Tomasz G. Smolinski, Mariofanna G. Milanova, and Aboul-Ella Hassanien (Eds.)

Computational Intelligence in Biomedicine and Bioinformatics

# Studies in Computational Intelligence, Volume 151

## Editor-in-Chief

Prof. Janusz Kacprzyk
Systems Research Institute
Polish Academy of Sciences
ul. Newelska 6
01-447 Warsaw
Poland
*E-mail:* kacprzyk@ibspan.waw.pl

Further volumes of this series can be found on our homepage:
springer.com

Vol. 130. Richi Nayak, Nikhil Ichalkaranje
and Lakhmi C. Jain (Eds.)
*Evolution of the Web in Artificial Intelligence Environments,*
2008
ISBN 978-3-540-79139-3

Vol. 131. Roger Lee and Haeng-Kon Kim (Eds.)
*Computer and Information Science,* 2008
ISBN 978-3-540-79186-7

Vol. 132. Danil Prokhorov (Ed.)
*Computational Intelligence in Automotive Applications,* 2008
ISBN 978-3-540-79256-7

Vol. 133. Manuel Graña and Richard J. Duro (Eds.)
*Computational Intelligence for Remote Sensing,* 2008
ISBN 978-3-540-79352-6

Vol. 134. Ngoc Thanh Nguyen and Radoslaw Katarzyniak (Eds.)
*New Challenges in Applied Intelligence Technologies,* 2008
ISBN 978-3-540-79354-0

Vol. 135. Hsinchun Chen and Christopher C. Yang (Eds.)
*Intelligence and Security Informatics,* 2008
ISBN 978-3-540-69207-2

Vol. 136. Carlos Cotta, Marc Sevaux
and Kenneth Sörensen (Eds.)
*Adaptive and Multilevel Metaheuristics,* 2008
ISBN 978-3-540-79437-0

Vol. 137. Lakhmi C. Jain, Mika Sato-Ilic, Maria Virvou,
George A. Tsihrintzis, Valentina Emilia Balas
and Canicious Abeynayake (Eds.)
*Computational Intelligence Paradigms,* 2008
ISBN 978-3-540-79473-8

Vol. 138. Bruno Apolloni, Witold Pedrycz, Simone Bassis
and Dario Malchiodi
*The Puzzle of Granular Computing,* 2008
ISBN 978-3-540-79863-7

Vol. 139. Jan Drugowitsch
*Design and Analysis of Learning Classifier Systems,* 2008
ISBN 978-3-540-79865-1

Vol. 140. Nadia Magnenat-Thalmann, Lakhmi C. Jain
and N. Ichalkaranje (Eds.)
*New Advances in Virtual Humans,* 2008
ISBN 978-3-540-79867-5

Vol. 141. Christa Sommerer, Lakhmi C. Jain
and Laurent Mignonneau (Eds.)
*The Art and Science of Interface and Interaction Design (Vol. 1),*
2008
ISBN 978-3-540-79869-9

Vol. 142. George A. Tsihrintzis, Maria Virvou, Robert J. Howlett
and Lakhmi C. Jain (Eds.)
*New Directions in Intelligent Interactive Multimedia,* 2008
ISBN 978-3-540-68126-7

Vol. 143. Uday K. Chakraborty (Ed.)
*Advances in Differential Evolution,* 2008
ISBN 978-3-540-68827-3

Vol. 144. Andreas Fink and Franz Rothlauf (Eds.)
*Advances in Computational Intelligence in Transport, Logistics,*
*and Supply Chain Management,* 2008
ISBN 978-3-540-69024-5

Vol. 145. Mikhail Ju. Moshkov, Marcin Piliszczuk
and Beata Zielosko
*Partial Covers, Reducts and Decision Rules in Rough Sets,* 2008
ISBN 978-3-540-69027-6

Vol. 146. Fatos Xhafa and Ajith Abraham (Eds.)
*Metaheuristics for Scheduling in Distributed Computing*
*Environments,* 2008
ISBN 978-3-540-69260-7

Vol. 147. Oliver Kramer
*Self-Adaptive Heuristics for Evolutionary Computation,* 2008
ISBN 978-3-540-69280-5

Vol. 148. Philipp Limbourg
*Dependability Modelling under Uncertainty,* 2008
ISBN 978-3-540-69286-7

Vol. 149. Roger Lee (Ed.)
*Software Engineering, Artificial Intelligence, Networking and*
*Parallel/Distributed Computing,* 2008
ISBN 978-3-540-70559-8

Vol. 150. Roger Lee (Ed.)
*Software Engineering Research, Management and*
*Applications,* 2008
ISBN 978-3-540-70774-5

Vol. 151. Tomasz G. Smolinski, Mariofanna G. Milanova
and Aboul-Ella Hassanien (Eds.)
*Computational Intelligence in Biomedicine and Bioinformatics,*
2008
ISBN 978-3-540-70776-9

Tomasz G. Smolinski
Mariofanna G. Milanova
Aboul-Ella Hassanien
(Eds.)

# Computational Intelligence in Biomedicine and Bioinformatics

## Current Trends and Applications

 Springer

Dr. Tomasz G. Smolinski
Department of Biology
Emory University
1510 Clifton Rd. NE
Atlanta, Georgia 30322
USA
Email: tsmolin@emory.edu

Professor Mariofanna G. Milanova
Department of Computer Science
University of Arkansas at Little Rock
2801 S. University Ave.
Little Rock, Arkansas 72204
USA
Email: mgmilanova@ualr.edu

Professor Aboul-Ella Hassanien
Department of Quantitative Methods and
Information Systems
College of Business and Administration
Kuwait University
P.O. Box 5486
Safat, 13055
Kuwait
Email: abo@cba.edu.kw
and
Department of Information Technology
Faculty of Computers and Information
Cairo University
5 Ahamed Zewal Street
Orman, Giza
Egypt
Email: a.hassanien@fci-cu.edu.eg

ISBN 978-3-540-70776-9          e-ISBN 978-3-540-70778-3

DOI 10.1007/978-3-540-70778-3

Studies in Computational Intelligence          ISSN 1860949X

Library of Congress Control Number: 2008930467

*Typeset & Cover Design:* Scientific Publishing Services Pvt. Ltd., Chennai, India.

Printed in acid-free paper
9 8 7 6 5 4 3 2 1
springer.com

To
    Jon and Tara

            Tomasz G. Smolinski

To
    my family

            Mariofanna G. Milanova

To
    my family

            Aboul-Ella Hassanien

# Preface

For the past decade or so, Computational Intelligence (CI) has been an extremely "hot" topic amongst researchers working in the fields of biomedicine and bioinformatics. There are many successful applications of CI in such areas as computational genomics, prediction of gene expression, protein structure, and protein-protein interactions, modeling of evolution, or neuronal systems modeling and analysis. However, there still are many problems in biomedicine and bioinformatics that are in desperate need of advanced and efficient computational methodologies to deal with tremendous amounts of data so prevalent in those kinds of research pursuits. Unfortunately, scientists in both these fields are very often unaware of the abundance of computational techniques that could be put to use to help them analyze and understand the data underlying their research inquiries. On the other hand, computational intelligence practitioners are often unfamiliar with the particular problems that their algorithms could be successfully applied for. The separation between the two worlds is partially caused by the use of *different languages* in these two spheres of science, but also by a relatively small number of publications devoted solely to the purpose of facilitating the exchange of new computational algorithms and methodologies on one hand, and the needs of the realms of biomedicine and bioinformatics on the other.

In order to help fill the gap between the scientists on both sides of this spectrum, we have solicited contributions from researchers actively applying computational intelligence techniques to important problems in biomedicine and bioinformatics. The purpose of this book is to provide an overview of powerful state-of-the-art methodologies that are currently utilized for biomedicine- and/or bioinformatics-oriented applications, so that researchers working in those fields could learn of new methods to help them tackle their problems. On the other hand, we also hope that the CI community will find this book useful by discovering a new and intriguing area of applications.

We have divided the book into three major parts. Part I, *Techniques and Methodologies*, contains a selection of contributions that provide a review of several theories and methods that could be (or to some extent already are) of great benefit to practitioners in the fields of biomedicine and bioinformatics dealing with problems of data exploration and mining, search-space exploration, optimization, etc.

In Chapter 1, *Aboul ella Hassanien et al.*, present an overview of selected CI techniques including Artificial Neural Networks (ANN), Particle Swarm Optimization (PSO), Genetic Algorithms (GA), Fuzzy Sets (FS), and Rough Sets (RS) and discuss their applications to several problems in bioinformatics and computational biology, including gene selection, DNA fragment assembly, multiple sequence alignment, protein structure prediction, and human genetics.

Chapters 2, by *Christopher M. Taylor and Arvin Agah*, and 3 by *Ray R. Hashemi, Alexander A. Tyler, Azita A. Bahrami*, deal with several methodologies widely used for mining of, quite broadly speaking, "bio-data." The discussed techniques and approaches include rough and fuzzy sets, rule induction, and genetic algorithms. The authors provide intuitive examples to explain the presented methods and review several real-life applications in the areas of analysis of gene and protein expression, genome annotation, and mutations in cancer.

Concluding this part of the book is Chapter 4, by *Nikola Kasabov et al.*, in which the authors propose a novel ontology-based decision support framework and a development platform for the creation of global knowledge representation for local and personalized modeling and decision support in biomedical and bioinformatics applications. The authors discuss a case study on brain-gene-disease ontology, where they derive from existing data and local profiles of patients a set of 12 genes related to the central nervous system cancer. Through this ontology analysis, these genes are found to be related to different functions, areas, and other diseases of the brain.

Part II of this book, *Computational Intelligence in Biomedicine*, contains a collection of contributions on current state-of-the-art biomedical applications of CI.

Opening this section of the book is Chapter 5, by *Samuel Neymotin et al.*, which presents a time-domain algorithm to facilitate data-mining of large electroencephalogram/electrocorticogram datasets to identify the occurrence of spike-wave or other activity patterns. The authors successfully used the proposed algorithm to identify and classify activity from both simulated and experimental seizures.

Chapter 6, by *Frank-Michael Schleif et al.*, treats on recent extensions of Self-Organizing Maps (SOM) as universal tools in the light of clinical proteomics. The authors consider extensions of the standard SOMs and Learning Vector Quantization (LVQs) for handling of more general metrics and propose a semi-supervised approach and a fuzzy classification scheme based on prototypes for classification of spectra.

Chapters 7, by *Andrzej Obuchowicz et al.*, and 8, by *Dongqing Chen et al.*, focus on state-of-the-art image processing techniques used in clinical applications. The first chapter presents an overview of CI methods applied to image segmentation for cytopathology and the second introduces a novel anisotropic

3D surface evolution model for detecting protrusion shape based colonic polyps on curved surfaces.

Chapter 9, by *Feng Chu et al.*, describes a study on an application of a fuzzy neural network (FNN) for cancer classification in microarray gene expression data. The authors use three well-known microarray databases, i.e., the lymphoma data set, the small round blue cell tumor (SRBCT) data set, and the ovarian cancer data set to test their approach. The results indicate that the FNN classifier not only improves the accuracy of cancer classification problem, but also helps find a better relationship between important genes and development of cancers.

Finally, in Chapter 10, by *Benjamin Haibe-Kains et al.*, the authors present a retrospective clinical study where the adoption of computational intelligence approaches for performing knowledge extraction from gene expression data enabled an improved oncological clinical analysis.

Part III, *Computational Intelligence in Bioinformatics*, opens with Chapter 11, by *Vitoantonio Bevilacqua et al.*, which presents an overview of artificial immune systems (AIS) in bioinformatics. The chapter describes how AIS have been successfully used in computational biology problems and gives readers further hints about possible implementations in unexplored fields.

Chapters 12, by *Heitor Silvério Lopes*, and 13, by *Xiao-min Hu et al.*, treat on applications of biology-inspired methods to the protein folding problem (PFP). The first chapter provides a review of current developments in the area of application of Evolutionary Algorithms (EA) to the PFP. The author discusses several computational approaches for the PFP, from molecular dynamics and approximation algorithms to several implementations of EAs. The second chapter presents an application of Ant Colony Optimization (ACO) to the flexible PFP. The authors test their proposed algorithm on benchmark two-dimensional hydrophobic-polar (2D-HP) protein sequences and compare its performance with some other well-known methods.

In Chapter 14, *Kirt M. Noël and Kay C. Wiese*, advocate considering stemloops as sequence signals for finding ribosomal RNA genes. The authors developed an algorithm to identify stem-loops along a genomic sequence which are similar to those found in rRNA secondary structures. The described results are encouraging and demonstrate that stem-loops indeed have the potential to act as sequence signals to discover rRNA genes.

Chapter 15, by *Radhakrishnan Nagarajan*, deals with the analysis of Genechip oligonucleotide microarrays, in which gene expression estimation is given as a complex combination of atomic entities on the array called probes. The study investigates qualitative similarities in the distributional signatures and local correlation structures/patchiness between the perfect-match (PM) and mis-match (MM) probe intensities. The results presented raise fundamental concerns in interpreting Genechip oligonucleotide microarray data.

In Chapter 16, *Vijayaraj Nagarajan and Mohamed O. Elasri* describe and discuss methods and tools used to predict structure and function of a putative protein sequence (Msa) with unknown function. The authors address the advantages and limitations of these approaches by using the Msa protein from the human pathogen *Staphylococcus aureus* as a case study.

Finally, in Chapter 17, *Deyu Zhou et al.* discuss the importance for biomedical researchers to be able to retrieve and mine specific knowledge from huge quantity of published articles with high efficiency. The authors provide a road map to the various information extraction methods in biomedical domain, such as protein name recognition and discovery of protein-protein interactions, and review current work and challenges in biomedical information extraction.

The editors are very grateful to the authors of the contributions included in this volume and to the referees for their tremendous service by critically reviewing the chapters. We would especially like to thank Prof. Janusz Kacprzyk, Editor-in-chief of the series *"Studies in Computational Intelligence,"* Dr. Thomas Ditzinger, Senior In-house Editor, and Ms. Heather King, Editorial Assistant of Springer Verlag, Germany, for their help, editorial assistance, and excellent cooperation. We sincerely hope that this book will prove useful to researchers working in biomedicine and bioinformatics as well as computational intelligence and that it will facilitate a productive dialog between the communities and result in fruitful collaborations and scientific advancements on both sides.


January 2008                                    Tomasz G. Smolinski
                                               Mariofanna G. Milanova
                                               Aboul-Ella Hassanien

# Contents

# List of Contributors

**Ajith Abraham**
Center for Quantifiable Quality of
Service in Communication Systems,
Norwegian University of Science and
Technology, Trondheim, Norway
ajith.abraham@ieee.org

**Arvin Agah**
Department of Electrical Engineering
and Computer Science
The University of Kansas
Lawrence, KS 66045, USA
agah@ku.edu

**Roberto T. Alves**
Federal Technological
University of Paraná
3165 Curitba, Brazil
r.t.alves@gmail.com

**Azita A. Bahrami**
Department of Information
Technology,
Armstrong Atlantic State University
Savannah, GA 31419, USA
Azita.Bahrami@armstrong.edu

**Lubica Benuskova**
Knowledge Engineering and Discovery
Research Institute, KEDRI,
Auckland University of Technology
Auckland 1142, New Zealand

**Vitoantonio Bevilacqua**
Polytechnic of Bari
70125 Bari, Italy
bevilacqua@poliba.it

**Gianluca Bontempi**
Machine Learning Group,
Université Libre de Bruxelles
Brussels, Belgium

**Dongqing Chen**
Computer Vision & Image Processing
(CVIP) Laboratory,
Department of Electrical & Computer
Engineering,
University of Louisville
Louisville, KY 40292, USA
dqchen@cvip.louisville.edu

**Feng Chu**
School of Electrical and
Electronic Engineering,
Nanyang Technological University,
Singapore 639798

**A. Deelder**
Biomolecular Mass
Spectrometry Unit,
Department of Surgery,
Leiden University Medical Center
Leiden, The Netherlands

**Myriam Delgado**
Federal Technological
University of Paraná
3165 Curitba, Brazil

**Mauro Delorenzi**
Bioinformatics Core Facility,
Institut Suisse de Recherche
Expérimentale sur le Cancer
Lausanne, Switzerland

**Christine Desmedt**
Functional Genomic Unit,
Institut Jules Bordet
Brussels, Belgium

**Gerald W. Dryden**
Division of Gastroenterology and
Hepatology, Department of Medicine,
University of Louisville
Louisville, KY 40292, USA

**Mohamed O. Elasri**
Department of Biological Sciences,
The University of Southern
Mississippi
Hattiesburg, MS 39406, USA
mohamed.elasri@usm.edu

**Robert L. Falk**
Department of Medical Imaging
Jewish Hospital & St. Mary's
Healthcare
Louisville, KY 40202, USA

**Aly A. Farag**
Computer Vision & Image Processing
(CVIP) Laboratory,
Department of Electrical & Computer
Engineering,
University of Louisville
Louisville, KY 40292, USA
farag@cvip.louisville.edu

**Farideh Fazayeli**
School of Electrical and
Electronic Engineering,
Nanyang Technological University,
Singapore 639798
FARI0004@ntu.edu.sg

**Alex A. Freitas**
Computing Laboratory
University of Kent
Canterbury, CT2 7NF, UK
a.a.freitas@kent.ac.uk

**Paulo Gottgtroy**
Knowledge Engineering and Discovery
Research Institute, KEDRI,
Auckland University of Technology
Auckland 1142, New Zealand

**Benjamin Haibe-Kains**
Machine Learning Group,
Université Libre de Bruxelles
Brussels, Belgium
Functional Genomic Unit,
Institut Jules Bordet
Brussels, Belgium
bhaibeka@ulb.ac.be

**Barbara Hammer**
Computer Science Department,
Technical University Clausthal
Clausthal-Zellerfeld, Germany
hammer@in.tu-clausthal.de

**Ray R. Hashemi**
Department of Computer Science,
Armstrong Atlantic State University
Savannah, GA 31419, USA
Ray.Hashemi@armstrong.edu

**Aboul-Ella Hassanien**
Information Technology Department,
FCI, Cairo University
Orman, Giza, Egypt
Information System Department,
CBA, Kuwait University,
Safat, 13055, Kuwait
a.hassanien@fci-cu.edu.eg

**M. Sabry Hassouna**
Computer Vision & Image Processing
(CVIP) Laboratory,
Department of Electrical & Computer
Engineering,
University of Louisville
Louisville, KY 40292, USA

**Ilkka Havukkala**
Knowledge Engineering and Discovery
Research Institute, KEDRI,
Auckland University of Technology
Auckland 1142, New Zealand
ilkka.havukkala@aut.ac.nz

**Yulan He**
Informatics Research Centre,
The University of Reading
Reading, RG6 6BX, UK
y.he@reading.ac.uk

**Maciej Hrebień**
Institute of Control and
Computation Engineering,
University of Zielona Góra
Zielona Góra, Poland
M.Hrebien@issi.uz.zgora.pl

**Ying-jie Hu**
Knowledge Engineering and Discovery
Research Institute, KEDRI,
Auckland University of Technology
Auckland 1142, New Zealand

**Xiao-min Hu**
Department of Computer Science,
SUN Yat-sen University
Guangzhou, 510275, China

**Vishal Jain**
Knowledge Engineering and Discovery
Research Institute, KEDRI,
Auckland University of Technology
Auckland 1142, New Zealand

**Nikola Kasabov**
Knowledge Engineering and Discovery
Research Institute, KEDRI,
Auckland University of Technology
Auckland 1142, New Zealand
nkasabov@aut.ac.nz

**Chee Keong Kwoh**
School of Computer Engineering,
Nanyang Technological University
Singapore 639798
asckkwoh@ntu.edu.sg

**Yun Li**
Department of Electronics and
Electrical Engineering,
University of Glasgow
Glasgow G12 8LT, Scotland, UK

**Sherene Loi**
Peter MacCallum Cancer Center
East Melbourne, Victoria, Australia

**Heitor Silvério Lopes**
Bioinformatics Laboratory
Federal University of Technology –
Paraná
80230-901 Curitiba, Brazil
hslopes@utfpr.edu.br

**William W. Lytton**
Department Biomedical Engineering,
SUNY Downstate Medical Center
Brooklyn, NY 11203, USA
billl@neurosim.downstate.edu

**Stephen MacDonell**
Knowledge Engineering and Discovery
Research Institute, KEDRI,
Auckland University of Technology
Auckland 1142, New Zealand

**Karen A. Manning**
Department of Anatomy,
University of Wisconsin
Madison, WI 53705, USA
kamannin@wisc.edu

**Andrzej Marciniak**
Institute of Control and
Computation Engineering,
University of Zielona Góra
Zielona Góra, Poland
A.Marciniak@issi.uz.zgora.pl

**Giuseppe Mastronardi**
Polytechnic of Bari
70125 Bari, Italy

**Filippo Menolascina**
Polytechnic of Bari
70125 Bari, Italy
National Cancer Institute
'Giovanni Paolo II'
70126 Bari, Italy
f.menolascina@ieee.org

**Mariofanna G. Milanova**
Computer Science Department,
University of Arkansas at Little Rock
Little Rock, AR 72204, USA
mgmilanova@ualr.edu

**Radhakrishnan Nagarajan**
University of Arkansas for
Medical Sciences
Little Rock, AR 72205, USA
nagarajanradhakrish@uams.edu

**Vijayaraj Nagarajan**
Department of Biological Sciences,
The University of Southern
Mississippi
Hattiesburg, MS 39406, USA
vijayaraj.nagarajan@usm.edu

**Samuel Neymotin**
Department Biomedical Engineering,
SUNY Downstate Medical Center
Brooklyn, NY 11203, USA
samn@neurosim.downstate.edu

**Giuseppe Nicosia**
University of Catania
95125 Catania, Italy
nicosia@dmi.unict.it

**Tomasz Nieczkowski**
Institute of Control and
Computation Engineering,
University of Zielona Góra
Zielona Góra, Poland
T.Nieczkowski@issi.uz.zgora.pl

**Kirt M. Noël**
McKesson Medical Imaging
Richmond, BC V6X 3G5, Canada
kirtnoel@gmail.com

**Andrzej Obuchowicz**
Institute of Control and
Computation Engineering,
University of Zielona Góra
Zielona Góra, Poland
A.Obuchowicz@issi.uz.zgora.pl

**Angelo Paradiso**
National Cancer Institute
'Giovanni Paolo II'
70126 Bari, Italy

**Russel Pears**
Knowledge Engineering and Discovery
Research Institute, KEDRI,
Auckland University of Technology
Auckland 1142, New Zealand

**Elaine Rush**
Knowledge Engineering and Discovery
Research Institute, KEDRI,
Auckland University of Technology
Auckland 1142, New Zealand

**Frank-Michael Schleif**
Medical Department,
University Leipzig
Leipzig, Germany
schleif@informatik.uni-leipzig.de

**Tomasz G. Smolinski**
Department of Biology,
Emory University
Atlanta, GA 30322, USA
tsmolin@emory.edu

**Qun Song**
Knowledge Engineering and Discovery
Research Institute, KEDRI,
Auckland University of Technology
Auckland 1142, New Zealand
qun.song@aut.ac.nz

**Christos Sotiriou**
Functional Genomic Unit,
Institut Jules Bordet
Brussels, Belgium

**Christopher M. Taylor**
Department of Electrical Engineering
and Computer Science
The University of Kansas
Lawrence, KS 66045, USA

**Alex Tjahjana**
Knowledge Engineering and Discovery
Research Institute, KEDRI,
Auckland University of Technology
Auckland 1142, New Zealand

**R. Tollenaar**
Biomolecular Mass
Spectrometry Unit,
Department of Surgery,
Leiden University Medical Center
Leiden, The Netherlands

**Stefania Tommasi**
National Cancer Institute
'Giovanni Paolo II'
70126 Bari, Italy

**Alexander A. Tyler**
University of Arkansas for Medical
Sciences,
Area Health Education
Center-Pine Bluff, 4010 Mulberry
Street, Pine Bluff,
AR 71603, USA
alexandertyler@yahoo.com

**Daniel J. Uhlrich**
Department of Anatomy,
University of Wisconsin
Madison, WI 53705, USA
duhlrich@wisc.edu

**Martijn van der Werff**
Biomolecular Mass
Spectrometry Unit,
Department of Surgery,
Leiden University Medical Center
Leiden, The Netherlands
M.P.J.van_der_Werff@lumc.nl

**Anju Verma**
Knowledge Engineering and Discovery
Research Institute, KEDRI,
Auckland University of Technology
Auckland 1142, New Zealand

**Thomas Villmann**
Medical Department,
University Leipzig
Leipzig, Germany
villmann@informatik.uni-leipzig.de

**Lipo Wang**
School of Electrical and
Electronic Engineering,
Nanyang Technological University,
Singapore 639798
elpwang@ntu.edu.sg

**Kay C. Wiese**
School of Computing Science,
Simon Fraser University
Surrey, BC V3T 0A3, Canada
wiese@cs.sfu.ca

**Wei Xie**
Institute for Infocomm Research
Singapore 119613

**Jun Zhang**
Department of Computer Science,
SUN Yat-sen University
Guangzhou, 510275, China
junzhang@ieee.org

**Deyu Zhou**
Informatics Research Centre,
The University of Reading
Reading, RG6 6BX, UK
d.zhou@reading.ac.uk

# List of Referees

**Arvin Agah**
Department of Electrical Engineering
and Computer Science
The University of Kansas
Lawrence, KS 66045, USA
agah@ku.edu

**James Archibald**
Brigham Young University
Provo, UT 84602, USA
jka@ee.byu.edu

**Joong-Hwan Baek**
School of Electronics, Telecommunica-
tion, and Computer Engineering,
Korea Aerospace University,
Koyang City, South Korea
jhbaek@kau.ac.kr

**Vitoantonio Bevilacqua**
Polytechnic of Bari
70125 Bari, Italy
bevilacqua@poliba.it

**Grzegorz M. Boratyn**
National Center for
Biotechnology Information
U.S. National Library of Medicine
Bethesda, MD 20894, USA
boratyng@ncbi.nlm.nih.gov

**Dongqing Chen**
Computer Vision & Image Processing
(CVIP) Laboratory
Department of Electrical and
Computer Engineering
University of Louisville
Louisville, KY 40292, USA
dqchen@cvip.louisville.edu

**Carlos A. Coello Coello**
CINVESTAV-IPN
Evolutionary Computation Group,
Departamento de Computación,
Av. IPN No. 2508
Col. San Pedro Zacatenco
México, D.F. 07360, Mexico
ccoello@cs.cinvestav.mx

**Zhihua Cui**
State Key Laboratory for Manufac-
turing Systems Engineering,
Xi'an Jiaotong University
Xi'an, Shaanxi, 710049, China
cuizhihua@gmail.com

**Christine Decaestecker**
Laboratory of Image Synthesis and
Analysis,
Université Libre de Bruxelles
Brussels, Belgium
cdecaes@ulb.ac.be

**Anca Doloc-Mihu**
Center for Advanced
Computer Studies,
University of Louisiana at Lafayette
Lafayette, LA 70503, USA
anca@louisiana.edu

**Aly A. Farag**
Computer Vision & Image Processing
(CVIP) Laboratory
Department of Electrical & Computer
Engineering
University of Louisville
Louisville, KY, 40292, USA
farag@cvip.louisville.edu

**Gary Fogel**
Natural Selection, Inc.
San Diego, CA 92121, USA
gfogel@natural-selection.com

**Adam E. Gaweda**
Kidney Disease Program
University of Louisville
Louisville, KY 40202, USA
adam.gaweda@louisville.edu

**Samik Ghosh**
Biological Networking
Research Group,
Department of Computer Science and
Engineering,
The University Of Texas at Arlington
Arlington, TX 76019, USA
fsghosh@cse.uta.edu

**Cengiz Günay**
Department of Biology,
Emory University
Atlanta, GA 30322, USA
cgunay@emory.edu

**Benjamin Haibe-Kains**
Machine Learning Group,
Université Libre de Bruxelles
Brussels, Belgium

Functional Genomic Unit,
Institut Jules Bordet
Brussels, Belgium
bhaibeka@ulb.ac.be

**Andrew Hamilton-Wright**
School of Rehabilitation Therapy,
Queen's University
Computing and Information Science,
University of Guelph
Mathematics and Computer Science,
Mount Allison University
Kingston, ON K7K 1T3, Canada
andrewhw@ieee.org

**Barbara Hammer**
Computer Science Department,
Technical University Clausthal
Clausthal-Zellerfeld, Germany
hammer@in.tu-clausthal.de

**Ray R. Hashemi**
Department of Computer Science,
Armstrong Atlantic State University
Savannah, GA 31419, USA
Ray.Hashemi@armstrong.edu

**Aboul-Ella Hassanien**
Information Technology Department,
FCI, Cairo University
Orman, Giza, Egypt
Information System Department,
CBA, Kuwait University,
Safat, 13055, Kuwait
a.hassanien@fci-cu.edu.eg

**Ilkka Havukkala**
Knowledge Engineering and Discovery
Research Institute, KEDRI,
Auckland University of Technology
Auckland 1142, New Zealand
ilkka.havukkala@aut.ac.nz

**Timothy J. Hickey**
Michtom School of Computer Science,
Brandeis University

Waltham, MA 02254, USA
tim@cs.brandeis.edu

Falk Huettmann
University of Alaska, EWHALE Lab,
Institute of Arctic Biology,
Department of Biology & Wildlife
Fairbanks, AK 99775, USA
fffh@uaf.edu

Wit Jakuczun
Nencki Institute of
Experimental Biology
Pasteura 3 St.,
02-093 Warsaw, Poland
WLOG Solutions
Harfowa 1A/25 St.,
02-389 Warsaw, Poland
W.Jakuczun@wlogsolutions.com

Fatma Gürel Kazancı
Department of Biology,
Emory University
Atlanta, GA 30322, USA
fgurelk@emory.edu

John T. Langton
Michtom School of Computer Science,
Brandeis University
Waltham, MA 02254, USA
psyc@cs.brandeis.edu

Dah-Jye Lee
Department of Electrical and
Computer Engineering,
Brigham Young University
Provo, UT 84602, USA
djlee@ee.byu.edu

Heitor Silvério Lopes
Bioinformatics Laboratory
Federal University of Technology –
Paraná
80230-901 Curitiba, Brazil
hslopes@utfpr.edu.br

Tom McTavish
Computational Bioscience Program,
University of Colorado
Health Sciences Center
P.O. Box 6511, Mail Stop 8303
Aurora, CO 80045, USA
Thomas.McTavish@uchsc.edu

Filippo Menolascina
Polytechnic of Bari
70125 Bari, Italy
National Cancer Institute 'Giovanni
Paolo II'
70126 Bari, Italy
f.menolascina@ieee.org

Mariofanna G. Milanova
Computer Science Department,
University of Arkansas at Little Rock
Little Rock, AR 72204, USA
mgmilanova@ualr.edu

Radhakrishnan Nagarajan
University of Arkansas for
Medical Sciences
Little Rock, AR 72205, USA
nagarajanradhakrish@uams.edu

Vijayaraj Nagarajan
Department of Biological Sciences,
The University of Southern
Mississippi
Hattiesburg, MS 39406, USA
vijayaraj.nagarajan@usm.edu

Samuel Neymotin
Dept. Biomedical Engineering,
SUNY Downstate Medical Center
Brooklyn, NY 11203, USA
samn@neurosim.downstate.edu

F. G. (Pat) Patterson, Jr.
Institute of Intelligent Systems,
FedEx Institute of Technology,
The University of Memphis
Memphis, TN 38152, USA
drfgp2@gmail.com

**Lech Polkowski**
Polish-Japanese Institute of
Information Technology
Koszykowa 86,
02008 Warszawa, Poland
polkow@pjwstk.edu.pl

**Astrid A. Prinz**
Department of Biology,
Emory University
Atlanta, Georgia 30322, USA
aprinz@emory.edu

**Rafał Scherer**
Department of Computer Engineering,
Częstochowa University of Technology
36 Armii Krajowej Ave.,
42-200 Częstochowa, Poland

**Frank-Michael Schleif**
Medical Department,
University Leipzig
Leipzig, Germany
schleif@informatik.uni-leipzig.de

**Jahangheer Shaik**
Department of Electrical and
Computer Engineering,
The University of Memphis
Memphis, TN 38152, USA
jshaik@memphis.edu

**Tomasz G. Smolinski**
Department of Biology,
Emory University
Atlanta, GA 30322, USA
tsmolin@emory.edu

**Ryszard Tadeusiewicz**
AGH University of Science and
Technology
30 Mickiewicza Ave.,
30-059 Krakow, Poland
rtad@agh.edu.pl

**David Windridge**
School of Electronics and
Physical Sciences,
University of Surrey
Guildford, Surrey, GU2 7XH,
United Kingdom
D.Windridge@surrey.ac.uk

**Ying Xie**
Department of Computer Science and
Information Systems,
Kennesaw State University
Kennesaw, GA 30144, USA
yxie2@kennesaw.edu

**Jun Zhang**
Department of Computer Science,
SUN Yat-sen University
Guangzhou, 510275, China
junzhang@ieee.org

**Bai-Tao Zhou**
School of Electronics, Telecommunica-
tion, and Computer Engineering,
Korea Aerospace University
Koyang City, South Korea
zhou@kau.ac.kr

# Part I

# Techniques and Methodologies

# 1

# Computational Intelligence in Solving Bioinformatics Problems: Reviews, Perspectives, and Challenges

Aboul-Ella Hassanien[1,2], Mariofanna G. Milanova[3], Tomasz G. Smolinski[4], and Ajith Abraham[5]

[1] Information Technology Department, FCI, Cairo University
   5 Ahamed Zewal Street, Orman, Giza, Egypt
[2] Information System Department, CBA, Kuwait University, Kuwait
   `a.hassanien@fci-cu.edu.eg, abo@cba.edu.kw`
[3] Computer Science Department, University of Arkansas at Little Rock
   2801 S. University Ave. Little Rock, Arkansas 72204, USA
   `mgmilanova@ualr.edu`
[4] Biology Department, Emory University
   1510 Clifton Rd. NE, Atlanta, Georgia 30322, USA
   `tsmolin@emory.edu`
[5] Center for Quantifiable Quality of Service in Communication Systems
   Norwegian University of Science and Technology,
   O.S. Bragstads plass 2E, N-7491 Trondheim, Norway
   `ajith.abraham@ieee.org, abraham.ajith@acm.org`

**Summary.** This chapter presents a broad overview of Computational Intelligence (CI) techniques including Artificial Neural Networks (ANN), Particle Swarm Optimization (PSO), Genetic Algorithms (GA), Fuzzy Sets (FS), and Rough Sets (RS). We review a number of applications of computational intelligence to problems in bioinformatics and computational biology, including gene expression, gene selection, cancer classification, protein function prediction, multiple sequence alignment, and DNA fragment assembly. We discuss some representative methods to provide inspiring examples to illustrate how CI could be applied to solve bioinformatic problems and how bioinformatics could be analyzed, processed, and characterized by computational intelligence. Challenges to be addressed and future directions of research are presented. An extensive bibliography is also included.

## 1.1 Introduction

The past few decades have seen a massive growth in biological information gathered by the related scientific communities. A deluge of such information coming in the form of genomes, protein sequences, gene expression data and so on have led to the absolute need for effective and efficient computational tools to store, analyze and interpret the multifaceted data. Bioinformatics and computational biology involve the use of techniques including applied mathematics, informatics, statistics, computer science, artificial intelligence, chemistry, and biochemistry

T.G. Smolinski et al. (Eds.): Comp. Intel. in Biomed. & Bioinform., SCI 151, pp. 3–47, 2008.
springerlink.com                                    © Springer-Verlag Berlin Heidelberg 2008

to solve biological problems usually on the molecular level. Research in computational biology often overlaps with systems biology. Major research efforts in the field include sequence alignment, gene finding, genome assembly, protein structure alignment, protein structure prediction, prediction of gene expression and protein-protein interactions, and the modeling of evolution [128]. Hence, in other words, bioinformatics can be described as the application of computational methods to make biological discoveries [10]. The ultimate attempt of the field is to develop new insights into the science of life as well as creating a global perspective, from which the unifying principles of biology can be derived [5]. There are at least 26 billion base pairs (bp) representing the various genomes available on the server of the National Center for Biotechnology Information (NCBI) [27]. Besides the human genome with about 3 billion bp, many other species have their complete genome available there. Cohen [23] explained the needs of biologists to utilize and help interpret the vast amounts of data that are constantly being gathered in genomic research. He also pointed out the basic concepts in molecular cell biology, and outlined the nature of the existing data, and illustrated the algorithms needed to understand cell behavior.

Bioinformatics involve the creation and advancement of algorithms using techniques including computational intelligence, applied mathematics and statistics, informatics, and biochemistry to solve biological problems usually on the molecular level. Major research efforts in the field include sequence analysis, gene finding, genome annotation, protein structure alignment analysis and prediction, prediction of gene expression, protein-protein docking/interactions, and the modeling of evolution.

Bioinformatics and computational biology are concerned with the use of computation to understand biological phenomena and to acquire and exploit biological data, increasingly large-scale data [38]. Methods from bioinformatics and computational biology are increasingly used to augment or leverage traditional laboratory and observation-based biology. These methods have become critical in biology due to recent changes in our ability and determination to acquire massive biological data sets, and due to the ubiquitous, successful biological insights that have come from the exploitation of those data. This transformation from a data-poor to a data-rich field began with DNA sequence data, but is now occurring in many other areas of biology [27].

Computational intelligence is a well-established paradigm, where new theories with a sound biological understanding have been evolving. The current experimental systems have many of the characteristics of biological computers ("brains") and are beginning to be built to perform a variety of tasks that are difficult or impossible to do with conventional computers. Computational intelligence methods are now being applied to problems in molecular biology and bioinformatics [70]. To name a few, Tasoulis et al. [104] present an application of neural networks, evolutionary algorithms, and clustering algorithms to DNA microarray experimental data analysis; Liang and Kelemen [60] propose a time lagged recurrent neural network with trajectory learning for identifying and classifying gene functional patterns from the heterogeneous nonlinear time series fmicroarray experiments. Reader may refer to [51, 22] for an extensive review of various computational intelligence techniques applied to different bioinformatics problems. Defining computational intelligence is not an easy task. In a nutshell, which becomes quite apparent in light of the current research pursuits, the area is heterogeneous with a combination of such technologies as neural networks, fuzzy systems, evolutionary computation, swarm intelligence, and probabilistic reasoning. The recent trend is to integrate different components to take advantage of complementary features and to develop a synergistic system [51]. Hybrid architectures like neuro-fuzzy systems, evolutionary-fuzzy systems, evolutionary-neural networks, evolutionary neuro-fuzzy systems, rough-neural, rough-fuzzy, etc. are widely applied for real world problem solving [1, 2, 46].

The objective of this book chapter is to present to the computational intelligence and bioinformatics research communities the state of the art computational intelligence applications to bioinformatics processing and motivate research in new trend-setting directions. Hence, we review and discuss in the following sections some representative methods to provide inspiring examples to illustrate how CI techniques could be applied to solve bioinformatics problems and how bioinformatics could be analyzed, processed, and characterized by computational intelligence. These representative examples include (i) CI in gene expression and clustering, (ii) rough discretization of gene expression, (iii) CI in protein sequence classification, (iv) CI in gene selection, (v) CI in cancer classification and the DNA fragment assembly problem, and (vi) CI in the multiple sequence alignment problem.

To provide useful insights for CI applications in bioinformatics, we structure the rest of this chapter as follows. Section 1.2 introduces some fundamental aspects and key components of modern computational intelligence including Artificial Neural Networks (ANN) , Rough Sets (RS), Fuzzy Sets (FS), Particle Swarm Optimization (PSO), and Genetic Algorithms (GA). Section 1.3 reviews some published papers on using computational intelligence in Gene Expression. A review of the current literature on CI-based approaches in Protein Sequence Classification problems is provided in Section 1.4. Section 1.5 discusses some successful work to illustrate how CI could be applied to Gene Selection problems. Applications of computational intelligence in DNA Fragment Assembly, Multiple Sequence Alignment Problems (MSA), and Protein Structure Prediction are reviewed in Sections 1.6, 1.7 and 1.8, respectively. An example of applications of CI in the field of human genetics, in the form of genetic programming neural networks, is presented in Section 1.9. CI in Microarray Classification is discussed and reviewed in Section 1.10. Conclusions, Challenges, and Future Directions are addressed in Section 1.11.

## 1.2 Computational Intelligence: Overview

In the following subsections, we present an overview of selected modern computational intelligence techniques including artificial neural networks, fuzzy sets, particle swarm optimization, genetic algorithms, and rough sets.

### 1.2.1   Artificial Neural Networks (ANN)

Artificial neural networks have been developed as generalizations of mathematical models of biological nervous systems. In a simplified mathematical model of the neuron, synapses are represented by connection weights that modulate the effect of the associated input signals, and the nonlinear characteristic exhibited by neurons is represented by a transfer function. There are many transfer functions developed to process the weighted and biased inputs, among which four basic and widely adopted in the field transfer functions are illustrated in Figure 1.1.

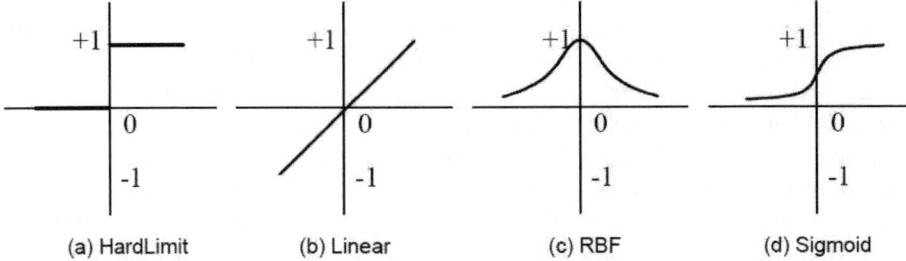

(a) HardLimit       (b) Linear       (c) RBF       (d) Sigmoid

**Fig. 1.1.** Basic transfer functions

The neuron impulse is computed as the weighted sum of the input signals, transformed by the transfer function. The learning capability of an artificial neuron is achieved by adjusting the weights in accordance to the chosen learning algorithm. Most applications of neural networks fall into the following categories: (1) *Prediction*: Use the input values to predict some output; (2) *Classification*: Use the input values to determine the classification of the input; (3) *Data Association*: Similar to classification, but also recognizes data containing errors; and (4) *Data conceptualization*: Analyze the inputs so that grouping relationships can be inferred.

### Neural Network Architecture

The behavior of the neural network depends largely on the interaction between the different neurons. The basic architecture consists of three types of neuron layers: input, hidden, and output layers.

In feed-forward networks the signal flow is from input to output units strictly in a feed-forward direction. The data processing can extend over multiple (layers of) units, but no feedback connections are present, that is, connections extending from outputs of units to inputs of units in the same layer or previous layers. Recurrent networks contain feedback connections. Contrary to feed-forward networks, the dynamical properties of such networks are important. In some cases, the activation values of the units undergo a relaxation process such that the network will evolve to a stable state in which these activations do not change anymore.

In other applications, the changes of the activation values of the output neurons are significant, such that the dynamical behavior constitutes the output of the network. There are several other neural network architectures (Elman network, adaptive resonance theory maps, competitive networks etc.) depending on the properties and requirement of the application.

Reader may refer to [13] for an extensive overview of the different neural network architectures and learning algorithms. A neural network has to be configured such that the application of a set of inputs produces the desired set of outputs. Various methods to set the strengths of the connections exist. One way is to set the weights explicitly, using a priori knowledge. Another way is to train the neural network by feeding it teaching patterns and letting it change its weights according to some learning rule. The learning situations in neural networks may be classified into three distinct sorts. These are supervised learning, unsupervised learning, and reinforcement learning. In supervised learning, an input vector is presented at the inputs together with a set of desired responses, one for each node, at the output layer. A forward pass is done and the errors or discrepancies, between the desired and actual response for each node in the output layer, are found. These are then used to determine weight changes in the network according to the prevailing learning rule. The term 'supervised' originates from the fact that the desired signals on individual output nodes are provided by an external teacher. The best-known examples of this technique occur in the backpropagation algorithm, the delta rule, and perceptron rule. In unsupervised learning (or self-organization) an output unit is trained to respond to clusters of patterns within the input. In this paradigm the system is supposed to discover statistically salient features of the input population. Unlike the supervised learning paradigm, there is no a priori set of categories into which the patterns are to be classified; rather the system must develop its own representation of the input stimuli. Reinforcement learning is learning what to do–how to map situations to actions–so as to maximize a numerical reward signal. The learner is not told which actions to take, as in most forms of Machine Learning (ML), but instead must discover which actions yield the most reward by trying them. In the most interesting and challenging cases, actions may affect not only the immediate reward, but also the next situation and, through that, all subsequent rewards. These two characteristics, trial-and-error search and delayed reward are the two most important distinguishing features of reinforcement learning.

### 1.2.2 Rough Sets (RS)

Rough set theory [83, 84, 86, 82] is a methodology fairly new to the medical domain capable of dealing with uncertainty in data. It is used to discover data dependencies, evaluate the importance of attributes, discover the patterns of data, reduce redundant objects and attributes, seek the minimum subset of attributes, recognize and classify objects. Moreover, it is being used for extraction of rules from databases. Rough sets have proven useful for representation of vague regions in spatial data. One advantage of rough sets is creation of readable if-then rules. Such rules have a potential to reveal new patterns in the data material. Furthermore, they also collectively function as a classifier for unseen data. Unlike other computational intelligence techniques, rough set analysis requires no external parameters and uses only the information presented in the given data. One of the nice features of rough sets theory is that its can tell whether the data is complete or not based on the data itself. If the data is incomplete, the theory can suggest more information about the objects needed to be collected in order to build a good classification model. On the other hand, if the data is complete, rough sets can determine whether there is any redundant information in the data and find the minimum data needed for classification. This property of rough sets is very important for applications where domain knowledge is very limited or data collection is very expensive/laborious because it makes sure the data collected is good enough to build a good classification model without sacrificing the accuracy of the classification model or wasting time and effort to gather extra information about the objects [83, 84, 86, 82].

In rough sets theory, the data is collected in a table, called decision table. Rows of the decision table correspond to objects, and columns correspond to attributes. In the data set, we assume that class labels to indicate the class to which each example belongs are given. We call the class label the decision attribute and the rest of the attributes the condition attributes. Rough sets theory defines three regions based on the equivalent classes induced by the attribute values Lower approximation, upper approximation, and the boundary. Lower approximation contains all the objects which are classified surely based on the data collected, and upper approximation contains all the objects which can be classified probably, while the boundary is the difference between the upper approximation and the lower approximation. Thus we can define a rough set as any set represented through its lower and upper approximations. On the other hand, indiscernibility notion is fundamental to rough set theory. Informally, two objects in a decision table are indiscernible if one cannot distinguish between them on the basis of a given set of attributes. Hence, indiscernibility is a function of the set of attributes under consideration. For each set of attributes we can thus define a binary indiscernibility relation, which is a collection of pairs of objects that are indistinguishable from each other. An indiscernibility relation partitions the set of cases or objects into a number of equivalence classes. An equivalence class of a particular object is simply the collection of objects that are indiscernible to the object in question. Here we provide an explanation of the basic framework of rough set theory, along with some of the key definitions. A review of this basic material can be found in sources such as [83, 84, 86, 82, 77, 125] and many others.

### 1.2.3   Fuzzy Logic (FL) and Fuzzy Sets (FS)

Zadeh [121] introduced the concept of fuzzy logic to present vagueness in linguistics, and further implement and express human knowledge and inference capability in a natural way. Fuzzy logic starts with the concept of a fuzzy set. An

FS set is a set without a crisp, clearly defined boundary. It can contain elements with only a partial degree of membership. A Membership Function (MF) is a curve that defines how each point in the input space is mapped to a membership value (or degree of membership) between 0 and 1. The input space is sometimes referred to as the universe of discourse. Let $X$ be the universe of discourse and $x$ be a generic element of $X$. A classical set $A$ is defined as a collection of elements or objects $x \in X$, such that each $x$ can either belong to or not belong to the set $A$, $A \subseteq X$. By defining a characteristic function (or membership function) on each element $x$ in $X$, a classical set $A$ can be represented by a set of ordered pairs $(x, 0)$ or $(x, 1)$, where 1 indicates membership and 0 non-membership. Unlike conventional set mentioned above, fuzzy set expresses the degree to which an element belongs to a set. Hence the characteristic function of a fuzzy set is allowed to have value between 0 and 1, denoting the degree of membership of an element in a given set. If $X$ is a collection of objects denoted generically by $x$, then a fuzzy set $A$ in $X$ is defined as a set of ordered pairs:

$$A = \{(x, \mu_A(x)) \mid x \in X\} \tag{1.1}$$

$\mu_A(x)$ is called the membership function of linguistic variable $x$ in $A$, which maps $X$ to the membership space $M$, $M = [0, 1]$, where $M$ contains only two points, 0 and 1, $A$ is crisp, and $\mu_A(x)$ is identical to the characteristic function of a crisp set. Triangular and trapezoidal membership functions are the simplest functions formed using straight lines. Some of the other shapes are Gaussian, generalized bell, sigmoidal, and polynomial based curves.

Figure 1.2, illustrates the shapes of two commonly used MFs. The most important thing to realize about fuzzy logical reasoning is the fact that it is a superset of standard Boolean logic.

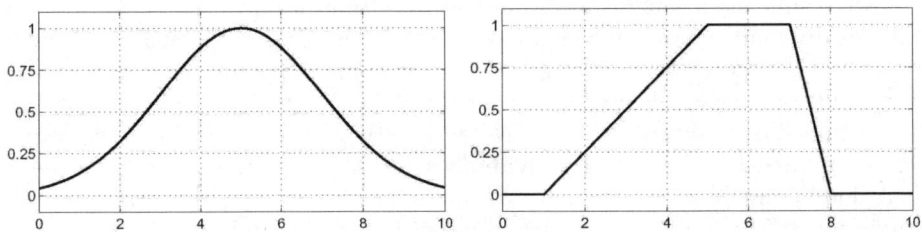

**Fig. 1.2.** Shapes of two commonly used MFs

## 1.2.4 Evolutionary Algorithms (EA)

Evolutionary Algorithms are adaptive methods, which may be used to solve search and optimization problems, based on the genetic processes of biological organisms. Over many generations, natural populations evolve according to the principles of natural selection and "survival of the fittest," first clearly stated by Charles Darwin in *The Origin of Species*. By mimicking this process, evolutionary algorithms are able to 'evolve' solutions to real world problems, if they have been suitably encoded [30]. Usually grouped under the term Evolutionary Algorithms (EA) or Evolutionary Computation (EC), we find the domains of genetic algorithms [43, 35], evolution strategies [8], evolutionary programming [32], genetic programming [57], and learning classifier systems [15]. They all share a common conceptual base of simulating the evolution of individual structures via processes of selection, mutation, and reproduction. The processes depend on the perceived performance of the individual structures as defined by the environment (problem).

EAs deal with parameters of finite length, which are coded using a finite alphabet, rather than directly manipulating the parameters themselves. This means that the search is unconstrained neither by the continuity of the function under investigation, nor the existence of a derivative function.

Genetic Algorithm (GA) is assumed that a potential solution to a problem may be represented as a set of parameters. These parameters (known as genes) are joined together to form a string of values (known as a chromosome). A gene (also referred to a feature, character or detector) refers to a specific attribute that is encoded in the chromosome. The particular values the genes can take are called its alleles. The position of the gene in the chromosome is its locus. Encoding issues deal with representing a solution in a chromosome and unfortunately, no one technique works best for all problems. A fitness function must be devised for each problem to be solved. Given a particular chromosome, the fitness function returns a single numerical fitness or figure of merit, which will determine the ability of the individual, which that chromosome represents. Reproduction is the second critical attribute of GAs where two individuals selected from the population are allowed to mate to produce offspring, which will comprise the next generation. Having selected two parents, their chromosomes are recombined, typically using the mechanisms of crossover and mutation.

There are many ways in which crossover can be implemented. In a single point crossover two chromosome strings are cut at some randomly chosen position, to produce two 'head' segments, and two 'tail' segments. The tail segments are then swapped over to produce two new full-length chromosomes. Crossover is not usually applied to all pairs of individuals selected for mating. Another genetic operation is mutation, which is an asexual operation that only operates on one individual. It randomly alters each gene with a small probability. Traditional view is that crossover is the more important of the two techniques for rapidly exploring a search space. Mutation provides a small amount of random search, and helps ensure that no point in the search space has a zero probability of being examined.

If the GA has been correctly implemented, the population will evolve over successive generations so that the fitness of the best and the average individual in each generation increases towards the global optimum. Selection is the survival of the fittest within GAs. It determines which individuals are to survive to the next generation. The selection phase consists of three parts. The first part involves determination of the individual's fitness by the fitness function. A fitness function must be devised for each problem; given a particular chromosome, the fitness function returns a single numerical fitness value, which is proportional to the ability, or utility, of the individual represented by that chromosome. For many problems, deciding upon the fitness function is very straightforward, for example, for a function optimization search; the fitness is simply the value of the function. Ideally, the fitness function should be smooth and regular so that chromosomes with reasonable fitness are close in the search space, to chromosomes with slightly better fitness. However, it is not always possible to construct such ideal fitness functions. The second part involves converting the fitness function into an expected value followed by the last part where the expected value is then converted to a discrete number of offsprings. Some of the commonly used selection techniques are roulette wheel and stochastic universal sampling. Genetic programming applies the GA concept to the generation of computer programs. Evolution programming uses mutations to evolve populations. Evolution strategies incorporate many features of the GA but use real-valued parameters in place of binary-valued parameters. Learning classifier systems use GAs in machine learning to evolve populations of condition/action rules.

### 1.2.5 Particle Swarm Optimization (PSO)

Swarm intelligence [54] is a collective behavior of intelligent agents in decentralized systems. Although there is typically no centralized control dictating the behavior of the agents, local interactions among them often cause a global pattern to emerge. Most of the basic ideas are derived from real swarms in the nature including ant colonies, bird flocking, honeybees, bacteria and microorganisms, etc. Ant Colony Optimization (ACO), have already been applied successfully to solve several engineering optimization problems. Swarm models are population-based and the population is initialized with a set of potential solutions. These individuals are then manipulated (optimized) over many iterations using several heuristics inspired from the social behavior of insects in an effort to find the optimal solution. Ant colony algorithms are inspired by the behavior of natural ant colonies, which solve their problems by multi agent cooperation using indirect communication through modifications in the environment. Ants release a certain amount of pheromone (hormone) while walking, and each ant prefers (probabilistically) to follow a direction, which is rich of pheromone. This simple behavior explains why ants are able to adjust to changes in the environment, such as optimizing shortest path to a food source or a nest. In ACO, ants use information collected during past simulations to direct their search and this information is available and modified through the environment. Recently ACO algorithms have also been used for clustering data sets [51].

The concept of particle swarms, although initially introduced for simulating human social behaviors, has become very popular these days as an efficient search and optimization technique. The Particle Swarm Optimization (PSO) [53], as it is called now, does not require any gradient information of the function to be optimized, uses only primitive mathematical operators, and is conceptually very simple. Since its advent in 1995, PSO has attracted the attention of many researchers all over the world resulting in a huge number of variants of the basic algorithm and many parameter automation strategies.

The canonical PSO model consists of a swarm of particles, which are initialized with a population of random candidate solutions [53]. They move iteratively through the $d$-dimension problem space to search for new solutions, where the fitness, $f$, can be calculated as the certain qualities measure. Each particle has a position represented by a position-vector $\mathbf{x}_i$ ($i$ is the index of the particle), and a velocity represented by a velocity-vector $\mathbf{v}_i$. Each particle remembers its own best position so far in a vector $\mathbf{x}_i^{\#}$, and its $j$-th dimensional value is $x_{ij}^{\#}$. The best position-vector among the swarm so far is then stored in a vector $\mathbf{x}^*$, and its $j$-th dimensional value is $x_j^*$. During the iteration time $t$, the update of the velocity from the previous velocity to the new velocity is determined by (1.2). The new position is then determined by the sum of the previous position and the new velocity by (1.3).

$$v_{ij}(t+1) = wv_{ij}(t) + c_1 r_1 (x_{ij}^{\#}(t) - x_{ij}(t)) + c_2 r_2 (x_j^*(t) - x_{ij}(t)). \qquad (1.2)$$

$$x_{ij}(t+1) = x_{ij}(t) + v_{ij}(t+1). \qquad (1.3)$$

where $w$ is called as the inertia factor, $r_1$ and $r_2$ are the random numbers, which are used to maintain the diversity of the population, and are uniformly distributed in the interval [0,1] for the $j$-th dimension of the $i$-th particle. $c_1$ is a positive constant, called the coefficient of the self-recognition component, $c_2$ is a positive constant, called the coefficient of the social component. From (1.2), a particle decides where to move next, considering its own experience, which is the memory of its best past position, and the experience of its most successful particle in the swarm. In the particle swarm model, the particle searches the solutions in the problem space with a range $[-s, s]$ (If the range is not symmetrical, it can be translated to a corresponding symmetrical range.) In order to guide the particles effectively in the search space, the maximum moving distance during one iteration must be clamped in between the maximum velocity $[-v_{max}, v_{max}]$ given in (1.4):

$$v_{ij} = sign(v_{ij}) min(|v_{ij}|, v_{max}). \qquad (1.4)$$

The value of $v_{max}$ is $p \times s$, with $0.1 \leq p \leq 1.0$ and is usually chosen to be $s$, i.e. $p = 1$. The end criteria are usually one of the following:

- Maximum number of iterations: the optimization process is terminated after a fixed number of iterations.
- Number of iterations without improvement: the optimization process is terminated after a fixed number of iterations without any improvement.
- Minimum objective function error: the error between the obtained objective function value and the best fitness value is less than a pre-fixed anticipated threshold.

## 1.3   CI in Gene Expression

Gene expression refers to a process through which the coded information of a gene is converted into structures operating in the cell. It provides the physical evidence that a gene has been *turned on* or activated. Expressed genes include those that are transcribed into mRNA and then translated into protein and those that are transcribed into RNA but not translated into protein (e.g., transfer and ribosomal RNAs) [64, 71]. The expression levels of thousands of genes can be measured at the same time using the modern microarray technology [87, 127]. DNA microarrays usually consist of thin glass or nylon substrates containing specific DNA gene samples spotted in an array by a robotic printing device. Researchers spread fluorescently labeled mRNA from an experimental condition onto the DNA gene samples in the array. This mRNA binds (hybridizes) strongly with some DNA gene samples and weakly with others, depending on the inherent double helical characteristics. A laser scans the array and sensors to detect the fluorescence levels (using red and green dyes), indicating the strength with which the sample expresses each gene. The logarithmic ratio between the two intensities of each dye is used as the gene expression data.

In this section, we provide a substantial review of the state of the art research, which focuses on the application of computational intelligence to different bioinformatics related Gene Expression problems. We also discuss some representative methods to provide inspiring examples to illustrate how CI could be applied to resolve bioinformatics Gene Expression problems and how Gene Expression problems could be analyzed, processed, and characterized by computational intelligence.

### 1.3.1   Gene Expression Data Clustering

In the field of pattern recognition, clustering [48] refers to the process of partitioning a dataset into a finite number of groups according to some similarity measure. Currently, it has become a widely used process in microarray engineering for understanding the functional relationship between groups of genes. Clustering was used, for example, to understand the functional differences in cultured primary epatocytes relative to the intact liver [9]. In another study, clustering techniques were used on gene expression data for tumor and normal colon tissue probed by oligonucleotide arrays [4].

A number of clustering algorithms, including hierarchical clustering [113, 97], Principle Component Analysis (PCA) [119, 89], genetic algorithms [59], and artificial neural networks [42, 101, 107], have been used to cluster gene expression data. However, in 2002, Yuhui et al. [120] proposed a new approach to analysis of gene expression data using Associative Clustering Neural Network (ACNN). ACNN dynamically evaluates similarity between any two gene samples through the interactions of a group of gene samples. It exhibits more robust performance than the methods with similarities evaluated by direct distances, which has been tested on the leukemia data set. The experimental results demonstrate that ACNN is superior in dealing with high dimensional data (7,129 genes).

The performance can be further enhanced when some useful feature selection methodologies are incorporated. The study has shown ACNN can achieve 98.61% accuracy on clustering the Leukemias data set with correlation analysis.

Herrero et al. [42] used the Self-Organizing Tree Algorithm (SOTA) for analysis of gene expression data coming from DNA array experiments, using an unsupervised neural network. DNA array technologies allow monitoring thousands of genes rapidly and efficiently. One of the interests of these studies is the search for correlated gene expression patterns, and this is usually achieved by clustering them. The result of the algorithm is a hierarchical cluster obtained with the accuracy and robustness of a neural network. SOTA clustering confers several advantages over classical hierarchical clustering methods. The clustering process is performed from top to bottom, i.e. the highest hierarchical levels are resolved before going to the details of the lowest levels. The growing can be stopped at the desired hierarchical level. Moreover, a criterion to stop the growing of the tree, based on the approximate distribution of probability obtained by randomisation of the original data set, is provided. In addition, obtaining average gene expression patterns is a built-in feature of the algorithm. Different neurons defining the different hierarchical levels represent the averages of the gene expression patterns contained in the clusters.

Xiao et al. [116] proposed a new clustering approach based on the synergism of the PSO and Self Organizing Maps (SOM). The authors achieved promising results by applying the hybrid SOM-PSO algorithm over the gene expression data of yeast and rat hepatocytes. We will briefly discuss their approach in the following paragraphs. The idea of the SOM [56] stems from the orderly mapping of information in the cerebral cortex. With SOMs, high dimensional datasets are projected onto a one- or two-dimensional space. Typically, a SOM has a two dimensional lattice of neurons and each neuron represents a cluster. The learning process of a SOM is unsupervised. All neurons compete for each input pattern and the neuron that is chosen for the input pattern wins.

In the approach proposed by Xiao et al., PSO is used to evolve the weights for the SOM. In the first stage of the hybrid SOM/PSO algorithm, a SOM is used to cluster the dataset. Authors used a SOM with conscience at this step. Conscience directs each component that takes part in competitive learning toward having the same probability to win. Conscience is added to the SOM by assigning each output neuron a bias. The output neuron must overcome its own bias to win. The objective is to obtain a better approximation of pattern distribution. The SOM normally runs for 100 iterations and generates a group of weights. In the second stage, PSO is initialized with the weights produced by the SOM in the first stage. Then a *gbest* PSO is used to refine the clustering process. Each particle consists of a complete set of weights for the SOM. The dimension of each particle is the number of input neurons of the SOM times the number of output neurons of the SOM. The objective of PSO is to improve the clustering result by evolving the population of particles.

Microarrays have recently made it possible to monitor the activity of thousands of genes simultaneously. They offer new insights into the biology of a cell.

However, the data produced by microarrays poses several challenges to over-come. One major task in the analysis of microarray data is to reveal structures despite a large noise component in the data. Futschik and Kasabov [33] used Fuzzy C-Means (FCM) clustering to achieve a robust analysis of gene expression time-series. Authors address the issues of parameter selection and cluster valid-ity. Using statistical models to simulate gene expression data, they show that FCM can detect genes belonging to different classes.

Chinatsu and Hanai [7] applied the Fuzzy Adaptive Resonance Theory (Fuzzy ART) [106] to gene clustering of DNA microarray data and their result indicate that the methodology may be more suitable for biological applications than most other methods including hierarchical clustering, k-means clustering, and SOM. In addition, the authors compared their technique with the fuzzy c-means clustering method and obtained comparable results.

Okada et al. [79] point out that although hierarchical clustering has been extensively used in analyzing patterns in microarray gene expression data, its biological interpretation is not easy. The authors propose a novel algorithm that automatically finds biologically interpretable cluster boundaries in hierarchical clustering by referring to gene annotations stored in public genome databases. In addition, the proposed algorithm has a new function of generating a set of clusters that are independent of each other with respect to the distributions of gene functions. The authors claim that this function would enable investigators to efficiently identify non-redundant and biologically-independent clusters.

## An Evolutionary Rough C-Means Clustering

Cluster analysis [104] is one key step in understanding how the activity of genes varies during biological processes and is affected by disease states and cellular environments. In particular, clustering can be used either to identify sets of genes according to their expression in a set of samples [26, 113], or to cluster samples into homogeneous groups that may correspond to particular macroscopic phenotypes [36]. The latter is in general more difficult, but is very valuable in clinical practice.

Several clustering algorithms have been developed and applied in bioinformat-ics problems, however, most of them cannot process objects in hybrid numeri-cal/nominal feature space or with missing values. In most of them, the number of clusters should be manually determined and the clustering results are sensi-tive to the input order of the objects to be clustered. These limit applicability of the clustering and reduce the quality of clustering. To solve this problem, an improved clustering algorithm based on rough set and entropy theory was presented by Chun-Bao et al. [19]. The approach aims at avoiding the need to pre-specify the number of clusters, and clustering in both numerical and nominal feature space with the similarity introduced to replace the distance index.

At the same time, rough sets are used to represent clusters in terms of upper and lower approximations. However, the relative importance of these approxi-mation parameters, as well as a threshold parameter, need to be tuned for good partitioning. The evolutionary rough c-means algorithm employs GAs to tune these parameters. The Davies-Bouldin index is used as the fitness function to be minimized. Various values of c are used to generate different sets of clusters, and GA is employed to generate the optimal partitioning [100].

Lingras [62] argued that incorporation of rough sets into k-means clustering requires the addition of the concept of lower and upper bounds. Calculation of the centroids of clusters from conventional k-means needs to be modified to include the effects of lower as well as upper bounds. The modified centroid calculations for rough sets are then given by:

$$cen_j = W_{low} \times \frac{\sum_{v \in \underline{R}(x)}}{|\underline{R}(x)|} + w_{up} \times \frac{\sum_{v \in (\overline{BN_R(x)})}}{|\overline{BN_R(x)}|} \tag{1.5}$$

Where $1 \leq j \leq m$. The parameters $w_{low}$ and $w_{up}$ correspond to the relative importance of lower and upper bounds, and $w_{low} + w_{up} = 1$. If the upper bound of each cluster were equal to its lower bound, the clusters would be conventional clusters. Therefore, the boundary region $\overline{BN_R(x)}$ will be empty, and the second term in the equation will be ignored. Thus, the above equation will reduce to conventional centroid calculations. The next step in the modification of the k-means algorithms for rough sets is to design criteria to determine whether an object belongs to the upper or lower bound of a cluster, for more details refer to [62]. The main steps of the algorithm are provided below.

---

**Algorithm 1.** Rough C-Means Algorithm
---
1: Set $x_i$ as an initial means for the c clusters.
2: Initialize the population of particles encoding parameters threshold and $w_{low}$
3: Initialize each data object $x_k$ to the lower approximation or upper approximation of clusters $c_i$ by computing the difference in its distance by:

$$diff = d(x_k, cen_i) - d(x_k, cen_j), \tag{1.6}$$

Where $cen_i$ and $cen_j$ are the cluster centroid pairs.
4: **if** diff $< \delta$ **then**
5:     $x_k \in$ the upper approximation of the $cen_i$ and $cen_j$ clusters and can not be in any lower approximation.
6:     Else
7:     $x_k \in$ lower approximation of the cluster $c_i$ such that distance $d(x_k, cen_i)$ is is minimum over the c clusters.
8: **end if**
9: Compute a new mean using equation (1.5)
10: **repeat**
11:     statements 3–9
12: **until** convergence i.e. there is no more new assignments

---

### 1.3.2   Rough Sets and DNA Microarray Technology

Biological research is currently undergoing a revolution. With the advent of microarray technology the behavior of thousands of genes can be measured simultaneously. This capability opens a wide range of research opportunities in

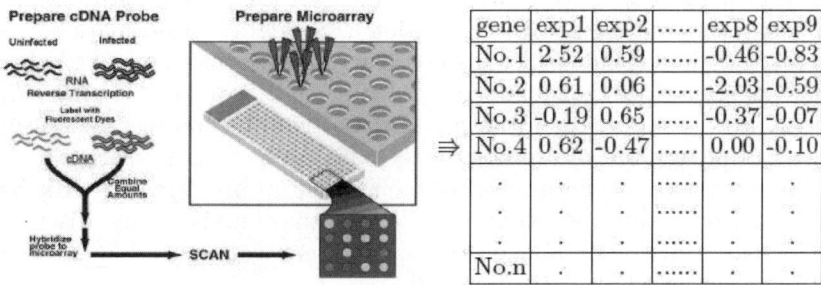

| gene | exp1 | exp2 | ...... | exp8 | exp9 |
|------|------|------|--------|------|------|
| No.1 | 2.52 | 0.59 | ...... | -0.46 | -0.83 |
| No.2 | 0.61 | 0.06 | ...... | -2.03 | -0.59 |
| No.3 | -0.19 | 0.65 | ...... | -0.37 | -0.07 |
| No.4 | 0.62 | -0.47 | ...... | 0.00 | -0.10 |
| . | . | . | ...... | . | . |
| . | . | . | ...... | . | . |
| . | . | . | ...... | . | . |
| No.n | . | . | ...... | . | . |

**Fig. 1.3.** Microarray production process:Microarrays provide the gene expression data. A sample of 9 experiments from Synovial Sarcoma data is illustrated, n=5,520 genes in this data set [37, 96].

biology, but the technology generates a vast amount of data that cannot be handled manually. Computational analysis is thus a prerequisite for the success of this technology, and research and development of computational tools for microarray analysis are of great importance [68]. The DNA microarray technology provides enormous quantities of biological information about genetically conditioned susceptibility to diseases [11]. The data sets acquired from microarrays refer to genes via their expression levels. Microarray production starts with preparing two samples of mRNA, as illustrated by Figure 1.3. The sample of interest is paired with a healthy control sample. The fluorescent red/green labels are applied to both samples. The procedure of samples mixing is repeated for each of thousands of genes on the slide. Fluorescence of red/green colors indicates to what extent the genes are expressed. The gene expressions can be then stored in numeric attributes, coupled with, e.g., clinical information about the patients [11].

One application of microarray technology is cancer studies, where supervised learning may be used for predicting tumor subtypes and clinical parameters. Herman et al. [68] present a general rough set approach for classification of tumor samples analyzed with microarrays. This approach is tested on a data set of gastric tumors, and authors develop classifiers for six clinical parameters. This research included only 2,504 genes out of a total of at least 30,000 genes in the human genome. Some of the genes that were not included in their study may have a connection to the parameters. In addition, their results show that it is possible to develop classifiers with a small number of tumor samples, and that rough set based methods may be well suited for this task. They believe that rough set based learning combined with feature selection may become an important tool for microarray analysis.

**Rough Discretization**

Microarray measurements are real numbers that have to be discretized before a learning algorithm can be applied on the them. It has been shown that the quality of a learning algorithm is dependent on the selected strategy used for real data discritization [25]. Discretization uses a data transformation procedure that involves finding cuts which divide the data values into intervals. Values lying within an interval are then mapped to the same 'label' value. Performing this process will lead to reduction in the size of the attributes value set and ensure that the rules that are mined are not too specific. Lots of discretization algorithms have been developed and applied in bioinformatics problems [68]. Examples of utilized discretization algorithms include frequency binning, naïve discretization, entropy-based discretization, discriminant discretization, and Boolean reasoning/rough set based discretization [68].

Here we demonstrate some reported examples of using discretization techniques in bioinformatics problems. Many successful work towards this issue has been addressed and discussed. For example, the rough sets with Boolean reasoning (RSBR) algorithm proposed by Zhong et al. [124, 40] was used for discretization of continuous-valued attributes. The main advantage of RSBR is that it combines discretization of real valued attributes and classification. The main steps of the RSBR discretization algorithm are provided below.

---

**Algorithm 2.** RSBR Discretization Algorithm

---

Input: Information system table $(S)$ with real valued attributes $A_{ij}$ and $n$ is the number of inter values for each attribute.

Output: Information table $(ST)$ with discretized real valued attribute

1: **for** $A_{ij} \in S$ **do**
2:      Define a set of Boolean variables as follows:

$$B = \{\sum_{i=1}^{n} C_{ai}, \sum_{i=1}^{n} C_{bi} \sum_{i=1}^{n} C_{ci}, ..., \sum_{i=1}^{n} C_{ni}\} \qquad (1.7)$$

3: **end for**
     Where $\sum_{i=1}^{n} C_{ai}$ corresponds to a set of intervals defined on the variables of attributes $a$
4: Create a new information table $S_{new}$ by using the set of intervals $C_{ai}$
5: Find the minimal subset of $C_{ai}$ that discerns all the objects in the decision class $D$ using the following formula:

$$\Upsilon^u = \wedge\{\Phi(i,j) : d(x_i \neq d(x_j)\} \qquad (1.8)$$

     Where $\Phi(i,j)$ is the number of minimal cuts that must be used to discern two different instances $x_i$ and $x_j$ in the information table.

---

Among further research directions, there is hybridization of rough set reduction framework with gene clustering. For example, in [37] authors used self-organizing maps to calculate the entropy distance for roughly discretized data. In another example, Ślęzak and Wróblewski [95] adapt the rough set-based approach to deal with gene expression data, where the problem is a huge amount of genes (attributes) $a \in A$ versus small amount of experiments (objects) $u \in U$. They perform gene reduction using standard rough set methodology based on approximate decision reducts applied against specially prepared data. In addition, the authors used rough discretization algorithm - Every pair of objects $(x, y) \in U \times U$ yields a new object, which takes values "$\geq a(x)$" if and only if $a(y) \geq a(x)$; and "$\leq a(x)$" otherwise; over original genes-attributes $a \in A$. In this way: 1) They work with desired, larger number of objects improving credibility of the obtained reducts; 2) They produce more decision rules, which vote during classification of new observations; 3) They avoid an issue of discretization of real-valued attributes, difficult and leading to unpredictable results in case of any data sets having much more attributes than objects. The authors illustrated their method by analysis of gene expression data related to breast cancer.

Another example given by Ślęzak and Wróblewski [96] extends the standard rough set-based approach to deal with huge amounts of numeric attributes versus a small amount of available objects. The authors transform the training data using a novel way of non-parametric discretization, called roughfication (in contrast to fuzzification known from fuzzy logic). Given roughfied data, they apply standard rough set attribute reduction and then classify the testing data by voting among the obtained decision rules. Roughfication enables to search for reducts and rules in the tables with the original number of attributes and far larger number of objects. It does not require expert knowledge or any kind of parameter tuning or learning. The authors illustrate it by analysis of gene expression data, where the number of genes (attributes) is enormously large with respect to the number of experiments (objects).

Given thousands of attributes against hundreds of objects, we face a *few-objects-many-attributes* problem, recognized as one of the main data mining challenges [118]. Moreover, in the case of gene expression, rough set based methods usually require discretization (cf. [76])–replacing the original values with the codes of intervals defined over attribute ranges. This additionally increases the amount of possible solutions of the optimization problem, now reformulated as searching for optimal subsets of attributes (genes) coupled with their optimal interval settings. Such a huge space of parameters, given too small samples of objects, leads to data overfitting (cf. [118]) and yields a kind of unreliability of the rough set techniques applied so far (cf. [109]). Ślęzak and Wróblewski [96] report an alternative method, illustrated by Figure 1.4. They call it rough discretization (or roughfication, compared to fuzzification).

As has been reported, e.g., [68], some discretization methods seem to work better than others for the problem of gene expression classification. Frequency binning and entropy-based discretization gave good results. Discretization based on linear discriminant analysis was also useful. The entropy-based method appeared to handle skewed class distributions better than the other methods. Boolean reasoning discretization had often a poor performance and behaved differently from the rest of the discretization methods. The AUC had a tendency to increase with additional genes. It is likely that this is due to the global nature of this method. The method considers all attributes at once when it creates cuts. The feature selection method, on the other hand, selects genes individually such that each selected gene may be a good classifier in itself. So, it is more appropriate to

|     | a | b | c | d |
|-----|---|---|---|---|
| u1  | 3 | 7 | 3 | 0 |
| u2  | 2 | 1 | 0 | 1 |
| u3  | 4 | 0 | 6 | 1 |
| u4  | 0 | 5 | 1 | 2 |

|         | a* | b* | c* | d* |
|---------|----|----|----|----|
| (u1,u1) | 1+ | 1+ | 1+ | 0 |
| (u1,u2) | 1– | 1– | 1– | 1 |
| (u1,u3) | 1+ | 1– | 1+ | 1 |
| (u1,u4) | 1– | 1– | 1– | 2 |
| (u2,u1) | 2+ | 2+ | 2+ | 0 |
| (u2,u2) | 2+ | 2+ | 2+ | 1 |
| (u2,u3) | 2+ | 2– | 2+ | 1 |
| (u2,u4) | 2– | 2+ | 2+ | 2 |
| (u3,u1) | 3– | 3+ | 3– | 0 |
| (u3,u2) | 3– | 3+ | 3– | 1 |
| (u3,u3) | 3+ | 3+ | 3+ | 1 |
| (u3,u4) | 3– | 3+ | 3– | 2 |
| (u4,u1) | 4+ | 4+ | 4+ | 0 |
| (u4,u2) | 4+ | 4– | 4– | 1 |
| (u4,u3) | 4+ | 4– | 4+ | 1 |
| (u4,u4) | 4+ | 4+ | 4+ | 2 |

$$POS(a^*,b^*) = POS(a^*,b^*,c^*)$$

$$POS(a^*) \subset POS(a^*,b^*,c^*)$$

$$POS(b^*) \subset POS(a^*,b^*,c^*)$$

IF a≥3 AND b≥7 THEN d=0
IF a≥3 AND b<7 THEN d=1
IF a≥2 AND b<1 THEN d=1
IF a<2 AND b≥1 THEN d=2
IF a≥4 AND b≥0 THEN d=1
IF a≥0 AND b<5 THEN d=1

**Fig. 1.4.** Rough discretization [96]. Top: A sample with 3 numeric attributes and 3 decision classes. Right: Its roughfied version. Middle: Some positive regions for the roughfied table. Bottom: Rules induced by reduct $a^*, b^*$.

make cuts individually for each gene. The Boolean reasoning approach is consequently less suited for this problem, but it may yield a good performance in other situations.

## 1.4  CI in Protein Sequence Classification

The problem of protein sequence classification is a crucial task in the interpretation of genomic data. Many high-throughput systems were developed with the aim of categorizing proteins based only on their sequences. However, modeling how proteins have evolved can also help in the classification task of sequenced data. Hence the phylogenetic analysis has gained importance in the field of protein classification. Busa-Fekete et al. [16] provide an overview about the problem of protein sequence classification area and propose two algorithms that are well suited to this scope. The two algorithms are based on a weighted binary tree representation of protein similarity data. The first one is called TreeInsert which assigns the class label to the query by determining a minimum cost necessary to insert the query in the (precomputed) trees representing the various classes. Then the TreeNN algorithm assigns the label to the query based on an analysis of the query's neighborhood within a binary tree containing members of the known classes. The two algorithms were tested in combination with various sequence similarity scoring methods (BLAST, Smith-Waterman, Local Alignment Kernel as well as various compression-based distance scores) using a large number of classification tasks representing various degrees of difficulty. They reported that, at the expense of a small computational overhead, both TreeNN and TreeInsert exceed the performance of simple similarity search (1NN) as determined by ROC analysis, at the expense of a modest computational overhead. Combined with a fast tree-building method, both algorithms are suitable for web-based server applications.

Mapping the pathways that give rise to metastasis is one of the key challenges of breast cancer research. Recently, several large-scale studies have shed light on this problem through analysis of gene expression profiles to identify markers correlated with metastasis. Han-Yu Chuang et al. [21] apply a protein-network-based approach that identifies markers not as individual genes but as subnetworks extracted from protein interaction databases. The resulting subnetworks provide novel hypotheses for pathways involved in tumor progression. Although genes with known breast cancer mutations are typically not detected through analysis of differential expression, they play a central role in the protein network by interconnecting many differentially expressed genes. Authors find that the subnetwork markers are more reproducible than individual marker genes selected without network information, and that they achieve higher accuracy in classification of metastatic versus non-metastatic tumors.

As shown in Figure 1.5, the subnetwork markers were significantly more reproducible between data sets than were individual marker genes selected without network information (12.7 versus 1.3%). In terms of biological function, extracellular signal-regulated kinase 1 (MAPK3) was reproducible as a central node in subnetworks identified from both data sets (Figure 1.5C versus Figure 1.5D. Figure 1.5E and 1.5F illustrate two other subnetworks that were discriminative in both data sets, although there was less consistency in the expression levels of genes comprising these subnetworks. For instance, PKMYT1 is significantly differentially expressed in van de Vijver et al [110] but not in Wang et al. [112] (Figure 1.5E; diamond versus circle), whereas CD44 is significantly differentially expressed in Wang et al. [112] but not in van de Vijver et al. [110] (Figure 1.5F). However, by aggregating the expression ratios of these genes with their network neighbors, the subnetworks containing these genes are found to be significant in both data sets.

Classification of protein sequences into families is an important tool in the annotation of structural and functional properties to newly discovered proteins. Mohamed et al [72] present a classification system using pattern recognition techniques to create a numerical vector representation of a protein sequence and then classify the sequence into a number of given families. Authors introduce the

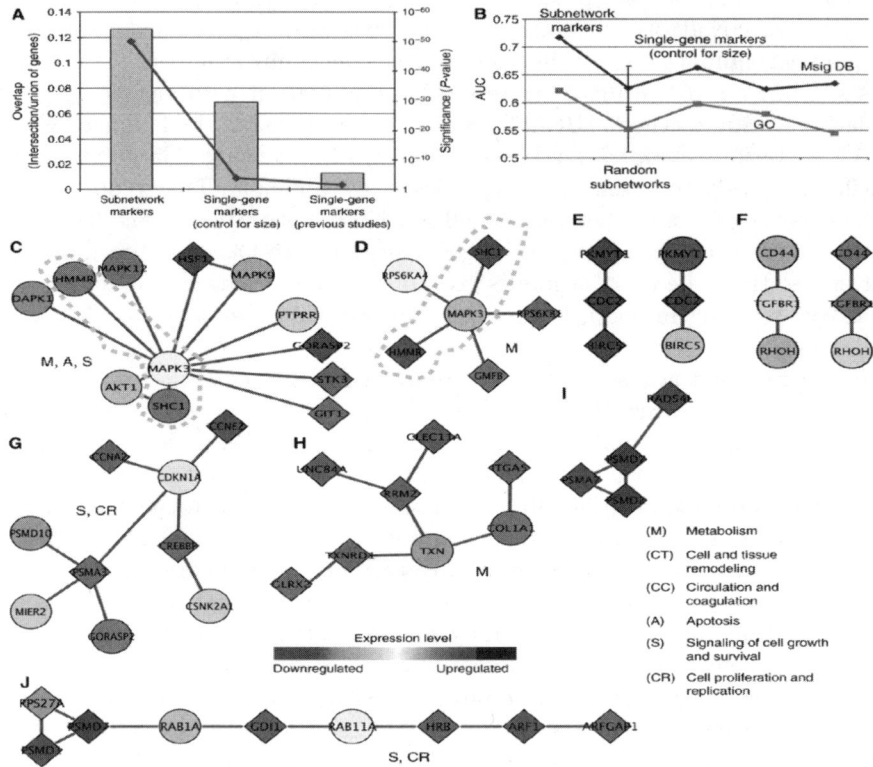

**Fig. 1.5.** Subnetwork markers across data sets [21]

use of fuzzy ARTMAP classifiers and show that coupled with a genetic algorithm based feature subset selection, the system is able to classify protein sequences with an accuracy of 93%. This accuracy is compared with numerous other classification tools and demonstrates that the fuzzy ARTMAP is suitable due to its high accuracy, quick training times, and ability for incremental learning.

Building improved intelligent protein sequence classification systems for effectively searching large biological database is significant for developing competitive pharmacological products. Wang et al [111] describe a methodology for constructing a neural protein classifier with various input features, rather than to train a neural classifier based on a given neural network architecture and some available data. A set of fuzzy classification rules with confidence factors can be extracted directly from the generalized radial basis function (GRBF) networks. The initial fuzzy rule set is refined using a new objective function, which compromises between misclassification rate and generalization capability, and GA programming. Their results compared favorably with other standard machine learning techniques.

## 1.5  CI in Gene Selection

Selecting informative and discriminative genes from huge microarray gene expression data is an important and challenging bioinformatics research topic. There have been many successful projects in this area reported in the literature. For example, Fernando et al. [29] demonstrate how a supervised fuzzy pattern algorithm can be used to perform DNA microarray data reduction over real data. The benefits of their method can be employed to find biologically significant insights relating to meaningful genes in order to improve previous successful techniques. Experimental results on acute myeloid leukemia diagnosis show the effectiveness of the proposed approach.

A new method combining correlation based clustering and rough sets attribute reduction for gene selection from gene expression data is proposed by Lijun et al [99]. Correlation based clustering is used as a filter to eliminate the redundant attributes, then the minimal reduct of the filtered attribute set is reduced by rough sets. Three different classification algorithms are employed to evaluate the performance of the proposed method. High classification accuracies achieved on two public gene expression data sets show that the introduced method is successful for selecting high discriminative genes for classification task. The experimental results indicate that rough sets based methods have the potential to become a useful tool in bioinformatics.

The approach to cancer classification based on selected gene expression data, rather than all the genes in the dataset, is important for efficient cancer diagnosis. Dingfang et al. [58] present a gene selection method, called RMIMR, which searches for the subset through maximum relevance and maximum positive interaction of genes. Compared to the classical methods based on statistics, information theory, and regression, this method led to significantly improved classification in experiments on 4 gene expression datasets.

Banerjee et al. [12] used an evolutionary rough feature selection algorithm for classifying microarray gene expression patterns. Since the data typically consist of a large number of redundant features, an initial reduction of the attributes is done to enable faster convergence. Rough set theory is employed to generate reducts, which represent the minimal sets of nonredundant features capable of discerning between all objects, in a multiobjective framework. The effectiveness of the algorithm is demonstrated on three cancer datasets.

Zhang et al. [123] present recent Support Vector Machine (SVM) classification approaches for gene selection, cancer classification, and functional gene classification, followed by analysis on the advantages and limitations of SVM on these applications.

Li et al. [59] introduced a multivariate approach that selects a subset of predictive genes jointly for sample classification based on expression data. They tested the algorithm on colon and leukemia data sets. The authors examined the sensitivity, reproducibility and stability of gene selection/sample classification to the choice of parameters of the algorithm. They used hybrid method that uses a genetic algorithms and the K-Nearest Neighbor (KNN) to identify genes that can jointly discriminate between different classes of samples (e.g. normal versus tumor). The genes identified are subsequently used to classify independent test set samples. The authors reported that the GA/KNN method is capable of selecting a subset of predictive genes from a large noisy data set for sample classification. It is a multivariate approach that can capture the correlated structure in the data.

Yuanchen et al. [41] proposed a fuzzy-granular method for the gene selection task. Firstly, genes are grouped into different function granules with the fuzzy c-means algorithm (FCM). And then informative genes in each cluster are selected with the signal-to-noise metric (S2N). With fuzzy granulation, information loss in the process of gene selection is decreased. As a result, more informative genes for cancer classification are selected and more accurate classifiers can be modeled. The simulation results on two publicly available microarray expression datasets show that the proposed method is more accurate than traditional algorithms for cancer classification.

## Gene Selection Using Neural Networks

Accurate diagnosis and classification are the key issues for the optimal treatment of cancer patients. Several studies demonstrate that cancer classification can be estimated with high accuracy, sensitivity, and specificity from microarray-based gene expression profiling using artificial neural networks.

Huang and Liao [45] introduced a comprehensive study to investigate the capability of the probabilistic neural networks (PNN) associated with a feature selection method, the so-called signal-to-noise statistic, in cancer classification. The signal-to-noise statistic, which represents the correlation with the class distinction, is used to select the marker genes and trim the dimension of data samples for the PNN. The experimental results show that the association of the probabilistic neural network with the signal-to-noise statistic can achieve superior classification results for two types of acute leukemias and five categories of embryonal tumors of central nervous system with satisfactory computation speed. Furthermore, the signal-to-noise statistic analysis provides candidate genes for future study in understanding the disease process and the identification of potential targets for therapeutic intervention.

Fogel [31] highlights recent advancements in the coupling evolutionary computation with artificial neural networks for microarray class prediction and discovery. The combination of these methods holds great promise for automated feature selection and data analysis. Neural networks have been noted elsewhere in the literature as particularly useful for microarray data clustering and classification. For instance, Khan et al. [55] developed a method of classifying cancers to specific diagnostic categories based on their gene expression signatures using artificial neural networks. The authors trained the ANNs using a small, round blue-cell tumors (SRBCTs) as a model. These cancers belong to four distinct diagnostic categories and often present diagnostic dilemmas in clinical practice. The ANNs correctly classified all samples and identified genes most relevant to the classification. Expression of several of these genes has been reported in SR-BCTs, but most have not been associated with these cancers. To test the ability

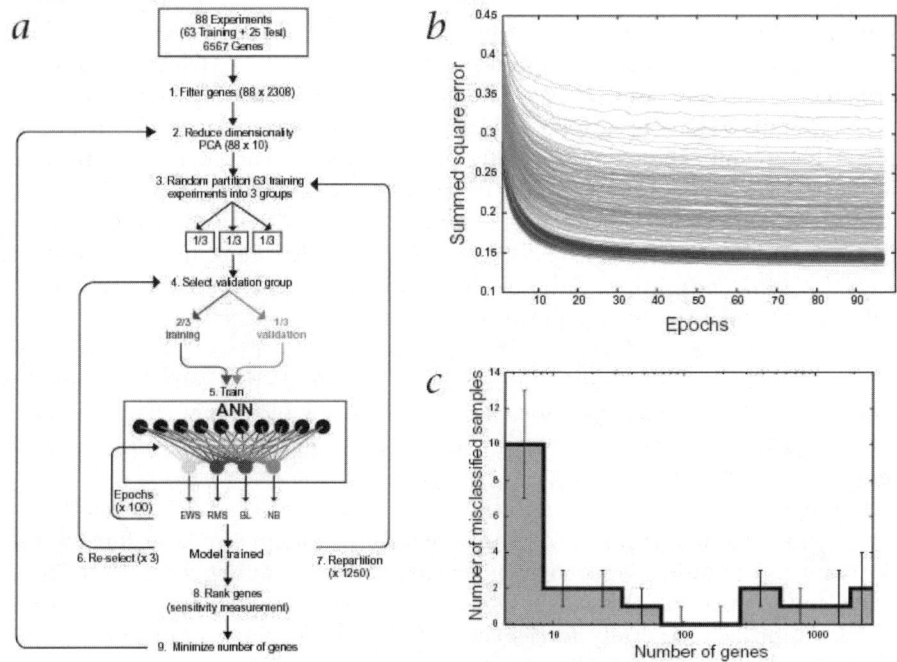

**Fig. 1.6.** Classification and diagnostic prediction of cancers using gene expression profiling and artificial neural networks [55]

of the trained ANN models to recognize SRBCTs, they analyzed additional blind samples that were not previously used for training, and correctly classified them in all cases. This study demonstrates the potential applications of these methods for tumor diagnosis and the identification of candidate targets for therapy.

As an illustrated in Figure 1.6a, the entire data-set of all 88 experiments was first quality filtered (1) and then the dimensionality was further reduced by principal component analysis (PCA) to 10 PC projections (2), from the original 6,567 expression values. Next, the 25 test experiments were set aside and the 63 training experiments were randomly partitioned into 3 groups (3). One of these groups was reserved for validation and the remaining 2 groups for calibration (4). ANN models were then calibrated using for each sample the 10 PC values as input and the cancer category as output (5). For each model, the calibration was optimized with 100 iterative cycles (epochs). This was repeated using each of the 3 groups for validation (6). The samples were again randomly partitioned and the entire training process repeated (7). For each selection of a validation group one model was calibrated, resulting in a total of 3750 trained models. Once the models were calibrated they were used to rank the genes according to their importance for the classification (8). The entire process (2–7) was repeated using only top ranked genes (9). The 25 test experiments were subsequently classified using all the calibrated models. Figure 1.6b presents monitoring of the calibration of the models. The average classification error per sample (using a summed square error function) is plotted during the training iterations (epochs) for both the training and the validation samples. A pair of lines, dark (training) and light (validation), represents one model. The decrease in the classification errors with increasing epochs demonstrates the learning of the models to distinguish these cancers. The results shown are for 200 different models, each corresponding to a random partitioning of the data. All the models performed well for both training and validation as demonstrated by the parallel decrease (with increasing epochs) of the average summed square classification error per sample. In addition, there was no sign of over-training: if the models begin to learn features in the training set, which are not present in the validation set, this would result in an increase in the error for the validation at that point and the curves would no longer remain parallel. Figure 1.6c shows minimizing the number of genes. The average number of misclassified samples for all 3,750 models is plotted against increasing number of used genes. The misclassifications were minimized to zero using the 96 highest ranked genes [31, 55].

While it is clear that neural network methods are well suited to microarray analysis, their proper training and optimization is a prerequisite for superior performance. A standard approach to neural network training is the use of back-propagation to optimize the weight assignments for a fixed neural network topology. This approach generally forces the user to choose the appropriate number of features to use and a fixed neural network topology. Backpropagation itself can also lead to suboptimal weight assignment if there are many local optima in the search space. Optimizing neural networks with stochastic optimization methods such as evolutionary computation, however, can outperform these classic methods by avoiding local optima and simultaneously identifying the most appropriate features to use for prediction [31].

In another study, Hwang et al. [47] applied neural networks in classification of patient samples using gene expressions levels. Here all gene expression levels are fed to the neural tree as input and the output is a binary classification. Through a structural learning process, essential genes for cancer classification are included into the neural tree and less important genes are weeded out automatically. In neural tree learning, all gene expression levels were linearly scaled into the interval [0.01, 0.99]. For the output value of neural tree learning, one was set to 0.01 and the other one to 0.99. Using this setup, their predicted accuracy was 86% and the number of genes selected was 16. Gene selection using Feed Forward Back Propagation Neural Network as a classifier is illustrated in Figure 1.7.

Francesca et al. [92] proposed a new gene selection method for analyzing microarray experiments pertaining to two classes of tissues and for determining relevant genes characterizing differences between the two classes. The new technique is based on Switching Neural Networks (SNN), learning machines that assign a relevance value to each input variable, and adopts Recursive Feature Addition (RFA) for performing gene selection. The performances of SNN-RFA are evaluated by considering its application on two real and two artificial gene

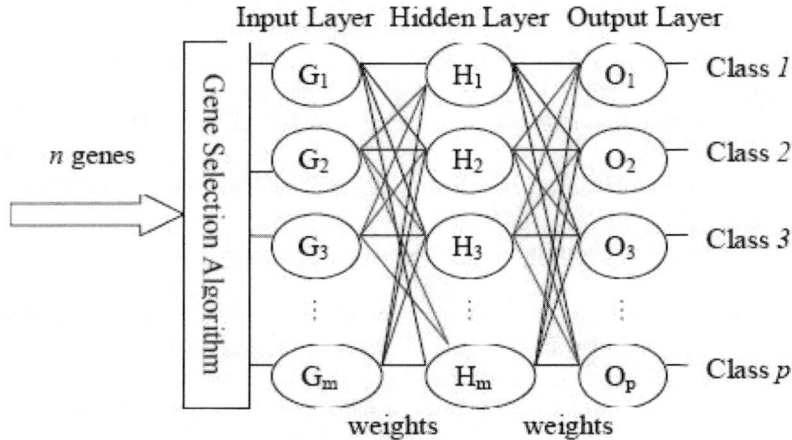

**Fig. 1.7.** Gene selection using Neural Network as classifere [81]

expression datasets generated according to a proper mathematical model that possesses biological and statistical plausibility.

Gene selection algorithms for cancer classification, based on the expression of a small number of biomarker genes, have been the subject of considerable research in recent years [81]. For instance, Feng et al. [20] use a t- test-based feature selection method to choose some important genes from thousands of genes. After that, authors classify the microarray data sets with a Fuzzy Neural Network (FNN). The FNN combines important features of initial fuzzy model self-generation, parameter optimization, and rule-base simplification. They applied the FNN to three well-known gene expression data sets, i.e., the lymphoma data set (with 3 sub-types), small round blue cell tumor (SRBCT) data set (with 4 sub-types), and the liver cancer data set (with 2 classes, i.e., non-tumor and hepatocellular carcinoma (HCC)). Their results in all the three data sets show that the FNN can obtain 100% accuracy with a much smaller number of genes in comparison with previously published methods. They reported that in view of the smaller number of genes required by the FNN and its high accuracy,the FNN classifier not only helps biological researchers differentiate cancers that are difficult to be classified using traditional clinical methods, but also helps biological researchers focus on a small number of important genes to find the relationships between those important genes and the development of cancers (see also [117]).

## 1.6 CI in DNA Fragment Assembly (FA)

The fragment assembly problem (FAP) deals with sequencing of DNA. Currently strands of DNA, longer than approximately 500 base pairs, cannot be sequenced very accurately. As a consequence, in order to sequence larger strands of DNA, they are first broken into smaller pieces. The FAP is then to reconstruct the original molecule's sequence from the smaller fragment sequences. FAP is basically a permutation problem, similar in spirit to the TSP, but with some important differences (circular tours, noise, and special relationships between entities) [94]. Meksangsouy and Chaiyaratana [67] attempted to solve the DNA fragment reordering problem with the ant colony system. The authors investigated two types of assembly problems: single-contig and multiple-contig problems. The simulation results indicate that the ant colony system algorithm outperforms the nearest neighbor heuristic algorithm when multiple-contig problems are considered.

The DNA fragment assembly is a problem to be solved in the early phases of the genome project and thus is critical since the other steps depend on its accuracy. This is an NP-hard combinatorial optimization problem which is growing in importance and complexity as more research centers become involved on sequencing new genomes. Various heuristics, including computational intelligence algorithms, have been designed for solving the fragment assembly problem, but since this problem is a crucial part of any sequencing project, better assemblers are needed. Here we demonstrated some reported examples of using the CI techniques in DNA Fragment Assembly problem.

Wannasak et al. [114] present the use of a combined ant colony system (ACS) and nearest neighbour heuristic (NNH) algorithm in DNA fragment assembly. The assembly process can be treated as combinatorial optimization where the aim is to find the right order of each fragment in the ordering sequence that leads to the formation of a consensus sequence that truly reflects the original DNA strands. The assembly procedure proposed is composed of two stages: fragment assembly and contiguous sequence (contig) assembly. In the fragment assembly stage, a possible alignment between fragments is determined where the fragment ordering sequence is created using the ACS algorithm. The resulting contigs are then assembled together using the NNH rule. Their results indicate that in overall the performance of the combined ACS/NNH technique is superior to that of a standard sequence assembly program (CAP3), which is widely used by many genomic institutions.

Angeler et al. [6] describes an alternative approach to the fragment assembly problem. The key idea is to train a recurrent neural network (RNN) to track a sequence of bases constituting a given fragment and to assign to the same cluster all sequences which are well tracked by this network. The authors make use of a 3-layer Recurrent Perceptron and examine both edited sequences from an ftp site and artificial fragments from a common simulation software. The clusters they obtain exhibit interesting properties in terms of error filtering, stability and self consistency; they define as well, with a certain degree of approximation, a metric on the fragment set. The proposed assembly algorithm is susceptible to becoming an alternative method with the following properties: (i) high quality of the rebuilt genomic sequences, (ii) high parallelizability of the computing process with consequent drastic reduction of the running time.

## 1.7  CI in Multiple Sequence Alignment (MSA)

Sequence Alignment (SA) refers to the process of arranging the primary sequences of DNA, RNA, or protein to identify regions of similarity that may be a consequence of functional, structural, or evolutionary relationships between the sequences. Given two sequences $X$ and $Y$, a pair-wise alignment indicates positions of each sequence that are considered to be functionally or evolutionarily related. From a family $S = (S_0, S_1, \ldots, S_{N-1})$ of $N$ sequences, we would like to find out common patterns of this family. Since aligning each pair of sequences from $S$ separately often does not reveal the common information, it is necessary to perform multiple sequence alignment (MSA). A multiple sequence alignment (MSA) is a sequence alignment of three or more biological sequences, generally protein, DNA, or RNA. In general, the input set of query sequences are assumed to have an evolutionary relationship by which they share a linkage and are descended from a common ancestor.

To evaluate the quality of an alignment, a popular choice is to use the SP (sum-of-pairs) score method [63]. The SP score basically sums the substitution scores of all possible pair-wise combinations of sequence characters in one column of a multiple sequence alignment. Assuming $c_i$ representing the $i^{th}$ character of a given column in the sequence matrix and match $(c_i, c_j)$ denoting the comparing score between characters $c_i$ and $c_j$, the score of a column may be computed using the formula:

$$SP = (c_1, c_2, \ldots, c_N) = \sum_{i=1}^{N-1} \sum_{j=i+1}^{N} match(c_i, c_j) \tag{1.9}$$

Progressive alignment is a heuristic widely used in MSA, but it does not guarantee optimality [28]. ClustalW [105] is another popular program that improved the algorithm presented by Feng and Doolittle [28]. The main shortcoming of ClustalW is that once a sequence has been aligned, that alignment can never be modified even if it conflicts with sequences added later.

Recently, Chen et al. [18] took a serious attempt to solve the classical MSA problem by using a partitioning approach coupled with the Ant Colony Optimization (ACO) algorithm. The algorithm consists of three stages. At first, a genetic algorithm is employed to find out the near optimal cut-off points in the original sequences from where they must be partitioned vertically. In this way a partitioning method is continued recursively to reduce the original problem to multiple smaller MSA problems until the lengths of the subsequences are all less than an acceptable threshold. Next, an ant colony system is used to align each small subsection derived from the previous step. The ant system consists of $N$ ants each of which represents a solution of alignment. Each ant searches for an alignment by moving on the sequences to choose the matching characters. Let the N sequences be $S = S_0, S_1, \ldots, S_{N-1}$. In that case an artificial ant starts from $S_0[0]$, the first character of $S_0$, and selects one character from each of the sequences of $S_1, \ldots, S_{N-1}$ matching with $S_0[0]$. From the sequence $S_i, i = 1, 2, \ldots, n_1$, the ant selects a character $S_i[j]$ by a probability determined by the matching score with $S_0[0]$, deviation of its location from $S_0[0]$ and pheromones trail on the logical edge between $S_i[j]$ and $S_0[0]$.In addition, an ant may choose to insert an empty space according to a predetermined probability. Next, the ant starts from $S_0[1]$, selects the characters of $S_1, \ldots, S_{N-1}$ matching with $S_0[1]$ to form the second path. Similarly, starting from $S_0[2], \ldots, S_0[|S_0| - 1]$, the ant can form other paths. Here $|S_0|$ indicates the number of characters in the sequence $|S_0|$.

To evaluate an alignment represented by a set of paths, the positions of characters not selected by the ants are calculated first by aligning them to the right and adding gaps to the left. Next their SP (sum-of-pairs) score is using relation (1.9). Finally, a solution to the MSA is obtained by concatenating the results from smaller sub-alignments. The Divide-Ant-MSA algorithm outperformed the SAGA [78], a leading MSA program based on genetic algorithms, in terms of both speed and accuracy especially for longer sequences.

Rasmussen and Krink [88] focussed on a new PSO based training method for Hidden Markov Models (HMMs) in order to solve the MSA problem. The authors showed how a combination of PSO and evolutionary algorithms can generate better protein sequence alignments than with more traditional HMM training methods, such as Baum-Welch [98] and simulated annealing [39].

Genetic algorithm is one of the important and successful approaches in MSA. Zhang and Huang [122] propose an improved GA method, multiple small-popsize initialization strategy (MSPIS) and hybrid one-point crossover scheme (HOPCS) based GA, which can search the solution space in a very efficient manner. The experimental results show that this improved approach can obtain a better result compared with traditional GA approach in aligning multiple protein sequences problem.

DNA matching is a crucial step in sequence alignment. Since sequence alignment is an approximate matching process there is a need for good approximation algorithms. The process of matching in sequence alignment is generally finding longest common subsequences. However, finding the longest common subsequence may not be the best solution for either a database match or an assembly. An optimal alignment of subsequences is based on several factors, such as quality of bases, length of overlap, etc. Factors such as quality indicate if the data is an actual read or an experimental error. Fuzzy logic allows tolerance of inexactness or errors in sub sequence matching. Nasser et al. [75] propose fuzzy logic for approximate matching of subsequences. Fuzzy characteristic functions are derived for parameters that influence a match. Authors develop a prototype for a fuzzy assembler. The assembler is designed to work with low quality data, which is generally rejected by most of the existing techniques. Authors test the assembler on sequences from two genome projects, namely Drosophila melanogaster and Arabidopsis thaliana. Their results are compared with other assemblers. The fuzzy assembler successfully assembled sequences and performed similar and in some cases better than existing techniques.

In multiple DNA sequence alignment, some researchers used divide-and-conquer techniques to cut the sequences for the sake of decreasing complexity. Because the cutting points of sequences of the existing methods are fixed at the middle or near-middle points, the performance of sequence alignment of the existing methods is not good enough. Chen et al. [17] present a new method for multiple DNA sequence alignment using genetic algorithms and divide-and-conquer techniques to choose optimal cut points of multiple DNA sequences. Their experimental results show that the proposed method is better than the existing methods for dealing with multiple DNA sequence alignment.

The similarity judgement of two sequences is often decomposed in similarity judgements of the sequence events with an alignment process. However, in some domains like speech or music, sequences have an internal structure which is important for intelligent processing like similarity judgements. In an alignment task, this structure can be reflected more appropriately by using two levels instead of aligning event by event. This idea is related to the structural alignment framework by Markman and Gentner [34]. Weyde and Klaus [115] introduce a method to align sequences by modeling the segmenting and matching of groups in an input sequence in relation to a target sequence, detecting variations or errors. This is realized as an integrated process, using a neuro-fuzzy system. The selection of segmentations and alignments is based on fuzzy rules which allow the integration of expert knowledge via feature definitions, rule structure, and rule weights. The rule weights can be optimized effectively with an algorithm adapted from neural networks. Thus the results from the optimization process are still interpretable. The system has been implemented and tested successfully in a sample application for the recognition of musical rhythm patterns.

Hiroshi [66] proposes a new method for efficient finding of the biologically optimal alignment of multiple sequences. A key technique used in his method is *deterministic annealing* that attempts to find the global optimum in a parameter space through the annealing process. The author proposes a new simple probabilistic model for the usually time-consuming iterative process of deterministic annealing. Probabilistic parameters of his model are trained from a given sequences based on the deterministic annealing and Expectation Maximization algorithm. When a new sequence is given, this sequence is aligned by parsing it using the trained model. Experimental results show that the proposed method gives a better performance than other competing methods, like a profile hidden markov models, and is time-efficient.

## 1.8   CI in Protein Structure Prediction (PSP)

Protein Structure Prediction (PSP) is one of the most important goals pursued by bioinformatics and theoretical chemistry. Its aim is prediction of the three-dimensional structure of proteins from their amino acid sequences, sometimes including additional relevant information such as the structures of related proteins [128]. In other words, it deals with the prediction of a protein's tertiary structure from its primary structure. Protein structure prediction is of high importance in medicine (e.g., in drug design) and biotechnology (e.g., in the design of novel enzymes). There have been many successful research projects focusing on this problem. For example, Tang et al. [102] address a problem of predicting protein homology between given two proteins. They propose a learning method that combines the idea of association rules with their previous method called Granular Support Vector Machines (GSVM), which systematically combines a SVM with granular computing. The method, called GSVM-AR, uses association rules with high enough confidence and significant support to find suitable granules to build a GSVM with good performance. The authors compared their method with SVM by KDDCUP04 protein homology prediction data. From the experimental results, GSVM-AR showed significant improvement compared to a single SVM.

The interface between combinatorial optimization and fuzzy sets-based methodologies is the subject of a very active and increasing research. In this context, Balnco et al. [14] describe a fuzzy adaptive neighborhood search (FANS) optimization heuristic that uses a fuzzy valuation to qualify solutions and adapts its behavior as a function of the search state. FANS may also be regarded as a local search framework. The authors show an application of this fuzzy sets-based heuristic to the protein structure prediction problem in two aspects: (1) to analyze how the codification of the solutions affects the results and (2) to confirm that FANS is able to obtain as good results as a genetic algorithm. Both results shed some light on the application of heuristics to the protein structure prediction problem and show the benefits and power of combining basic fuzzy sets ideas with heuristic techniques.

Solving the structure prediction problem for complex proteins is difficult and computationally expensive. Tantar et al. [103] propose a bicriterion parallel hybrid genetic algorithm in order to efficiently deal with the problem using a computational grid. The use of a near-optimal metaheuristic, such as a GA, allows a significant reduction in the number of explored potential structures. However, the complexity of the problem remains prohibitive as far as large proteins are concerned, making the use of parallel computing on the computational grid essential for its efficient resolution. A conjugated gradient-based Hill Climbing local search is combined with the GA in order to intensify the search in the neighborhood of its provided configurations. Authors consider two molecular complexes: (1) the tryptophan-cage protein (Brookhaven Protein Data Bank ID 1L2Y) and (2) a-cyclodextrin. The experimentation results obtained on a computational grid show the effectiveness of their approach.

Predicting the three-dimensional structure of proteins from their linear sequence is one of the major challenges in modern biology. It is widely recognized that one of the major obstacles in addressing this question is that the *standard* computational approaches are not powerful enough to search for the correct structure in the huge conformational space. Genetic algorithms, a cooperative computational method, have been successful in many difficult computational tasks. Thus it is not surprising that in recent years several studies were performed to explore the possibility of using genetic algorithms to address the protein structure prediction problem. Ron Roger [108] reviewed a general framework of how genetic algorithms can be used for structure prediction problem. Using this framework, significant studies that were published in recent years are discussed and compared. Applications of genetic algorithms to the related question of protein alignments are also mentioned. The rationale of why genetic algorithms are suitable for protein structure prediction is presented, and future improvements that are still needed are discussed.

The understanding of protein structures is vital to determine the function of a protein and its interaction with DNA, RNA, and enzymes. The information about its conformation can provide essential information for drug design and protein engineering. While there are over a million known protein sequences, only a limited number of protein structures are experimentally determined. Hence, prediction of protein structures from protein sequences using computer programs is an important step to unveil proteins' three dimensional conformation and functions. As a result, prediction of protein structures has profound theoretical and practical influence over biological study. Pan [80] shows how to use machine learning methods with various advanced encoding schemes and classifiers improve the accuracy of protein structure prediction. The explanation of how a decision is made is also important for improving protein structure prediction. The reasonable interpretation is not only useful to guide the "wet experiments," but also the extracted rules are helpful to integrate computational intelligence with symbolic AI systems for advanced deduction. The author also presents some preliminary results using SVM and decision tree for rule extraction and prediction interpretation.

## 1.9   CI in Human Genetics

One goal of genetic epidemiology is to identify genes associated with common, complex multifactorial diseases. Success in achieving this goal will depend on a research strategy that recognizes and addresses the importance of interactions among multiple genetic and environmental factors in the etiology of diseases such as essential hypertension [50, 73, 91]. The identification of genes that influence the risk of common, complex disease primarily through interactions with other genes and environmental factors remains a statistical and computational challenge in genetic epidemiology. This challenge is partly due to the limitations of parametric statistical methods for detecting genetic effects that are dependent solely or partially on interactions. Recently, Marylyn et al. [74] took a serious attempt to introduce a genetic programming neural network (GPNN) as a method for optimizing the architecture of a neural network to improve the identification of genetic and gene-environment combinations associated with a disease risk. This empirical studies suggest GPNN has excellent power for identifying gene-gene and gene-environment interactions. In [91] Marylyn et al. continued their study to compare the power of GPNN to stepwise logistic regression (SLR) and classification and regression trees (CART) for identifying gene-gene and gene-environment interactions. SLR and CARTare standard methods of analysis for genetic association studies. Using simulated data,authors show that GPNN has higher power to identify gene-gene and gene-environment interactions than SLR and CART. These results indicate that GPNN may be a useful

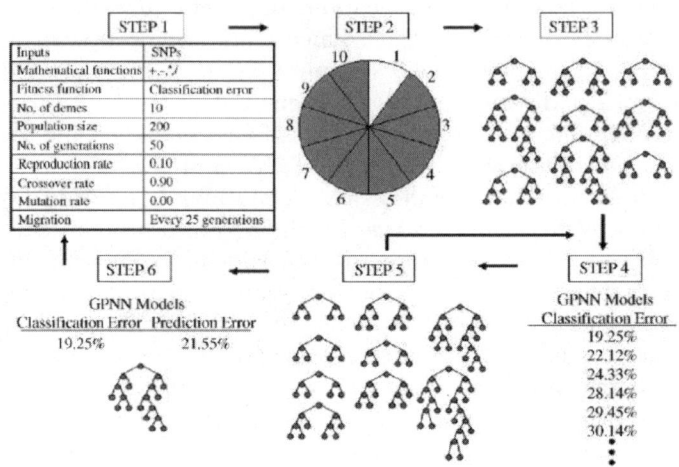

**Fig. 1.8.** The steps of the GPNN algorithm [91]

pattern recognition approach for detecting gene-gene and gene-environment interactions in studies of human disease. We will briefly discuss their approach in the following paragraphs. Their method contains six steps as shown in Figure 1.8 and described in brief as follows.

- *Step-1: Set of GPNN parameters.* GPNN has a set of parameters that must be initialized before beginning the evolution of NN models. These include an independent variable input set, a list of mathematical functions, a fitness function, and finally the operating parameters of the GP. These operating parameters include number of demes (or populations), population size, number of generations, reproduction rate, crossover rate, mutation rate, and migration [90].
- *Step-2: Divide the data based on cross validation.* The data are divided into 10 equal parts for 10-fold cross-validation. Here, we will train the GPNN on 9/10 of the data to develop an NN model. They test this model on the 1/10 of the data left out to evaluate the predictive ability of the model.
- *Step-3: Generate an initial population.* Training of the GPNN begins by generating an initial population of random solutions. Each solution is a binary expression tree representation of an NN.
- *Step-4:GPNN evaluation.* Each GPNN is evaluated on the training set and its fitness recorded.
- *Step-5: The best solutions selection.* The best solutions are selected for crossover and reproduction using a fitness-proportionate selection technique, called roulette wheel selection, based on the classification error of the training data.
- *Step-6: Classification and prediction error.* Classification error is defined as the proportion of individuals where the disease status was incorrectly

specified. A predefined proportion of the best solutions are directly copied (reproduced) into the new generation. Another proportion of the solutions is used for crossover with other best solutions. The new generation, which is equal in size to the original population, begins the cycle again. T'his continues until some criterion is met at which point the GPNN stops.

Another work introduced by Alison et al [74] which developed a grammatical evolution neural network (GENN) approach that accounts for the drawbacks of GPNN. In this study, they show that this new method has high power to detect gene-gene interactions in simulated data. They also, compare the performance of GENN to GPNN, a traditional Back-Propagation Neural Network (BPNN) and a random search algorithm. GENN outperforms both BPNN and the random search, and performs at least as well as GPNN. This study demonstrates the utility of using GE to evolve NN in studies of complex human disease.

## 1.10  CI in Microarray Classification

A DNA microarray (also commonly known as DNA chip or gene array) is a collection of microscopic DNA spots attached to a solid surface, such as glass, plastic, or silicon chip, forming an array for the purpose of expression profiling, monitoring expression levels for thousands of genes simultaneously. Microarrays provide a powerful basis to monitor the expression of thousands of genes, in order to identify mechanisms that govern the activation of genes in an organism. Short DNA patterns (or binding sites near the genes) serve as switches that control gene expression. Therefore, similar patterns of expression correspond to similar binding site patterns. A major cause of coexpression of genes is their sharing of the regulation mechanism (coregulation) at the sequence level. Clustering of coexpressed genes into biologically meaningful groups helps in inferring the biological role of an unknown gene that is coexpressed with a known gene(s). Cluster validation is essential, from both the biological and statistical perspectives, in order to biologically validate and objectively compare the results generated by different clustering algorithms.

Microarray classification has a broad variety of biomedical applications. Support Vector Machines (SVM) have emerged as a powerful and popular classifier for microarray data. At the same time, there is increasing interest in the development of methods for identifying important features in microarray data. Many of these methods use SVM classifiers either directly in the search for good features or indirectly as a measure of dissociating classes of microarray samples. Peterson and Thaut [85] present study that describes empirical results in model selection for SVM classification of DNA microarray data. Authors demonstrate that classifier performance is very sensitive to the SVM's kernel and model parameters. They also demonstrate that the optimal model parameters depend on the cardinality of feature subsets and can influence the evolution of a genetic search for good feature subsets. Their results suggest that application of SVM classifiers to microarray data should include careful consideration of the space of possible SVM parameters. The results also suggest that feature selection search and model selection should be conducted jointly rather than independently.

Tasoulis et al. [104] study and compare various computational intelligence approaches such as neural networks, evolutionary algorithms, and clustering algorithms, then they demonstrate their applicability as well as their weaknesses and shortcomings to efficient DNA microarray data analysis.

Heterogeneous types of gene expressions may provide a better insight into the biological role of gene interaction with the environment, disease development, and drug effect at the molecular level. Liang and Kelemen [60] proposed a Time Lagged Recurrent Neural Network with trajectory learning for identifying and classifying the gene functional patterns from the heterogeneous nonlinear time series microarray experiments. The proposed procedures identify gene functional patterns from the dynamics of a state-trajectory learned in the heterogeneous time series and the gradient information over time. Also, the trajectory learning with back-propagation through time algorithm can recognize gene expression patterns varying over time. This may reveal much more information about the regulatory network underlying gene expressions. The analyzed data were extracted from spotted DNA microarrays in the budding yeast expression measurements, produced by Eisen et al. [26]. The gene matrix contained 79 experiments over a variety of heterogeneous experiment conditions. The number of recognized gene patterns in our study ranged from two to ten and were divided into three cases. Optimal network architectures with different memory structures were selected based on Akaike and Bayesian information criteria using two-way factorial design. The optimal model performance was compared to other popular gene classification algorithms, such as nearest neighbor, support vector machine, and self-organized maps. The reliability of the performance was verified with multiple iterated runs.

Efficient and reliable methods that can find a small sample of informative genes amongst thousands are of great importance. In this area, much research is devoted to combining advanced search strategies (to find subsets of features), and classification methods [44]. Juliusdottir et al. [49] investigate a simple evolutionary algorithm/classifier combination on two microarray cancer datasets, where this combination is applied twice–once for feature selection, and once for further selection and classification. Their contribution are: (further) demonstration that a simple EA/classifier combination is capable of good feature discovery and classification performance with no initial dimensionality reduction; demonstration that a simple repeated EA/K-NN approach is capable of competitive or better performance than methods using more sophisticated preprocessing and classifier methods; new and challenging results on two public datasets with clear explanation of experimental setup; review material on the EA/K-NN area; and specific identification of genes that their work suggests are significant regarding colon cancer and prostate cancer.

Lin et al. [61] propose a genetic algorithm with silhouette statistics as discriminant function (GASS) for gene selection and pattern recognition. The proposed method evaluates gene expression patterns for discriminating heterogeneous cancers. Distance metrics and classification rules have also been analyzed to design a GASS with high classification accuracy. Moreover, the proposed method is compared to previously published methods. Various experimental results show that their method is effective for classifying the NCI60, the GCM and the SR-BCTs datasets. Moreover, GASS outperforms other existing methods in both the leave-one-out cross-validations and the independent test for novel data.

Identification of the short DNA sequence motifs that serve as binding targets for transcription factors is an important challenge in bioinformatics. Unsupervised techniques from the statistical learning theory literature have often been applied to motif discovery, but effective solutions for large genomic datasets have yet to be found. Mahonya et al. [65] present three self-organizing neural networks that have applicability to the motif-finding problem. The core system in this study is a previously described SOM-based otif-finder named SOMBRERO. The motif-finder is integrated in this work with a SOM-based method that automatically constructs generalized models for structurally related motifs and initializes SOMBRERO with relevant biological knowledge. A self-organizing tree method that displays the relationships between various motifs is also presented in this work, and it is shown that such a method can act as an effective structural classifier of novel motifs. The performance of the three self-organizing neural networks is evaluated and analyzed using various datasets.

## 1.11 Conclusions, Challenges, and Future Directions

Computational Intelligence (CI) has increasingly gained attention in bioinformatics research and computational biology. With the availability of different types of CI algorithms, it has become common for researchers to apply the off-shelf systems to classify and mine their databases. At present, with various intelligent methods available in the literature, scientists are facing difficulties in choosing the best method that could be applied to a specific data set. Researchers need tools, which present the data in a comprehensible fashion, annotated with context, estimates of accuracy, and explanation. The terms bioinformatics and computational biology mean about the same. Recently, however, the US National Institutes of Health (NIH) [126] came up with slightly different definitions, which for the convenience of the reader are repeated below. *Bioinformatics*: Research, development, or application of computational tools and approaches for expanding the use of biological, medical, behavioral, or health data, including those to acquire, store, organize, archive, analyze, or visualize such data. *Computational biology*: The development and application of data-analytical and theoretical methods, mathematical modeling, and computational simulation techniques to the study of biological, behavioral, and social systems.

The goal of motif finding is to detect novel, over-represented unknown signals in a set of sequences. Most widely used algorithms for finding motifs obtain a generative probabilistic representation of the over-represented signals and try to discover profiles that maximize information content score. The major difficulty for these algorithms arises from the fact that the best motif corresponds to the global maximum of a non-convex continuous function. Algorithms like Expectation Maximization (EM) and Gibbs sampling are very sensitive to the initial guesses and only converge to the nearest local maximum. A challenge here is to develop a novel optimization framework that searches the neighborhood regions of the initial alignments in a systematic manner to explore the neighborhood profiles. Algorithms like PSO could lead to new and interesting avenues of research.

The problem of cancer classification is another challenge. It has been divided into two related but separate challenges: class prediction and class discovery [31]. Class prediction refers the assignment of samples to one of several previously defined classes. Class discovery refers to defining a previously unrecognized tumor subtype(s) in expression data. Both of these tasks are challenging and require computational assistance. Class prediction via cluster analysis is typically used to infer the function of novel genes by grouping them with genes of well-known functionality in gene expression profiling. Genes that show similar activity patterns are often related functionally and are controlled by the same mechanisms of regulation. A major obstacle to the eventual utility of microarrays is the lack of efficient methods for cataloging the data into coexpressed groups. A new way of processing numeric data with large number of attributes versus low number of objects turns out to be well-suited to the gene expression data. Furthermore, tumors are not identical–even when they occur in the same organ, and patients may need different treatments depending on their particular subtype of cancer. Identification of tumor subgroups is therefore important for diagnosis and design of medical treatment. Most medical classification systems for tumors are currently based on clinical observations and the microscopical appearance of the tumors. These observations are not informative with regard to the molecular characteristics of the cancer. The genes, whose expression levels are associated with the tumor subtypes, are largely unknown. A better understanding of the cancer could be achieved if these genes were identified. Furthermore, the disease may manifest itself earlier on the molecular level than on a clinical level. Hence, gene expression data from microarrays may enable prediction of tumor subtype and outcome at an earlier stage than clinical examination. Thus microarray analysis may allow earlier detection and treatment of the disease, which again may increase the survival rate.

Most universities and companies have the same reasons for pursuing biomarker research: better diagnosis and better treatment for patients. According to Lynn Rutkowski, co-leader of clinical translational medicine at Wyeth Company (a global leader in pharmaceuticals, consumer health care products, and animal health care products), "*You need a strategy in place, so you have time to do the research you need to fill in gaps and get biomarkers you have confidence in. There are so many technologies emerging. The moment you commit to one, there is another right behind it.*" Both companies and researchers have already considered a new approach of combining imaging technology text mining and biomarkers discovery as a possible solution in future biometric research. For example, Wyeth Company is investing almost $86 million for biomarker discovery, including ten in cardiovascular and metabolic disease, four in inflammation and seven in neuro-science. This company has developed new markers using the 'combine' approach. In stroke, for example, in addition to imaging technology, Wyeth has used re-habilitation tools to measure patients' responses. A robotic instrumentation has been used for therapy that can also provide a quantitative assessment of motor-function recovery. Another example includes Alzheimer's disease (AD). AD has 11 compounds in development. One of these compounds is FK962. The company's long-term strategy involves molecular markets, structural and functional brain imaging, and physiological, behavioral, and associative learning tests.

Another challenge is to combine gene expression research with noninvasive imaging techniques. Eran Segal [93] and his collaborators hypothesized that the global gene expression patterns of human cancers may systematically correlate with their dynamic imaging features [93]. To address the challenges of relating gene expression to imaging, the researches followed a three step methodology and created an association map between imaging features on tree-phase contrast enhanced CT scans and gene expression patterns of 28 human hepatocellular carcinomas (HCC). First, the researchers defined and quantified 138 *units of distinctiveness* named *traits* present in one or more HCCs. Second, the module networks algorithm was implemented. The algorithm systematically search for associations between expression levels of 6,732 well-measured genes determined by mycroarary analysis and combinations of imaging traits. Third, the statistical significance of the association map was validated by comparison with permuted data sets, and by testing the prediction of the association map in an independent set of tumors.

Paralleling the diversity of genetic and protein activities pathologic human tis-sues also exhibit diverse radiographic features. It is proven that dynamic imaging trails in noninvasive computer tomography (CT) systematically correlate with the global gene expression profiles. For example: the association map of imaging traits and gene expression revealed that a large fraction of the gene expression program can be reconstructed from a small number of image trails. The expres-sion variation in 6,732 genes was captured by 116 gene modules, each of which was associated with specific combination of imaging trails. For each module, the presence or absence of combination of imaging traits explained the aggregate ex-pression level of genes within the module. The combinations of relevant imaging trials are depicted in decision trees: each split in the tree is specified by variation of an imaging trait, each terminal leaf in the tree is a cluster of samples that share a similar expression pattern of module genes. Thus the association map al-lowed the user to reconstruct the relative expression level of a gene (by mapping it to a module) in a given HCC sample (by mapping it to a cluster) Across all 116 gene modules capturing 6,732 genes in the presented model, the difference in the level of expression of member genes from their cognate module averages is 1.36- 1.33 fold. Thus the expression level of individual genes can be reconstructed from imaging features with an average deviation of about twofold, within the ex-perimental determination level allowed by microarray analysis. The experiment shows that only 8 imaging traits are sufficient to reconstruct the variation of all 116 gene modules [93].

The term cyber-infrastracture has been established by US National Science Foundation (NSF) to address the needs for new mechanisms of information handling and exchange. Eric Neumann, Director of Clinical Semantics Group at MIT, has presented the following project as an example of text mining research: NeuroCommons is a project within Science Commons at MIT. This project is using text mining to extract neuro-molecular relations from text mining, representing them as RDF (Resource Description Framework). SWAN (Semantic Web Applications in Neuromedicine) is an NIH-funded project that allows scientists to directly annotate knowledge onto findings using RDF. The user interface consists of a SPARQL–a query page that permits a wide variety of questions regarding genes, neurological diseases, microanatomy, and publications. Examples include: "Find all publications with neural dendrites in their description;" "Show all genes expressed in brain region CA1 involved in signal transduction;" "Find all papers on Parkinson's disease that involve gene products localized in the nucleus;" etc. Results can be formatted as tables. In RDF additional tools can process the data for enhanced scientific view. Tool such as Google can also be applied to the output from a query. The future of cyberinfrastarcture for bioinformatics and biomedical research is becoming a reality: a connected research community more effectively utilizing data and computational resources from different areas.

Also, intelligent support is essential for managing and interpreting this great amount of information. One of the well-known constraints specifically related to microarray data is the large number of genes in comparison with the small number of available experiments. In this context, the ability of design methods capable of overcoming current limitations of state-of-the-art algorithms is crucial to the development of successful applications.

A combination of computational intelligence techniques in application to bioinformatics and computational biology has become one of the most important areas of research in intelligent information processing [24]. Neural networks show their strong ability to solve complex problems for many bioinformatics problems. From the perspective of specific rough sets approaches that can be applied, exploration into possible applications of hybridizing rough sets with other intelligent systems like neural networks, genetic algorithms, fuzzy logic, etc. to bioinformatics and computational biology could lead to new and interesting avenues of research. Moreover, algorithms like PSO or ACO and their variants involve a large degree of randomness and different runs of the same program may yield different results; so it is necessary to incorporate problem specific domain knowledge in the Swarm Intelligence tools to reduce randomness and computational time and current research should progress in this direction as well.

The main purpose of this chapter was to present to the CI and bioinformatics and computational biology research communities the state of the art in CI applications to bioinformatics and computational biology, and to inspire further research and development on new applications and new concepts in new trend-setting directions and in exploiting computational intelligence.

# References

1. Abraham, A.: Intelligent systems: Architectures and perspectives, recent advances in intelligent paradigms and applications. In: Abraham, A., Jain, L., Kacprzyk, J. (eds.) Studies in Fuzziness and Soft Computing, pp. 1–35. Springer, Heidelberg (2002)
2. Abraham, A.: Nature and scope of AI techniques. In: Sydenham, P., Thorn, R. (eds.) Handbook for Measurement Systems Design, pp. 893–900. John Wiley and Sons Ltd., Chichester (2005)
3. Alba, E., Luque, G.: A New Local Search Algorithm for the DNA Fragment Assembly Problem. In: Cotta, C., van Hemert, J. (eds.) EvoCOP 2007. LNCS, vol. 4446, pp. 1–12. Springer, Heidelberg (2007)
4. Alon, U., et al.: Broad patterns of gene expression revealed by clustering analysis of tumor and normal colon tissues probed by oligonucleotide arrays. Proc. Natl. Acad. Sci. USA, Cell Biology 96, 6745–6750 (1999)
5. Altman, R.B., Valencia, A., Miyano, S., Ranganathan, S.: Challenges for intelligent systems in biology. IEEE Intelligent Systems 16(6), 14–20 (2001)
6. Angeleri, E., Apolloni, B., de Falco, D., Grandi, L.: DNA Fragment assembly using neural prediction techniques. Intl. J. Neural Systems 9(6), 523–544 (1999)
7. Arima, C., Hanai, T.: Gene expression analysis using Fuzzy k-Means Clustering. Genome Informatics 14, 334–335 (2003)
8. Back, T.: Evolutionary Algorithms in Theory and Practice: Evolution Strategies, Evolutionary Programming, Genetic algorithms. Oxford University Press, Oxford (1996)
9. Baker, T.K., et al.: Temporal gene expression analysis of monolayer cultured rat hepatocytes. Chem. Res. Toxicol. 14(9), 1218–1231 (2001)
10. Baldi, P., Brunak, S.: Bioinformatics: The Machine Learning Approach. MIT Press, Cambridge (1998)
11. Baldi, P., Hatfield, G.W.: DNA Microarrays and Gene Expression: From Experiments to Data Analysis and Modeling. Cambridge University Press, Cambridge (2002)
12. Banerjee, M., Mitra, S., Banka, H.: Evolutionary rough feature selection in gene expression data. IEEE Transactions on Systems, Man, and Cybernetics, Part C: Applications and Reviews 37(4), 622–632 (2007)
13. Bishop, C.M.: Neural Networks for Pattern Recognition. Oxford University Press, Oxford (1995)
14. Blanco, A., Pelta, D.A., Verdegay, J.L.: Applying a fuzzy sets-based heuristic to the protein structure prediction problem. Intl. J. Intelligent Systems 17(7), 629–643 (2002)
15. Bull, L., Kovacs, T. (eds.): Foundations of Learning Classifier Systems. Studies in Fuzziness and Soft Computing, 183 (2005)
16. Busa-Fekete, R., Kocsor, A., Pongor, S.: Tree-Based Algorithms for Protein Classification. Studies in Computational Intelligence 94, 165–182 (2008)
17. Chen, S.-M., Lin, C.-H., Chen, S.-J.: Multiple DNA sequence alignment based on genetic algorithms and divide-and-conquer techniques. Intl. J. Applied Science and Engineering 3(2), 89–100 (2005)

18. Chen, Y., Pan, Y., Chen, L., Chen, J.: Partitioned optimization algorithms for multiple sequence alignment. In: Proc. 20th Intl. Conf. on Advanced Information Networking and Applications, pp. 618–622 (2006)
19. Chena, C.-B., Wang, L.-Y.: Rough set-based clustering with refinement using Shannon's entropy theory. Computers and Mathematics with Applications 52(10-11), 1563–1576 (2006)
20. Chu, F., Xie, W., Wang, L.: Gene selection and cancer classification using a fuzzy neural network. In: Proc. IEEE Annual Meeting of Fuzzy Information, pp. 555–559 (2004)
21. Chuang, H.-Y., Lee, E., Liu, Y.-T., Lee, D., Ideker, T.: Network-based classification of breast cancer metastasis. Molecular Systems Biology 3(140) (2007)
22. Cios, K.J., Mamitsuka, H., Nagashima, T., Tadeusiewicz, R.: Computational intelligence in solving bioinformatics problems. Artificial Intelligence in Medicine 35(1-2), 1–8 (2005)
23. Cohen, J.: Bioinformatics: An introduction for computer scientists. ACM Computing Surveys 36(2), 122–158 (2004)
24. Das, S., et al.: Swarm Intelligence Algorithms in Bioinformatics. Studies in Computational Intelligence 94, 113–147 (2008)
25. Dougherty, J., Kohavi, R., Sahami, M.: Supervised and unsupervised discritization of continuous features. In: Proc. XII Intl. Conf. on Machine Learning, pp. 294–301 (1995)
26. Eisen, M.B., Spellman, P.T., Brown, P.O., Botstein, D.: Cluster analysis and display of genome-wide expression patterns. PNAS 95(25), 14863–14868 (1998)
27. Ezziane, Z.: Applications of artificial intelligence in bioinformatics: A review. Expert Systems with Applications 30, 2–10 (2006)
28. Feng, D.F., Doolittle, R.F.: Progressive sequence alignment as a prerequisite to correct phylogenetic trees. J. Mol. Evol. 25, 351–360 (1987)
29. Fernando, D., Fdez-Riverola, F., Glez-Pea, D., Corchado, J.M.: Using fuzzy patterns for gene selection and data reduction on microarray data. In: Corchado, E., Yin, H., Botti, V., Fyfe, C. (eds.) IDEAL 2006. LNCS, vol. 4224, pp. 1087–1094. Springer, Heidelberg (2006)
30. Fogel, D.B.: Evolutionary Computation: Toward a New Philosophy of Machine Intelligence. IEEE Press, Los Alamitos (1999)
31. Fogel, G.B.: Gene expression analysis using methods of computational intelligence. Pharmaceutical Discovery 5(8), 12–18 (2005)
32. Fogel, L.J., Owens, A.J., Walsh, M.J.: Artificial Intelligence Through Simulated Evolution. John Wiley & Sons, Chichester (1967)
33. Futschik, M.E., Kasabov, N.K.: Fuzzy clustering of gene expression data. In: Proc. 2002 IEEE Intl. Conf. on Fuzzy Systems, pp. 414–419 (2002)
34. Gentner, D., Markman, A.B.: Structure mapping in analogy and similarity. American Psychologist 52(1), 45–56 (1997)
35. Goldberg, D.E.: Genetic Algorithms in Search, Optimization, and Machine Learning. Addison-Wesley Publishing, Reading (1989)
36. Golub, T., et al.: Molecular classification of cancer: Class discovery and class prediction by gene expression monitoring. Science 286(5439), 531–537 (1999)
37. Gruźdź, A., Ihnatowicz, A., Ślęzak, D.: Interactive Gene Clustering: A Case Study of Breast Cancer Microarray Data. Information Systems Frontiers 8(1), 21–27 (2006)
38. Gusfield, D.: Introduction to the IEEE/ACM transactions on computational biology and bioinformatics. IEEE/ACM Transactions on Computational Biology and Bioinformatics 1(1), 2–3 (2004)

39. Hamam, Y., Al-Ani, T.: Simulated annealing approach for Hidden Markov Models. In: Proc. 4th WG-7.6 Working Conf. on Optimization-Based Computer-Aided Modeling and Design, ESIEE, France (1996)
40. Hassnein, A.-E., Abdelhafez, M., Own, H.: Rough sets data analysis: A case of Kuwaiti diabetic children patients. In: Advances in Fuzzy Systems (in press)
41. He, Y., Tang, Y., Zhang, Y.-Q., Sunderraman, R.: Fuzzy-granular gene selection from microarray expression data. In: Proc. 6th IEEE Intl. Conf. on Data Mining - Workshops, pp. 153–157 (2006)
42. Herrero, J., Valencia, A., Dopazo, J.: A hierarchical unsupervised growing neural network for clustering gene expression patterns. Bioinformatics 17(2), 126–136 (2001)
43. Holland, J.: Adaptation in Natural and Artificial Systems. University of Michigan Press (1975)
44. Hong, J.-H., Cho, S.-B.: The classification of cancer based on DNA microarray data that uses diverse ensemble genetic programming. Artificial Intelligence in Medicine 36, 43–58 (2006)
45. Huang, C.-J., Liao, W.-C.: A comparative study of feature selection methods for probabilistic neural networks in cancer classification. In: Proc. 15th IEEE Intl. Conf. on Tools with Artificial Intelligence, p. 451 (2003)
46. Hunga, C.-M., Huanga, Y.-M., Changb, M.-S.: Alignment using genetic programming with causal trees for identification of protein functions. Nonlinear Analysis 65, 1070–1093 (2006)
47. Hwang, K.B., Cho, D.Y., Wook Park, S.W., Kim, S.D., Zhang, B.Y.: Applying machine learning techniques to analysis of gene expression data: Cancer diagnosis. In: Proc. 1st Conf. on Critical Assessment of Microarray Data Analysis (2000)
48. Jain, A.K., Murty, M.N., Flynn, P.J.: Data clustering: A review. ACM Computing Surveys 31(3), 264–323 (1999)
49. Juliusdottir, T., Keedwell, E., Corne, D., Narayanan, A.: Two-phase EA/k-NN for feature selection and classification in cancer microarray datasets. In: Proc. 2005 IEEE Symposium on Computational Intelligence in Bioinformatics and Computational Biology, pp. 1–8 (2005)
50. Kardia, S.L.R.: Context-dependent genetic effects in hypertension. Curr. Hypertens. Rep. 2, 32–38 (2000)
51. Kelemen, A., Abraham, A., Chen, Y. (eds.): Computational Intelligence in Bioinformatics. Studies in Computational Intelligence. Springer, Heidelberg (2008)
52. Kennedy, J., Eberhart, R.: Particle swarm optimization. In: Proc. IEEE Intl. Conf. on Neural Networks, pp. 1942–1948 (1995)
53. Kennedy, J.: Small worlds and mega-minds: Effects of neighborhood topology on particle swarm performance. In: Proc. 1999 Congress of Evolutionary Computation, pp. 1931–1938 (1999)
54. Kennedy, J., Eberhart, R., Shi, Y.: Swarm Intelligence. Morgan Kaufmann Academic Press, San Francisco (2001)
55. Khan, J., et al.: Classification and diagnostic prediction of cancers using gene expression profiling and artificial neural networks. Nat. Med. 7(6), 673–679 (2001)
56. Kohonen, T.: Self-organizing maps. Springer, Heidelberg (1995)
57. Koza, J.R.: Genetic Programming. MIT Press, Cambridge (1992)
58. Li, D., Zhang, W.: Gene selection using rough set theory. In: Wang, G.-Y., Peters, J.F., Skowron, A., Yao, Y. (eds.) RSKT 2006. LNCS (LNAI), vol. 4062, pp. 778–785. Springer, Heidelberg (2006)

59. Li, L., Weinberg, C.R., Darden, T.A., Pedersen, L.G.: Gene selection for sample classification based on gene expression data: Study of sensitivity to choice of parameters of the GA/KNN method. Bioinformatics 17, 1131–1142 (2001)
60. Liang, Y., Kelemen, A.: Time course gene expression classification with time lagged recurrent neural network. Studies in Computational Intelligence 94, 149–163 (2008)
61. Lin, T.-C., et al.: Pattern classification in DNA microarray data of multiple tumor types. Pattern Recognition 39(12), 2426–2438 (2006)
62. Lingras, P.: Applications of rough set based k-means, Kohonen SOM, GA Clustering. In: Peters, J.F., Skowron, A., Marek, V.W., Orłowska, E., Słowiński, R., Ziarko, W. (eds.) Transactions on Rough Sets VII. LNCS, vol. 4400, pp. 120–139. Springer, Heidelberg (2007)
63. Lipman, D.J., Altschul, S.F., Kececioglu, J.D.: A tool for multiple sequence alignment. Proc. Natl. Acad. Sci. USA 86, 4412–4415 (1989)
64. Luscombe, N.M., Greenbaum, D., Gerstein, M.: What is Bioinformatics? A proposed definition and overview of the field. Yearbook of Medical Informatics, 83–100 (2001)
65. Mahonya, S., Benosa, P.V., Smithd, T.J., Goldend, A.: Self-organizing neural networks to support the discovery of DNA-binding motifs. Neural Networks 19, 950–962 (2006)
66. Mamitsuka, H.: Finding the biologically optimal alignment of multiple sequences. Artificial Intelligence in Medicine 35(1-2), 9–18 (2005)
67. Meksangsouy, P., Chaiyaratana, N.: DNA fragment assembly using an ant colony system algorithm. In: Proc. Congress on Evolutionary Computation (2003)
68. Midelfart, H., Komorowski, J., Nørsett, K., Yadetie, F., Sandvik, A.K., Lægreid, A.: Learning rough set classifiers from gene expressions and clinical data. Fundamenta Informaticae 53, 155–183 (2002)
69. Mitra, S.: An evolutionary rough partitive clustering. Pattern Recognition Letters 25, 1439–1449 (2004)
70. Mitra, S., Hayashi, Y.: Bioinformatics with soft computing. IEEE Transactions on Systems, Man, and Cybernetics, Part C: Applications and Reviews 36, 616–635 (2006)
71. Mitra, S., Banka, H., Paik, J.H.: Evolutionary fuzzy biclustering of gene expression data. In: Yao, J., Lingras, P., Wu, W.-Z., Szczuka, M.S., Cercone, N.J., Ślęzak, D. (eds.) RSKT 2007. LNCS (LNAI), vol. 4481, pp. 284–291. Springer, Heidelberg (2007)
72. Mohamed, S., Rubin, D., Marwala, T.: Multi-class Protein Sequence Classification Using Fuzzy ARTMAP. In: Proc. IEEE Intl. Conf. on Systems, Man, and Cybernetics, pp. 1676–1681 (2006)
73. Moore, J.H., Williams, S.M.: New strategies for identifying gene-gene interactions in hypertension. Ann. Med. 34, 88–95 (2002)
74. Motsinger, A.A., Dudek, S.M., Hahn, L.W., Ritchie, M.D.: Comparison of Neural Network Optimization Approaches for Studies of Human Genetics. In: Rothlauf, F., Branke, J., Cagnoni, S., Costa, E., Cotta, C., Drechsler, R., Lutton, E., Machado, P., Moore, J.H., Romero, J., Smith, G.D., Squillero, G., Takagi, H. (eds.) EvoWorkshops 2006. LNCS, vol. 3907, pp. 103–114. Springer, Heidelberg (2006)
75. Nasser, S., Vert, G.L., Nicolescu, M., Murray, A.: Multiple Sequence Alignment using Fuzzy Logic. In: Proc. IEEE Symposium on Computational Intelligence and Bioinformatics and Computational Biology, pp. 304–311 (2007)

76. Nguyen, H.S.: Approximate Boolean reasoning: Foundations and applications in data mining. In: Peters, J.F., Skowron, A. (eds.) Transactions on Rough Sets V. LNCS, vol. 4100, pp. 334–506. Springer, Heidelberg (2006)
77. Ning, S., Ziarko, W., Hamilton, J., Cercone, N.: Using rough sets as tools for knowledge discovery. In: Proc. 1st Intl. Conf. on Knowledge Discovery and Data Mining, pp. 263–268 (1995)
78. Notredame, C., Higgins, D.G.: SAGA: sequence alignment by genetic algorithm. Nucleic Acids Research 24(8), 1515–1524 (1996)
79. Okada, Y., et al.: Knowledge-assisted recognition of cluster boundaries in gene expression data. Artificial Intelligence in Medicine 35(1-2), 171–183 (2005)
80. Pan, Y.: Protein structure prediction and understanding using machine learning methods. In: Proc. IEEE Intl. Conf. on Granular Computing, pp. 13–20 (2005)
81. Paul, T.K.: Gene expression based cancer classification using evolutionary and non-evolutionary methods. Technical Report No. 041105A1, Dept. of Frontier Informatics, University of Tokyo, Japan (2004)
82. Pawlak, Z.: Rough sets. Intl. J. Comp. Inform. Science 11, 341–356 (1982)
83. Pawlak, Z.: Rough Sets – Theoretical Aspects of Reasoning About Data. Kluwer, Dordrecht (1991)
84. Pawlak, Z., Grzymala-Busse, J., Slowinski, R., Ziarko, W.: Rough sets. Communications of the ACM 38(11), 88–95 (1995)
85. Peterson, D.A., Thaut, M.H.: Model and feature selection in microarray classification Peterson. In: Proc. IEEE Symposium on Computational Intelligence in Bioinformatics and Computational Biology, pp. 56–60 (2004)
86. Polkowski, L.: Rough Sets: Mathematical Foundations. Physica-Verlag, Heidelberg (2003)
87. Quackenbush, J.: Computational analysis of microarray data. National Review of Genetics 2, 418–427 (2001)
88. Rasmussen, T.K., Krink, T.: Improved Hidden Markov Model training for multiple sequence alignment by a particle swarm optimization-evolutionary algorithm hybrid. BioSystems 72, 5–17 (2003)
89. Raychaudhuri, S., Stuart, J.M., Altman, R.B.: Principal components analysis to summarize microarray experiments: Application to sporulation rime series. In: Proc. Pacific Symposium on Biocomputing, pp. 452–463 (2000)
90. Ritchie, M.D., et al.: Optimization of neural network architecture using genetic programming improves detection of gene-gene interactions in studies of human diseases. BMC Bioinformatics 4(28) (2003)
91. Ritchie, M.D., et al.: Genetic programming neural networks: A powerful bioinformatics tool for human genetics. Applied Soft Computing 7, 471–479 (2007)
92. Ruffino, F., Costacurta, M., Muselli, M.: Evaluating switching neural networks for gene selection. In: Masulli, F., Mitra, S., Pasi, G. (eds.) WILF 2007. LNCS (LNAI), vol. 4578, pp. 557–562. Springer, Heidelberg (2007)
93. Segal, E., et al.: Decoding global gene expression programs in liver cancer by noninvasive imaging. Nature Biotechnology 25, 675–680 (2007)
94. Setubal, J., Meidanis, J.: Introduction to Computational Molecular Biology. Intl Thomson Publishing (1999)
95. Ślęzak, D., Wróblewski, J.: Rough Discretization of Gene Expression Data. In: Proc. 2006 Intl. Conf. on Hybrid Information Technology, pp. 265–267 (2006)
96. Ślęzak, D., Wróblewski, J.: Roughfication of numeric decision tables: The case study of gene expression data. In: Yao, J., Lingras, P., Wu, W.-Z., Szczuka, M.S., Cercone, N.J., Ślęzak, D. (eds.) RSKT 2007. LNCS (LNAI), vol. 4481, pp. 316–323. Springer, Heidelberg (2007)

97. Spellman, E.M., Brown, P.L., Brown, D.: Cluster analysis and display of genome-wide expression patterns. Proc. Natl. Acad. Sci. USA 95, 14863–14868 (1998)
98. Stolcke, A., Omohundro, S.: Hidden Markov Model induction by Bayesian model merging. NIPS 5, 11–18 (1993)
99. Sun, L., Miao, D., Zhang, H.: Gene selection with rough sets for cancer classification. In: Proc. 4th Intl. Conf. on Fuzzy Systems and Knowledge Discovery, pp. 167–172 (2007)
100. Sushmita, M.: An evolutionary rough partitive clustering. Pattern Recognition Letters 25, 1439–1449 (2004)
101. Tamayo, P., et al.: Interpreting patterns of gene expression with self organizing maps: Methods and applications to hematopoietic differentiation. PNAS 96, 2907–2912 (1999)
102. Tang, Y., Jin, B., Zhang, Y.-Q.: Granular support vector machines with association rules mining for protein homology prediction. Artificial Intelligence in Medicine 35(1-2), 121–134 (2005)
103. Tantar, A.A., Melab, N., Talbi, E.G., Parent, B., Horvath, D.: A parallel hybrid genetic algorithm for protein structure prediction on the computational grid. Future Generation Computer Systems 23(3), 398–409 (2007)
104. Tasoulis, D.K., Plagianakos, V.P., Vrahatis, M.N.: Computational intelligence algorithms and DNA microarrays. Studies in Computational Intelligence 94, 1–31 (2008)
105. Thompson, J.D., Higgins, D.G., Gibson, T.J.: CLUSTAL W: Improving the sensitivity of progressive multiple sequence alignment through sequence weighting, position specific gap penalties and weight matrix choice. Nucleic Acids Research 22(22), 4673–4680 (1994)
106. Tomida, S., Hanai, T., Honda, H., Kobayashi, T.: Gene expression analysis using Fuzzy ART. Genome Informatics 12, 245–246 (2001)
107. Toronen, P., Kolehmainen, M., Wong, G., Castren, E.: Analysis of gene expression data using self-organizing maps. FEBS letters 451, 142–146 (1999)
108. Unger, R.: The genetic algorithm approach to protein structure prediction. Structure and Bonding 110, 153–175 (2004)
109. Valdes, J.J., Barton, A.J.: Relevant attribute discovery in high dimensional data: Application to breast cancer gene expressions. In: Wang, G.-Y., Peters, J.F., Skowron, A., Yao, Y. (eds.) RSKT 2006. LNCS (LNAI), vol. 4062, pp. 482–489. Springer, Heidelberg (2006)
110. van de Vijver, M.J., et al.: A gene-expression signature as a predictor of survival in breast cancer. N. Engl. J. Med. 347, 1999–2009 (2002)
111. Wang, D., Lee, N.K., Dillon, T.S.: Extraction and optimization of fuzzy protein sequences classification rules using GRBF neural networks. Neural Information Processing - Letters and Reviews 1(1), 53–57 (2003)
112. Wang, Y., et al.: Gene-expression profiles to predict distant metastasis of lymph-node-negative primary breast cancer. Lancet 365, 671–679 (2005)
113. Wen, X., et al.: Large scale temporal gene expression mapping of cns development. Proc. Natl. Acad. Sci. USA, Neurobiology 95, 334–339 (1998)
114. Wetcharaporn, W., Chaiyaratana, N., Tongsima, S.: DNA fragment assembly by ant colony and nearest neighbour heuristics. In: Rutkowski, L., Tadeusiewicz, R., Zadeh, L.A., Żurada, J.M. (eds.) ICAISC 2006. LNCS (LNAI), vol. 4029, pp. 1008–1017. Springer, Heidelberg (2006)
115. Weyde, T., Dalinghaus, K.: A neuro-fuzzy system for sequence alignment on two levels. Mathware and Soft Computing XI(2-3), 197–210 (2004)

116. Xiao, X., Dow, E.R., Eberhart, R.C., Miled, Z.B., Oppelt, R.J.: Gene clustering using self-organizing maps and particle swarm optimization. In: Proc. 17th Intl. Symposium on Parallel and Distributed Processing (2003)
117. Xie, W., Chu, F., Wang, L.: Fuzzy neural network applications for gene selection and cancer classification. In: Proc. Artificial Intelligence and Soft Computing (2004)
118. Yang, Q., Wu, X.: Challenging problems in data mining research. Intl. J. Information Technology and Decision Making 5(4), 597–604 (2006)
119. Yeung, K.Y., Ruzzo, W.L.: Principal component analysis for clustering gene expression data. Bioinformatics 17, 763–774 (2001)
120. Yuhui, Y., Lihui, C., Goh, A., Wong, A.: Clustering gene data via associative clustering neural network. In: Proc. 9th Intl. Conf. on Information Processing, pp. 2228–2232 (2002)
121. Zadeh, L.A.: Fuzzy sets. Information and Control 8, 338–353 (1965)
122. Zhang, G.-Z., Huang, D.-S.: Aligning multiple protein sequence by an improved genetic algorithm. In: Proc. IEEE Intl. Joint Conf. on Neural Networks, pp. 1179–1183 (2004)
123. Zhang, J., Lee, R., Wang, Y.J.: Support vector machine classifications for microarray expression data set. In: Proc. 5th Intl. Conf. on Computational Intelligence and Multimedia Applications, pp. 67–71 (2003)
124. Zhang, Q.: An approach to rough set decomposition of incomplete information systems. In: Proc. 2nd IEEE Conf. on Industrial Electronics and Applications, pp. 2455–2460 (2007)
125. Ziarko, W.: Variable precision rough sets model. J. Computer and Systems 46(1), 39–59 (1993)
126. NIH: http://www.bisti.nih.gov (last accessed December 2007)
127. Special Issue on Bioinformatics. IEEE Computer 35 (July 2002)
128. http://en.wikipedia.org/wiki/DNA_microarray (last accessed December 2007)

# 2

# Data Mining and Genetic Algorithms: Finding Hidden Meaning in Biological and Biomedical Data

Christopher M. Taylor and Arvin Agah

Department of Electrical Engineering and Computer Science,
The University of Kansas, Lawrence, Kansas 66045, USA

**Summary.** The amount of biological and biomedical data being accumulated continues to grow at incredible rates. Having tools that can search through these enormous databases is of critical importance to the advancement of research. Data mining is a field of research in Computer Science that specializes in examining large collections of data and extracting patterns that occur within the data. One useful technique for performing data mining is through a genetic algorithm, a process that mimics evolution. This chapter highlights data mining and the genetic algorithm technique, and it also lists many applications where data mining tools have been beneficial to biological and biomedical researchers, and lists some of the available data mining tools.

## 2.1 Introduction

Before 1994, predicting premature births was something of a shot in the dark. It involved a lot of manual techniques and guess work, and the results were accurate only about 38% of the time–less accurate than flipping a coin. However, a paper published in 1994 [42] improved the accuracy of these predictions to a colossal 88%, shattering the predictive power of earlier methods. This astounding accuracy did not result from any new medical discoveries—in fact, no additional laboratory work or testing of any kind was performed. Instead, the researchers gathered data from three different large databases representing a mix of high-risk and low-risk pregnant women in the United States. The data contained 214 pieces of information on 18,890 patients. Using this collection of patient information and the known results of each pregnancy, the researchers then employed a technique known as Data Mining to search the data and find patterns.

Data Mining offers a fresh perspective to research problems. Where most research is done with a hypothesis-driven approach, data mining approaches the problem from a different direction. Rather than test a hypothesis developed by an expert, data mining derives hypotheses from the data itself, letting the data essentially play the role of the expert. Sometimes human experts form opinions, whether consciously or not, about the cause of a problem and their opinions drive research and the development of predictive models. Sometimes these opinions are erroneous or flawed and can lead research down the wrong path, resulting in inaccurate predictive models. Data mining offers researchers the opportunity to

T.G. Smolinski et al. (Eds.): Comp. Intel. in Biomed. & Bioinform., SCI 151, pp. 49–68, 2008.
springerlink.com

examine data to see what insights the data might provide. By letting the data speak for itself, new strides are being made in numerous research fields.

This chapter will provide an overview of data mining and genetic algorithms. First, some basic concepts of Set Theory are provided to help readers understand how data mining techniques describe patterns that lie within data. Some metrics are then introduced to measure the quality of the patterns found in the data. A portion of this chapter is dedicated to genetic algorithms and their ability at finding solutions to very complicated problems, such as data mining. Toward the end of the chapter, several examples of how data mining has benefited research in biology are highlighted. The chapter concludes with a list of several data mining tools that can benefit researchers.

## 2.2   Data Mining

Data mining is the process of finding patterns that lie within large collections of data. Data mining is discovery-driven rather than assumption-driven [30]. As the process searches through the data, patterns are automatically extracted.

In general, data mining objectives can be placed into two categories: descriptive and predictive [11]. The goal of descriptive data mining is to find general patterns or properties that lie within the data set. This often involves aggregate functions such as mean, variance, count, sum, etc. In other words, descriptive data mining reports patterns about the data itself. Predictive data mining, however, attempts to infer meaning from the data in order to create a model that can be used to predict future data. This is often done by using data with known results, and analyzing the properties those data elements have in common. The common properties should be a reasonable predictor for the given result.

Another important concept is the difference between supervised and unsupervised learning. Supervised learning takes place when data has been pre-classified. In other words, the items in the data have already been placed into groups or been assigned some value or result. For example, in a database about house values, each item in the database will contain values such as the number of bedrooms, square footage, etc. In supervised learning, each house will also be assigned a monetary value, either by an expert or by the amount for which the house actually sold. The goal of the data mining process is then to find the patterns that result in a given value. Unsupervised learning, on the other hand, occurs when data is not pre-classified. In these cases, the data mining process cannot make value judgments. It can find correlations within data, but it is not able to make any inferences about what those patterns might mean. Thus, unsupervised learning is descriptive while supervised learning is predictive. In order to make predictions, data must be classified, or given value, by some outside source. In many cases, data can be classified using experimental or observable results, such as which patients had a recurrence of cancer. In other cases, data are classified by an expert, such as data using appraised property values for houses.

Some of the difficulties involved in data mining are problems with the data itself. When collecting data from various sources, they often have different formats, collect different pieces of information, and have different protocols regarding the data. For example, one set of records might contain a person's age, while a similar set of records from another source might not. Further compounding the problem, even within the same source, data can be erroneous or even missing. Perhaps the ages for some, but not all, of the people are recorded in the database. This can make finding patterns in the data very difficult when records are incomplete or inconsistent. Thus, a major task in data mining is in how to handle these anomalies that occur within data.

There are many different approaches to data mining, and the different techniques vary as widely as the data they are used to analyze. There are data mining techniques implementing neural networks [26], clustering algorithms [7], genetic algorithms [37], data visualization [40], and even hybrid approaches that attempt to utilize the strengths of multiple techniques, such as combining neural networks with genetic algorithms [38]. Rather than describe the various approaches, this chapter will present the important concepts, applications, and tools, with a focus on genetics algorithms.

## 2.3 Concepts

Several key concepts are essential to the field of data mining.

### 2.3.1 Set Theory

One fundamental concept is that of Set Theory. Set Theory can be defined as the mathematical science of the infinite [20]. It studies the properties of sets, which are then used to formalize all the concepts of mathematics. An important notion of Set Theory is membership. For example, if $a$ is a member of $A$ (denoted as $a \in A$), then the set $A$ contains $a$ as one of its members. A member is also referred to as an element. All members of a set share similar properties as defined by the set. For example, if set $A$ is defined as the set of all mammals, then for any $a$, if $a$ is a mammal, then $a$ is an element of $A$. Thus, a dog would be a member of $A$, but a worm would not. All elements of $A$ would share the properties that define a thing as being a mammal, such as using lungs to breathe air. Other properties do not necessarily hold, however. While some members of $A$ live on land, it is not reasonable to conclude that all members of A are land-dwellers. Thus, the definition for any particular set is very important because the definition of the set describes the common properties of its members. The term *cardinality* refers to the number of elements in a set. If a set contains six elements, then that set has the cardinality of six.

Another important concept of Set Theory is that of the subset. For a set $A$, if all members of $A$ are also members of set $B$, then $A$ is a subset of $B$ (denoted as $A \subset B$). Every set can be broken into smaller sets, even if the smaller set contains no elements. The set without any elements is called the "empty set."

If set $B$ is the set of all living things, then set $B$ could be broken into subsets of animals and plants. The animals set could be further divided into mammals, birds, reptiles, etc. The set mammals could be divided into subsets of dogs, cats, humans, etc. Depending on how these sets were defined, the same element could actually be a member of several different sets. For example, a dog could be a member of the sets mammals, dogs, pets, and canines. A cat could also be a member of the sets mammals and pets but that would not make it as a dog. It would mean that cats and dogs do have some common properties.

Anything that is not a member of a set is said to be in the *compliment* of the set. For any $c$ such that $c$ is not a member of $A$ ($c \notin A$), then $c$ is a member of the compliment of $A$ ($c \in \overline{A}$). This concept can also be referred to as not or negation. Thus, if $a$ is not a member of $A$ then it is a member of $\overline{A}$.

There are some important operations that can occur between sets to define a new set, namely, *intersection* and *union*. These operations can also be thought of as logical *and* and *or*, respectively. The intersection of two sets is the set of elements that are common to both sets. Set $C$ is the intersection of sets $A$ and $B$ if all elements of $C$ are also elements of $A$ and $B$ (denoted as $A \cap B$). For example, if $A$ is the set of people who wear glasses and B is the set of people under age 20, the the intersection of the two sets would be the set of people under age 20 who wear glasses. The intersection of two sets can be empty, meaning that there are no elements that occur in both sets. The intersection can also be referred to as the AND operator.

The union of two sets is the set of elements that occur in at least one of the sets. Set $C$ is the union of sets $A$ and $B$ if all elements of $C$ are also elements of $A$ or $B$ (denoted as $A \cup B$). The union operation is an inclusive or, meaning that the elements of $C$ must occur in $A$, or $B$, or both. There is another operation called an exclusive or in which the elements must occur in one or the other set, but cannot occur in both (sometimes expressed as $A \oplus B$). If $A$ is defined as the set of all canines and $B$ is defined as the set of all pets, the the set $A \cup B$ would include cats (which are pets), dogs (which are pets and canines), and wolves (which are canines). The set $A \oplus B$ would include cats and wolves, but not dogs. The union is also referred to as the OR operator and exclusive or is referred to as the XOR operator.

## Crisp Sets

The sets used in classical set theory are often called *crisp sets*. This is because the size of the sets, the number of elements, and the identity of the individual elements are all very well defined. An example of a crisp set could be the number of books on a specific bookshelf. Membership in the set is easily determined—if a book is on the bookshelf, then it is in the set; books not on the shelf are not in the set. The cardinality (number of elements in the set) can be easily calculated by counting the books on the shelf. Questions about membership are also easily answered (e.g., Does the set contain War and Peace? or What are the titles of all books in the set?). However, there are many real-world situations which are difficult to define in terms of crisp sets.

**Rough Sets**

Because crisp sets are sometimes difficult to represent precisely, they are sometimes defined using two *rough sets*. A rough set approximates the upper and lower bounds of a set, also known as its upper and lower approximation, that is otherwise difficult to define precisely. The upper and lower bounds are sets themselves. If $X$ is a set that is difficult to define as a crisp set, then set $A$ can be the lower bound for set $X$; and set $B$ can be the upper bound for set $X$. The definition for set $A$ would be such that all elements of $A$ are definitely elements of set $X$. In other words, $A \subseteq X$. Set $B$ is given a much broader definition so that it contains at least some of the elements in $X$ so that $B \cap X \neq \emptyset$ [27].

Rough sets are useful when dealing with uncertainty or ambiguity. For example, if it is unclear as to which set a particular element should belong, a rough set can be created to include the unknown element. Rough sets can also be used to make space for elements which do not fit well into a crisp set, but must be placed into a single set. For example, considering two bookshelves where one is comprised entirely of books about mathematics and the other only contains philosophical texts, if a history book has just been purchased and must be placed on a shelf, with which books should it be included? It does not belong on either shelf according to their current definitions. Regardless of which shelf it is placed on, the definition for the books on that shelf becomes a rough set (either math and history books, or philosophy and history books).

**Fuzzy Sets**

Sometimes there are elements that only partially belong to a set and the definition of membership becomes a little blurry, which is where fuzzy sets are useful. Fuzzy sets allow an individual element to partially belong to multiple sets at the same time, a concept that cannot be handled with crisp sets. Each element in a fuzzy set is given a membership value indicating the degree to which the element belongs in the set [34]. For example, if $A$ is the set of weekdays and $B$ is the set of weekend days, the to which set does Friday belong? In crisp sets, Friday is a weekday because the weekend only consists of Saturday and Sunday. However, to many people Friday is part of the weekend. Using fuzzy sets, Friday could be assigned to set $A$ with 60% membership and to set $B$ with 40% membership. Thus, fuzzy sets allow for partial membership and can even be used to make claims such as element $x$ is more of a member of the set than element $y$.

### 2.3.2 Decision Tables

Decision tables are similar to tables in a database. Each row in the table is called a *tuple* and represents one specific item such as a house, an employee, a business, a car, etc. Each column in the table is an attribute and is used to describe each tuple in the table. What distinguishes a decision table from a database table is that the decision table has some *decision* (or classification) associated with each tuple. The attributes can be thought of as a condition, with the decision

**Table 2.1.** Decision table indicating conditions for reimbursement from insurance

| Tuple # | Deductible Met | Type of Visit | Participating Physician | Reimbursement |
|---|---|---|---|---|
| 1 | yes | office | yes | 90 |
| 2 | yes | office | no | 50 |
| 3 | yes | hospital | no | 80 |
| 4 | yes | kab | no | 70 |
| 5 | no | office | yes | 0 |
| 6 | no | office | no | 0 |
| 7 | no | hospital | no | 0 |
| 8 | no | lab | no | 0 |

being associated with that condition. Table 2.1 is an example of a decision table indicating the conditions for which a patient will receive different percentages of reimbursement from their health insurance provider.

There are eight tuples in Table 2.1, each representing a different condition under which a patient might apply for a reimbursement from the insurance company. Each tuple can take on a different value for each attribute. As seen in the table, there are three types of visits: *office*, *hospital*, and *lab*. The other two attributes only have two possible values of *yes* or *no*. Thus, each tuple can be described by using the values of the attributes. The decision in this table is the *reimbursement* attribute. The combination of an attribute with a value is called a feature [22]. For example, $(type\_of\_visit, office)$ is a feature—it specifies that the type of visit was an office visit.

Some of the combinations of attributes in Table 2.1 are missing, such as a hospital visit that is also a participating physician. In order to cover all possible combinations of features in the first three attributes of the table, only 12 tuples would be required. Real world data, however, can have millions or even billions of tuples.

### 2.3.3  Rule Induction

Rule induction is the process of taking the data and searching for meaningful patterns that can be described in terms of features. The result is a set of rules that describe the patterns in the data. Each rule consists of two parts: the antecedent, and the consequent [33]. The antecedent, or left-hand side, of the rule is the condition that must be met for the rule to be applicable. It is the if part of the rule. The consequent, or right-hand side, of the rule is the action or decision that follows if the antecedent is true. If the antecedent is true, then the consequent follows. A sample rule from Table 2.1 is:

$(deductible\_met, no) \rightarrow (reimbursement, 0\%)$

This rule can be read as "if the deductible is not met then there is 0% reimbursement." If the antecedent of $(deductible\_met, no)$ is true, then the consequent of $(reimbursement, 0\%)$ is true. This does not hold when reversed, however—the antecedent does not follow from the consequent—receiving 0% reimbursement does not necessarily mean that the deductible was not met because there are other conditions when a 0% reimbursement might result. The example rule describes tuples 5, 6, 7, and 8 as to how those conditions are reimbursed. It does not cover tuples 1 through 4 because they have (*deductible_met, yes*). More rules need to be induced to cover the first four tuples. Thus, by creating a suitable set of rules, all of the patterns in the decision table can be described.

A consequent can be described by a single set in the antecedent, i.e., one feature can be used to describe a result. However, this occurs very rarely in the real-world data. It usually requires a combination of sets to provide an adequate classification that does not include members from other consequents. These sets have to be separated by set operators. The set operators describe how the sets relate to one another. Without set operators, the meaning of the rule is ambiguous. For example: (*deductible_met, yes*) (*participating_physician, yes*) → (*reimbursement*, 90%) does not have a set operator. From Table 2.1, it is apparent that for the consequent to be true, a tuple must be a member of both sets in the antecedent, requiring that an AND operator be placed between the two sets. If an OR operator had been used then tuples 1 through 5 would be members of the antecedent. Tuples 1 through 4 are members of (*deductible_met, yes*) and tuple 5 is a member of (*participating_physician, yes*). Regardless of the antecedent, only tuple 1 would be a member of the consequent. The AND operator makes the rule correct. The OR operator would make the rule correct only one-fifth of the time. However, there are times when the OR operator is more appropriate: (*age*, 15)*OR*(*age*, 17) → (*age_group, teen*) Changing this OR to an AND would mean that there must be a tuple that has ages of both 15 and 17 in order to be a teen.

The four most common set operators used in data mining are:

- Y AND Z: must be a member of both sets Y and Z.
- Y OR Z: is a member of at least one of the sets Y and Z.
- Y XOR Z: is a member of only one of the sets Y and Z.
- NOT Z: is not a member of Z.

Using sets and set operators, complex statements can be constructed about the data. This is necessary because the patterns within the data rarely allow for a single set to be used as the antecedent for a rule. Statements can be constructed with sets and set operators that allow rules to make unique classifications.

### 2.3.4 Confusion Matrices and Quality Metrics

When discussing the results of a data mining model, the two common measures are sensitivity and specificity. Sensitivity (often called the true positive rate) measures the percentage correctly identified as positive out of the total number of positives. Specificity (often called the true negative rate) measures the percentage correctly identified as negative out of the total number of negatives.

A method to discuss sensitivity and specificity is to use a confusion matrix [41]. A confusion matrix is an $L \times L$ matrix, where $L$ is the number of different label values. Table 2.2 is an example of a confusion matrix.

In the confusion matrix, quadrant $a$ denotes those tuples that were predicted as negative and were actually negative. Quadrant $b$ is composed of those tuples that were predicted as positive but were actually negative, also known as *false positives*. Quadrant $c$ is those tuples which were classified as negative but were actually positive, also known as *false negatives*. Quadrant $d$ contains the tuples that were classified as positive and were actually positive [22]. Thus, using the confusion matrix, it is easier to define a few metrics:

$$Accuracy = (a + d)/(a + b + c + d)$$

$$Sensitivity(TruePositiveRate) = d/(c + d)$$

$$Specificity(TrueNegativeRate) = a/(a + b)$$

$$Precision = d/(b + d)$$

$$FalsePositiveRate = b/(a + b) = 1 - Specificity$$

$$FalseNegativeRate = c/(c + d) = 1 - Sensitivity$$

- Accuracy is the percentage of tuples that are correctly classified out of all the tuples that are given a classification.
- Sensitivity (sometimes called recall) is the percentage of positive classification.
- Specificity is the percentage of negative classification.
- Precision indicates the number of exceptions to a rule. For example, a precision of 4/5 indicates that there is 1 exception to the rule.
- False Positive Rate is the percentage of tuples that are classified as positive but in reality are negative.
- False Negative Rate is the percentage of tuples that are classified as negative but in reality are positive.

There is one more metric which is useful when discussing the quality of a data mining model, and that is coverage. The coverage of a model is the proportion of the data for which there is a rule. Thus, a model which has 90% coverage

**Table 2.2.** A $2 \times 2$ confusion matrix

|  | Predicted Negative | Predicted Positive |
|---|---|---|
| Actual Negative | $a$ | $b$ |
| Actual Positive | $c$ | $d$ |

provides rules which classify 90% of the tuples. Coverage only means that a classification is made. It does not measure the accuracy of the classification.

Using these seven metrics, the results of a data mining model can be evaluated. This allows for reasonable comparisons to be made between different models.

One popular method of comparing the quality of different models is the Receiver Operating Characteristics curve, also known as the ROC curve [15]. The vertical axis of the ROC curve represents the sensitivity of the model, and the horizontal axis represents the false positive rate. The ROC curve is useful for determining cutoff points in testing so that true positive results are maximized while the false positive results are minimized. The ROC curve illustrates a challenge shared by researchers and data miners—the more sensitive a test or model becomes, more false positives are also likely to result.

## 2.4 Genetic Algorithms

Genetic algorithms, or GAs, are based upon evolutionary principles of natural selection, mutation, and survival of the fittest [12]. GAs are different from most computer algorithms, which have well-defined techniques for coming up with solutions to problems. The genetic algorithm approach is to randomly generate a large number of potential solutions in a search space and then evolve a solution that is suitable for the problem. Genetic algorithms are a useful technique for performing data mining. However, GAs are not the only technique used for data mining, nor are they exclusive to data mining, having been applied to numerous other problems as well.

### 2.4.1 Genetic Algorithm Techniques

One of the big keys to a successful genetic algorithm is in the development of a good fitness function. The fitness function determines how the algorithm evaluates each potential solution and defines the problem to be solved. For example, if the purpose of the genetic algorithm is to design a car, then the fitness function will provide a means for evaluating the fitness of a car design.

When developing a genetic algorithm, one must decide how each solution will be represented in the algorithm. For simplicity, a string of bits is most often used. The bits can be used to represent any part of the solution. Taking the car example, some of the bits might represent the color of the car, others the size of the wheels, and others the gas mileage of the car. It is the responsibility of the fitness function to understand what the bits mean and how to use them to evaluate the fitness of each potential solution.

During the first iteration of a GA, it generates an initial population of potential solutions. Usually this is done randomly. Each member of the population is then examined and its fitness is evaluated and recorded. Once each member of the population has been evaluated, then the next generation is produced from the current generation. There are many ways of creating the next generation, but the two most popular techniques involve "crossover" and "mutation." In crossover, two members of the population are chosen at random with higher probability given to the more fit members of the population. These two members are then combined to produce two offsprings. This is usually performed by selecting a position in the bit sequence and exchanging the two sequences after that position, which is called the crossover point. For example, given the following two members of a population: 111010100 and 100101000, if these were crossed over at position 4, the resulting offsprings would be: 111001000 and 100110100.

This process results in two new members in the next generation. In theory, these two new members should be reasonably more fit because they likely came from fit members in the previous population. Each member of the population is assigned a biased probability of selection. Because of this increased probability of selection, the most fit members of a population are more likely to be selected for crossover. However, there is always a possibility that a less fit member will be selected instead.

Usually, the crossover process continues until the size of the next generation is the same as the population size of the previous generation. After crossover has taken place, mutation is then applied to each member of the population. A typical mutation function is to assign a probability for flipping each bit. Thus, if a 10% value for mutation is assigned, then each bit of each member of the population has a 10% chance of being flipped, i.e., a 0 becomes a 1, and a 1 becomes a 0. After mutation is completed, each member of the new generation is evaluated for fitness and the process repeats for another generation. This process of evolving new populations continues until some stopping criterion is met. The stopping criteria could be when: (a) the overall fitness of the population reaches a certain value, (b) the overall fitness over several generations fails to change more than a specified threshold, or (c) a certain number of generations have been evaluated.

There are different ways to perform crossover and mutation. Genetic operators can also differ between GAs. The choice of genetic operators depends on the problem to be solved and the fitness function used to evaluate potential solutions. The genetic operators should reflect ways in which a member of the population could potentially become more fit; and thus a better solution to the problem. Regardless of the approach, the overall process remains the same: generate an initial population, evaluate the members of the population, generate a new population based upon the more fit members of the previous generation, and repeat the process until a certain stop criteria is achieved.

### 2.4.2   Applications of Genetic Algorithms

Genetic algorithms are powerful search tools. By search it is meant that GAs are capable of pouring through a large number of potential solutions to find good solutions. Scheduling has been an area where genetic algorithms have proven very useful. The GA searches the space of potential schedules and finds those schedules which are most effective, and maximize the desired criteria, such as minimizing idle time. For example, GAs are used by some airlines to schedule their flights [12]. It was reported that an application of GAs to a financial problem—tactical asset allocation and international equity strategies—resulted in an 82% improvement in portfolio value over a passive benchmark model, and a 48% improvement over a non-GA model used to improve the passive benchmark [12]. GAs have also been applied to problems such as protein motif discovery through multiple sequence alignment [23], and obtaining neural network topologies [36]. More information on genetic algorithms can be found in [19].

### 2.4.3 Genetic Algorithms for Data Mining

There are currently two different approaches to rule discovery in data mining using genetic algorithms, namely, the Michigan approach and the Pittsburg approach [2]. The Michigan approach represents a rule set by the entire population, with each member of the population representing a single rule. The Pittsburg approach represents an entire rule set with a single representation, thus each member of the population represents an entire set of potential rules for describing the data.

Data sets are either single-class or multi-class. This refers to the number of possible values for the decision class. If there is only one decision value, all of the rules will describe that value—thus it is a single-class set. In a multi-class set, there are multiple decision values in the data (but each rule describes a single value). Table 2.1 is a multi-class set because there are multiple values for the "reimbursement" decision. The Michigan method is useful for multi-class problems, but it suffers in that there is no way to ensure a high coverage of the data by evaluating a single rule at a time. The Pittsburg approach has been used to learn rules for a single class. In order to induce rules for multiple classes, the algorithm needs to be run multiple times.

In their experiments, [2] used the Pittsburg approach to great success. They compared the results of their data mining model, DMEL, developed by a genetic algorithm, to the results developed by C4.5, a very popular and well-known data mining algorithm that uses decision trees [29]. The experiment included seven different data sets that were diverse in nature. In each instance, the genetic algorithm produced more accurate results than C4.5, ranging from as little as 0.3% up to a 13% improvement in accuracy [2].

In their work, [16] concluded that genetic algorithms are well suited to undirected data mining, but can also be used for directed data mining. Undirected data mining is the most common form, where the program looks for patterns and describes them. In directed data mining, the user specifies the type of information in which they are interested. Using the Michigan approach, GA-MINER was able to find interesting, non-trivial rules within the data sets used for the experiment. They also posed an idea for "hypothesis refinement" in which the user could "seed" the genetic algorithm with a rule or set of rules which the GA can use as an initial population. In this way, the GA can refine the initial hypotheses to produce a better model for the data.

Thus, genetic algorithms can be extremely useful in data mining. They can start with a random set of hypotheses about the patterns within the data. They can then evolve these hypotheses and refine them until they reflect the real patterns that lie within the data. They can even start with a given set of hypotheses and attempt to refine them to reflect the patterns within the data.

## 2.5   Data Mining in Biology

There are numerous examples where data mining has benefited research in biology. Several of them are included in this section. The first two examples show how generating a common database has helped open the door for data mining and how it has helped research in those areas. Examples three through eight show how data mining has helped in the study of genomics and proteomics. Lastly, example nine shows how data mining can be used with many diverse data sets to find patterns within the data and hopefully give researchers new insights.

### 2.5.1   Measuring Biodiversity

Researchers with the Global Mountain Biodiversity Assessment GMBA [18] are encouraging the data mining of the various geo-referenced archive databases on mountain organisms. These databases contain geographical coordinates and altitude specifications with respect to where each organism was observed or collected. Using this geo-referenced data, biological and geophysical information can be linked together and used to test evolutionary and ecological theories across the worlds mountain ranges. The ability to separate global from regional geographic features offers new perspectives in how species adapt. In June 2006, GBMA held a meeting to promote data mining by bringing together database experts with biological experts.

### 2.5.2   Analysis of Protein Expression

A large number of proteins expressed in bacterial hosts form inclusion bodies, where proteins interact with each other, making it necessary for researchers to be able to isolate proteins for study. This process involves separating out the proteins in the inclusion body and then refolding the protein of interest back into its correct form. Thus, there is a need for highly efficient methods of protein folding that minimize miss-folding and other reactions that limit research [8].

Typically, refolding experiments are done on an individual protein basis and are published in a non-standardized fashion, making data mining for a particular method practically impossible. Instead, researchers have had to perform exhaustive manual searches through the literature. This led [8] to create a relational database for protein refolding methods. The REFOLD database encourages a standard for reporting methods, and establishes a central repository that can be easily searched. Researchers can deposit new methodologies into the database, thus allowing for dissemination of their results quickly throughout the scientific community.

By creating a standard for reporting of methods and results, as well as a central repository for those methods and results, researchers can now quickly gain access to the information they need. This also opens the door for opportunities in further data mining to see what successful refolding techniques have in common and could perhaps also assist in the prediction of protein structure.

### 2.5.3  Genome Annotation

Following the significant contributions of genome sequencing projects, there has been an increasing need for applications capable of examining these incredibly long sequences for biological data. There is an abundance of genomic information, but gathering knowledge from that information is a major challenge for bioinformatics today. In order to address this challenge, [6] developed a hybrid Bayesian statistical method to supply protein functional annotation in the yeast Saccharomyces cerevisiae. Their method involves an integration of micro-array profiles, protein complexes, and large-scale biological data such as yeast two-hybrid data, a method used to test for protein interaction [17]. Their approach quantified the relationship between functional similarity and the actual sequences themselves. This was done through a functional linkage graph, where each node of the graph represents a protein, with links to other proteins. Each link has a weight associated with the Bayesian probability that the two proteins have similar functionality. Their predictions also included evolution information and protein subcellular localization information. This technique was able to assign function to 1,802 out of 2,280 (79%) unannotated proteins in yeast.

### 2.5.4  Analysis of Gene Expression

Analyzing micro-array data is a critical task for many researchers. Often, the most important goal is to identify small sets of genes that share coherent expression (all similarly up-regulated or down-regulated) across a limited number of the tested conditions. It is also important to discover which conditions are the ones related to the expression of these gene sets, and how these conditions relate to conditional covariates such as disease diagnosis or prognosis. Researchers have developed a data analysis package that facilitates visualization and subsequent data mining of the independent sources of significant variation present in gene micro-array expression datasets [32]. They applied their work to two public datasets, highlighting sets of genes most affected by specific subsets of conditions (e.g. tissues, treatments, samples, etc.). Statistically significant associations for highlighted gene sets were shown via global analysis for Gene Ontology term enrichment. Together with covariate associations, the tool provides a basis for building testable hypotheses about the biological or experimental causes of observed variation.

This resulted in an unsupervised data mining technique for diverse micro-array expression datasets that is distinct from major methods now in routine use. Test cases, based on publicly available gene annotations, appear to identify numerous sets of biologically relevant genes. In instances where there are many diverse conditions (tens to hundreds of different tissues or cell types), a difficult situation for many clustering and ordering algorithms, this technique has proven especially effective. This approach also shows promise in other domains such as multi-spectral imaging datasets.

## 2.5.5    Analysis of Regulation

Micro-array experimentation often yields thousands of results that need to be analyzed. It is difficult for biologists to interpret and assimilate these results due to the sheer volume of measurements taken per experiment. A different type of micro-array, called a yeast deletion array, allows researchers to examine the effects of some reporter system when each of the approximately 5,000 genes on the array is knocked out [10]. The Aryl Hydrocarbon Receptor (AHR) is a protein that can act as a transcription factor. When a cell is exposed to various toxic chemicals, the AHR system turns on the expression of certain genes. One of the tasks of the KDD Cup Challenge in 2002 was to discover which genes played a role in the performance of the AHR signaling pathway. Data miners were given a large volume of information, including the known function of proteins expressed by many of the genes, the interaction between proteins, and over 15,000 journal abstracts with annotations to the genes associated with the abstracts. None of the teams entered in the competition had better than 70% accuracy, indicating that this is a very difficult problem for data mining, but an impossible problem without it, given the sheer volume of information involved.

## 2.5.6    Analysis of Mutations in Cancer

It is believed that there is a link between a family history of cancer and various genetic mutations. In research performed by [13], data mining was used to see if there was a correlation between specific mutations and related forms of cancer. They examined patients who had undergone gene testing and tested positive for mutations associated with cancer and who also had a family history of cancer. The features they chose to examine were:

1. Itemized cancers among the patients relatives
2. Relationship of cancer-affected relatives to the patient
3. Age of onset of cancers
4. Evidence of vertical transmission
5. Evidence of cancer in same generation
6. Repetition of identical cancers in the family
7. Level of overall cancer occurrences

Their data set of patients focused on those with confirmed mutations in genes associated with breast or ovarian cancers. The rules that resulted from data mining all showed an early age of onset for the cancers. Nearly all of the rules showed an intense family tree of cancer, with three or more cases of cancer within one generation of each other.

While this approach did not yield any new scientific knowledge (there was already research that established a link between these mutations and the cancers), the fact that the results are supported by the existing literature gives validity to the use of data mining in establishing links between specific mutations and hereditary forms of cancer. Thus, data mining has the potential to tremendously improve the treatment of cancer by finding more correlations that can be used in genetic screening for cancer mutations.

### 2.5.7  Prediction of Protein Structure

Analyzing proteins based upon their amino acid sequences has been the focus of many research efforts. The function of a protein is determined by its structure, which is determined by the amino acids that comprise the protein. Thus, by being able to predict the structure of an amino acid sequence, the function of the protein can then be inferred by comparing it to other proteins with similar structure.

[5] used a data mining algorithm called LEM2 ("Learning by Examples Module") to examine protein families for motifs (amino acid sequences that characterize a family of proteins). They examined ten different families of proteins, using 40 examples of each protein from different species. The proteins were then broken down into overlapping subsequences containing three amino acids each. Ignoring the order of the sequences, LEM2 was able to induce rules that could uniquely classify proteins into the correct families. The resulting model was 96% accurate but was only able to make predictions for 74% of the data. Thus, it found potential motifs for the families, but there was a gap in the coverage of the model—it was unable to make a decision for 26% of the proteins investigated. Further work is being done to see if the predictive model could be improved by taking into account the order of the amino acid subsequences.

### 2.5.8  Comparative Genomics

The entire genomes of many organisms have been sequenced. By using data mining techniques on these sequences, different kingdoms of life can be compared and contrasted, potentially giving insight into how the organisms evolved differently. In a study by [24] data mining tools were used to compare the kingdoms of archaea, eubacteria, and eukaryota. The results showed common features and different patterns in the protein evolution of the organisms. The researchers used principle component analysis of the various proteins and discovered that a majority of the proteins clustered closely together, with only a few outliers. Within these outlying proteins are the likely reasons for difference between these organisms.

### 2.5.9  Analysis of Data Sets

At the heart of data mining is the idea that patterns lie within data, and that by uncovering these patterns that lie within previously collected data, new knowledge can be acquired. The number of medical records residing in hospital databases is astronomical, with a wealth of information that remains largely untapped. This is the driving force behind Arcanum [37], a data mining system that uses genetic algorithms to manipulate rules that can be used to describe patterns within any set of data. Arcanum has been used to examine the recurrence of breast cancer after different kinds of treatment and to develop rules for predicting the malignancy of cancer tumors. Future tasks for Arcanum include mining large databases of patient records in hopes of discovering causes and/or treatments for particular forms of illness.

## 2.6  Data Mining Tools

There are numerous data mining tools and resources available on the Internet, including many which specialize in bioinformatics applications. A few of these tools are highlighted in this section. Because there are many different domains and problems within biological and biomedical research, the tools used to analyze these problems are also very different from one another.

### 2.6.1  Protein Prospector

The Protein Prospector [28] is a data mining tool that allows users to search protein databases using mass spectrometry data. With the numerous pre-filters, users can limit searches to within a particular species, a series of accession numbers, or can even pre-filter the searches based upon protein names.

### 2.6.2  BLAST

The Basic Local Alignment Search Tool (BLAST) [3] is used for comparing genomic or proteomic sequences against public databases in order to find matches. It is very useful for helping researchers to identify a protein or a genetic sequence, or to find similar sequences in other organisms, thus giving the researcher insight into potential functions for the protein sequence. There are several different versions of BLAST which allow for different kinds of searches, such as matching a protein to the DNA sequence that codes the protein.

### 2.6.3  Fasta

Fasta [14] is a sequence search program similar to BLAST that allows researchers to match proteins to proteins, DNA to DNA, or even to match protein sequences with the DNA sequence that codes for the protein. Fasta is able to perform matches quickly because it uses many customizable parameters to limit the number of potential matches upfront.

### 2.6.4  CDART

The Conserved Domain Architecture Retrieval Tool (CDART) [4] allows the users to enter a protein sequence or accession number. It then lists the functional domains that comprise the protein and searches for other proteins with similar domain structures. Unlike BLAST and Fasta, CDART does not perform a direct sequence similarity search. Instead, it looks at functional annotations and retrieves proteins that have similar conserved domains.

### 2.6.5  VAST

The VAST Search [39] tool performs a structure to structure similarity search. It compares actual 3D coordinates of a protein structure to the 3D structure of other proteins. The results are shown graphically, so that the users can compare the images of the structures and even superimpose one structure on the other to examine similarities and differences.

### 2.6.6   Tanagra

Tanagra [35] is a free data mining software package. It is open source, so users can add their own algorithms or modules to it. The source code is written in Delphi 6. It implements construction of decision trees, clustering, factorial analysis, feature selection, and many other algorithms of use.

### 2.6.7   Other Resources

Other Websites offer free tools, or links to tools, including [1] which contains an extensive list of tools ranging from sequence analysis to phylogeny and protein splicing, [25] which includes a list of data mining tools available from the National Center of Biotechnology Information (NCBI), and [9] where their Gene Workbench is a free software download that provides data management and analysis tools.

## 2.7   Conclusion

This chapter has provided an overview of data mining and genetic algorithms and how they pertain to biological and biomedical research. Data mining is unlike traditional forms of research in that it is data-driven rather than hypothesis-driven. It searches through data to find the hidden knowledge buried within. As such, data mining is domain independent—it can be used in many different fields using many different types of data. Because it is data-driven, data mining does not require expertise within the field from whence the data originate. However, for the results to be useful, experts must examine the outcomes of data mining to determine if any useful knowledge has been gained. Thus, data mining encourages interdisciplinary studies between data miners and experts of the field being studied. As the amount of biological data continues to grow, it is increasingly important for researchers to have access to the tools that data mining provides.

The examples of data mining applications in biology show the importance of generating central repositories for biological data so that data mining can be performed. Collecting the diverse forms of data together into a standardized format can be a difficult task, but researchers are seeing the benefits that can result when the vast data available is gathered and mined. Other examples showed the success of data mining in genomics and proteomics, where there are tremendous amounts of high-throughput data that require analysis.

Because data mining is proving to be very valuable to researchers, and because there is such a vast amount of information needing to be mined, there are many data mining tools available. Some tools are very specialized, while others are generalized. A select list of some of the data mining tools relating to biology have been provided, including some Web sites that contain links to large numbers of such tools.

The amount of data available to researchers is enormous and continues to grow. In fact, biological data is growing at an exponential rate as more and more researchers contribute their findings. Data mining is currently one of the best hopes for handling such vast amounts of information. There is also tremendous potential to find patterns and to make discoveries that have gone overlooked in this wealth of information. Data need to be collected and mined and then presented to the experts to see if perhaps some new discovery or lead can be uncovered. It is possible that the secrets to curing some of the world's diseases already lie within the data currently accumulated, just like the secret to helping prevent premature births was waiting in the information previously collected. Through data mining, it is entirely possible that more can be learned from what is already known.

# References

1. 123Genomics (2007), http://www.123genomics.com/files/analysis.html (retrieved June 2007)
2. Au, W.-H., Chan, K.C.C., Yao, X.: A novel evolutionary data mining algorithm with applications to churn prediction. IEEE Transactions on Evolutionary Computation 7(6), 532–545 (2003)
3. BLAST (2007), http://www.ncbi.nlm.nih.gov/BLAST (retrieved June 2007)
4. CDART (2007), http://www.ncbi.nlm.nih.gov/Structure/lexington/lexington.cgi?cmd=rps (retrieved June 2007)
5. Chen, X.-W., Taylor, C.M.: Predicting protein function using sequence information (manuscript in preparation, 2007)
6. Chen, Y., Xu, D.: Global protein function annotation through mining genome-scale data in yeast Saccharomyces Cerevisiae. Nucleic Acids Research 32(21), 6414–6424 (2004)
7. Cheng, C.H., Fu, A.W.-C., Zhang, Y.: Entropy-based subspace clustering for mining numerical data. Knowledge Discovery and Data Mining, 84–93 (1999)
8. Chow, M.K., Amin, A.A., Fulton, K.F., Fernando, T., Kamau, L., Batty, C., Louca, M., Ho, S., Whisstock, J.C., Bottomley, S.P., Buckle, A.M.: The REFOLD Database: A tool for the optimization of protein expression and refolding. Nucleic Acids Research 34(D), 207–212 (2006)
9. CLC bio (2007), http://www.clcbio.com (retrieved June 2007)
10. Craven, M.: The genomics of a signaling pathway: A KDD cup challenge task. SIGKDD Explorations 4(2), 97–98 (2003)
11. De Raedt, L., Blockeel, H., Dehaspe, L., Van Laer, W.: Three companions for data mining in first order logic. In: Dzeroski, S., Lavrac, N. (eds.) Relational Data Miningm. Springer, Heidelberg (2001)
12. Dulay, N.: Genetic Algorithms. Surprise 96 Journal On-Line (2005), http://wwwhomes.doc.ic.ac.uk/nd/surprise_96/journal/vol4/tcw2/report.html (retrieved April 2005)
13. Evans, S., Lemon, S.J., Deters, C., Fusaro, R.M., Durham, C., Snyder, C., Lynch, H.T.: Using data mining to characterize DNA mutations by patient clinical features. In: Proc. AMIA Annual Fall Symposium, Nashville, Tennessee, October 1997, pp. 253–257 (1997)
14. Fasta (2007), http://www.ebi.ac.uk/fasta33 (retrieved June 2007)

15. Fawcett, T.: Using rule sets to maximize ROC performance. In: Proc. 2001 IEEE International Conference on Data Mining (ICDM 2001), Washington, DC, pp. 131–138 (2001)
16. Flockhart, I.W., Radcliffe, N.J.: A Genetic algorithm-based approach to data mining. In: Proc. Second International Conference on Knowledge Discovery and Data Mining (KDD 1996), Portland, Oregon, August 1996, pp. 299–302 (1996)
17. Gietz, R.D., Triggs-Raine, B., Robbins, A., Graham, K., Woods, R.: Identification of proteins that interact with a protein of interest: Applications of the yeast two-hybrid system. Molecular and Cellular Biochemistry 172, 67–79 (1997)
18. Global Mountain Biodiversity Assessment, Mountain Biodiversity Data Mining (2007), http://gmba.unibas.ch/research/datamining.htm (retrieved June 2007)
19. Goldberg, D.: Genetic Algorithms in Search, Optimization, and Machine Learning. Addison-Wesley Publishing, Inc., Reading (1989)
20. Jech, T.: Set Theory. In: Zalta, E.N. (ed.) The Stanford Encyclopedia of Philosophy (2002), http://plato.stanford.edu/contents.html
21. Kohavi, R.: The power of decision tables. In: Lavrač, N., Wrobel, S. (eds.) ECML 1995. LNCS, vol. 912. Springer, Heidelberg (1995)
22. Kohavi, R., Provost, F.: Glossary of Terms. Machine Learning, Special Issue on Applications of Machine Learning and the Knowledge Discovery Process 30, 271–274 (1998)
23. Mendez, J., Falcon, A., Lorenzo, J.: A procedure for biological sensitive pattern matching in protein sequences. In: Proc. First Iberian Conference on Pattern Recognition and Image Analysis, Mallorca, Spain, June 2003, pp. 547–555 (2003)
24. Nandi, T.B., Rao, C., Ramachandran, S.: Comparative genomics using data mining tools. J. BioSciences 27(1), 15–25 (2002)
25. NCBI (2007), http://www.ncbi.nlm.nih.gov/Tools/ (retrieved June 2007)
26. Ofran, Y., Rost, B.: Predicted Protein-Protein Interaction Sites From Local Sequence Information (2003), http://citeseer.ist.psu.edu/ofran03predicted.html (retrieved October 2007)
27. Pawlak, Z.: Rough sets present state and further prospects. In: Proc. Third International Workshop on Rough Set and Soft Computing, San Jose, California, pp. 72–76 (1994)
28. Protein Prospector (2007), http://prospector.ucsf.edu/ (retrieved June 2007)
29. Quinlan, R.: C4.5: Programs for Machine Learning. Morgan Kaufmann, San Diego (1993)
30. Radivojevic, Z., Cvetanovic, M., Milutinovic, V., Sievert, J.: Data mining: A brief overview and recent IPSI Research. Annals of Mathematics, Computing, and Teleinformatics 1(1), 84–90 (2003)
31. Rakotomalala, R.: TANAGRA: A free software for research and academic purposes. In: Proc. EGC 2005, RNTI-E-3, vol. 2, pp. 697–702 (2005) (in French)
32. Roden, J.C., King, B.W., Trout, D., Mortazavi, A., Wold, B.J., Hart, C.E.: Mining Gene expression data by interpreting principal components. BMC Bioinformatics 7(194) (2006)
33. Siler, W.: Rule-Based Reasoning: Antecedent and Consequent, Building Fuzzy Expert Systems (2005), http://members.aol.com/wsiler/chap03.htm (retrieved November 2005)
34. Straccia, U.: A fuzzy description logic. In: Proc. 15th National Conference on Artificial Intelligence (AAAI 1998), Madison, Wisconsin, pp. 594–599 (1998)
35. Tanagra Project (2007), http://eric.univ-lyon2.fr/ ricco/tanagra/en/tanagra.html (retrieved June 2007)

68      C.M. Taylor and A. Agah

36. Taylor, C.M., Agah, A.: Evolving Neural Network Topologies for Object Recognition. In: Proc. 6th International Symposium on Soft Computing for Industry (ISSCI 2006), Budapest, Hungary, July 2006, ISSCI-71, pp. 1–6 (2006)
37. Taylor, C.M.: An enhanced genetic algorithm with direct manipulation of sets for data mining. Ph.D. Dissertation, University of Kansas (2008)
38. Valdes, J.J., Mateescu, G.: Time series model mining with similarity-based neuro-fuzzy networks and genetic algorithms: A parallel implementation. In: Alpigini, J.J., Peters, J.F., Skowron, A., Zhong, N. (eds.) RSCTC 2002. LNCS (LNAI), vol. 2475. Springer, Heidelberg (2002)
39. VAST Search (2007), http://www.ncbi.nlm.nih.gov/Structure/VAST/vastsearch.html (retrieved June 2007)
40. Vesanto, J.: SOM-based data visualization methods. Intelligent Data Analysis 3, 111–126 (1999)
41. Wikipedia, Confusion Matrix (2007), http://en.wikipedia.org/wiki/Confusion_matrix (retrieved June 2007)
42. Woolery, L.K., Grzymala-Busse, J.: Machine learning for an expert system to predict preterm birth risk. J. American Medical Informatics Association 1(6), 439–446 (1994)

# 3

# The Use of Rough Sets as a Data Mining Tool for Experimental Bio-data

Ray R. Hashemi[1], Alexander A. Tyler[2], and Azita A. Bahrami[3]

[1] Department of Computer Science, Armstrong Atlantic State University, Savannah, GA 31419, USA
Ray.Hashemi@armstrong.edu
[2] University of Arkansas for Medical Sciences, Area Health Education Center-Pine Bluff, 4010 Mulberry Street, Pine Bluff, AR 71603, USA
alexandertyler@yahoo.com
[3] Department of Information Technology, Armstrong Atlantic State University, Savannah, GA 31419, USA
Azita.Bahrami@armstrong.edu

**Summary.** The Rough Sets methodology has great potential for mining experimental data. Since its introduction by Pawlak, it has received a lot of attention in the computing community. However, due to the mathematical nature of the Rough Sets methodology, many experimental scientists lacking sufficient mathematical background have been hesitant to use it. The goal of this chapter is twofold: (1) to introduce "Rough Sets" methodology (along with one of its derivatives, "Modified Rough Sets") in a non-mathematical fashion hoping to share the potentials of this approach with a larger group of non-computationally-oriented scientists (Mining of one specific form of implicit data within a bio-dataset is also discussed), and (2) to apply this methodology to a dataset of children with and without Attention Deficit/Hyperactivity Disorder (ADHD), to demonstrate the usefulness of the approach in patient differentiation. Discriminant Analysis statistical approach as well as the ID3 approach were also applied to the same dataset for comparison purposes to find out which approach is most effective.

## 3.1 Introduction

Data Mining provides methodologies for finding patterns of interest in a given dataset, experimental bio-dataset or otherwise. There are six general types of patterns of interest and they are discovered from a given dataset by Association Analysis, Classification and Prediction, Cluster Analysis, Outlier Analysis, Evolution Analysis, and Data Dependency Analysis [1]. In this chapter, the focus will be on Classification and Prediction. Classification is the process of finding a model (or function) that is able to describe and distinguish data classes or concepts within the dataset. Prediction is the process of using such a model to predict the class of those objects that are new to the dataset and whose class labels are unknown [2, 3, 4].

There are many methodologies that can be used to find a classification model for a dataset. Some of these methodologies are: Rough Sets [5], Neural Networks [6, 7, 8, 9], Genetic Algorithms [10, 11, 12], Fuzzy Logic [13, 14], Decision trees

T.G. Smolinski et al. (Eds.): Comp. Intel. in Biomed. & Bioinform., SCI 151, pp. 69–91, 2008.
springerlink.com

[15, 16], Statistical Methods [17, 18], and many different Hybrid methodologies [11, 14, 19, 20, 21, 22, 23, 24]. For all of these methods, the goal is to mine a dataset for "implicit" data . By implicit data, it is meant that the data that are hidden in the semantics of the observations. One specific type of "implicit" data is the relationship between "cause" and "effect" in some clinical and experimental studies. For example, it is desirable to mine datasets to determine a model that captures the relationship(s) between the toxic effect of a chemical agent (a dependent variable) and the conditions under which the subject was exposed to the toxic agent (independent variable). This specific type of implicit data provides the basis for building diagnostic, extrapolative, pedagogical, and many other predictive systems with a variety of applications.

Based on previous observations it is clear that the Rough Sets approach has great potential for helping to find relationships between "cause" and "effect" in empirical datasets. This potential stems from two facts: (1) a bio-dataset may be presented as a special type of information system founded on Rough Sets methodology; and (2) the results of data mining using this approach can be expressed in the form of 'If .... Then' statements or rules such that they have tangible meaning(s).

Since the Rough Sets approach was introduced by Pawlak [5], it has received a lot of attention in the computing community [21, 23, 25, 26, 27, 28, 29, 30, 31, 32, 33]. However, due to the mathematical nature of the Rough Sets approach, many experimental scientists lacking sufficient mathematical background have been reluctant to use it. In this chapter, the Rough Sets methodology (along with one of its derivatives, "Modified Rough Sets") is described in a non-mathematical fashion in the hopes that non-computationally-oriented scientists can also benefit from its capabilities. In addition, the usefulness of the approach and its effectiveness are compared with the more familiar statistical approach of Discriminant Analysis and ID3.

## 3.2   Literature Review

There is a larg number of articles about Rough Sets in the literature. The articles cover a vast spectrum of research about different aspects of Rough Sets. One end of the spectrum deals with the basic foundation of Rough sets and the other end deals with hybridization of Rough sets with other soft computing models such as Fuzzy Sets, Neural Networks, and Evolutionary algorithms [34, 35, 36, 37, 38]. Invariably, the articles within the two ends use mathematical language in order to explain Rough Sets and its features, abilities, and shortcommings. Should an article delve into the paths of a particular aspect of Rough Sets, the mathematical language used become substantially heavier. The authors searched the existing literature in the hope of finding some articles that have explained Rouh Sets and its characteristics in a non-matematical language, but no such articles were found.

## 3.3    The Rough Sets Methodology

Suppose a group of human subjects have been tested with a variety of behavioral tasks [39]. In addition, suppose that the IQ of each subject has been measured independently. As a result, each subject's record will be composed of a set of behavioral task results (Conditional attributes, or simply Conditions) and IQ score (Decision attribute, or simply Decision). Figure 3.1 shows an example of such a record for which the set of conditions are: Short-term memory task [Delayed Matching-to-Sample (DMTS) task] Accuracy, Learning task [Incremental Repeated Acquisition (IRA) task] Accuracy, Motivation task [Progressive Ratio (PR) task] Response Rate, and Time Estimation task [Temporal Response Differentiation (TRD) task] Response Rate. The decision is IQ. Each condition may have the possible values of "Above Average", "Average", or "Below average" and the possible values for the decision are "High", "Average", and "Low".

| Conditional attributes | | | | Decision's attribute |
|---|---|---|---|---|
| **DMTS Accuracy** | **IRA Accuracy** | **PR Response Rate** | **TRD Response Rate** | **IQ** |
| Above Average | Average | Above Average | Above Average | High |

**Fig. 3.1.** An example of a subject's record

Suppose a set of behavioral task results and IQ data are collected from a variety of subjects. This dataset can be visualized as shown in Figure 3.2-a (the large rectangle represents the entire dataset that is composed of records from 42 subjects, the small rectangles). The goals here are (a) to establish the relationships (if any) between the behavioral task results and IQ (i.e., between the set of conditions and the decision), and (b) to express the relationships in the form of *'If .... Then'* rules. To meet these goals, the "Rough Sets" methodology is introduced in a non-mathematical fashion.

Depending on what the intention is, the subjects can be categorized into *classes* or *partitions*. All of the subjects who have the same values for their conditions constitute a class. Suppose that, after the classification process is completed, there are four classes in the dataset, class 1 to class 4. The classes of the dataset are visualized in Figure 3.2-b: the rectangles with the same shaded patterns collectively represent one class. The four classes 1, 2, 3, and 4 contain data for 26, 9, 5, and 2 subjects, respectively. Since each subject's record represents a data summary for that subject, the word "subject" and the term "subject's record" are used interchangeably in this chapter.

All partition subjects are the ones about whom the same decsions are made. Since there are only three possible values for the decision, the number of partitions could be three. For example, all the subjects with IQ = "High" constitute one partition. And let the subjects surrounded by the thick solid lines, Figure 3.2-c, represent this partition. The number of subjects in this partition is 20.

A closer look at this partition reveals that not all subjects in class 3 (dark black rectangles), for example, are contained within this partition. That is, the subjects

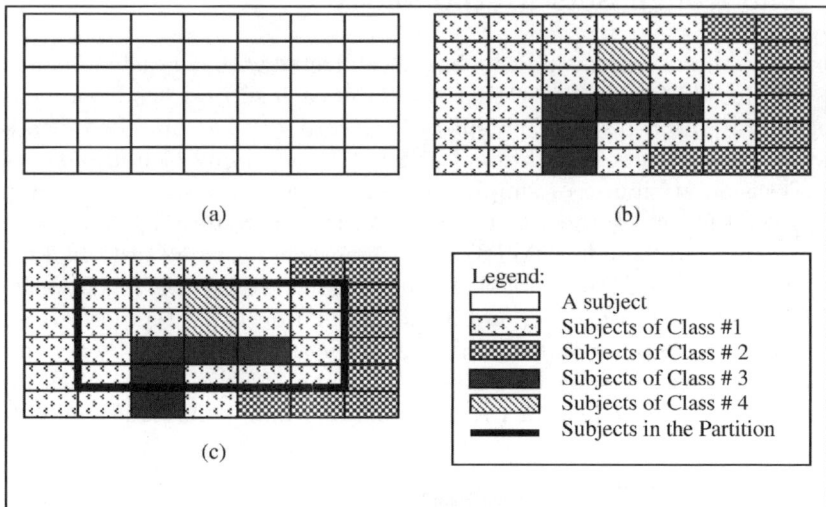

**Fig. 3.2.** Visualization of a dataset: (a) subjects, (b) classes and (c) a partition which includes only subjects with High IQ

in class 3 have the same conditions' values, but not all of them have the same IQ values. This kind of observation is often made throughout experimental sciences: in any given study within a class of subjects that share similar characteristics (conditions), a group of subjects that behaves differently from the rest of the subjects in the class can often be identified. As observed here, one subject in class 3 does not have the same IQ value as the rest of the subjects in that class.

In general, a class may be "totally", "partially", or "not-at-all" contained in a specific partition. Here, Class 4 is "totally" contained in the partition while Class 1 and 3 are "partially" contained in the partition and class 2 is "not-at-all" contained in the partition. The subjects of those classes that are totally contained within the partition are said to constitute the *lower approximation space* of the High IQ partition. These subjects exhibit a strong relationship between their conditions and their decision. The subjects in class 4, thus, define the lower approximation space of the High IQ partition, Figure 3.3-a. The subjects of all the classes that are either "totally" or "partially" contained in the partition make the *upper approximation space* of the High IQ partition, Figure 3.3-b.

### 3.3.1 Rough Sets

A mathematical set, which we refer to as a "traditional" set, is a well-defined collection of objects called members or elements. A well-defined set means that, for any given object, it either belongs to the set or it does not belong to the set. In other words, a given object has only two choices: being "inside" or "outside" of the set. With this in mind, let us look at the High IQ partition as a traditional set and let us look at each class as an object. Except for object 4 (class 4) that

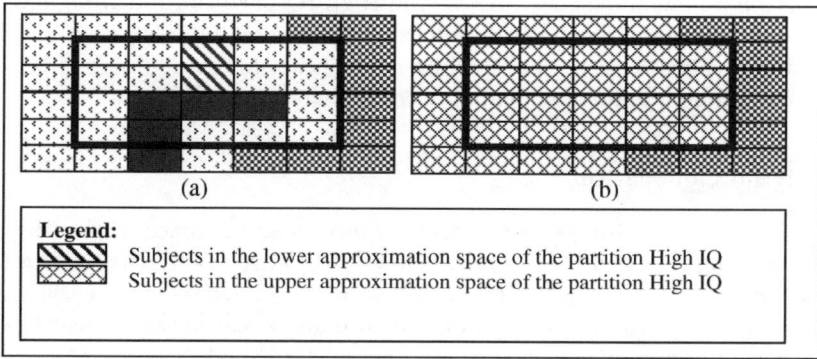

**Fig. 3.3.** Visualization of the approximation spaces for high IQ: (a) The lower approximation space and (b) the upper approximation space

is inside the set, objects 1 and 3 (classes 1 and 3) are partially inside the set and object 2 (class 2) is outside of the set. Thus, in reference to the High IQ set, objects can either be "inside", "outside", or "partially inside" of the set. These three possibilities (instead of the traditional two) contradict the definition of a traditional set. Therefore, the High IQ set is not a traditional set and it is called a "Rough" set. To be more precise, we may say that the High IQ partition is represented by a Rough Set.

A Rough Set has two boarders, Figure 3.4-a. The inner boarder signifies the lower approximation space and the outer boarder signifies the upper approximation space of the Rough Set. The classes that are partially in the Rough Set make the *boundary* of the Rough Set. Any subject of the classes that make the boundary *possibly* resides inside the Rough Set. But any subject of the classes in the lower approximation space of the Rough Set *certainly* resides inside the Rough Set. For the High IQ Rough Set the inner and outer boarders are shown in Figure 3.4-b.

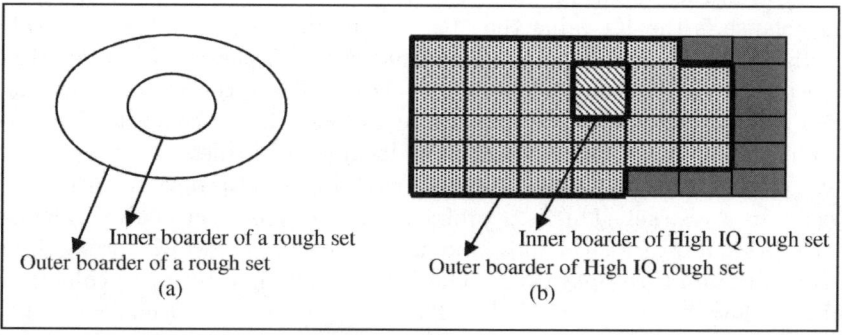

**Fig. 3.4.** Visualization of the High IQ Rough Set: (a) a Rough Set and its boarders and (b) the inner and outer boarders for High IQ Rough Set

It must be remembered that, just as the High IQ partition is a Rough Set, so are the partitions for the other two values of IQ (Average IQ partition and Low IQ partition.) Thus, here we are dealing with three Rough Sets and this is the reason that the methodology is called "Rough Sets" (plural).

### 3.3.2   Rule Generation

Suppose one of the subjects in the lower approximation space of the High IQ Rough Set is represented by the data shown in Figure 3.1. A rule in the form of *If...Then...* can be generated from this subject. The *If* clause of the rule is composed of the conditions' values of the record and because the subject belongs to the lower approximation space of the "High" IQ Rough Set the *Then* clause is IQ = "High", (see Figure 3.5).

```
If      DMTS Accuracy = Above Average  &
        IRA Accuracy = Average  &
        PR Response Rate = Above Average  &
        TRD Response Rate = Above Average  &
Then    IQ = High
```

**Fig. 3.5.** A local certain rule for High IQ subjects

A rule generated from a subject that belongs to the lower approximation space of the High IQ Rough Set is called a *local certain rule*. The term "local" means that the rule is generated from one subject's record. The term "certain" means that the subject belongs to the lower approximation spce. That is, any new subject that satisfies the *If* clause of a local certain rule, will *certainly* be categorized as a subject with the IQ value of the *Then* clause. If a rule is derived from a subject in the boundary of the High IQ Rough Set, then the rule will be called a *local possible rule*. Again, the "If" clause of a local possible rule is borrowed from the conditions' values of the subject in the boundary, while the "Then" clause is the IQ value that belongs to the Rough Set. A new subject that satisfies the *If* clause of a local possible rule, will *possibly* be categorized as a subject with the IQ value of the *Then* clause. There is a chance that the IQ value in the *Then* clause of the rule can be different from the actual IQ value of the subject. This is not the case with the local certain rules.

A large rule set may contain some rules that are not appropriate for prediction of objects in a test set. This is a problem that may be rectified by calculating every rule's significance level and removing those rules for which the significance level is less than a threshold value. There are many algorithms for calculating a significance level for individual rules. These algorithms are based on the quality of approximation, statistics, or both. However, the details of such approaches are beyond the scope of this chapter but can be obtained from [29, 40, 41, 42].

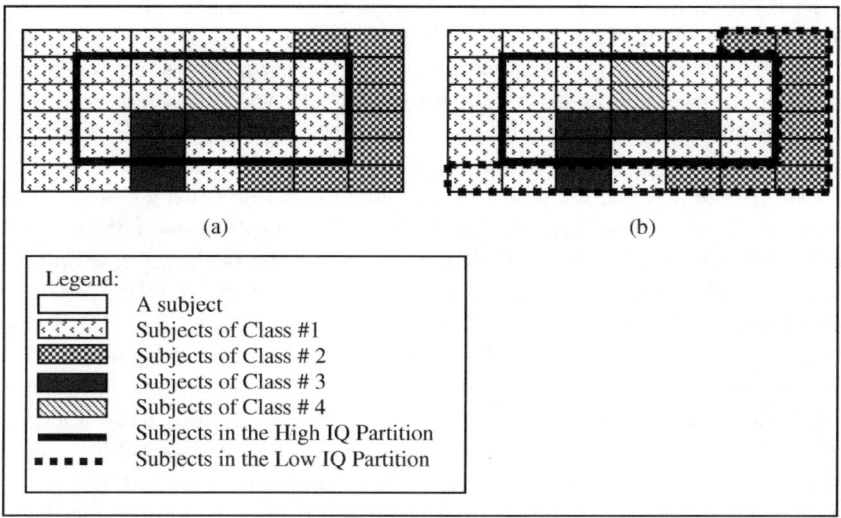

(a)                                                    (b)

Legend:
☐         A subject
▨         Subjects of Class #1
▨         Subjects of Class # 2
■         Subjects of Class # 3
▧         Subjects of Class # 4
▬▬▬       Subjects in the High IQ Partition
■ ■ ■ ■   Subjects in the Low IQ Partition

**Fig. 3.6.** Visualization of classes and partitions: (a) classes and High IQ Partition and (b) Low IQ Partition

| Local Rules | IRA Accuracy | PR Response Rate | TRD Response Rate | IQ |
|---|---|---|---|---|
| Certain | Below Average | Below Average | Average | low |
| | Average | Above Average | Average | Average |
| | Below Average | Above Average | Average | Average |
| | Above Average | Average | Below Average | High |
| Possible | Below Average | Below Average | Below Average | Low |
| | Below Average | Below Average | Below Average | Average |
| | Average | Above Average | Above Average | Average |
| | Average | Above Average | Above Average | High |
| | Above Average | Average | Average | High |
| | Above Average | Average | Average | Average |

**Fig. 3.7.** The tabular presentation of a set of local certain and possible rules

### 3.3.3   Globalization of Rules

A copy of Figure 3.3-c is shown in Figure 3.6-a. To refresh the reader's memory, Figure 3.6-a shows four different classes (each class is shaded differently) and a partition for IQ = "High" (the realm of the partition is shown with a thick line.) A new partition for IQ = "Low" along with the High IQ partition is shown in Figure 3.6-b using a thick broken line.

Let us concentrate on the subjects of class 3. Since the subjects of this class are partially in the High IQ partition, all the subjects are in the boundary of the High IQ Rough Set. By the same analogy, all the subjects of class 3 are also in the boundary of the Low IQ Rough Set.

Since all the subjects in class 3 have the same decision values, and they belong to both boundaries of Low and High IQ values, two local possible rules can be generated. Both local possible rules have the same conditions' values but different IQ values. Such a case is never true for a set of local certain rules.

An example of a set of local certain and local possible rules generated from a new dataset in which each record has only three conditional attributes of IRA Accuracy, PR Response Rate, and TRD Response Rate, is shown in tabular form in Figure 3.7. The first row of this table is read as follows: "*If* (IRA Accuracy = "Below Average" & PR Response Rate = "Below Average" & TRD Response Rate = "Average") *Then* IQ = "Low". Concentrating only on the conditions, Figure 3.8, reveals that there are no duplications of the conditions' values among the local certain rules, but there exist such duplications among the local possible rules. As part of the process of globalization, the local possible rules with duplicated conditions' values are collapsed into one rule, Figure 3.9.

Close inspection of the rule set in Figure 3.9 reveals that none of the rules have a value of "Below Average" for the PR Response Rate condition and a

| Local Rules | IRA Accuracy | PR Response Rate | TRD Response Rate |
|---|---|---|---|
| Certain | Below Average | Below Average | Average |
| | Average | Above Average | Average |
| | Below Average | Above Average | Average |
| | Above Average | Average | Below Average |
| Possible | Below Average | Below Average | Below Average |
| | Below Average | Below Average | Below Average |
| | Average | Above Average | Above Average |
| | Average | Above Average | Above Average |
| | Above Average | Average | Average |
| | Above Average | Average | Average |

**Fig. 3.8.** The tabular presentation of the rules without the decision attribute

| Local Rules | IRA Accuracy | PR Response Rate | TRD Response Rate |
|---|---|---|---|
| Certain | Below Average | Below Average | Average |
| | Average | Above Average | Average |
| | Below Average | Above Average | Average |
| | Above Average | Average | Below Average |
| Possible | Below Average | Below Average | Below Average |
| | Average | Above Average | Above Average |
| | Above Average | Average | Average |

**Fig. 3.9.** The collapsed local possible rules

| Local Rules | IRA Accuracy | PR Response Rate | TRD Response Rate |
|---|---|---|---|
| Certain | - | Below Average | Average |
| | Average | - | Average |
| | Below Average | Above Average | - |
| Possible | - | Average | Below Average |
| | Below Average | - | Below Average |
| | Average | - | Above Average |
| | - | Average | Average |

**Fig. 3.10.** The dropped conditions from Figure 3.9

| Local Rules | IRA Accuracy | PR Response Rate | TRD Response Rate | IQ |
|---|---|---|---|---|
| Certain | - | Below Average | Average | low |
| | Average | - | Average | Average |
| | Below Average | Above Average | - | Average |
| | - | Average | Below Average | High |
| Possible | Below Average | - | Below Average | Low |
| | Below Average | - | Below Average | Average |
| | Average | - | Above Average | Average |
| | Average | - | Above Average | High |
| | - | Average | Average | High |
| | - | Average | Average | Average |

**Fig. 3.11.** Expansion of Figure 3.10 to include the decision attribute and duplicated local possible rules

value of "Average" for the TRD Response Rate condition except for the first rule. In other words, the first rule may be differentiated from the rest of the rules by using only the PR Response Rate and the TRD Response Rate conditions' values and, thus, the first condition is "dropped". Therefore, the first rule may be expressed as: If (PR Response Rate = "Below Average" & TRD Response Rate = "Average") Then IQ = "Low". The new form of the local certain rule is called a *global certain rule*. The term "global" means that the minimum number of conditions for the given rule is kept while preserving its uniqueness within the rule set. The adopted method for generating global rules using the local rules is called the *dropping condition* [43]. Figure 3.10 illustrates the dropped conditions and they are shown by a dash (-) symbol. Now, Figure 3.10 can be expanded to include the decision attribute and duplicated local possible rules, Figure 3.11, from which the final set of Global certain and possible rules are generated and displayed in Figure 3.12.

Before we leave this section, an important point needs to be discussed. Let us look at the local certain rule number four in Figure 3.9 one more time. This rule is globalized in Figure 3.10 by keeping the conditions PR Response Rate and TRD Response Rate. Keeping the PR Response Rate and TRD Response

**Global Certain Rules:**
 If (PR Response Rate = "Below Average" & TRD Response Rate = " Average")
  Then IQ = "low"
 If (IRA Accuracy = "Average" & TRD Response Rate = "Average")
  Then IQ = "Average"
 If (IRA Accuracy = "Below Average" & PR Response Rate = "Above Average")
  Then IQ = "Average"
 If (PR Response Rate = "Average" & TRD Response Rate = "Below Average")
  Then IQ = "High"

**Global Possible Rules:**
 If (IRA Accuracy = "Below Average" & TRD Response Rate = "Below Average")
  Then (IQ = "Low" OR IQ = "Average")
 If (IRA Accuracy = "Average" & TRD Response Rate = "Above Average")
  Then (IQ = "Average" OR IQ = "High")
 If (PR Response Rate = "Average" & TRD Response Rate = " Average")
  Then (IQ = "Average" OR IQ = "High")

**Fig. 3.12.** The list of global certain and possible rules for the rule set of Figure 3.7

Rate is but one of the possible choices. Another choice is keeping IRA Accuracy and TRD Response Rate. Therefore, the local certain rule number four will be transformed into two following global certain rules:

- If (PR Response Rate = "Average" & TRD Response Rate = "Below Average") Then IQ = "High"
- If (IRA Accuracy = "Above Average" & TRD Response Rate = "Below Average") Then IQ = "High"

The domain expert may dismiss none or one of these rules. The point is that for every local certain and possible rule, there may exist more than one global rule and Figure 3.10 does not show all of these global rules.

### 3.3.4   Information Systems

In Rough Set nomenclature, an *information system* is a set of records for a number of subjects. Each record in this system is described with a set of attributes. Data for the attributes are discrete. That is, data are represented in categorical values such as "1", "2", . . . and/or linguistic values such as "high", "low", etc. No limitation is placed on the number of discrete values that can be used for a given attribute. To use the information system in a bio-data mining process, one may name the attributes that represent dependent variables as *conditions* and the independent variable as *decision*. Doing so, technically, turns the information system into a *decision table* in which each record has two parts: Conditions part and Decision part, Figure 3.13. However, we continue using the term information system throughout the chapter. An information system may be representative

| Subjects | Conditions | | | | Decision |
|---|---|---|---|---|---|
| | DMTS Accuracy | IRA Accuracy | PR Response Rate | TRD Response Rate | IQ |
| s1 | Below Average | Below Average | Below Average | Below Average | Low |
| s2 | Below Average | Above Average | Average | Average | High |
| s3 | Above Average | Average | Above Average | Above Average | Average |
| s4 | Above Average | Average | Above Average | Above Average | High |
| s5 | Average | Average | Above Average | Average | Average |
| s6 | Above Average | Below Average | Above Average | Average | Average |
| s7 | Above Average | Above Average | Average | Below Average | High |
| s8 | Below Average | Below Average | Below Average | Below Average | Low |
| s9 | Below Average | Below Average | Below Average | Average | Low |
| s10 | Below Average | Above Average | Average | Average | Average |
| s11 | Below Average | Below Average | Below Average | Below Average | Average |

**Fig. 3.13.** An Information System

of a typical dataset obtained from experimentation. Three following questions are needed to be answered in reference to this decision table:

1- Is there any redundant condition in the system?
2- What are the local certain and local possible rules for the system?
3- What are the global certain and global possible rules for the system?

The answers to these questions are discussed below.

## Reduction of Information Systems

Let us concentrate on only the conditions of the above information system. In this system, subjects s1, s8, and s11 have the same conditions' values. Also, subjects s2 and s10 have the same values for their conditions. In addition, the conditions' values for subjects s3 and s4 are the same. In other words, s1, s8, and s11 make one class, s2 and s10 make another class, s3 and s4 make a third class and each remaining subject in the system makes a class by itself, for a total of 7 classes. The number of classes, along with their contents is referred to, in this chapter, as the *internal structure* of the system.

If we remove a condition from the information system and the information system's internal structure remains the same, then the removed condition is a *redundant condition*. For example, the DMTS Accuracy is a redundant condition, but the TRD Response Rate is not. To explain further, when the TRD Response Rate condition is removed, subjects s1, s8, s11, and s9 will have the same conditions' values, whereas prior to removal of this condition, only s1, s8, and s11 had the same conditions' values. Therefore, the internal structure of the information system changed with the removal of the TRD Response Rate condition because the TRD Response Rate is not redundant. Using the same analogy, it can be seen that the DMTS Accuracy condition is redundant. In addition, the conditions PR Response Rate and IRA Accuracy are likewise redundant.

Checking for condition redundancy takes place in the *presence of all conditions* except the condition whose redundancy is being assessed. After all redundant conditions are identified, one of the redundant conditions is selected randomly and permanently removed from the information system. In this manner, the system becomes a new information system. For this new information system, the process of identifying all redundant conditions is repeated, followed by the random removal of one of the newly identified redundant conditions, and, thus, the creation of a new information system. This process is repeated until an information system is obtained whose set of conditions can not be reduced further. This final set of conditions is referred to as a *reduct* of the original set of conditions. Since redundant conditions are randomly selected for removal from the information system, the condition set of an information system may have more than one possible reduct. Reducts enable researchers to identify the minimum set of independent variables for a given study.

For the conditions of the information system shown in Figure 3.13, there are three reducts:

- Reduct 1 includes conditions (IRA Accuracy, PR Response Rate, TRD Response Rate)
- Reduct 2 includes conditions (DMTS Accuracy, PR Response Rate, TRD Response Rate)
- Reduct 3 includes conditions (DMTS Accuracy, IRA Accuracy, TRD Response Rate)

The information system composed of the first reduct is shown in Figure 3.14. The analysis of all the possible reducts, for a given information system, will reveal and quantify the significance of each individual independent variable (condition) in the system. For example, the condition TRD Response Rate is the most significant condition in the information system, because it appears in all three reducts. Analysis of reducts is beyond the scope of this chapter. However, a thorogh discussion of the topic may be found in [44, 45].

The reader needs to be reminded that there are other algorithms for finding the reducts of an information system including but not limited to the algorithms based on Boolean reasoning [46] and the attributes' significance [47].

Finding a reduct of an information system is significant because reducing the number of independent variables may shorten the length of data analysis and minimize the presence of artifacts in the outcome of such analysis. In addition, if the experiment is to be conducted again or by another group, then the length of the experiment may be shortened and the introduction of noise into the dataset and the cost may be reduced. Furthermore, the rules generated from a reduct are shorter and usually easier to interpret.

## Deriving the Local Certain and Local Possible Rules from a Reduct

The reduced information system, Figure 3.14, is ready for classification and partitioning. The reader needs to be reminded that all the subjects in one class have the same conditions' values and all the subjects in one partition have the same decision value. The number of classes for the reduced information system remains the same as the number of classes in the original information system, seven. The subjects in each class are as follow: class 1: (s1, s8, and s11), class 2: (s2 and s10),

| Subjects | Conditions | | | Decision |
| | IRA Accuracy | PR Response Rate | TRD Response Rate | IQ |
|---|---|---|---|---|
| s1 | Below Average | Below Average | Below Average | Low |
| s2 | Above Average | Average | Average | High |
| s3 | Average | Above Average | Above Average | Average |
| s4 | Average | Above Average | Above Average | High |
| s5 | Average | Above Average | Average | Average |
| s6 | Below Average | Above Average | Average | Average |
| s7 | Above Average | Average | Below Average | High |
| s8 | Below Average | Below Average | Below Average | Low |
| s9 | Below Average | Below Average | Average | Low |
| s10 | Above Average | Average | Average | Average |
| s11 | Below Average | Below Average | Below Average | Average |

**Fig. 3.14.** A reduct of the information system of Figure 3.13

class 3: (s3 and s4), class 4: (s5), class 5: (s6), class 6: (s7), and class 7: (s9). The number of partitions for the reduced information system is three, because the IQ has three possible values. Partitions 1, 2, and 3 represent IQ values of "Low", "Average" and "High", respectively. The subjects in each partition are as follow:

Partition 1: (s1, s8, and s9),
Partition 2: (s3, s5, s6, s10, and s11), and
Partition 3: (s2, s4, and s7).

To build the lower approximation space, the upper approximation space, and the boundary for the above partitions, the reader needs to be reminded again that the following rules are used:

- If the subjects of a class are totally in a given partition, then the subjects of the class become a part of the lower approximation space of that partition.
- If the subjects of a class are totally or partially in a given partition, then the subjects of that class become a part of the upper approximation space of the partition.
- If the subjects of a class are partially in a given partition, then the subjects of that class become a part of the boundary of the partition.

The application of the above rules may generate an empty lower approximation space, an empty upper approximation space, or an empty boundary for a given partition. Following the above rules, the lower approximation space, upper approximation space, and boundary for each partition of the reduced information system are displayed in Figure 3.15.

The local certain and local possible rules for partition 1 (Low IQ) are derived from the subjects in the lower approximation space and boundary of the partition 1, respectively. These rules are:

Local certain rule:

If (IRA Accuracy = "Below Average" & PR Response Rate = "Below Average" & TRD Response Rate = "Average") Then IQ = "Low"

| Partition | Details |
|---|---|
| 1 | Lower Approximation Space: (s9) |
| | Upper Approximation Space: (s1, s8, s9, and s11) |
| | Boundary: (s1, s8, and s11) |
| 2 | Lower Approximation Space: (s5, and s6) |
| | Upper Approximation Space: (s1, s2, s3, s4, s5, s6, s8, s10, and s11) |
| | Boundary: (s1, s2, s3, s4, s8, s10, and s11) |
| 3 | Lower Approximation Space: (s7) |
| | Upper Approximation Space: (s2, s3, s4, s7, and s10) |
| | Boundary: (s2, s3, s4, and s10) |

**Fig. 3.15.** Subjects in the lower approximation space, upper approximation space, and boundary for the three partitions of the reduct of Figure 3.14

Local possible rule:

If (IRA Accuracy = "Below Average" & PR Response Rate = "Below Average" & TRD Response Rate = "Below Average") Then IQ = "Low

Following the same analogy the local certain rules for "Low", "Average" and "High" values of IQ are shown in Figure 3.16.

**Deriving the Global Rules**

The globalization of the local certain and possible rules takes place as described in section 3.3. The set of local certain and possible rules of Figure 3.16, in tabular format, is the same as the rule set in Figure 3.7. Therefore, the final set of Global certain and possible rules for the information system of Figure 3.14 is the same as the rule set displayed in Figure 3.12.

Considering the nature of the local and global possible rules, one may choose to dismiss them all together and only use the local and global certain rules to classify the new subjects.

**Discussion**

Let us look at Figure 3.6 one more time and concentrate again on subjects of class 3. It was previously concluded that the subjects of this class are in the boundaries of both High and Low IQ Rough Sets, thus, two local possible rules can be generated. Both local possible rules have the same conditions' values but different IQ values. These local possible rules would not seem to have a lot of use in real life unless we logically dismiss one of them. The main question is which one of the two rules is a candidate for dismissal? It is logical to dismiss the local possible rule with decision IQ = "Low" because only one subject of the class is in the partition Low IQ and 4 out of 5 subjects of class 3 are within the partition High IQ. The influence of the dismissed rule on the surviving rule manifests itsel in form of a probability $(4/5 = 0.8)$ that will be assigned to the decision value of the surviving rule and the surviving rule will no longer be a

| Partition | Rules |
|---|---|
| 1<br>(IQ = "Low") | Local certain:<br>• If (IRA Accuracy = "Below Average" & PR Response Rate = "Below Average" & TRD Response Rate = "Average") Then IQ = "Low"<br>Local Possible:<br>• If (IRA Accuracy = "Below Average" & PR Response Rate = "Below Average" & TRD Response Rate = "Below Average") Then IQ = "Low" |
| 2<br>(IQ = "Average") | Local certain:<br>• If (IRA Accuracy = "Average" & PR Response Rate = "Above Average" & TRD Response Rate = "Average") Then IQ = "Average"<br>• If (IRA Accuracy = "Below Average" & PR Response Rate = "Above Average" & TRD Response Rate = "Average") Then IQ = "Average"<br>Local Possible:<br>• If (IRA Accuracy = "Below Average" & PR Response Rate = "Below Average" & TRD Response Rate = "Below Average") Then IQ = "Average"<br>• If (IRA Accuracy = "Above Average" & PR Response Rate = "Average" & TRD Response Rate = "Average") Then IQ = "Average"<br>• If (IRA Accuracy = "Average" & PR Response Rate = "Above Average" & TRD Response Rate = "Above Average") Then IQ = "Average" |
| 3<br>(IQ = "High") | Local Certain:<br>• If (IRA Accuracy = "Above Average" & PR Response Rate = "Average" & TRD Response Rate = "Below Average") Then IQ = "High"<br>Local Possible:<br>• If (IRA Accuracy = "Average" & PR Response Rate = "Above Average" & TRD Response Rate = "Above Average") Then IQ = "High"<br>• If (IRA Accuracy = "Above Average" & PR Response Rate = "Average" & TRD Response Rate = "Average") Then IQ = "High" |

**Fig. 3.16.** Local certain and local possible rules for "Low", "Average", and "High" values of IQ

local possible rule. This method of handling local possible rules forms the basis of a new concept, Modified Rough Sets, that is introduced in [12] and described in the following subsection.

### 3.3.5 Modified Rough Sets

In reference to the reduced information system of Figure 3.14, one can examine all of the classes of the information system and identify those classes in which the subjects of the class do not all share the same decision values (these classes end up in the boundaries of more than one Rough Sets.) For each class, one of the decision values in that class is designated as the *dominant* decision of

| Class | Subjects of the class | Dominant Decision | Probability of the Dominant Decision |
|---|---|---|---|
| 1 | s1, s8, and s11 | Low | 0.66 |
| 2 | s2 and s10 | Average* | 0.5 |
| 3 | s3 and s4 | High* | 0.5 |

*Since there is a tie, the dominant decision is chosen by a domain expert.

**Fig. 3.17.** The dominant decision for the classes 1, 2, and 3 of Figure 3.14

| Subjects | Conditions' attributes | | | Dominant Decision | |
|---|---|---|---|---|---|
| | IRA Accuracy | PR Response Rate | TRD Response Rate | IQ | Probability |
| s1 | Below Average | Below Average | Below Average | Low | 0.66 |
| s2 | Above Average | Average | Average | Average | 0.5 |
| s3 | Average | Above Average | Above Average | High | 0.5 |
| s5 | Average | Above Average | Average | Average | 1 |
| s6 | Below Average | Above Average | Average | Average | 1 |
| s7 | Above Average | Average | Below Average | High | 1 |
| s9 | Below Average | Below Average | Average | Low | 1 |

**Fig. 3.18.** The information system of Figure 3.14 after enforcing dominant decisions

that class. Although the Bayes' theorem [12] is used to designate a dominant decision for a given class and identify a probability for the dominant decision, the dominant decision is, practically, the most common decision among the subjects of the class. If there is a tie between two or more than two decision values, then the dominant decision is either chosen randomly or a domain expert makes the choice. The ratio of the number of subjects with the dominant decision in the class to the total number of subjects in the class defines the probability for the dominant decision. After the dominant decision is identified, the decision values for all the subjects in that class are then changed to the dominant decision. By taking this action, the lower and upper approximation spaces for each Rough Set become the same and, thus, boundary of the set becomes empty.

For our example, the classes 1, 2, and 3 participate in the boundaries of the IQ Rough Sets. Therefore, the dominant decisions for these classes are identified and shown in Figure 3.17. The reader needs to be reminded that for those classes in which all the subjects have the same decision values, the dominant decision is the shared value and the probability of the dominant decision is 1.

Enforcing the dominant decisions causes the subjects of class 1, for example, to become exactly the same because they have the same conditions' values and the same dominant decision. Therefore, extra copies have to be removed from Figure 3.14 (Figure 3.18). As a result, the classes for the information system of Figure 3.18 are as follow: class 1: (s1), class 2: (s2), class 3: (s3), class 4: (s5), class 5: (s6), class 6: (s7), and class 7: (s9). The partitions and their lower approximation spaces are shown in Figure 3.19. (The upper approximation spaces are the same as the lower approximation spaces and, thus, the boundaries of the partitions are *always* empty.)

| Partition | Subjects of the partition | Lower Approximation Space |
|-----------|---------------------------|---------------------------|
| 1 | s1 and s9 | s1 and s9 |
| 2 | s2, s5, and s6 | s2, s5, and s6 |
| 3 | s3 and s7 | s3 and s7 |

**Fig. 3.19.** The partitions and lower approximation spaces for the information system of Figure 3.18

---

**Local Approximate Rules:**
If (IRA Accuracy ="Below Average" & PR Response Rate = "Below Average" &
   TRD Response Rate = " Below Average") Then IQ = "Low" and CF =.66
If (IRA Accuracy ="Above Average" & PR Response Rate = "Average" &
   TRD Response Rate = "Average") Then IQ = "Average" and CF = 0.5
If (IRA Accuracy ="Average" & PR Response Rate = "Above Average" &
   TRD Response Rate = " Above Average") Then IQ = "High" and CF = 0.5
If (IRA Accuracy =" Average" & PR Response Rate = "Above Average" &
   TRD Response Rate = "Average") Then IQ = "Average" and CF = 1
If (IRA Accuracy ="Below Average" & PR Response Rate = "Above Average" &
   TRD Response Rate = "Average") Then IQ = "Average" and CF = 1
If (IRA Accuracy ="Above Average" & PR Response Rate = "Average" &
   TRD Response Rate = " Below Average") Then IQ = "High" and CF = 1
If (IRA Accuracy ="Below Average" & PR Response Rate = "Below Average" &
   TRD Response Rate = "Average") Then IQ = "Low" and CF = 1

**Global Approximate Rules:**
If (PR Response Rate = "Below Average" & TRD Response Rate = "Below Average")
   Then IQ = "Low" and CF = 0.66
If (IRA Accuracy ="Above Average" & TRD Response Rate = "Average")
   Then IQ = "Average" and CF = 0.5
If (PR Response Rate = "Above Average" & TRD Response Rate = "Above
   Average") Then IQ = "High" and CF = 0.5
If (IRA Accuracy =" Average" & TRD Response Rate = "Average")
   Then IQ = "Average" and CF = 1
If (IRA Accuracy ="Below Average" & PR Response Rate = "Above Average")
   Then IQ = "Average" and CF = 1
If (IRA Accuracy ="Above Average" & TRD Response Rate = "Below Average")
   Then IQ = "High" and CF = 1
If (PR Response Rate = "Below Average" & TRD Response Rate = "Average")
   Then IQ = "Low" and CF = 1

**Fig. 3.20.** The local and global approximation rules for the information system of Figure 3.18

The rules that are generated for the information system of Figure 3.18 are called *local approximate rules*. Each local approximate rule has a *certainty factor (CF)* that is the same as the probability assigned to its decision value. If the dropping condition approach is applied to a set of local approximate rules, then the results obtained are referred to as *global approximate rules*. The local and global approximate rules for the information system of Figure 3.18 are shown in Figure 3.20. If a new subject satisfies the conditions' values of the first global approximate rule 1, for example, then the decision value for the new subject is "Low" with 66% certainty.

In creation of the *Modified Rough Sets* the inner and outer boarders of every Rough Set overlay each other. One may say that a Modified Rough Set becomes a traditional set because the set now has only one boarder. This is partially correct. There is still a difference between a traditional set and a Modified Rough Set and the difference lies in the fact that each subject in a Modified Rough Set is assigned a probability that may be different from the next subject. This is not the case for a traditional set. (However, there are similarities between the Modified Rough Set and the Fuzzy Set, but the comparison of these two approaches is beyond the scope of this chapter.)

To establish the usefulness of the Rough Sets and Modified Rough Sets approaches, a real dataset containing the results of an actual experiment with children diagnosed with or without Attention Deficit/Hyperactivity Disorder (ADHD) was used. The more traditional statistical approach of Discriminant Analysis was also applied to the same dataset to provide a comparison of the effectiveness of the Rough Sets and Modified Rough Sets approach.

## 3.4   Experimental Results and Discussion

A set of five behavioral tasks was performed by a group of 73 children at a large children hospital. The age range of the subjects was from 6 to 9 years. The number of conditions for each subject was 20 and these included subject gender and 19 different measures for the five behavioral tasks. The tasks utilized (and the cognitive functions they are thought to model) were: Conditioned Position Responding (color and position discrimination); Progressive Ratio (motivation); Temporal Response Differentiation (time estimation); Delayed Matching-to-Sample (short-term memory); and Incremental Repeated Acquisition (learning/indexlearning). The decision for each subject was either "positive ADHD" or "negative ADHD". The main objective of the experiment was to first establish the relationships or lack there of, between diagnosis of ADHD and the outcome of behavioral tests expressed in the form of rules and then predict the diagnosis of the subjects in the test set. The objective was attained using Cross-Validation process [48] that is explained below.

The number of subjects described as "positive ADHD" and "negative ADHD" were 34 and 39 respectively. The process of conditions reduction was applied to the information system and the resulting reduct had 9 conditions. From both the "positive ADHD" and "negative ADHD" groups, 15% of the subjects were randomly chosen for inclusion in a testing set (a total of 11 subjects). The remainder of the subjects (i.e., 62 subjects or 85% of the total subjects in the dataset) comprised a training set.

For the training set, the local and global certain rules were generated using the Rough Sets methodology. (The local and global possible rules were not generated.) Also, the local, and global approximate rules were generated using the

| No. of Classes | No. of Partitions | No. of Certain Rules | | No. of Approximate Rules | |
|---|---|---|---|---|---|
| | | Local | Global | Local | Global |
| 13 | 2 | 5 | 12 | 13 | 13 |

**Fig. 3.21.** Number of different types of rules generated from the training set

| Methodology | | % of correct classification of the testing set |
|---|---|---|
| **a)** | | |
| Rough Sets | Local Certain Rules | 60 |
| | Global Certain Rules | 64 |
| Modified | Local Approximate Rules | 64 |
| Rough sets | Global Approximate Rules | 73 |
| Discriminant Analysis | | 60 |
| ID3 | | 62 |
| **b)** * | | |
| Rough Sets | Local Certain Rules | 58 |
| | Global Certain Rules | 65 |
| Modified | Local Approximate Rules | 72 |
| Rough sets | Global Approximate Rules | 81 |
| Discriminant Analysis | | 60 |
| ID3 | | 64 |
| * The percentages reported in part b are the average of the diagnostic accuracy for 10 pairs of training and test sets. | | |

**Fig. 3.22.** Diagnostic accuracy for Rough Sets (using local and global certain rules), Modified Rough Sets (using local and global approximate rules) compared to discriminant analysis and ID3 on (a) the original dataset, and (b) random files generated by use of a resampling technique

Modified Rough Sets methodology. The number of certain and approximate rules along with the number of classes and partitions are shown in Figure 3.21.

The resulting rules were then used for the corresponding testing set to categorize subjects as either ADHD or not. The results are shown in Figure 3.22-a. To show the effectiveness of the approach we also used two approaches of Discriminant analysis and ID3 to perform the same task. The results are depicted also in Figure 3.22-a.

Since the dataset was small, 10 pairs of training and testing sets were generated from the original training and testing sets using the Random Resampling approach [49]. For each pair, the Cross-Validation process was repeated and the average of the results are shown in Figure 3.22-b.

The results indicate that Rough Sets and Modified Rough Sets can accommodate records for subjects whose conditions values are the same but whose decision values are different. Such capability is of a great import because in clinical situations, for example, there are (a) patients with the same symptoms but are diagnosed differently or (b) there are patients who react differently to the same drug despite the controlled experimental environment.

Where statistical models *remove* such conflicting records from bio-datasets, Rough sets and Modified Rough Sets fully *analyze* them. This ability is particularly valuable in biological data regarding both human and non-human (primate) subjects because when the number of available data points from an experiment is small, retaining every one of them is crucial and Modified Rough sets do, in fact, retain them.

## 3.5  Conclusion

The Rough Sets and Modified Rough Sets methodologies were fully introduced in a non-mathematical fashion so that researchers of non-mathematical backgrounds can also benefit from their capabilities. A comparison was then conducted among Rough Sets, Modified Rough Sets, Discriminant Analysis, and ID3. A dataset of children diagnosed with or without Attention Deficit/Hyperactivity Disorder (ADHD) was used for the comparison. The conclusion drawn from the analysis of the dataset is that first, the global rules have better predictive ability than do the local rules. Second, the performance of Modified Rough Sets is better than that of Discriminant Analysis, ID3, and Rough Sets. Third, the performance of Rough sets using global certain rules, is better or as good as that of Discriminant analysis and ID3.

## References

1. Han, J., Kamber, M.: Data mining, Concepts and Techniques, 2nd edn. Morgan Kaufmann, San Francisco (2005)
2. Hashemi, R., Bahar, M., Early, J., Tyler, A., Young, J.: A signature-based Liver Cancer Prediction System. In: Srimani, P. (ed.) Proc. 2005 Int. Conf. on Info Tech Coding and Computing (ITCC 2005), Las Vegas, Nevada, pp. 154–160 (2005)
3. Young, J., Tong, W., Fang, H., Xie, Q., Pearce, B., Hashemi, R., Beger, R., Cheeseman, M., Chen, J., Chang, Y., Kodell, R.: Building an Organ-Specific Carcinogenic Database for SAR Analyses. Toxicology and Environmental Health, Part A 67, 1363–1389 (2004)
4. Hashemi, R., Young, J.: The Prediction of Methylmercury Elimination Half-Life In Humans Using Animal Data: A Neural Network/Rough Sets Analysis. Toxicology and Environmental Health, Part A 66(23), 2227–2252 (2003)
5. Pawlak, Z.: Rough Classification. Man-Machine Studies 20, 469–483 (1984)
6. Fausett, L.: Fundamentals of Neural Networks: Architectures, Algorithms, and Applications. Prentice-Hall, Englewood Cliffs (1994)
7. Kennedy, M.P., Chua, L.: Neural Networks for Nonlinear Programming. IEEE Transac. Circuits Sys. CAS-35(5), 554–562 (1988)

8. Hashemi, R., Tyler, A., Slikker, W., Paule, M.: Profiling Through Kohonen Self-Organizing Map: The Effect of Birth Weight On The Performance Measure Of An Operant Test Battery. In: Dagli, C.H., Buczak, A.L., Ghosh, J., Embrechts, M.J., Ersoy, O. (eds.) Proc. Annie 1999 Smart Engineering System Design: Neural Networks, Fuzzy Logic, Evolutionary Programming, Complex Systems and Data Mining, St. Louis, MO, pp. 941–946 (1999)

9. Hashemi, R., Schafer, T., Hinson, W., Lay, J.: Identifying and Testing of Signatures for Non-Volatile Biomolecules Using The Tandom Mass Spectra. In: Proc. 1996 ACM Int. Symp. on Applied Computing, Philadelphia, PA, pp. 44–49 (1996)

10. Holland, J.: Genetic Algorithms. Scientific American 267(1), 66–72 (1992)

11. Smolinski, T., Boratyn, G., Milanova, M., Zurada, J., Wrobel, A.: Evolutionary Algorithms and Rough Sets-Based Hybrid Approach to Classificatory Decomposition of Cortical Evoked Potentials. In: Alpigini, J.J., Peters, J.F., Skowron, A., Zhong, N. (eds.) RSCTC 2002. LNCS (LNAI), vol. 2475, pp. 621–628. Springer, Heidelberg (2002)

12. Hashemi, R., Pearce, B., Arani, R., Hinson, W., Paule, M.: A Fusion of Rough Sets, Modified Rough Sets, and Genetic Algorithms for Hybrid Diagnostic Systems. In: Lin, T.Y., Cercone, N. (eds.) Rough Sets and Data Mining: Analysis of Imprecise Data, pp. 149–176. Kluwer Academic Publishers, Dordrecht (1997)

13. Zadeh, L.: A Fuzzy-Algorithmic Approach to the Definition of Complex or Imprecise Concepts. Man-Machine Studies 8, 249–291 (1976)

14. Hashemi, R., Choobineh, F., Slikker, W., Paule, M.: On Integration of Modified Rough Set and Fuzzy Logic in Classification. In: Proc. 3rd Int. Joint Conf. on Info. Sciences, Research Triangle Park, NC, pp. 255–258 (1997)

15. Quinlan, J.R.: Learning Efficient Classification Procedures and Their Application to Chess Endgames. In: Michalski, J.S., Carbonell, J.G., Mirchell, T.M. (eds.) Machine Learning: An Artificial Intelligence Approach, vol. 1, pp. 463–482. Morgan Kaufmann, Palo Alto (1983)

16. Hashemi, R., Le Blanc, L., Traywick, B.: A Decision Tree-Based Prediction of Philanthropic Giving. In: The 2007 Int. Conf. on Machine Learning and Applications (ICMLA 2007), Cincinnati, Ohio (submitted, 2007)

17. Razzaghi, M., Kodell, R.: Risk Assessment for Quantitative Responses Using a Mixture Model. Biometrics 56(2), 519–527 (2000)

18. Tomaszewski, P., Håkansson, J., Grahn, H., Lundberg, L.: Statistical Models vs. Expert Estimation for Fault Prediction in Modified Code - An Industrial Case Study Source. Systems and Software 8(8), 1227–1238 (2007)

19. Hashemi, R., Epperson, C.: The Integration of Rough Sets and Neural Network Paradigm. In: Predictive Systems. Proc. Annie 1999 Smart Engineering System Design: Neural Networks, Fuzzy Logic, Evolutionary Programming, Complex Systems and Data Mining, St. Louis, MO, pp. 499–504 (1999)

20. Hashemi, R., Danley, J., Bolan, B., Tyler, A., Slikker, W., Paule, M.: Info Granulation and Super Rules. In: Proc. 4th Int. Joint Conf. on Info Sciences. Research Triangle Park, NC, pp. 383–386 (1998)

21. Smolinski, T., Boratyn, G., Milanova, M., Buchanan, R., Prinz, A.: Hybridization of Independent Component Analysis, Rough Sets, and Multi-Objective Evolutionary Algorithms for Classificatory Decomposition of Cortical Evoked Potentials. In: Rajapakse, J.C., Wong, L., Acharya, R. (eds.) PRIB 2006. LNCS (LNBI), vol. 4146, pp. 174–183. Springer, Heidelberg (2006)

22. Boratyn, G., Smolinski, T., Zurada, J., Milanova, M., Bhattacharyya, S., Suva, L.: Hybridization of Blind Source Separation and Rough Sets for Proteomic Biomarker Indentification. In: Rutkowski, L., Siekmann, J.H., Tadeusiewicz, R., Zadeh, L.A. (eds.) ICAISC 2004. LNCS (LNAI), vol. 3070, pp. 486–491. Springer, Heidelberg (2004)
23. Hassanien, A.E., Slezak, D.: Rough Neural Intelligent Approach for Image Classification: A Case of Patients with Suspected Breast Cancer. Hybrid Intelligent Systems 3(4), 205–218 (2006)
24. Hassanien, A.E.: Fuzzy Rough Sets Hybrid Scheme for Breast Cancer Detection. Image Vision Computing, Electronic Edition 25(2), 172–183 (2007)
25. Deogun, J., Raghavan, V.V., Sever, H.: Rough Set Based Classification Methods and Extended Decision. In: Proc. 3rd Int. Workshop on Rough Sets and Soft Computing, San Jose, CA, pp. 302–309 (1994)
26. Grzymala-Busse, J.W.: LERS - A System for Learning from Examples based on Rough Sets, Intel Decision Supt: Handbook of Applications and Advances in Rough Set Theory, pp. 3–18. Kluwer Academic Publishing, Dordrecht (1992)
27. Lin, T.Y.: Neighborhood Systems-Information Granulation. In: Proc. 3rd Int. Joint Conf. on Info Sciences, Research Triangle Park, NC, vol. 3, pp. 161–164 (1997)
28. Hashemi, R., Epperson, C., Tyler, A., Young, J.: Knowledge Discovery from Sparse Pharmacokinetic Data. In: Proc. 2000 ACM Int. symp. on Applied Computing (SAC 2000), Como, Italy, pp. 75–79 (2000)
29. Pawlak, Z., Grzymala-Busse, J., Slowinski, R., Ziarko, W.: Rough Sets. Comm. of the ACM 38(11), 89–95 (1995)
30. Ras, Z.W., Koo, T.: Knowledge Discovery for Intelligent Query Answering. In: Proc. 3rd Int. Joint Conf. on Info Sciences, Research Triangle Park, NC, vol. 3, pp. 367–370 (1997)
31. Wong, S.K.M., Ziarko, W., Ye, R.L.: Comparison of Rough-set and Statistical Methods in Inductive Learning. Man-Machine Studies 24, 53–72 (1986)
32. Yao, Y.Y., Noroozi, N.: A Unified Model for Set-Based Computations. In: Proc. 3rd Int. Workshop on Rough Sets and Soft Computing, San Jose CA, pp. 236–243 (1994)
33. Ziarko, W.: Variable Precision Rough Set Model. Computer and Sys. Scie. 46, 39–59 (1993)
34. Moshkov, M.J., Skowron, A., Suraj, Z.: On Covering Attribute Sets by Reducts. In: Kryszkiewicz, M., Peters, J.F., Rybinski, H., Skowron, A. (eds.) Proc. Int. Conf. on Rough Sets and Intel. Sys. Paradigms, Warsaw, Poland, pp. 175–180 (2007)
35. Grzymala-Busse, J.W.: Mining Numerical Data - A Rough Set Approach. In: Kryszkiewicz, M., Peters, J.F., Rybinski, H., Skowron, A. (eds.) Proc. of Int. Conf. on Rough Sets and Intel. Sys. Paradigms, Warsaw, Poland, pp. 12–21 (2007)
36. Suraj, Z., Pancerz, K.: Flow Graphs as a Tool for Mining Prediction Rules of Changes of Components in Temporal Information Systems. In: Yao, J., Lingras, W.W., Szczuka, M.S., Cercone, N., Slezak, D. (eds.) Proc. of Int. Conf. on Rough Sets and Knowledge Tech., Toronto, Canada, pp. 468–475 (2007)
37. Renpu, L.R., Wang, Z.: Mining Classification Rules Using Rough Sets and Neural Networks. European J. Operat. Resch. 157(2), 439–448 (2004)
38. Stefanowski, J.: On Combined Classifiers, Rule Induction and Rough Sets. In: Peters, J.F., Skowron, A., Duntsch, I., Grzymala-Busse, J.W., Orlowska, E., Polkowski, L. (eds.) Transactions on Rough Sets VI, Part I, pp. 329–350 (2007)
39. Paule, M.: Analysis of Brain Function Using a Battery of Schedule-Controlled Operant Behaviors. In: Weiss, B., O'Donoghue, J. (eds.) Neurobehavioral Toxicity: Analysis and Interpretations, pp. 331–338. Raven Press, New York (1994)

40. Ruckert, U., Kramer, S.: A Statistical Approach to Rule Learning. In: Proc. 23rd Int. Conf. on Machine Learning, pp. 785–792. Carnegie Mellon University, Pittsburgh (2006)
41. Hassanien, A.E.: Rough Set Approach for Attribute Reduction and Rule Generation: a Case of Patients with Suspected Breast Cancer. J. of the Americ. Soc. for Info. Sci. and Tech. 55(11), 954–962 (2004)
42. Pawlak, Z.: Combining Rough Sets and Bayes' Rule. Comput. Intel. 17(3), 401–408 (2001)
43. Michalski, R.: A theory and Methodology of Inductive Learning. In: Michalski, J.S., Carbonell, J.G., Michell, T.M. (eds.) Machine Learning, pp. 83–134. Tioga Publication Co., Palo Alto (1983)
44. Riddle, D.L., Coovert, M.D., Elliott, L.R., Schiflett, S.G.: Potential Contributions of Rough Sets Data Analysis to Training Evaluations. Military Psychology 15(1), 41–58 (2003)
45. Maciag, T.J., Slezak, D., Hepting, D.H.: Consumer Modelling in Support of Interface Design. In: Proc. 2006 Int. Conf. on Hybrid Info. Tech (ICHIT 2006), Cheju Island, Korea, vol. 2, pp. 153–160 (2006)
46. Skowron, A., Rauszer, C.: The Discernibility Matrices and Functions in Decision systems. In: Slowinski, R. (ed.) Intelligent decision support. Handbook of applications and advances of the rough sets theory, pp. 311–362. Kluwer, Dordrecht (1992)
47. Sloinski, R., Zopounidis, C.: Application of Rough Set Approach to Evaluation of Bankrupcy Risk. Intel. Sys. in Account, Finance & Mngmt. 4, 27–41 (1995)
48. Kohavi, R.: A Study of Cross-Validation and Bootstrap for Accuracy Estimation and Model Selection. In: Proc. 14th Int. Joint Con. on Arti. Intel., vol. 2(12), pp. 1137–1143 (1995)
49. Efron, B., Tibshirani, R.: Bootstrap methods for standard errors, confidence intervals, and other measures of statistical accuracy. Statistic Sci. (1), 54–77 (1985)

# 4

# Integrating Local and Personalised Modelling with Global Ontology Knowledge Bases for Biomedical and Bioinformatics Decision Support

Nikola Kasabov, Qun Song, Lubica Benuskova, Paulo Gottgtroy, Vishal Jain, Anju Verma, Ilkka Havukkala, Elaine Rush, Russel Pears, Alex Tjahjana, Yingjie Hu, and Stephen MacDonell

Knowledge Engineering and Discovery Research Institute, KEDRI,
Auckland University of Technology, Auckland, New Zealand
nkasabov@aut.ac.nz
http://www.kedri.info

**Summary.** A novel ontology based decision support framework and a development platform are described, which allow for the creation of global knowledge representation for local and personalised modelling and decision support. The main modules are: an ontology module; and a machine learning module. Both modules evolve through continuous learning from new data. Results from the machine learning procedures can be entered back to the ontology thus enriching its knowledge base and facilitating new discoveries. This framework supports global, local and personalised modelling. The latter is a process of model creation for a single person, based on their personal data and the information available in the ontology. Several methods for local and personalised modelling, both traditional and new, are described. A case study is presented on brain-gene-disease ontology, where a set of 12 genes related to central nervous system cancer are revealed from existing data and local profiles of patients are derived. Through ontology analysis, these genes are found to be related to different functions, areas, and other diseases of the brain. Two other case studies discussed in the paper are chronic disease ontology and risk evaluation, and cancer gene ontology and prognosis.

## 4.1 Introduction

With the accumulation of both data and knowledge in the biomedical area and bioinformatics, it becomes eminent that these data and knowledge need to be organized in a more global knowledge repository and used in their complexity and richness for an efficient profiling, prognosis, diagnosis and decision support for every individual person who needs that. This task requires both adaptive, evolving knowledge repository systems and methods for local and personalised modeling in their integration and dynamic interaction.

To illustrate the problem above, let us take for example the brain in its multiple aspects of functioning and disease. The brain evolves its structure and functionality at different levels – Fig. 4.1: quantum-, molecular (genetic)-, single neuron-, ensemble of neurons-, cognitive-, evolutionary.

At the quantum level, particles (atoms, ions, electrons, etc.), that make every molecule in the material world, are moving continuously, being in several states at the

T.G. Smolinski et al. (Eds.): Comp. Intel. in Biomed. & Bioinform., SCI 151, pp. 93–116, 2008.

same time that are characterized by probability, phase, frequency, energy. At a molecular level, RNA and protein molecules evolve in a cell and interact in a continuous
way, based on the stored information in the DNA and on external factors, and affect
the functioning of a cell (neuron) under certain conditions. At the level of a neuron,
the internal information processes and the external stimuli cause the neuron to produce a signal that carries information to be transferred to other neurons. At the level
of neuronal ensembles, all neurons operate in a "concert", defining the function of the
ensemble, for instance perception of a spoken word. At the level of the whole brain,
cognitive processes take place, such as language and reasoning, and global information processes are manifested, such as consciousness. At the level of a population of
individuals, species evolve through evolution, changing the genetic DNA code for a
better adaptation.

The processes at each level from Fig.4.1 are very complex and difficult to understand, but much more difficult to understand is the interaction between the different
levels, e.g. gene- brain function-disease (Benuskova and Kasabov 2007). It may be
that understanding the interaction through its modeling would be a key to understanding each level of processing and perhaps the brain as a whole.

| |
|---|
| 6.  Evolutionary (population/generation) processes |
| 5. Brain cognitive processes |
| 4. System information processing (e.g. neural ensemble) |
| 3. Information processing in a cell (neuron) |
| 2. Molecular information processing (genes, proteins) |
| 1. Quantum information processing |

**Fig. 4.1.** Levels of information processing in the brain and the interaction between the levels
(from Kasabov 2002, 2007a)

The enormous amount of information so far related to the brain processes at the
different levels from Fig.4.1 need to be globally structured and made accessible for
the purpose of a better decision support for every individual person and for the purpose of new knowledge discovery and a better understanding.

Similar problems relate to cancer diseases; to chronic diseases, such as diabetes,
and to many more unsolved medical and health problems.

This chapter suggests integrating local and personalised modelling methods with a
global ontology knowledge and data repository for a better personalised decision
support, for new knowledge discovery and for a better understanding. Section 4.2
presents several methods for local and personalised modelling. Section 4.3 contains
an introduction to the Ontology systems. Sections 4.4, 4.5, 4.6 and 4.7 contain case
studies on brain-gene-disease ontology and inference; chronic disease ontology and
risk analysis; cancer gene ontology and profiling.

## 4.2 Local and Personalised Modelling

Contemporary medical and bioinformatics decision support systems use both inductive and transductive reasoning to derive global, local, and personalised models for
the prediction of a person's risk or outcome of disease (Kasabov 2007a, b). While
inductive modelling results in the incremental creation of a global model where new,
unlabeled data is "mapped" through a recall procedure, transductive inference methods estimate the value of a potential model (function) only in a single point of the
space (the new data vector) utilizing additional information related to this point (Vap-
niak 1998). This approach seems to be more appropriate for clinical and medical
applications of learning systems, where the focus is not on the model, but on the individual patient. And it is not so important what the global error of a global model over
the whole problem space is, but rather the accuracy of prediction for any individual
patient. Each individual data vector (e.g. a patient in the medical area) may need an
individual, local model that best fits the new data, rather than a global model, where
new data are matched without taking into account any specific information about
these data.

In transductive modelling, for every new input vector $x_i$ that needs to be processed
for a prognostic task, the closest $N_i$ examples, that form a data set $D_i$, are derived from
an existing data set D. A new model $M_i$ is dynamically created from these samples to
approximate the function in the point $x_i$. The system is then used to calculate the output value $y_i$ for this input vector $x_i$.

A simple transductive inference method is the k-nearest neighbour method (k-NN),
where the output value $y_i$ for the new vector $x_i$ is calculated as the average of the output values of the k-nearest samples from the data set $D_i$. In a weighted k-NN method
(WKNN) the output $y_i$ is calculated based on the distance of the k-NN samples to $x_i$:

$$y_i = \frac{\sum_{j=1}^{Ni} w_j y_j}{\sum_{j=1}^{Ni} w_j} \tag{4.1}$$

Where: $y_j$ is the output value for the sample $x_j$ from $D_i$ and $w_j$ is its weight measured
as:

$$w_j = \max(\boldsymbol{d}) - [d_j - \min(\boldsymbol{d})] \tag{4.2}$$

In Eq. (4.2), the vector $\boldsymbol{d} = [d_1, d_2, \ldots d_{Ni}]$ is defined as the distances between the
new input vector $x_i$ and the input vectors of the nearest samples $(x_j, y_j)$ for $j = 1$ to $N_i$;
$\max(\boldsymbol{d})$ and $\min(\boldsymbol{d})$ are the maximum and minimum values in $\boldsymbol{d}$ respectively.

In the WWKNN method (Kasabov 2007a, b) not only the nearest samples are
weighted based on their distance to the new sample, but the contribution of each of
the variables is weighted based on their importance for the local area where the new
sample belongs. The WWKNN algorithm is given in Appendix A.

Recently, two other methods for personalised modelling were proposed: Transductive Neural Fuzzy Inference System – NFI (Song and Kasabov 2005) and Transductive

Neural Fuzzy Inference System with Weighted Data Normalization – TWNFI (Song and Kasabov 2006).

NFI is a dynamic neural-fuzzy inference system with a local generalization, in which, either Zadeh-Mamdani or Takagi-Sugeno type fuzzy inference engine is used. Gaussian fuzzy membership functions are applied in each fuzzy rule for both the antecedent and the consequent parts (Zadeh-Mamdani type) or for antecedent part only (Takagi-Sugeno type). A back-propagation learning algorithm is used for optimizing the parameters of the fuzzy membership functions. An additional learning function is derived for the Takagi-Sugeno model. The distance between vectors x and y is measured in NFI as the normalized Euclidean distance defined as follows:

$$\|x - y\| = \frac{1}{P}\left[\sum_{j=1}^{P}|x_j - y_j|^2\right]^{\frac{1}{2}} \tag{4.3}$$

To partition the input space for creating fuzzy rules and obtaining initial values of fuzzy rules, the ECM (Evolving Clustering Method) is applied (Kasabov and Song 2002) and the cluster centres and cluster radiuses are respectively taken as initial values of the centres and widths of the Gaussian membership functions (for both Zadeh-Mamdani and Takagi-Sugeno types). The data in a cluster are used for creating a linear function (Takagi-Sugeno type fuzzy inference).

TWNFI (Song and Kasabov 2006) is a transductive weighted data neuro-fuzzy inference method, similar to NFI, but the input variables in the Eq.(4.3) are weighted based on their importance for the problem, derived through the back-propagation or an evolutionary optimization algorithm. The TWNFI algorithm is given in Appendix B.

Transductive, personalised modelling is suitable when small data bases are available for a problem. In case of large data bases, a global model, which consists of many local models, may be more appropriate to derive and update on new data and to use for a personalised profiling and prognosis. This is called here *local modelling*.

Evolving connectionist systems (ECOS) are neural network models that develop, evolve their structure - nodes (neurons) and connections between them, through supervised or un-supervised incremental learning from data samples (Kasabov 2001, 2002, 2007a). One of the ECOS models – DENFIS, a dynamic evolving neuro-fuzzy inference system is a fuzzy inference system that first evolves fuzzy rules from data through the evolving clustering algorithm ECM, and then incrementally modifies a local function to approximate the data in this cluster (Kasabov and Song 2002). The cluster and the function associated with it form a Takagi-Sugeno fuzzy rule. For every test input vector $x_i$, several neighbouring fuzzy rules are activated together to infer the output value $y_i$. Different fuzzy membership functions can be used in the rules. ECM is a connectionist clustering algorithm, where the evolved nodes represent cluster centres of samples in the input space. The number of the centres is not specified or fixed and depends on a maximum radius of the clusters, which is either defined or derived from the data. In the clustering procedure, data samples are allocated to rule nodes based on the similarity between the samples and the nodes calculated in the input space.

The distance between samples and rule nodes can be measured in different ways. The most popular measurement is the normalized Euclidean distance. In case of missing values for some of the input variables, a partial normalized Euclidean distance can be used which means that only the existing values for the variables in a current sample ($\mathbf{x}$, $\mathbf{y}$) are used for the distance measure between this sample and an existing rule node N ($W_{1N}$):

$$d(S,N) = \frac{1}{P}\left[\sum_{j=1}^{P}\left|x_j - W_{IN\ (j)}\right|^2\right]^{\frac{1}{2}} \tag{4.4}$$

For all P input variables $x_i$ that have a defined value in the input vector x and in an already established connection $W_{1N(j)}$ to the cluster node N.

At any time of an ECOS' continuous, incremental learning from data, fuzzy inference rules can be derived from the ECOS structure. Each rule associates a cluster area from the input variable space to a local output function applied to the data in this cluster, for example,

IF [an input vector x is in cluster $N_{cj}$ (cluster center $N_j$ and a cluster radius $R_j$)]

THEN [the local output function is fc, with $N_{jex}$ samples in the cluster approximated by this function].

In case of DENFIS, first order local fuzzy rule models are derived incrementally from data, for example:

IF [the value of $x_1$ is in the area defined by a Gaussian membership function with a center at 0.2, and a deviation of 0.12) AND (the value of $x_2$ is in the area defined by a Gaussian membership function with a center at 0.7, and a deviation of 0.20 respectively]

THEN [the output value y is calculated by the formula y= 0.51+ 3.9 $x_1$+ 1.45 $x_2$].

In another ECOS model – EFuNN (Kasabov 2001, 2002, 2007a) the following local rules are derived to represent a cluster of data:

IF [an input vector x is in an input cluster $NI_{cj}$ (cluster center $NI_j$; cluster radius $R_j$)]

THEN [the output y is in output cluster $NO_{cj}$ (cluster center $NO_j$; cluster radius $E_j$)], with $N_{jex}$ samples approximated by this rule].

The Evolving Classification Model (ECF) is a simplified version of EFuNN, where the local rules are of the following form:

IF [an input vector x is in an input cluster $NI_{cj}$ (cluster center $NI_j$; cluster radius $R_j$)]

THEN [the output belongs to class $NO_{cj}$], with $N_{jex}$ samples approximated by this rule].

The above described techniques for local and personalised modelling are part of a modelling environment NeuCom (www.theneucom.com) and have been widely used so far (Kasabov 2007a). Here, we apply them on ontology structured data to derive local and personalised profiles, where the results are entered back to the ontology to enrich its knowledge repository and to facilitate new discoveries.

## 4.3 An Introduction to Ontology Systems for Information and Knowledge Representation

In modern computer science, ontology is a data model that represents a set of concepts, information and data within a domain, for example the domain of brain (see Fig.4.1), and the relationships between those concepts. Ontology is used to reason and make inferences about the objects within that domain (Gruber 1993).

Ontology is generally written as a set of definitions of formal vocabulary of objects and relationships in the given domain. It supports the sharing and reuse of formally represented knowledge among systems (Chandrasekaran et al 1999; Fensel 2004). In recent years, ontologies have been adopted in many business and scientific communities as a way to share, reuse and process domain knowledge (Fensel 2004). As a database technology, ontologies are commonly coded as triple stores (subject, relationship, object), where a network of objects is formed by relationship linkages, as a way of storing semantic information (Owens 2005; Berners Lee et al 2001).

Several medical ontologies (Pisanelli 2004), including Open Bio-medical Ontology OBO (http://www.bioontology.org/) and the Gene Ontology have been created (http://www.geneontology.org/) (Ashburner et al 2000). The goal of a Biomedical Ontology is to allow scientists to create, disseminate, and manage biomedical information and knowledge in machine-processable form for accessing and using this biomedical information in research.

The Gene Ontology (GO) project provides a controlled vocabulary to describe gene and gene product attributes in any organism. The GO project is an effort to address the need for consistent descriptions of gene products in different databases. The project began as collaboration between three model organism databases, FlyBase (Drosophila), the Saccharomyces Genome Database (SGD), and the Mouse Genome Database (MGD), in 1998. Since then, the GO Consortium has grown to include many databases, including several of the world's major repositories for plant, animal (mouse, rat), human, and microbial genomes. But this is still not all. According to the 2007 update of the world-wide molecular database collection, there are 968 freely available gene/protein related databases (Galperin 2007). Since 2004, a total of 110-170 databases have been added each year (Galperin 2005, 2006, 2007). Therefore intelligent integration of relevant knowledge needs to be embodied in any biodata ontology that deals with personalised decision support.

Disease Ontology is a controlled medical vocabulary designed to facilitate the mapping of diseases and associated conditions to particular medical codes such as ICD9CM, SNOMED and others (http://diseaseontology.sourceforge.net/). The Disease Ontology can also be used to associate model organism phenotypes to human disease as well as medical record mining.

Simultaneously with the emerging need for standardized nomenclatures and concept ontologies for biosciences, the new science of systems biology has emerged. It is needed for the grand unification of biological (and medical) knowledge for basic and applied research. Importantly, systems biology is the ultimate tool for describing metabolic and genetic networks interacting with environmental variables to produce phenotypes of all organisms, including health and disease in individuals. Systems biology knowledge is essential for both personalised medicine and molecular epidemiology studies of human diseases in stratified populations (Nicholson 2006). In such systems, biological knowledge needs to be represented, stored and analyzed in a standardized ontological framework, so that data from different domains of biology and medicine can be properly integrated.

A standardized ontology framework makes data easily available for advanced methods of analysis, including artificial intelligence algorithms, that can tackle the multitude of large and complex datasets for clustering, classification, and rule inference for biomedical and bioinformatics applications.

The challenge is to create computational platforms that dynamically integrate ontology and a set of efficient machine learning methods, including new methods for personalised modelling that would manifest a better accuracy at a personal level and facilitate new discoveries.

## 4.4  An Integrated Famework and a Platform for Ontology-Based Local and Personalised Modelling and Knowledge Discovery

The framework and a software platform presented here bring together ontology knowledge repository and machine learning techniques to facilitate sophisticated adaptive data and information storage, retrieval, modelling, and knowledge discovery.

The framework utilizes ontology based data, as well as new knowledge inferred from the data embedded in the ontology. The platform allows for the adaptation of an existing knowledge base to new data sources and through entering results from machine learning and reasoning models. A generic diagram of the framework is shown in Fig. 4.2. It consists of two main modules: an ontology knowledge and data repository module; and a machine learning module. There is an interface module between the two modules that is specific for every application.

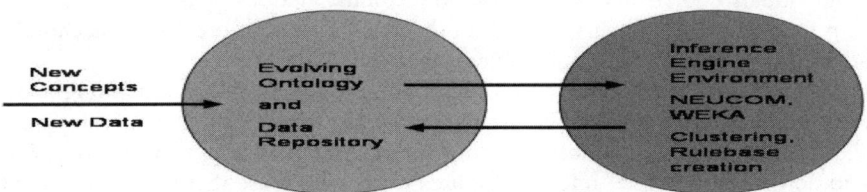

**Fig. 4.2.** The ontology-based personalised decision support framework consists of two interconnected parts: (1) An ontology/data base sub-system; (2) Machine learning sub-system

The general framework from Fig. 4.2 is implemented as a software platform characterized by the following characteristics:

- Protégé ontology development environment (http://protege.stanford.edu/)
- Data import module to enter external multimodal data into ontology
- Data retrieval module to search and retrieve relevant data from an ontology
- Machine inference module that includes local and personalised techniques such as the described in Section 4.2, included for example in a decision support environment NeuCom (www.theneucom.com).
- User-friendly interface modules that can be tailored to specific applications in different knowledge domains.
- A module for updating the ontology, based on classification and clustering results from the machine inference module.

A sample implementation schema of an ontology-based personalised decision support system in biomedicine is shown in Fig. 4.3 (Gottgtroy et al 2006).

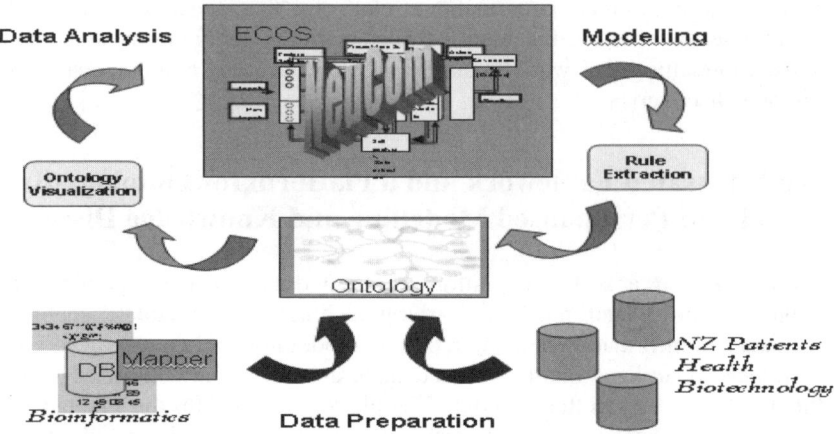

**Fig. 4.3.** A sample ontology-based decision support system. The inference engine at the top utilized data retrieved from Ontology in Protégé (from Gottgtroy 2006).

The system from Fig. 4.3 is able to combine data from numerous sources to provide individualized person/case reports and recommendations on actions/interventions to help modify the outcome in a desired direction based on previously accumulated information on input variables and outcomes in the database.

In keeping with our overall vision of bringing together machine learning and ontologies into a single integrated environment, we propose to use the patterns produced by the machine learning module to refine the structure of the existing ontology. One way to do that is to extract relevant features from a database that resides in an ontology, to create local profiles, and then to enter the extracted features and profiles back into the ontology in order to enrich it and to discover new relationships. Feature selection has long been known to be a key success factor in improving the accuracy of the classification/prediction process in machine learning (Witten and Frank 2000; Kasabov 2002, 2007a). Since ontologies link related concepts together, they can be used to extract a set of related features of different kind (e.g. clinical, genetic, cognitive, etc.) for a particular machine learning model. For example, in classifying whether patients are at high risk or low risk of contracting a heart disease, an ontology such as the Chronic Disease Ontology (CDO), described later in this chapter, can be used to determine all the currently known risk factors (encompassing the clinical, genomic and demographic data types). Since the predictors used are acknowledged to be the best that are currently known, we could expect performance to improve over uninformed or ad-hoc methods of feature selection only from a single database.

A major challenge is how to use the newly discovered knowledge to further evolve existing ontologies. In general, the knowledge extracted from machine learning methods can fall into three distinct categories; those that refer to:

1. concepts that already exist in the ontology
2. concepts not covered by the existing ontology
3. changes in  the nature of existing concepts in the ontology

In terms of category 1, no changes need to be made to the existing ontology. Categories 2 and 3 pose significant challenges as they could represent knowledge hitherto unknown to the knowledge engineer. A naïve approach of immediately refining the ontology may not be desirable, given that an ontology represents the collective wisdom and knowledge of world-class domain experts gained through their life experiences. A more prudent approach would be to monitor such knowledge over a period of time and only update the ontology when a clear and consistent trend emerges that shows that such knowledge persistently improves the accuracy of predictions on newly arriving data. The rank aggregation technique proposed by Domshlak (Domshlak, Gal and Roitman 2007) and the knowledge pattern technique proposed by Clark et al. (Clark, Thompson and Porter 2004) provides us with the right  tools for assessing when changes should be made to the existing ontology.

The problem of linking ontologies with machine learning systems requires building a specific interface. To enable a machine learning workbench to automatically obtain the right data, there should be shared contextual "understanding" between the learning system and the ontology itself, as each of them may have their own contextual meaning which may differ from one another. Thus the integration of these local contexts is yet another challenging issue, and as discussed by Maamar et al (2006) and Satyanarayanan (2001), should address the following issues: how can changes in a concept be detected; how should the context be found and stored within the systems/data; how should the context be taken into account; how should an inference engine obtain sufficient information to act in a context-aware manner. A further issue is the mutual trust between the system and user / data source; and whether the system retrieves accurate and relevant information.

Local and Transductive inference methods focus only on a small area of data space and its relevant information (Song and Kasabov 2006; Kasabov 2007a, b). Thus new incoming data will dynamically change the contextual meaning of the information within the database; especially when new data point is being introduced near the area of interest. This can lead to changes in how the data are being clustered, or the new data might strengthen a particular cluster. Either way, the changes will affect the ontology, because as the data change, the representation depicted by the ontology will need to be updated. Therefore, to accommodate the dynamics of the data, the ontology must be able to evolve. Evolving the ontology involves modifying the originally designed ontology based on the knowledge and new clustering discovered during the inference (Gottgtroy et al 2006).

In general, the ontology evolution process can be classified as conceptual changes and explication changes (Lenzerini, Milano and Poggi 2004). Conceptual changes deal and includes new concepts or relationships which are emerging, or flagging already existing concepts which display a diminishing level of support from new data streaming in ; while explication changes focus on the modifications in the description of the concepts, such as adding a new description or property of a concept. As a general guide, Uschold and Maedche offer good frameworks for ontology building and learning (Maedche 2002; Uschold and Grüninger 1996). However, in our case, we are mainly interested in the evaluation and refining of the frameworks; due to our concern in assuring that the evolved ontology will still reflect the real world which it represents, as well as the refining process in order to support its evolving nature.

In its early stage, the ontology evolution focuses on the ontology learning process by proving from the machine learning process. In its subsequent stages, the system will grow and further evolve. This includes the ability of the machine learning module to automatically select appropriate data from the database and for the ontology to detect newly emerging concepts or relationships. For instance, the patient's health and medical data stored in the database might be stored in several separate tables, thus the ontology and context mediation system will help the machine learning module (e.g. NeuCom) to collect data from relevant columns and tables based on the information and relationship described in the ontology. For example, if the user wants to perform chronic disease analysis, a context mediation system can be used to ensure that the system will collect all of the right information about chronic diseases, but not about kidney functions.

After the machine learning has performed its analysis on a given data set, and identified new relationships, these new findings will be fed back into the ontology and will be noted. However, this doesn't mean that this new relationship will be immediately acknowledged as new concepts. It will be noted as possible discovery but confirmed by further evidence to establish its status.

Sometimes, when we analyze a set of data using one methodology, for example numerical prediction, it may not show any new findings, but if we combine it with the result of some other methodology, such as pattern recognition or clustering on the same data set, the combined results may reveal new insights. These new insights can then be used to update and/or evolve the ontology. Therefore, the capability of the system to evolve is not just limited to a certain learning method, or findings.

The implementation of this technique will raise the issue of how one can be sure that the particular concepts or relationships already have enough evidence to be claimed as new findings. As we are using the rank and weighting methods to overcome this issue, we believe that by adopting the rank aggregation technique proposed by Domshlak (Domshlak, Gal and Roitman 2007) will help us ensure that the ontology evolution process will not go amiss. We also will adopt the knowledge pattern technique proposed by Clark et al. (Clark, Thompson and Porter 2004), to help us ascertain that the emerging concepts fit with certain knowledge patterns and are reliable new findings.

The platform described above can be used to create ontology and simulation systems for various bioinformatics and biomedical applications, such as:

- Brain-gene-disease repository and disease risk simulation (illustrated in section 4.5);
- Chronic disease (e.g. heart disease, obesity, diabetes) personal risk evaluation (illustrated in section 4.6);
- Diagnosis and risk assessment of multiple types of cancer on a genomic scale (illustrated in section 4.7);
- Kidney function prediction system (Marshal et al 2005);
- Longevity prediction for patients on haemodialysis (Song et al 2005);
- Environmental monitoring and prognostic systems (Kasabov 2007a);
- Many more.

## 4.5  Brain-Gene-Disease Ontology and Brain Cancer Gene Profiling

It is a long term goal to develop a global ontology and simulation system covering all functional levels in the brain from Fig. 4.1. Here we describe a smaller–scale brain-gene-disease ontology (BGO) and a simulation system developed with the use of the framework and the platform from Section 4.3.

The system supports computational neurogenetic modelling (CNGM), that is concerned with modelling and understanding the influence of genes upon brain functions (Anon 2005; Kasabov and Benuskova 2004; Benuskova and Kasabov 2007). The BGO and the CNGM simulation system integrates knowledge that comes from different

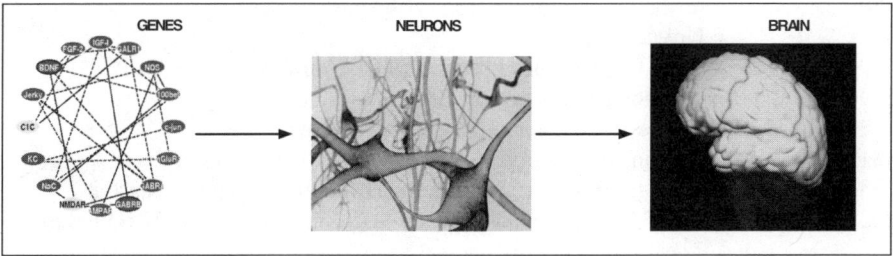

**Fig. 4.4.** CNGM is concerned with simulating complex relationships between genes, and their influence upon neurons and the brain, encompassed in a comprehensive ontology integrating hierarchical levels of organization from genes to neurons to neural networks to specific brain tissues (the pictures are snapshots from the BGO).

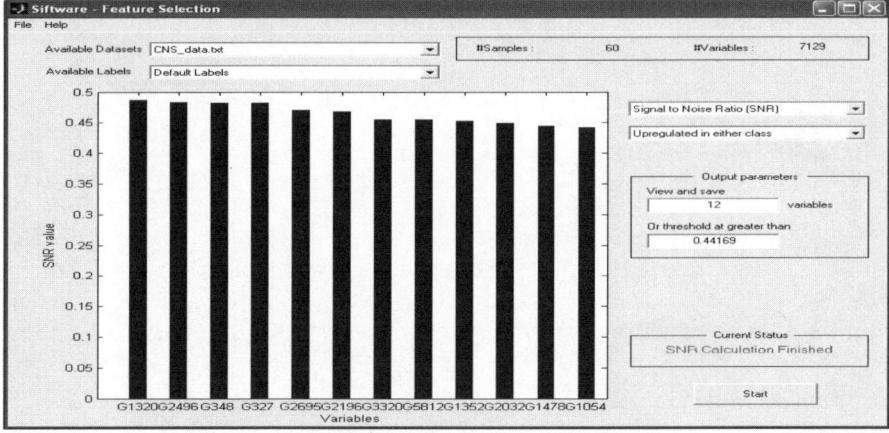

**Fig. 4.5.** The 12 genes selected as top discriminating genes from the Central Nervous System (CNS) cancer data that discriminates two classes – responders and non- responders to treatment (Pomeroy, Tamayo et al 2002). The NeuCom software system was used for the analysis using the Signal-to-Noise Ratio method.

disciplinary domains such as neuroscience, bioinformatics, genetics, computer and information sciences (Benuskova and Kasabov 2007). The scope of the phenomena included in the developed BGO is shown in Figure 4.4 (see www.kedri.info).

**Table 4.1.** The 12 selected most informative genes discriminating responders from non-responders in the treatment of the CNS cancer (dataset of Pomeroy et al (2002)), based on Signal-to-Noise Ratio method using NeuCom software

| | |
|---|---|
| G1320 | = FBN1 Fibrillin 1 (Marfan syndrome) |
| G2496 | = NTRK3 Neurotrophic tyrosine kinase, receptor, type 3 (TrkC) - (one of the 50 markers of survival identified by Pomeroy et al. 2002) |
| G348 | = probable ubiquitin carboxyl-terminal hydrolase |
| G327 | = Unknown product |
| G2695 | = TAR RNA binding protein (TRBP) mRNA |
| G2196 | = polyposis locus protein 1 |
| G3320 | = Leukotriene C4 synthase (LTC4S) gene |
| G5812 | = Elastin, Alt. Splice 2 |
| G1352 | = High mobility group protein (HMG-I(Y)) gene exons 1-8 |
| G2032 | = MMP2 Matrix metalloproteinase 2 (gelatinase A, 72kD gelatinase, 72kD type IV collagenase) |
| G1478 | = PCOLCE Procollagen C-endopeptidase enhancer |
| G1054 | = APOD Apolipoprotein D |

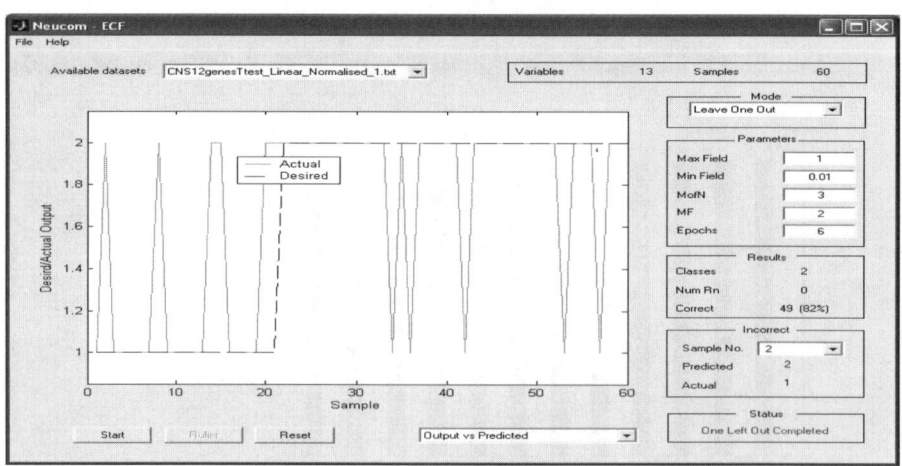

**Fig. 4.6.** A leave-one-out cross validation method is applied on the selected 12 genes (Table 4.1) to validate an ECF ECOS model on the 60 CNS cancer samples (Pomeroy and Tamayo et al 2002), where 60 models are created – each one trained on 59 samples and tested on the left out sample. The average accuracy over all 60 examples is 82%, where 49 samples are classified accurately and 11 incorrectly.

The BGO contains data, information, and knowledge about several levels of the functioning of the brain, including: molecular and gene information; 500 genes related to brain functions and diseases such as Epilepsy, Alzheimer, Mental retardation, Cancer; single neuron functions; functions of different areas of the brain.

How this information can be used to derive personalised profiles and new knowledge is illustrated on gene expression data of 60 samples of central nervous system (CNS) cancer (medulloblastoma) representing 39 child patients who survived the cancer after treatment, and 21 patients who did not respond to the treatment (Pomeroy et al 2002).

Fig. 4.5 and Table 4.1 show the selection of the top 12 genes out of 7129, as numbered in the original publication (Pomeroy et al 2002), based on 60 samples, using a signal to noise ratio method (SNR) in the software environment NeuCom. The selected

a

b

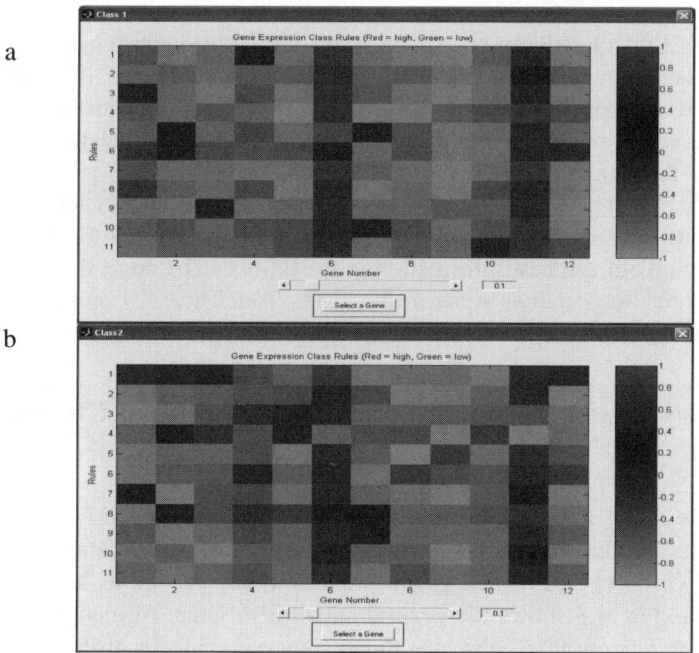

**Fig. 4.7.** For each class (1-not responding, and 2 – survive) 11 profiles are extracted that represent different expression patterns of the 12 genes from table 1 for each cluster of data consisting of one or more samples (individuals). Different profiles point to the heterogeneity of the gene expressions in the CNS cancer samples due to their different interaction for an individual or a group of persons (data from (Pomeroy, Tamayo et al. 2002)). (a) Class 1 – profiles of not responding to treatment individuals and groups; (b) Class 2 – profiles of survival individuals and groups. The analysis was done using a proprietary software system SIFTWARE (www.peblnz.com). Red indicates a highly expressed gene and green – low expressed on the scale of normalised values [1,-1].

small number of genes, out of thousands, can be further analyzed and modelled in terms of their interactions and relations to the functioning of neurons, neural networks, the brain and the CNS.

Local and personalised modelling on the extracted 12 genes using ECOS is illustrated in Figures 4.6 and 4.7, where a classification system is evolved using ECF and the local cluster profiles, 11 of them for each of the two classes, are shown. The profiles capture the interaction between genes for an individual person or for a group of per-sons clustered together. This suggests that an interesting interaction between genes defines survival of the CNS cancer, rather than a single gene alone.

Before the final classifier is evolved, leave-one-out cross-validation method is applied to validate the ECF model on the 60 samples, where 60 models are created – each one trained on 59 samples and tested on the left-out sample. The average accuracy over all 60 examples is 82% as shown in Fig. 4.6. A total of 49 samples are classified accurately, out of 60.

A final ECF ECOS classifier is trained on all 60 samples and local cluster profiles are extracted for each class – Fig.4.7. We can see that the profiles are different, which points to the heterogeneity of the CNS cancer across the population in the data set.

A new input vector will be mapped into a trained ECOS model recalling the closest local rules (profiles) and the class output will be calculated.

The extracted 12 genes and 22 patterns of their expression related to the two classes constitute new information, which can be entered back to the BGO. After that an analysis of both old and new information in the BGO can be done, that may further reveal new information, such as some of the genes from Table 4.1 and gene patterns from Fig. 4.7 being involved in some other diseases (multiple disease genes/patterns).

## 4.6  Ontology-Based Personalised Risk Evaluation of Chronic Disease

As another case study, a Protégé-based ontology is being developed for entering and linking concepts and data for various chronic diseases (Type 2 Diabetes; Cardio Vascular Disease; Obesity) and related genes and mutations, as well as health, diet and life history data. Fig. 4.8 shows the general network structure of main concept linkages in our prototype ontology. Fig. 4.9 shows a more detailed snapshot of a portion of the developed ontology, depicting diabetes related genes linked to specific types of mutations.

Health-related data from surveys of New Zealand individuals are being imported into the ontology. Examples of suitable data to include are in Table 4.2. These health related data can be used in the machine learning module to evaluate a patient's risk towards chronic diseases.

This is the framework into which information on individual patients for their disease symptoms, genetic mutations, diet, and life history details can be inputted, and risks, profiles, and recommendations derived.

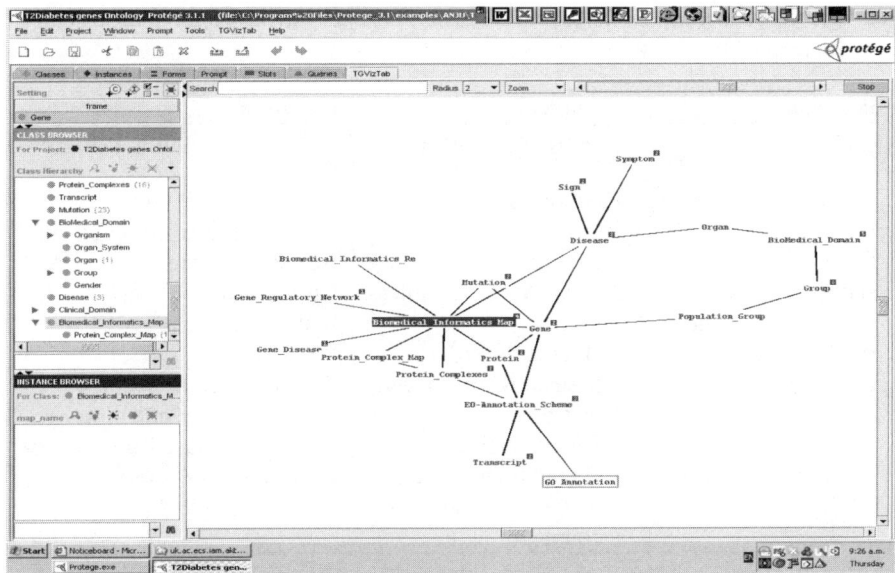

**Fig. 4.8.** Protégé display of the general ontological network for chronic disease concepts linking diseases, symptoms, genes, proteins etc

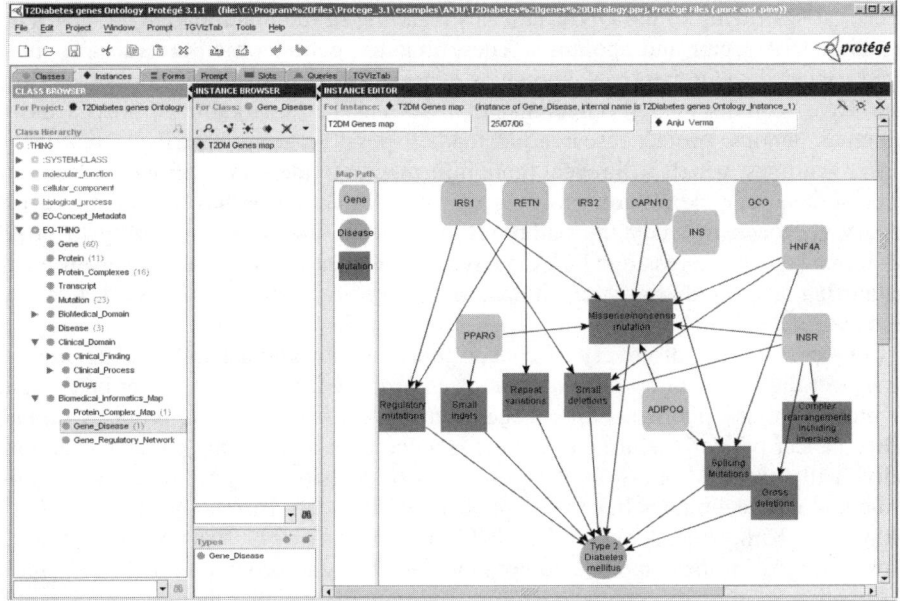

**Fig. 4.9.** Protégé ontology structure for genes and mutation types related to one chronic disease (Type 2 diabetes)

**Table 4.2.** National Nutrition Survey 1997 (NNS97) dataset variables used in the ontology for health risk prediction:

| Personal | Clinical | Nutrition |
|---|---|---|
| Age | Body mass index | Nutrients including Energy |
| Gender | Waist | Protein |
| Ethnicity | Blood pressure | Fat – saturated fat |
| Immigration | Pulse | Carbohydrates complex and |
| Social          economic | Blood: | free sugar Fibre |
| status | Total & HDL choles- | Cholesterol |
| Geographic location – | terol | Salt |
| rural – urban | Haemoglobin | Supplements |
| Occupation | Measures of inflam- | Dietary pattern: |
|  | mation: | Energy and nutrition re- |
|  | Serum ferritin | quirements based on activity |
|  | C reactive protein | pattern |
|  | White blood cell |  |
|  | count |  |

## 4.7 Ontology-Based Personalised Cancer Prognosis – Work in Progress

Cancer arises owing to the DNA damages resulting in the mutations in the genes that regulate DNA repair and apoptosis. These mutated genes, which are causally implicated cancer development, are known as cancer genes, and so far, more than 350 of them have been discovered (Greenman et al 2007). The remarkable progress achieved in human genome project research has made it possible to systematically resequence cancer genomes, which will reveal more information to identify other cancer genes. In recent years, gene expression profiling with DNA microarray has provided a revolutionary approach to study the pathology of cancer. Owing to the ability to profile differential gene expression, DNA microarray data have been extensively used for diagnosing and predicting the clinical outcomes in response to cancer treatment (Schena 2002).

Coupled with the discovery of cancer genes, a substantial number of approaches for cancer diagnosis and prognosis have been proposed. However, cancer heterogeneity prevents many published approaches and models from working properly on individual cancer patient's treatment. Cancer is a complex disease that patients need to be individually treated. Many models for clinical decision support based on various computational techniques have been published (Alizadeh, Eisen, Davis, Ma and et al 2000; Cho, Kim, Wee, Jeon and Lee 2003; Golub et al 1999; van't Veer et al 2002). However, most of these models do not have the ability to provide reliable and precise information to the patients who require individual therapy schemes (Navins et al 2003).

There have been several attempts to use ontology for cancer research. The National Cancer Institute Thesaurus (NCIT) is a biomedical ontology that provides consistent, unambiguous definitions for concepts and terminologies in cancer research domain

(Ceusters et al 2005). It opens a way to integrate various types of information through semantic relationships, including cancer related disease, findings, gene data, drugs, etc. NCI is also linked to other internal or external information resources, such as caCore, caBIO and Gene Ontology (GO). In the study proposed by Dameron et al (2006), it has been demonstrated that ontology is capable of automatically analyzing the grading of lung cancer.

One of the research goals in this project is to create an ontology for cancer data analysis that can be used to assist personalised cancer diagnosis and risk measurement based on gene data analysis. Such cancer diagnosis and prognosis ontology will be beneficial to both the business and research communities. This is based on the high financial cost and available data resources for cancer research using gene data analysis (Hermida et al 2006; Shipp et al 2006). The cancer diagnosis and prognosis ontology will help the scientists in providing the relationships, either evidential or predicted, between genes; therefore, the scientists can target their research appropriately. The other benefit is to avoid repeatedly re-discovering any relationships that have been already been made by other researchers.

The main advantage of this system is the evolving ontology and the use of a multi-model machine learning module. This module will have a personalised modelling system, which has been demonstrated to be efficient for clinical and medical applications of learning systems. It is efficient because it focuses not on the model, but on the individual samples (Nevins et al 2003; Song and Kasabov 2006; Song, Ma, and Kasabov 2005). To solve the cancer heterogeneity issue, the implementation of personalised modelling can efficiently diagnose cancers and predict clinical outcomes for cancer patients (see for example Fig. 4.7). Ultimately, the results of the personalised modelling will be used to evolve the ontology in such a way that it will be able to find any emerging patterns, as well as strengthening the existing ones.

The cancer diagnosis and prognosis ontology can also be a sophisticated platform to store, manage, and to share the large amount of data and insight collected from the last two decades of advances in cancer research. Moreover, with such a system, the disparate datasets and distinct computational models for cancer research can be integrated. Furthermore, such data can serve as a good base for our machine learning module to make its predictions and analysis.

One of the main notions in this research is to develop a model that can accurately diagnose cancer samples and predict clinical outcome based on gene data for future individual patients. A personalised cancer diagnosis and risk management system will be built based on gene data expression analysis. Novel personalised neuro-fuzzy inference techniques will be incorporated specifically for constructing the system for cancer diagnosis and prognosis. Based on the model obtained from learning procedure, an individual model (personalised model) will be created and used for every new input data vector (real cancer tissue samples) from clinical sources.

## 4.8   Conclusions and Directions for Further Research

This chapter presented a framework and a software platform for the integration of local and personalised modelling techniques with ontology information bases to achieve:

1. A better personal data processing, resulting in personalised models, profiles, risk evaluations, treatment.
2. New information creation
3. A better understanding of complex, unsolved so far problems in their multivariate interaction at different levels of the functioning a biological system

The presented framework needs to be developed further in terms of:

1. Efficient integration of old and new concepts and information in the ontology
2. Automated search for relevant data to a new person's data
3. Novel, more efficient methods for personalised feature selection and model optimization.
4. Multiple model creation for a single person and cross model analysis for the discovery of new interactions between variables from different models, e.g. genes related to different types of cancer.

The presented case studies will be further developed in terms of:

1. A continuous update of the BGO with new information made available related to any of the functional levels from Fig.1, including quantum and evolutionary level.
2. A continuous update of the chronic disease ontology.
3. The creation and the continuous update of a cancer ontology and prognostic personalised modelling system, to incorporate all available data related to all types of cancer for a cross model creation and new information discovery.

# References

Alizadeh, A.A., Eisen, M., Davis, R., Ma, C., et al.: Distinct types of diffuse large B-cell lymphoma identified by gene expression profiling. Nature 403(6769), 503–511 (2000)
Anon: The Nervous System. In: Genes and Disease, National Centre for Biotechnology Information (NCBI) (2005),
http://www.ncbi.nlm.nih.gov/books/bv.fcgi?rid=gnd.chapter.75
Ashburner, M., Ball, C.A., Blake, J.A., Botstein, D., Butler, H., Cherry, J.M.: Gene ontology: tool for the unification of biology. The Gene Ontology Consortium. Nature Genetics 25, 25–29 (2000)
Benuskova, L., Kasabov, N.: Computational Neurogenetic Modeling. Springer, New York (2007), http://www.springer.com/east/home/default?SGWID=5-40356-22-173696910-0
Benuskova, L., Jain, V., Wysoski, S.G., Kasabov, N.: Computational neurogenetic modeling: a pathway to new discoveries in genetic neuroscience. Intl. Journal of Neural Systems 16(3), 215–227 (2006)
Benuskova, L., Kasabov, N.: Modeling L-LTP based on changes in concentration of pCREB transcription factor. Neurocomputing 70(10-12), 2035–2040 (2007)
Berners-Lee, T., Hendler, J., Lassila, O.: The Semantic Web. Scientific American (May 17, 2001)
Ceusters, W., Smith, B., Coldberg, L.: A Terminological and Ontological Analysis of the NCI Thesaurus. Methods Inf Med 44(4), 498–507 (2005)
Chandrasekaran, B., Josephson, J.R., Benjamins, V.R.: What are ontologies, and why do we need them? Intelligent Systems and Their Applications 14, 20–26 (1999)

Cho, H.S., Kim, T.S., Wee, J.W., Jeon, S.M., Lee, C.H.: cDNA Microarray Data Based Classification of Cancers Using Neural Networks and Genetic Algorithms. Nanotech 1, 28–31
(2003)

Clark, P., Thompson, J., Porter, B.: Knowledge Patterns. In: Staab, S., Studer, R. (eds.) Handbook on Ontologies, pp. 191–208. Springer, Berlin (2004)

Dameron, l., Roques, E., Rubin, D., Marquet, G., Burgun, A.: Grading lung tumors using
OWL-DL based reasoning. In: The 9th International Protege Conference, Stanford, USA
(paper presented at, 2006)

Domshlak, C., Gal, A., Roitman, H.: Rank Aggregation for Automatic Schema Matching. IEEE
Transactions on Knowledge and Data Engineering 19(4), 538–553 (2007)

Fensel, D.: Ontologies: A Silver Bullet for Knowledge Management and Electronic Commerce,
2 ed. Springer, Heidelberg (2004)

Futschik, M., Reeve, A., Kasabov, N.: Evolving connectionist systems for knowledge discovery
from gene expression data of cancer tissue. Artificial Intelligence in Medicine 28, 165–189
(2003)

Galperin, M.Y.: The Molecular Biology Database Collection. Nucl. Acids. Res. 33, D5–D24
(update 2005)

Galperin, M.Y.: The Molecular Biology Database Collection. Nucl. Acids. Res. 34, D3–D5
(update 2006)

Galperin, M.Y.: The Molecular Biology Database Collection. Nucl. Acids. Res. 35, D3–D4
(update 2007)

Golub, T.R., Slonim, D.K., Tamayo, P., Huard, C., Gaasenbeek, M., Mersirov, J.P., et al.:
Molecular classification of cancer: class discovery and class prediction by gene expression
monitoring. Science 286, 531–537 (1999)

Gottgtroy, P., Kasabov, N., Macdonell, S.: Evolving Ontologies for Intelligent Decision Support. In: Fuzzy Logic And The Semantic Web, ch. 21, pp. 415–439. Elsevier, Amsterdam
(2006)

Greenman, C., Stephens, P., Smith, R., Dalgliesh, G., Hunter, C., Bignell, G., et al.: Patterns of
somatic mutation in human cancer genomes. Nature 446(7132), 153–158 (2007)

Gruber, T.R.: A translation approach to portable ontologies. Knowledge Acquisition 5, 199–
220 (1993)

Havukkala, I., Benuskova, L., Pang, S., Jain, V., Kroon, R., Kasabov, N.: Image and Fractal
Information Processing for Large-Scale Chemoinformatics. In: Rajapakse, J.C., Wong, L.,
Acharya, R. (eds.) PRIB 2006. LNCS (LNBI), vol. 4146, pp. 163–173. Springer, Heidelberg
(2006)

Hermida, L., Schaad, O., Demougin, P., Descombes, P., Primig, M.: MIMAS: an innovative
tool for network-based high density oligonucleotide microarray data management and annotation. BMC Bioinformatics 7(190) (2006)

Kasabov, N.: Adaptive learning system and method, Patent USA, PEBL, PCT WO 01/78003
(2001)

Kasabov, N.: Evolving Connectionist Systems: Methods and Applications in Bioinformatics.
In: Brain Study and Intelligent machines. Springer, London (2002)

Kasabov, N.: Adaptation and Interaction in Dynamical Systems: Modelling and Rule Discovery
Through Evolving Connectionist Systems. Applied Soft Computing 6(3), 307–322 (2006)

Kasabov, N.: Evolving Connectionist Systems: The Knowledge Engineering Approach, 2nd
edn. Springer, London (2007a)

Kasabov, N.: Global, local and personalised modelling and profile discovery in Bioinformatics:
An integrated approach. Pattern Recognition Letters 28(6), 673–685 (2007b)

Kasabov, N., Song, Q.: Transductive Neuro-Fuzzy Inference Method for Personalised Modelling, Patent, PCT WO 2005/048185 A1 (2005)

Kasabov, N., Futschik, M., Sullivan, M., Reeve, A.: Method and Medical Decision Support System Utilizing Gene Expression and Clinical Information, Patent USA, PEBL, PCT/US03/25563 (2003)

Kasabov, N., Reeve, A., Futschik, M., Sullivan, M., Guildford, P.: Medical Applications of Adaptive Learning Systems using Gene Expression Data, Patent USA, PEBL, PCT WO 03/079286 (2003)

Lenzerini, M., Milano, D., Poggi, A.: State of the art and state of the practice including initial possible research orientations (InterOP Report). UniRoma, Roma, Italy (2004)

Maamar, Z., Benslimane, D., Narendra, N.C.: What can context do for web services? Communications of the ACM 49(12), 98–103 (2006)

Maedche, A.: Ontology learning for the semantic Web. Kluwer Academic Publishers, Dordrecht (2002)

Marshall, M.R., Song, Q., Ma, T.M., MacDonell, S., Kasabov, N.: Evolving Connectionist System versus Algebraic Formulae for Prediction of Renal Function from Serum Creatinine. Kidney International 67, 1944–1954 (2005)

Nevins, J.R., Huang, E.S., Dressman, H., Pittman, J., Huang, A.T., West, M.: Towards integrated clinico-genomic models for personalized medicine: combining gene expression signatures and clinical factors in breast cancer outcomes prediction. Human Molecular Genetics 12(2), R153-R157 (2003)

Nicholson, J.K.: Global systems biology, personalized medicine and molecular epidemiology. Mol. Syst. Biol. 52(2) (2006)

Noy, N.F., McGuinness, D.L.: Ontology Development 101: A Guide to Creating Your First Ontology (Medical Informatics Technical Report No. SMI-2001-0880): Standford Knowledge Systems Laboratory (2001)

Owens, A.: Semantic Storage: Overview and Assessment. Technical Report IRP Report 2005, Electronics and Computer Science, U of Southampton (2005)

Pisanelli, D.M. (ed.): Ontologies in Medicine. IOS Press, Amsterdam (2004)

Pomeroy, S.L., Tamayo, P., Gaasenbeek, M., Sturla, L.M., et al.: Prediction of central nervous system embryonal tumour outcome based on gene expression. Nature 415(6870), 426 (2002)

Satyanarayanan, M.: Pervasive computing: vision and challenges. IEEE Personal Communications 8(4), 10–17 (2001)

Schena, M.: Microarray analysis. John Wiley & Sons, New York (2002)

Shegogue, D., Zheng, W.J.: Integration of the Gene Ontology into an object-oriented architecture. Bioinformatics 6(113), 1–14 (2005)

Shippy, R., Fulmer-Smentek, S., Jensen, R.V., Jones, W.D., Wolber, P.K., Johnson, C.D., et al.: Using RNA sample titrations to assess microarray platform performance and normalization techniques. Nature Biotechnology 24, 1123–1131 (2006)

Song, Q., Kasabov, N.: NFI: a neuro-fuzzy inference method for transductive reasoning. Fuzzy Systems, IEEE Transactions on 13(6), 799–808 (2005)

Song, Q., Kasabov, N.: TNFI: A Neuro-Fuzzy Inference Method for Transductive Reasoning. IEEE Transactions on Fuzzy Systems 13(6), 799–808 (2005)

Song, Q., Kasabov, N.: TWNFI - a transductive neuro-fuzzy inference system with weighted data normalisation for personalised modelling. Neural Networks 19(10), 1591–1596 (2006)

Song, Q., Ma, T., Kasabov, N. (eds.): Transductive Knowledge Based Fuzzy Inference System for Personalized Modeling. In: Wang, L., Jin, Y. (eds.) FSKD 2005. LNCS (LNAI), vol. 3614. Springer, Heidelberg (2005)

Song, Q., Kasabov, N., Ma, T., Marshall, M.: Integrating regression formulas and kernel functions into locally adaptive knowledge-based neural networks: a case study on renal function evaluation. In: Artificial Intelligence in Medicine (February 2006)

Uschold, M.F., Grüninger, M.: Ontologies: Principles, Methods and Applications. The Knowledge Engineering Review 11(2), 93–155 (1996)

Van de vijies, M.J., et al.: Gene expression profiling predicts clinical outcome of breast cancer. Nature 415(6871), 530–536 (2002)

Verma, A., Song, Q., Kasabov, N.: Developing "Evolving Ontology" for Personalised Risk Evaluation for Type-2 Diabetes Patients. In: 6th International Conference on Hybrid Intelligence, Auckland, New Zealand (2006)

Witten, I.H., Frank, E.: Data Mining: Practical Machine Learning Tools and Techniques with Java Implementations. Morgan Kaufmann, San Francisco (2000)

## Appendix A: The WWKNN Algorithm (from Kasabov, 2007 a,b)

Using the ranking of the variables in terms of a discriminative power within the neighborhood of K vectors, when calculating the output for the new input vector, is the main idea behind the WWKNN algorithm (Kasabov, 2007a,b), which includes one more weight vector to weigh the importance of the variables.

The Euclidean distance $d_j$ between a new vector $x_i$ and a neighboring one $x_j$ is calculated now as:

$$d_j = \text{sqr} \left[ \text{sum}_{l=1 \text{ to } v} \left( c_{i,l} \left( x_{i,l} - x_{j,l} \right) \right)^2 \right] \tag{A1}$$

where: $c_{i,l}$ is the coefficient weighing variable $x_l$ for in neighbourhood of $x_i$. It can be calculated using a Signal-to-Noise Ratio (SNR) procedure that ranks each variable across all vectors in the neighborhood set $D_i$ of Ni vectors:

$$\mathbf{C}i = (c_{i,1}, c_{i,2}, \ldots, c_{i,v}) \tag{A2}$$

$c_{i,l} = S_l / \text{sum}(S_l)$,   for: $l = 1,2,\ldots,v$, where:

$$S_l = \text{abs}\left(M_l^{(\text{class 1})} - M_l^{(\text{class 2})}\right) / \left(\text{Std}_l^{(\text{class 1})} + \text{Std}_l^{(\text{class2})}\right) \tag{A3}$$

Here $M_l^{(\text{class 1})}$ and $\text{Std}_l^{(\text{class 1})}$ are respectively the mean value and the standard deviation of variable $x_l$ for all vectors in $D_i$ that belong to class 1.

The new distance measure, that weighs all variables according to their importance as discriminating factors in the neighborhood area $D_i$, is the new element in the WWKNN algorithm when compared to the WKNN.

Using the WWKNN algorithm, a "personalised" profile of the variable importance can be derived for any new input vector, which represents a new piece of "personalised' information.

Weighted variables in personalized models is used in the TWNFI models (Transductive Weighted Neuro-Fuzzy Inference ) in (Song and Kasabov 2006), where a back propagation or an evolutionary optimization algorithm is applied.

There are several open problems related to transductive learning and reasoning, e.g. how to choose the optimal number of vectors in a neighborhood and the optimal number of variables, which for different new vectors may be different.

## Appendix B: TWNFI – Neural Fuzzy Inference System with Weighted Data Normalization (from Song and Kasabov 2006)

The distance between vectors x and y is measured in TWNFI in weighted normalized Euclidean distance defined as follows (the values are between 0 and 1):

$$\left\| x - y \right\| = \left[ \frac{1}{P} \sum_{j=1}^{P} w_j \left| x_j - y_j \right|^2 \right]^{\frac{1}{2}} \tag{B1}$$

Where: $x, y \in R^P$ and $w_j$ are weights.

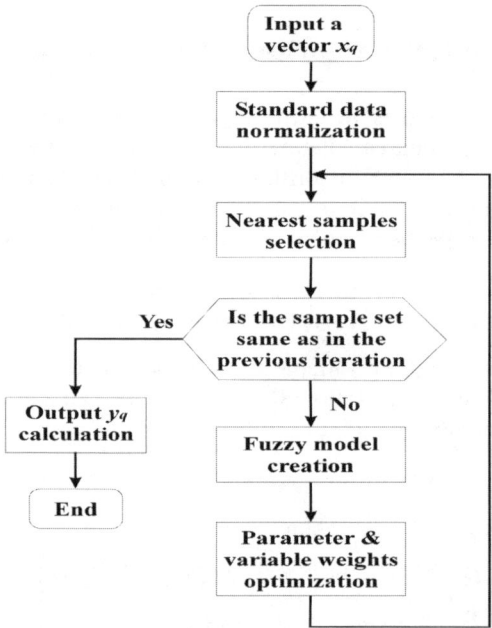

**Fig. B1.** A block diagram of the TWNFI

For each new data vector $x_q$ an individual model is created with the application of the following steps:

1. Normalize the training data set and the new data vector $x_q$ (the values are between 0 and 1) with value 1 as the initial input variable weights.
2. Search in the training data set in the input space to find $N_q$ training examples that are closest to $x_q$ using weighted normalized Euclidean distance defined as Eq. B1.
3. Calculate the distances $d_i$, $i = 1, 2, \ldots, N_q$, between each of these data samples and $x_q$. Calculate the vector weights $v_i = 1 - (d_i - \min(d))$, $i = 1, 2, \ldots, N_q$, $\min(d)$ is the minimum value in the distance vector $d = [d_1, d_2, \ldots, d_{Nq}]$.
4. Use a clustering algorithm to cluster and partition the input sub-space that consists of $N_q$ selected training samples.

5. Create fuzzy rules and set their initial parameter values according to the clustering procedure results; for each cluster, the cluster centre is taken as the centre of a fuzzy membership function (Gaussian function) and the cluster radius is taken as the width.
6. Apply the steepest descent method (back-propagation) to optimize the weights and parameters of the fuzzy rules in the local model $M_q$ following Eq. (B6 – B13).
7. Search in the training data set to find $N_q$ samples (the same to Step 2); if the same samples are found as the last search, the algorithm goes to Step 8, otherwise, to Step 3.
8. Calculate the output value $y_q$ for the input vector $x_q$ applying fuzzy inference over the set of fuzzy rules that constitute the local model $M_q$.
9. End of the procedure.

The weight and parameter optimization procedure is described below:

Consider the system having P inputs, one output and M fuzzy rules defined initially through the ECM clustering procedure, the $l$ -th rule has the form of:

$$R_l: \quad \text{If } x_1 \text{ is } F_{l1} \text{ and } x_2 \text{ is } F_{l2} \text{ and } \dots x_P \text{ is } F_{lP}, \text{ then } y \text{ is } G_l. \tag{B2}$$

Here, $F_{lj}$ are fuzzy sets defined by the following Gaussian type membership function:

$$GaussianMF = \alpha \exp\left[-\frac{(x-m)^2}{2\sigma^2}\right] \tag{B3}$$

and $G_l$ are of a similar type as $F_{lj}$ and are defined as

$$GaussianMF = \exp\left[-\frac{(y-n)^2}{2\delta^2}\right] \tag{B4}$$

Using the Modified Centre Average defuzzification procedure the output value of the system can be calculated for an input vector $x_i = [x_1, x_2, \dots, x_P]$ as follows:

$$f(x_i) = \frac{\sum_{l=1}^{M} \frac{G_l}{\delta_l^2} \prod_{j=1}^{P} \alpha_{lj} \exp\left[-\frac{w_j^2(x_{ij}-m_{lj})^2}{2\sigma_{lj}^2}\right]}{\sum_{l=1}^{M} \frac{1}{\delta_l^2} \prod_{j=1}^{P} \alpha_{lj} \exp\left[-\frac{w_j^2(x_{ij}-m_{lj})^2}{2\sigma_{lj}^2}\right]} \tag{B5}$$

Here, $w_j$ are weights of the input variables.

Suppose the TWNFI is given a training input-output data pair $[x_i, t_i]$, the system minimizes the following objective function (a weighted error function):

$$E = \frac{1}{2}v_i[f(x_i)-t_i]^2 \tag{B6}$$

($v_i$ are defined in Step 3)

The steepest descent algorithm (BP) is used then to obtain the formulas for the optimization of the parameters $G_l$, $\delta_l$, $\alpha_{lj}$, $m_{lj}$, $\sigma_{lj}$ and $w_j$ such that the value of E from Eq. (B6) is minimized:

$$G_l(k+1) = G_l(k) - \frac{\eta_G}{\delta_l^2(k)} v_i \Phi(\boldsymbol{x}_i)\left[ f^{(k)}(\boldsymbol{x}_i) - t_i \right] \tag{B7}$$

$$\delta_l(k+1) = \delta_l(k) -$$
$$\frac{\eta_\delta v_i \Phi(\boldsymbol{x}_i)}{\delta_l^3(k)}\left[ f^{(k)}(\boldsymbol{x}_i) - t_i \right]\left[ f^{(k)}(\boldsymbol{x}_i) - G_l(k) \right] \tag{B8}$$

$$\alpha_{lj}(k+1) = \alpha_{lj}(k) -$$
$$\frac{\eta_\alpha v_i \Phi(\boldsymbol{x}_i)}{\delta_l^2(k)\alpha_{lj}(k)}\left[ f^{(k)}(\boldsymbol{x}_i) - t_i \right]\left[ G_l(k) - f^{(k)}(\boldsymbol{x}_i) \right] \tag{B9}$$

$$m_{lj}(k+1) = m_{lj}(k) -$$
$$\frac{\eta_m w_j^2(k) v_i \Phi(\boldsymbol{x}_i)}{\delta_l^2(k)\,\sigma_{lj}^2(k)}\left[ f^{(k)}(\boldsymbol{x}_i) - t_i \right]\left[ G_l(k) - f^{(k)}(\boldsymbol{x}_i) \right]\left[ x_{ij} - m_{lj}(k) \right] \tag{B10}$$

$$\sigma_{lj}(k+1) = \sigma_{lj}(k) -$$
$$\frac{\eta_\sigma w_j^2(k) v_i \Phi(\boldsymbol{x}_i)}{\delta_l^2(k)\,\sigma_{lj}^3(k)}\left[ f^{(k)}(\boldsymbol{x}_i) - t_i \right]\left[ G_l(k) - f^{(k)}(\boldsymbol{x}_i) \right]\left[ x_{ij} - m_{lj}(k) \right]^2 \tag{B11}$$

$$w_j(k+1) = w_j(k) -$$
$$\frac{\eta_w w_j(k) v_i \Phi(\boldsymbol{x}_i)}{\delta_l^2(k)\,\sigma_{lj}^2(k)}\left[ f^{(k)}(\boldsymbol{x}_i) - t_i \right]\left[ f^{(k)}(\boldsymbol{x}_i) - G_l(k) \right]\left[ x_{ij} - m_{lj}(k) \right]^2 \tag{B12}$$

Here,

$$\Phi(\boldsymbol{x}_i) = \frac{\prod_{j=1}^{P} \alpha_{lj} \exp\left\{ -\frac{w_j^2(k)\left[ x_{ij} - m_{lj}(k) \right]^2}{2\sigma_{lj}^2(k)} \right\}}{\sum_{l=1}^{M} \frac{1}{\delta_l^2} \prod_{j=1}^{P} \alpha_{lj} \exp\left\{ -\frac{w_j^2(k)\left[ x_{ij} - m_{lj}(k) \right]^2}{2\sigma_{lj}^2(k)} \right\}} \tag{B13}$$

where $\eta_G$, $\eta_\delta$, $\eta_\alpha$, $\eta_m$, $\eta_\sigma$ $\eta_w$ and are learning rates for updating the parameters $G_l$, $\delta_l$, $\alpha_{lj}$, $m_{lj}$, $\sigma_{lj}$ and $w_j$ respectively.

In the TWNFI training–simulating algorithm, the following indexes are used:

- Training data samples:    $i = 1, 2, \ldots, N$;
- Input variables:    $j = 1, 2, \ldots, P$;
- Fuzzy rules:    $l = 1, 2, \ldots, M$;
- Training epochs:    $k = 1, 2, \ldots$.

# Part II

# Computational Intelligence in Biomedicine

# 5

# Data-Mining of Time-Domain Features from Neural Extracellular Field Data

Samuel Neymotin[1], Daniel J. Uhlrich[2], Karen A. Manning[2], and William W. Lytton[1]

[1] SUNY Downstate Medical Center, Dept. Biomedical Engineering, Brooklyn, NY
 {samn,billl}@neurosim.downstate.edu
[2] University of Wisconsin Dept. of Anatomy, Madison, WI
 {duhlrich,kamannin}@wisc.edu

**Summary.** Spike-wave and polyspike-wave activity in electroencephalogram are waveforms typical of certain epileptic states. Automated detection of such patterns would be desirable for automated seizure detection in both experimental and clinical venues. We have developed a time-domain algorithm denominated SPUD to facilitate data-mining of large electroencephalogram/electrocorticogram datasets to identify the occurrence of spike-wave or other activity patterns. This algorithm feeds into our enhanced Neural Query System [2, 12] database application to facilitate data-mining. We have used our algorithm to identify and classify activity from both simulated and experimental seizures.

## 5.1 Introduction

There are vast databases of existing neural data which are growing in number and size at an ever increasing rate. The data is of clinical and research significance, but without the proper tools to sift through it, it is of limited use [2]. Custom search, interpretation, and quantification methods are needed to optimize the use of the data. These methods and search tools should allow the investigator to visualize the available information in a clear and comprehensible manner, and allow him/her to find patterns that would be difficult to see without the aid of a computer. These types of methods come under the heading *data-mining*, informally defined as user-guided analysis of large datasets using automated techniques. Ideally, data-mining methods should be general enough to work for different types of neural data, *i.e.*, simulated in software, *in vivo*, and *in vitro* recordings, yet allow customizable options to find relevant and particular patterns.

We have developed an algorithm, named SPUD (mnemonic for slice, peak, up, down — explained below), which successfully extracts useful features from disparate types of neural data. It operates on signals in the time-domain, and the features it extracts remain in the time-domain, providing exact timing information for events of interest. This is in contrast to many popular signal analysis algorithms, such as wavelets [1] or FFT (Fast Fourier transform [5]), which

T.G. Smolinski et al. (Eds.): Comp. Intel. in Biomed. & Bioinform., SCI 151, pp. 119–140, 2008.
springerlink.com                                    © Springer-Verlag Berlin Heidelberg 2008

operate mainly in the frequency-domain, and therefore, lack precise timing information. For the detection of spike-wave (SW), algorithms operating in the frequency-domain may have difficulty, because the timing of SWs is extremely precise. For example, a population spike, which is very brief, may be followed by a wave, which is considerably longer-lasting. This type of pattern may provide conflicting information in the frequency-domain because spikes and waves are so close together in time.

The SPUD algorithm feeds the extracted features directly into the previously developed Neural Query System (NQS) [12]. We enhanced NQS to allow for standard structured query language (SQL [4]) queries by incorporating the open source database MySQL (http://www.mysql.com). The versatile search capabilities already provided by NQS, together with standard SQL, provides for highly efficient and customizable search directly from the NEURON simulation environment [10, 3]. This makes it convenient for a researcher to run simulations and analyze their results all in one familiar environment. Using the developed techniques, we were able to find SW and polyspike-wave (PSW) patterns in both simulated and *in vivo* recordings from Sprague Dawley rats.

## 5.2   Methods

### 5.2.1   Data-Mining Architecture

*Database setup*

The data of interest may either be read in directly to the database system or run through a feature extraction algorithm. The feature extraction algorithm simplifies the data by pulling out properties deemed to be useful for the particular application. It allows viewing of the data at a higher level of abstraction by ignoring irrelevant details and giving structure to the data. The feature extraction algorithm we used, SPUD, is described below. Other feature extraction algorithms fit into this pipeline and may be used in SPUD's place.

*Performing queries*

Once the database is set up, the user can perform searches/queries on the data. The query syntax is described below. The results returned by the database system can be displayed visually with NQS's (described below) graphical output system and/or read into data structures of the hoc programming language that is available with the NEURON simulation environment. This allows for a bidirectional flow of information between the user and the database. Fig. 5.1 shows the data-mining architecture developed.

### 5.2.2   SPUD Feature Extraction Algorithm

SPUD is a general algorithm for extracting information from signals in the time-domain. It works on both noisy and clean data. It was originally intended to

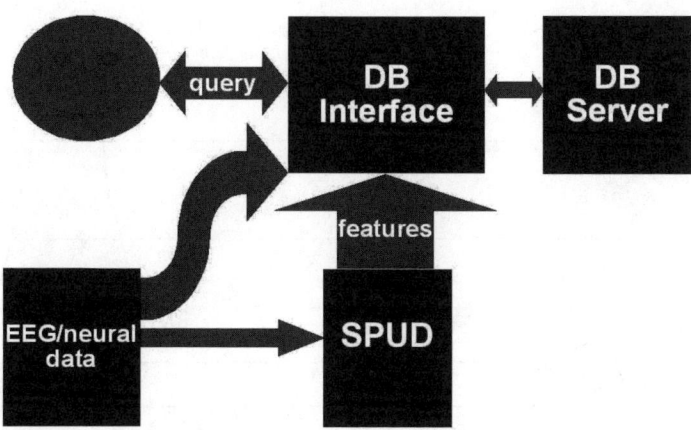

**Fig. 5.1.** Schematic of data-mining architecture. Arrows represent flow of information in a given direction. Bi-directional information transfer occurs between the user and the database (DB) system.

work on signals obtained from simulations in the NEURON environment, but performed equally well on experimental data obtained from *in vivo* recordings of rats with electrocorticographic leads. One key design goal was the ability of the algorithm to extract the information characterizing the morphology and precise time properties of an arbitrary signal.

*Steps of the algorithm*

The SPUD algorithm is used to build a database of *bump* characteristics in the time domain, where a bump is any deviation with return to the baseline.

The algorithm proceeds as follows:

1. Slice the EEG/ECoG trace horizontally at logarithmically spaced voltage levels. These levels are more finely spaced at lower voltage levels to capture small bumps.
2. Iterate over slice positions in the horizontal direction to find matched upward and downward crossing points.
3. Find peaks between crossing points to define slice traces as a triad: upward crossing; peak; downward crossing.
4. Group slices with common peaks to define individual bumps.
5. Extract information about each bump. Information may include *e.g.,* first derivative of upstroke, first derivative of downstroke, amplitude, duration, sharpness (defined below), start time, and others.

Fig. 5.2 shows bumps extracted from a sample trace.

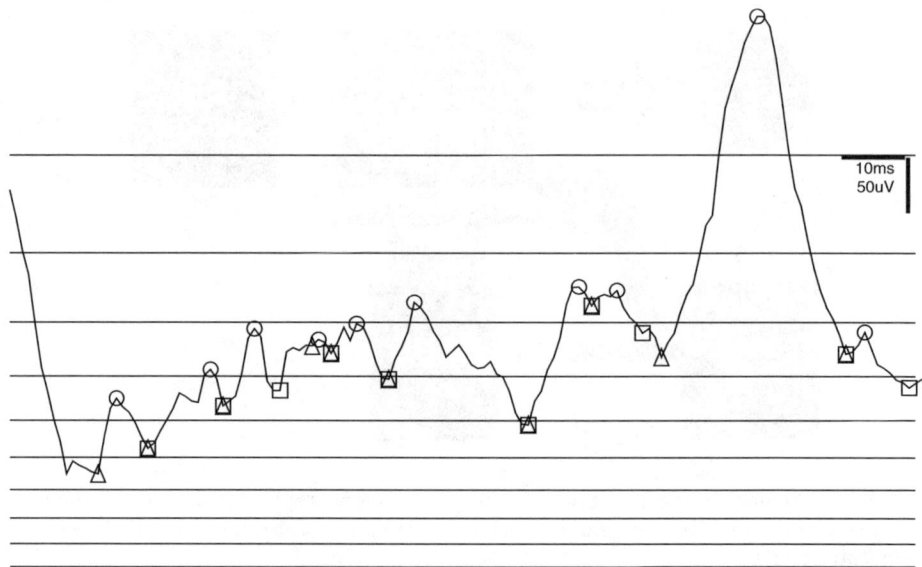

**Fig. 5.2.** Schematic showing different steps of SPUD feature extraction algorithm on recording from rat right occipital cortex. Horizontal lines are threshold locations. Open triangles, circles, and squares represent left/upward crossing position, peak position, and right/downward crossing position of bumps respectively. Note: some bumps starting and ending points are at the same time location and therefore a square and triangle marks them.

*Bumps*

The information extracted from the signal, which is defined as the sequence of sampled time-ordered pairs $(t_i, s_i)$, takes the form of *bumps*. A bump, $b$, is a section of the signal containing a deviation from the baseline value, $L_m \in L$ (set of threshold lines), followed by a return to that value. Given a series of samples in the signal, $s_i, s_{i+1}, s_{i+2}, \ldots s_{i+n}$ , occurring at times $t_i, t_{i+1}, t_{i+2}, \ldots t_{i+n}$ , and a fixed horizontal baseline, $L_m$, a bump begins at the first time signal pair, $(t_f, s_f)$, such that $s_f \geq L_m$. The bump consists of the time-ordered set of pairs $(t_k, s_k)$ such that $s_k \geq L_m$. The last pair, $(t_l, s_l)$, where $s_l \geq L_m$ but $s_{l+1} < L_m$, defines the end of the bump. The next bump in the set may begin at the next time the signal crosses $L_m$. For a bump $b$, $\forall (t_k, s_k) \in b, t_k \geq t_f, t_k \leq t_l, s_k \geq L_m$. For notational convenience $(t_f, s_f)$ , $(t_l, s_l)$ , and $(t_p, s_p)$ denote the first, last, and peak time/signal pairs in a bump. A given bump $B_i \in B$ 's start, end, and peak time will be denoted $B_{i_{tf}}$ , $B_{i_{tl}}$ , and $B_{i_{tp}}$ respectively.

Bumps have properties associated with them which may include width $(t_l - t_f)$, height from baseline $(\max(s_i - L))$, absolute peak value $(\max(s_i)$, first derivative $\left(\frac{s_i - s_{i-1}}{t_i - t_{i-1}}\right)$ at arbitrary positions within the bump, number of nested bumps, etc. The properties may be easily extended and determined algorithmically once the extents of the bump have been determined with the default algorithm. The shape of the bumps can be anything from a single action potential to the complex spiking patterns from extracellular recordings.

*Thresholding*

In order to extract only the *significant* bumps, a set of horizontal threshold lines, $L$, is used to define the baselines where bumps deviate from. This set of horizontal lines may be selected interactively by the user as it becomes apparent where significant bumps occur. Initially it can be set to either linear spacing, or logarithmic spacing. Further work needs to be done to algorithmically space the threshold lines in an optimal way depending on a given signal. The linear and logarithmic spacings allow for more control of bump extraction for different types of signals. For example, use of logarithmic spacing allows finer spacing at lower values of the signal, allowing for the extraction of smaller bumps. In general, the minimum value of the signal determines a value above which the first horizontal threshold line is placed and the maximum value determines a location below which the last threshold line is placed. These threshold lines must be monotonically increasing.

*Varying threshold lines*

There are several options for specifying how to space the threshold lines. Depending on the option chosen, the extracted bumps can be somewhat different. The default option is to space the lines logarithmically from 0.05 of the maximum amplitude in the trace to 0.95 of the maximum amplitude. This generally picks out the most significant bumps and smaller bumps that are placed near the bottom of the trace due to the finer spacing there. Another option which can yield better performance when the user's focus is on the middle of a trace, allows a *fan-out* logarithmic spacing whereby the start of a logarithmically spaced set of threshold lines extends upward and downward from 0.5 of the maximum amplitude of the trace. The most simple option, which does not allow for focusing in on any particular region is to space the threshold lines linearly. In general, logarithmic spacing starting at a certain vertical position will extract finer details in the surrounding region. The user is also given the option of specifying his own threshold lines to narrow in on activity of interest. Figure 5.3(A,B) shows the differences in resulting bumps using 2 different sets of threshold slices.

*Bump finding*

Once the threshold lines, $L$, have been determined, they are traversed from the lowest threshold line to the highest threshold line. Each threshold line, $L_m$, is traversed in the horizontal direction from the start of the signal, $(t_0, s_0)$, to the end of the signal, $(t_n, s_n)$, in a search for matched upward, peak, and downward crossing points. To start the search for upward crossing points, a point in the signal below the current threshold line $(s_i < L_m)$ is found. This point is the starting position for the search. The horizontal position is incremented until the signal crosses the threshold ($s_i \geq L_m$) line in an upward direction. This horizontal point is then noted as the left-most point of the first bump. The search then continues until a maxima is reached. This point is noted. Once again, the search continues until the signal passes the current threshold line in a downward direction ($s_i \leq L_m$) and the point noted as the right-most position of the current bump. In this fashion, each threshold line is traversed with a triad of upward, peak, and downward crossing points being extracted as the initial bumps.

*Overlapping bumps*

Since multiple threshold lines are used, the set of bumps found from the previous step, $B$, may be overlapping. This occurs with two bumps, $B_i$ and $B_j$, when $B_{i_{tf}} \geq B_{j_{tf}}$ and $B_{i_{tl}} \leq B_{j_{tl}}$. This means that $B_i$ is entirely contained within the time bounds of $B_j$. Overlapping also occurs when $B_{i_{tf}} \geq B_{j_{tf}}$, $B_{i_{tf}} \leq B_{j_{tl}}$, and $B_{i_{tl}} > B_{j_{tl}}$ . There are two options for dealing with these types of inevitabilities. The first, and simplest, is to allow overlapping bumps. Though this is not the default behavior of the algorithm, for certain situations, this is the desired result, *i.e.*, determining the # of nested bumps in a signal/time interval. Overlapping bumps may also help determine the bounds of low frequency, long duration bumps in the presence of high frequency components. For example, in fig. 5.2, the first 6 low amplitude bumps are high frequency fluctuations on a longer low frequency deflection. In other situations, overlapping bumps can have their starting and ending times adjusted. This is done by grouping threshold slices with common peaks to define individual bumps. First, the bumps are traversed in increasing time order and checked for overlaps on the left or right sides. A left overlap is defined as the start of the current bump, being between the start and the peak time of the previous bump ($B_{c_{tf}} > B_{c-1_{tf}}$ and $B_{c_{tf}} < B_{c-1_{tp}}$). In such a case, the threshold lines are traversed from $\min(L)$ to $\max(L)$ to find the lowest threshold containing a bump with the same peak time as the current bump's peak time and containing a starting time greater than the previous bump's ending time (all threshold lines have their associated bumps stored during initial bump extraction). The current bump's starting time is set as this threshold line's peak time. The same idea is used to check for and correct overlaps on the right side.

*Creeping*

At this point there is a non-default option allowing a bump $B_i \in B$ 's $B_{i_{tf}}$ and $B_{i_{tl}}$ and associated signal values to *creep* to a local minima. This means decreasing the starting time until it is a local minima, *i.e.*, having a signal value less than the surrounding signal values ($s_i < s_{i-1}, s_i < s_{i+1}$), and increasing the starting time in a similar fashion. In certain situations, this helps get a more accurate estimate of a bump's time bounds, i.e., if a threshold line is much higher than the local minima found by creeping.

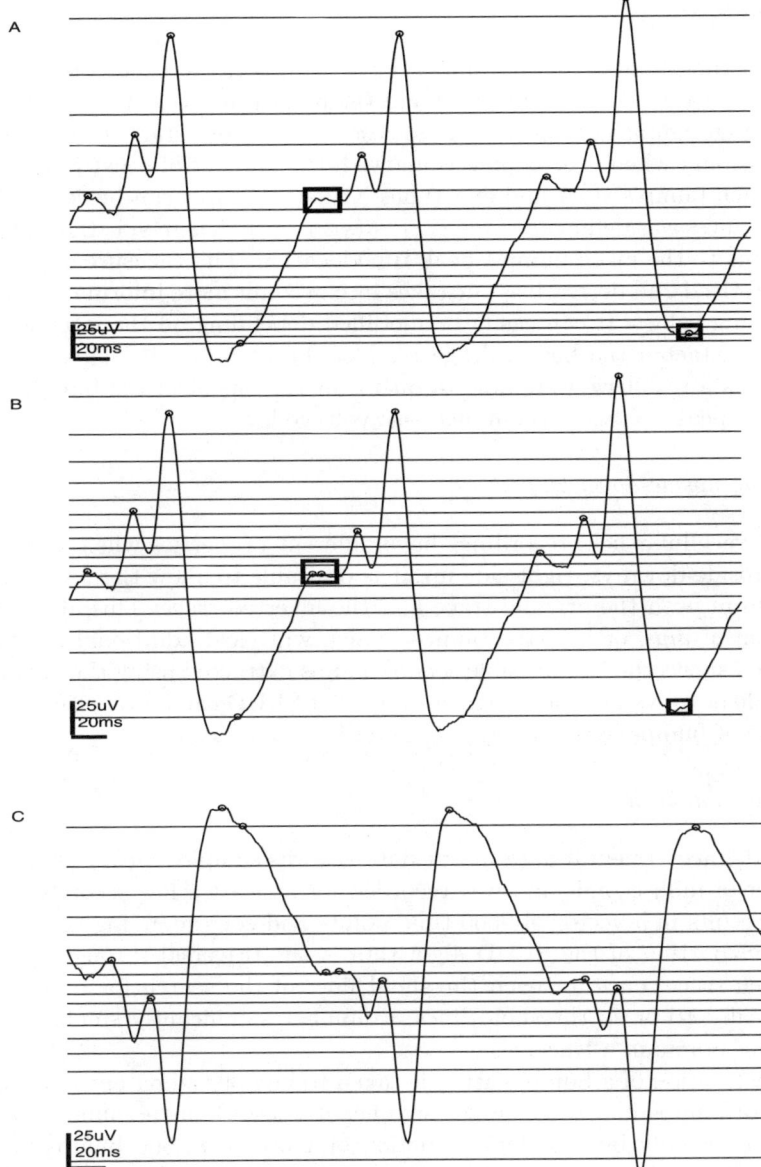

**Fig. 5.3.** Traces obtained by averaging all triplets of photic responses across 25 trials. Spacing the threshold slices (horizontal lines) differently may result in different bumps being extracted. Bump peaks represented as circles. Differences in extracted bumps between sets in A and B slices highlighted by rectangles. (A) Logarithmic spacing starting near bottom of trace. (B) Logarithmic spacing centered at middle (fan-out). (C) Dips extracted as bumps by flipping trace over x-axis.

*Bump properties*

Once the time bounds of the bumps have been set, the other properties of a bump are extracted. The values we had the algorithm extract were: start time, peak voltage amplitude, peak time location, time from start to finish (*width*), sharpness, and when overlapping is allowed, the number of other bumps nested within each bump's start and end times. Others features can be readily added. The sharpness was defined as $(s_i - s_{i-4}) - (s_{i+4} - s_i)$ , where $s_i$ is the sample corresponding to the given bump's peak time location. This measure approximates the discretized 2nd derivative, but takes into account more information from the surrounding values. It can easily be modified depending on the intended usage. We also extracted the base voltage level of the bump on its left and/or right side and as a result we were able to query for the height of the bump, which is defined as peak voltage level minus base voltage level.

*Extracting upside-down bumps*

Since extracellular field recordings have different properties when they are displayed upside-down vs. right-side up, it is desirable to allow the user to extract bumps from both the original trace and the inverted trace. This allows for the extraction of *dips*, or inverted bumps, which will yield additional information. Figure 5.3 shows the two possible sets of bumps extracted using the normal, and the upside down version of the trace. Note that SPUD can automatically extract both sets of bumps by setting a parameter.

*Run-time complexity*

A typical trace of neural data in our data set was sampled with a frequency on the order of milliseconds and was recorded for at least a few seconds, typically 9. This results in a vector of 9000 time points and corresponding voltage levels. The main portion of the SPUD algorithm is the traversal through each point in a given trace, once for each threshold line, in the search for bumps. There are typically 10 or so threshold lines, a number significantly smaller than the number of points in a trace.

The extraction of a bump feature is taken to be approximately constant. The number of bumps in a trace is also significantly less than the number of points in a trace, usually by a 30-fold reduction or more (*e.g.,* 300 bumps in a 9000 millisecond trace).

The main steps of the algorithm consist of iterating over the threshold lines for each time point and extracting bumps. Then for each bump, features are extracted which has a constant cost. The time complexity of the algorithm can therefore be approximated as $|L| \times |B| \times |S|$ where $|L|$ is the number of threshold lines, $|B|$ is the number of bumps found, and $|S|$ is the number of sampled points in a trace. When removing overlapping bumps, additional terms proportional to $|B|^2$ may be added to the run-time complexity. This is because, for a single given bump, $b \in B$, overlapping with another bump, the search for the modified starting/ending time is linear in $|B|$ in the worst case (the average will be far less). Since there are $|B|$ bumps, this results in an additional cost of $|B|^2$. The total run-time complexity will then be $|L| \times |S| \times |B| + |B|^2$ , with the dominating term being $|S|$. As a consequence the algorithm will have an $o(|S|)$ run-time. The results of a signal may be analyzed many times in a data-mining framework and consequently, the run time cost of bump extraction can be viewed as an amortized payment which has returns each time a query is performed by the user.

In runtime complexity, SPUD is competitive with FFT which is $\theta$ $(|S| \times \log(|S|))$ [11], and with the the fast wavelet transform using the lifting scheme, which is $o(|S|)$ when using finite filters [15].

### 5.2.3  NQS and MySQL

*Overview*

The next step in our data-mining pipeline was storing the extracted bumps and their associated information into the Neural Query System database as well as a MySQL table to allow data-mining of features.

*NQS*

The Neural Query System (NQS) is a relational database system built into the NEURON simulation environment. It allows storage of structured records and a flexible search syntax on these records. It works seamlessly with NEURON's data types, including Vectors, Lists, strings, scalars and neural or network models. It also allows storage of sub-tables within tables. This is very handy for storage of complex data types. NQS also provides for graphical display of data, aiding in finding patterns in the data. Using NEURON with NQS, it is possible to run large simulations and store partially analyzed data quickly and easily from one environment.

*MySQL*

The MySQL database system (http://www.mysql.com/) has been in use for over a decade and is open-source. It has a large user base familiar with Structured Query Language (SQL). Though NQS has an intuitive syntax and much in common with SQL, there are differences in the details of the two query languages. Requiring users who already know SQL to learn NQS syntax is a possible hindrance to fast learning of the data-mining system. Although NQS has a convenient front-end and runs quickly on fast machines, it does not have the benefit of more than a decade of open source contributions from a wide community of programmers. For these reasons, we decided to create an interface between NQS and MySQL that would maintain the strengths of both systems — maintaining the ease of use of NQS and SQL, as well as the optimizations available in MySQL.

*NQS/MySQL Interface*

The interface we created enhances NQS with the ability to perform standard SQL queries using MySQL's version of SQL. The interface was written with the MySQL C application programmer interface (API) being translated into blocks in NEURON's NMODL language [9]. NMODL allows C code to be compiled into the NEURON environment and then called once it is running. The API we wrote connects directly to a MySQL database server from within NEURON. This allows conversion of NQS databases to MySQL databases and vice versa. It also allows for SQL queries to be performed on MySQL databases and then reading the returned data into NEURON data types. Exchanging data and queries between NQS and a MySQL server becomes quite easy. Once the queries are done, the user may retrieve their results into an NQS table.

*Function overview*

The main interface consists of functions for connecting to the MySQL server as well as performing queries. Once the queries have been made, their results may be obtained with helper functions. These functions read the data directly into NEURON data structures. There are other helper functions made to help search tables in the database. A typical session in NEURON will consist of connecting to the MySQL server and exchanging queries and data back and forth. We also added some functions written in NEURON's scripting language, hoc, to allow for the conversion between NQS tables and MySQL tables easily, as well as a function to help create MySQL tables programmatically. These functions make it very easy to use MySQL and NQS together efficiently.

### 5.2.4   Simulation Setup

The simulations were run in NEURON using *tonic-clonic* simulation networks previously described [13]. The networks consisted of 1350 neurons comprising three types of cells: spontaneously active cells (drivers), inhibitory cells, and expressors (main excitatory cell population). 138,240 traces of extracellular field potentials were produced. Realistic patterns of activity, such as SW, were present (a sample trace with SW activity is shown in figure 5.4). The traces were then stored in the NQS system along with the 8,205,557 bumps and corresponding features extracted with the SPUD algorithm. The enhanced NQS system also converted the resulting database into a MySQL database and indexed each column to allow for real-time/interactive searching of the full dataset. Having to sift through such a massive set of data without the aid of a flexible data-mining tool would have prohibitive time costs and most likely introduce errors.

### 5.2.5   Rat Data Recording Setup

*Materials*

Data was acquired from normal adult Sprague Dawley rats using BrainWare (TDT), which employs an electrode channel window for electrophysiological

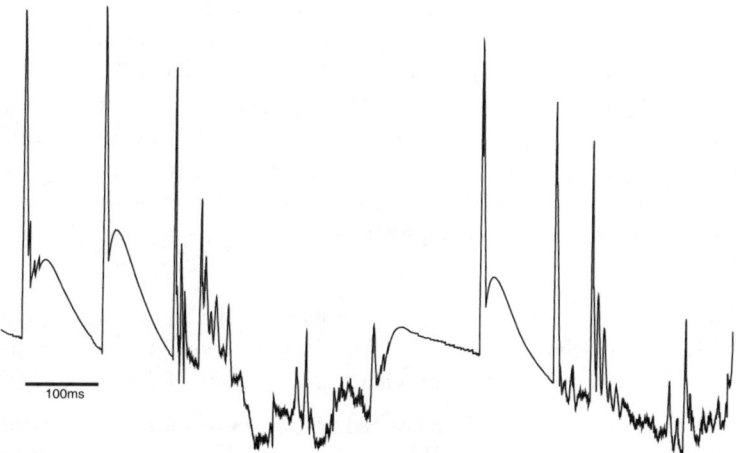

**Fig. 5.4.** Single trace from simulation. In this and following traces derived from simulation, voltage is in arbitrary units.

**Fig. 5.5.** ECoG response from photically-sensitized rat in response to a train of stroboscopic flashes presented at 8 Hz. Vertical lines represent occurrence of strobe flashes. The first flashes in the strobe train elicited a photic driving response that morphed into PSW activity.

**Fig. 5.6.** ECoG recording from right occipital cortex during a PTZ-induced seizure with SW discharges

recording. We recorded the electrocorticogram (ECoG) at a sampling rate of 25kHz from a stainless steel wire placed on the dural surface over right occipital cortex; a second lead located over the cerebellum served as the differential reference [16].

*Experimental conditions*

There were two experimental conditions used for recording. In all recordings, the rat was awake and moving freely.

*Epileptiform behavior via photic-induced sensitization.* The first experimental condition involved photic-induced sensitization following repeated exposure to trains of stroboscopic flashes [16]. One strobe train consisted of 17 flashes presented at 8Hz. Trains were presented every 30 seconds and 30-40 trains were presented during a daily recording session. On initial train presentations, the responses were small, showing at most an entrained sinusoidal driving response. However, over the course of three sessions of strobe exposure, the response grew in magnitude and acquired epileptiform characteristics like SW morphology (Fig. 5.5).

*Chemically-induced seizures.* The rat was injected with 24 mg/kg of pentylenetetrazol (PTZ), a convulsant agent [7]. We started recording as soon as the rat was injected. SW seizures appeared within 2-7 min of PTZ injection (Fig. 5.6).

*Artifact rejection.* In some of the recorded traces there were artifacts introduced by the recording equipment or sudden animal movement. These artifacts were generally a voltage value of 0 or amplifier maximal value. Therefore, before running our algorithms on these traces, we preprocessed them by setting outliers/artifacts to an average of the surrounding voltage levels programmatically. Outliers were considered as any value that differed from the median voltage level of that trace by a predetermined threshold. Our algorithms and tools were then tested on several hundred of these preprocessed traces.

## 5.3   Results

### 5.3.1   Simulated Data

*Data-mining for SWs*

Once the bumps and their corresponding information were stored in the database, we were able to data-mine for patterns of interest. We were able to use bump properties to classify the simulated EEG traces shown in figure 5.7A. A simple search looked for all traces with bumps having height above the 95th percentile, and sharpness above the 30th percentile. This was done by first creating a table in NQS storing the percentile values of several properties of interest such as height, sharpness, and number of bumps in a trace. Percentile tables were then created with a few function calls. Then the following NQS query was used to select all traces having more than 3 bumps above the 95th percentile of height, and 3 bumps above the 30th percentile of sharpness:

```
sq.select("HE95" , ">" ,3, "SH30", ">" ,3, "NUM", "<" ,10)
```

This query returned traces with spike-wave activity. Some results are shown in figure 5.7. If it is most important to identify all seizures, a liberal criteria is used in the search which will result in a potentially higher number of false positives. On the other hand, a more conservative select is used in order to collect a few clear seizures,

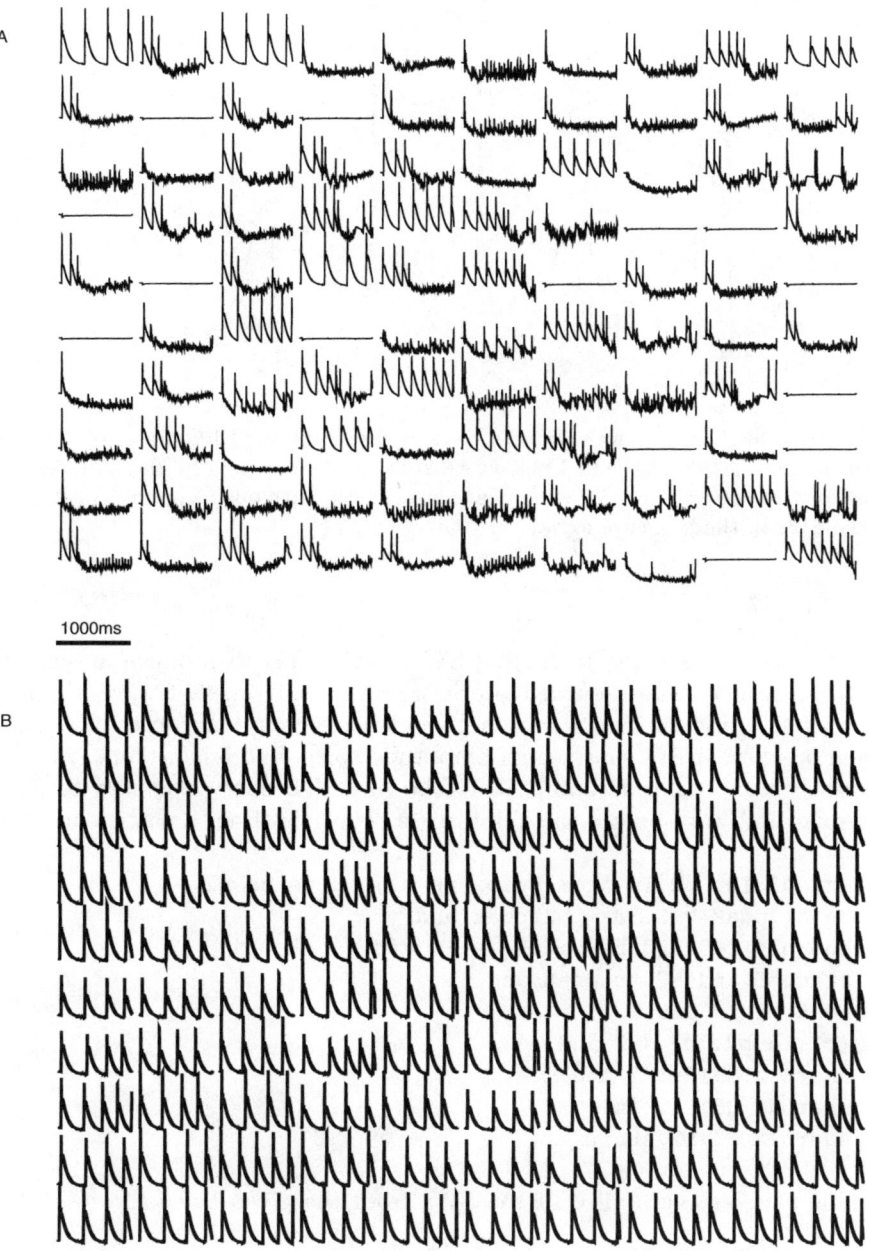

**Fig. 5.7.** Simulation traces. (A) 100 traces of 1000ms duration each, from tonic-clonic simulation described in methods section. There are many different types of patterns in these traces, including SW activity. (B) Some of the traces returned with the NQS *percentile-based* select described. These traces were selected from the data set of 138,240 simulation traces, some of which are shown in A.

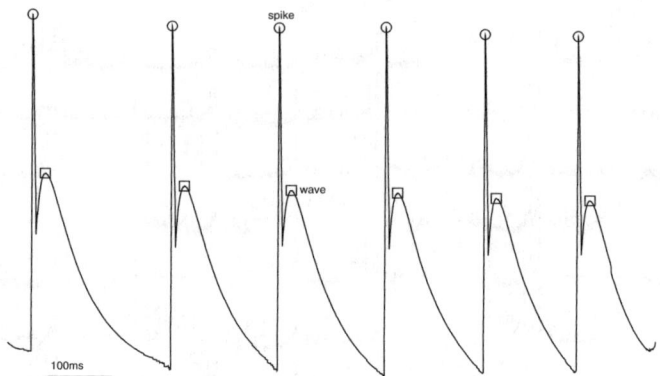

**Fig. 5.8.** This figure shows that all 6 SWs in the simulated EEG trace were returned with the SQL select mentioned below. All spike peaks returned by the SQL select are marked with open circles. All wave peaks are marked with open squares. As shown, all of the SWs in the trace are found.

*SQL SW select*

An alternative approach to finding SW activity was to produce an enhanced NQS/SQL query that returned times of spikes and times of waves. This was done by selecting all bumps from a given trace having two consecutive bumps, *a* and *b*, where *a* is a sharp, high amplitude spike, and *b* is a blunt, wide, and relatively low amplitude bump. The following SQL query was used to find all the spike and wave bumps satisfying these properties from a particular trace:

```
select b1.id, b2.id from bumps as b1, bumps as b2
where b1.peak > 1.33 * b2.peak and
b2.sharp < 200 and b2.width >= 15 and
b1.trace=6 and b2.trace=6 and
b1.id < b2.id and
abs(b1.id-b2.id) < 2 and
b1.sharp > 3312*1.5 and
b2.sharp < 3312/2 and
b1.sharp > 30*b2.sharp;
```

This query returns all 6 of the SWs from the given trace shown above in figure 5.8.

*System performance*

Using the SQL query above, the data-mining system was able to search through the entire bump database, which is on the order of 8e6 bumps, and return results in less than 1ms. This performance level is useful for a fast back and forth between investigator and the data-mining system.

### 5.3.2   Rat Data

*Classification algorithms*

To help find different types of activity patterns in the *in vivo* electrocorticogram recordings of rat epileptiform activity several types of classification algorithms were used on the bumps. One such application was in determining whether a PSW occurred during a PPR response. This involved clustering the time intervals based on their bump content. We also developed an algorithm to determine the time bounds of the PTZ induced seizures by analyzing the bump properties in a trace.

*Clustering with k-means*

Classifying time windows in traces into background activity vs. PSW is useful for detecting the onset and duration of seizures and/or epileptiform activity. We were able to use k-means clustering [14, 8] to classify intervals into two classes. K-means used 4-D vectors consisting of the un-normalized maximal bump width, height, peak, and sharpness from the bumps of a given interval. The time intervals were cut to be 125ms, which is the time between successive strobe flashes. Although PSWs typically occur in between strobe flashes, we did not limit the clustering solely to those intervals so that we would obtain accurate background vs. PSW activity. When we only used the time intervals between strobe flashes, the false positive/negative rates increased. Once the k-means clustering was performed, there were two classes - one with a high amplitude width, height, peak, and sharpness, and one with a significantly lower amplitude of these values. Though k-means has an element of randomness in it, running it thousands of times helps ensure an accurate result. This is done by choosing the cluster centroids that occur the majority of the time as the final cluster centroids. All time intervals are classified accordingly. The final centroids of the two classes came out with these ratios for the PSW/non-PSW values : peak  1.67, width  2.75, sharpness  1.77, height  3.3. It is clear from these ratios that the PSW intervals have significantly higher values of all the properties used for classification. This difference in magnitude of the properties allowed the algorithm to determine which class had PSWs and which did not.

The resultant clusters, shown in fig. 5.9, separating between PSW and non-PSW intervals correlated well with the results of the algorithm described below, providing a measure of cross-validation. The results were also visually inspected for accuracy. Though certain intervals are mis-classified, overall the method does well. Since the distinction between PSW and driving activity is sometimes blurry, errors may be reduced by thresholding the results of a fuzzy k-means algorithm [6]. However, there are certain sharp borders in this classification scheme: based on visual inspection, PSWs detected outside of the strobing windows are always false positives. An NQS select shows that there are 21 PSW cycles detected outside of this time interval (before 3000ms or after 5000ms) out of a total of 243 total PSW cycles detected. That is an approximately 8.6% false positive rate. False negative rates are more difficult to obtain due to the blurry border

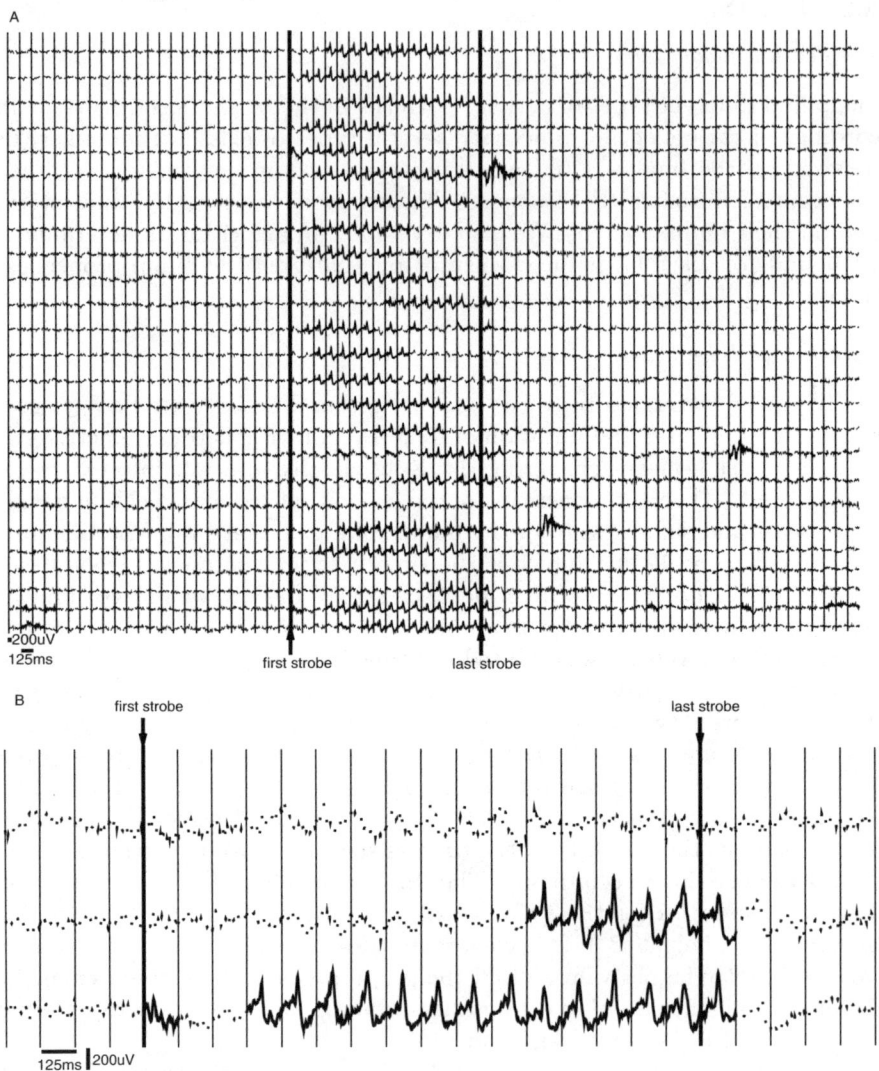

**Fig. 5.9.** K-means clustering on strobe data traces. Vertical lines indicate time windows for bump analysis. Each time window is 125ms. Thick, dark lines represent PSW type activity. Thinner lines indicate absence of PSWs. Length of each trace is 9000ms. PSW activity begins shortly after onset of strobing. Vertical solid lines represent time interval borders for clustering and 8Hz strobe times (between first and last strobe only). (A) Full set of results. (B) Zoom in of portion of traces.

between PSW and non-PSW responses (see discussion). More generally, the clustering of the time intervals may be viewed as another step in the data-mining pipeline, where the researcher filters the data sets to find ever more particular information. Though this method has the additional information of optimal time window length, it may be possible to dynamically determine this based on bump properties for other types of traces.

*Detecting time bounds of seizures*

To detect the time bounds of the PTZ-induced seizures, we could not rely on a clustering algorithm like k-means, because there were no predefined time windows that would be guaranteed to work. Therefore we used a strategy based on the fact that in general, the seizure regions of the trace consisted of *tall* and *wide* bumps with other small bumps in intermediate positions. The presence of seizures could then be indicated by positive deviations from the mean width and height of the bumps in a trace by a certain factor. The first step in detecting the seizure was iterating over the set of bumps and searching for tall and wide bumps. These bumps were designated as *seed* bumps. The qualifications for a seed bump that worked well for the PTZ seizures were that the height be greater than 4.5× the mean height and the width be greater than 3× the mean width. Once the seed bumps were found, we had a rough estimate of where seizure activity was likely to occur. We next had to find the extent of the seizure. To find where the seizure started and ended, we determined positions to the left and

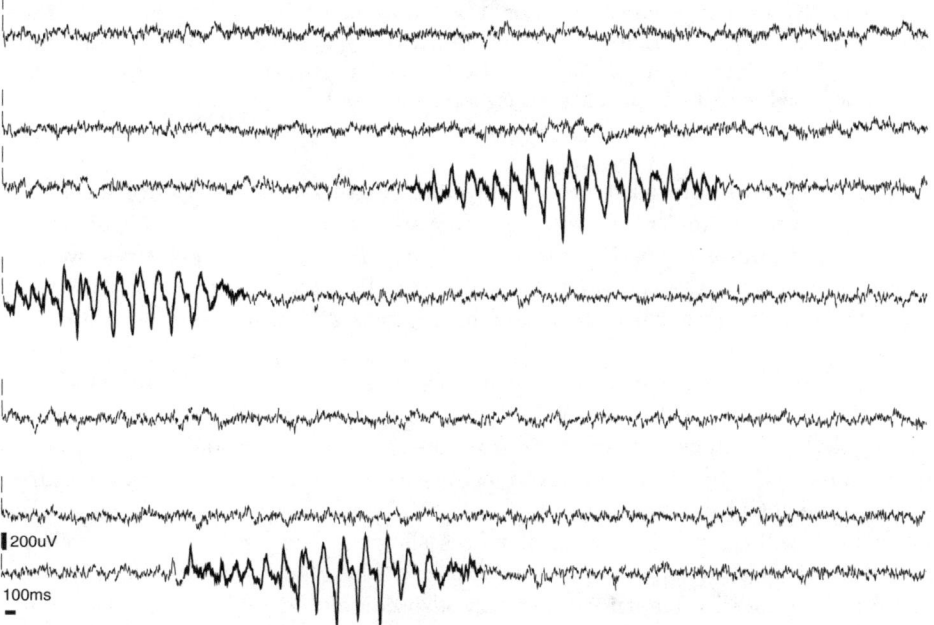

**Fig. 5.10.** Recordings from the rat's right occipital cortex after PTZ administration. Solid, bold black lines indicate extents of seizures detected algorithmically. Each trace is 9000ms.

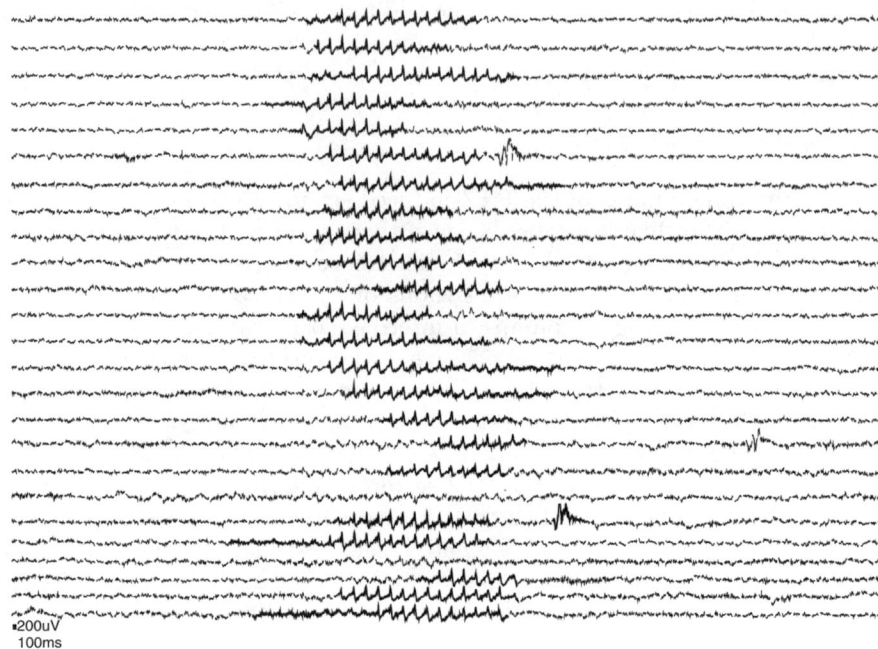

**Fig. 5.11.** Recordings from the rat's right occipital cortex during strobing trials. Dark black regions indicate extents of epileptiform activity detected algorithmically. Each trace is 9000ms. Note high correlation between epileptiform activity detected here and PSWs detected with k-means clustering (same data as Fig. 5.9).

right of the seed bump where the bumps returned to baseline size for a selected time duration (50 ms in the current analysis). This prevents misclassification due to small bumps occurring within a seizure. To avoid false positives, we also had an option of excluding very short seizures. This method was able to extract all of the significant seizures from our initial data set. A few sample results are shown in fig. 5.10.

This algorithm performed well on the photic-induced epileptiform activity as well (results shown in fig. 5.11), but we had to specify different parameters to the algorithm, such as the deviation from mean bump width and height required for a bump to qualify as a *seed* bump, as well as the minimum time span required for baseline activity, which was reduced to 12.5 ms. The fact that the algorithm performed well on various types of traces shows it has usefulness as a general detection algorithm for epileptiform activity. The results of this algorithm on the photic-induced epileptiform activity also correlated highly with the results of the k-means clustering. This lends support to both methods and to the feature extraction algorithm's robustness. The run-time complexity of this algorithm is essentially linear in the number of bumps in the time interval being analyzed, since it essentially traverses the bumps in order.

*Data-mining seizure properties*

Once we had the PTZ seizures extracted we used data-mining to look for different types of correlations between properties of bumps and properties of seizures. For

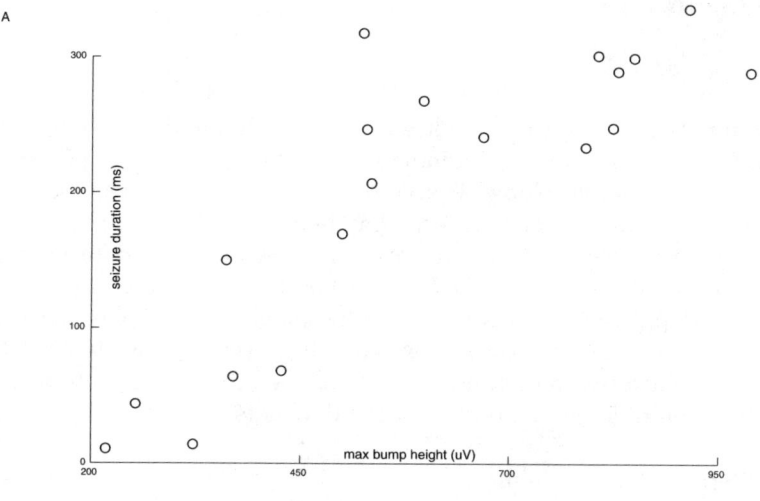

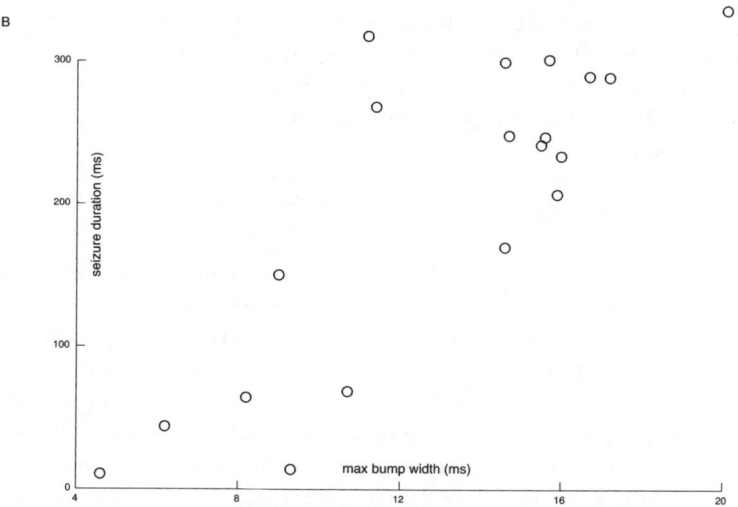

**Fig. 5.12.** Correlations between bump properties during a PTZ-induced seizure and the seizure properties. (A) Correlation between maximum bump height in a PTZ-induced seizure vs. that seizure's duration. Computed correlation coefficient was 0.856. (B) Correlation between maximum bump width in a PTZ-induced seizure vs. that seizure's duration. Computed correlation coefficient was 0.819.

example, we found that the maximum bump height and width of a PTZ-induced seizure had a positive correlation with the duration of the seizure, with values of 0.856 and 0.819 respectively. This is shown in a scatter plot in fig. 5.12.

## 5.4   Discussion

*Data-mining system*

The combination of the feature extraction algorithm and the NQS search software, provides a powerful data-mining tool for finding patterns of interest. The SPUD algorithm we developed is able to efficiently extract features of interest in the time-domain. We have shown that more complicated frequency-domain methods are not always needed. We have also shown that the data-mining framework can be used with various *ad hoc* techniques to interact with the feature database and pull out interesting and physiologically complex patterns. This was done with simple algorithms as well as with intuitive, easy to use NQS/SQL selects. Once patterns were pulled out, it was possible to find interesting correlations between different properties of the data sets.

*Quantification of neural data*

One of the goals of our quantitative analysis is to provide measures that could be used to define what is to be considered a seizure and what is not. There is no gold standard for seizure definition since there is a continuum of electrographic activity appearance resulting in a large grey area where abnormal activity might or might not be defined as epileptiform depending on the criteria used. In order to estimate classification accuracy and error rates in the future we would like to develop several criteria-sets that operate along different dimensions and use these for cross validation.

*Applications using extracted features*

Using the features extracted with the SPUD algorithm it was also possible to determine the time bounds of several different classes of activity including PTZ-induced seizures and photic-induced epileptiform activity. This will have both research and clinical usefulness. Clinical applications may involve a real-time seizure detector that extracts features from recordings of an epileptic patient and detects when they begin with algorithms similar to those here developed. Research applications include searching for the underlying features of bumps occurring pre, post, and during seizures. This may shed light on network and neuronal dynamics leading to seizure genesis. Future work will involve the comparison of different classes of epileptiform activity, *i.e.,* PTZ vs. photic-induced, by quantifying differences in the bumps and their associated properties. We also plan on quantifying and tracking seizure genesis by watching how bump properties change spatio-temporally.

Another future direction is extending the seizure detection algorithms developed to cover a wide class of seizures. Other types of neural activity may also be analyzed using the feature extraction algorithms developed. Future work may entail analyzing the correlations between activity properties in different brain regions as well as spatio-temporal patterns during specific behaviors. This may help uncover the interplay between different brain regions. More generally, the feature extraction methods developed will help organize the vast amounts of neuronal information currently available.

*Frequency-domain measures*

Frequency-domain measures from wavelet or Fourier analysis can be readily added to the databases to provide a still richer information matrix for pattern extraction. This added information may help in understanding neuronal data. It may also reduce false positive/negative rates in pattern extraction algorithms. Note that some of the attributes that we pull out in the time-domain, particularly bump duration and sharpness, are correlates of frequency. This is not a problem in data-mining: we have no need to eliminate redundancy and simply seek to include as many features as may be useful.

*Compression*

Due to the large size of raw trace databases, and potentially limited storage space, it may be desirable to store extracted features in place of full traces for some applications. Corresponding segmental traces could be stored with bump features to minimize loss of information. This can result in an order of magnitude compression in addition to permitting arbitrary activity-pattern searches to be done across large amounts of physiological data. Researchers will be more likely to store their data in central repositories if they can be readily stored and easily accessed.

## 5.5 Download

The software discussed is available at ModelDB:
   http://senselab.med.yale.edu/ModelDB

## Acknowledgments

The authors wish to thank Michael Hines and Ted Carnevale for continuing assistance with NEURON. The simulations and data analysis were performed at the Neurosim lab at SUNY Downstate and the authors would like to thank the Neurosim staff, particularly Larry Eberle, for long hours of advice and assistance. Simulation and data-mining were performed under NIH NS045612. Physiological recording experiments were conducted at the U. of Wisconsin and were supported by NSF grant IOB-0445606.

# References

1. Adeli, H., Zhou, Z., Dadmehr, N.: Analysis of EEG records in an epileptic patient using wavelet transform. J. Neurosci Methods 123(1), 69–87 (2003)
2. Ascoli, G.A., De Schutter, E., Kennedy, D.N.: An information science infrastructure for neuroscience. Neuroinformatics 1(1), 1–2 SPR (2003)
3. Carnevale, N.T., Hines, M.L.: The NEURON Book. Cambridge University Press, Cambridge (2006)
4. Chamberlin, D.D., Boyce, R.F.: SEQUEL: A structured English query language, International Conference on Management of Data. In: Proceedings of the 1974 ACM SIGFIDET (now SIGMOD) workshop on Data description, access and control, Ann Arbor, Michigan, pp. 249–264 (1974)
5. Cooley, J.W., Tukey, J.W.: An algorithm for the machine calculation of complex Fourier series. Math. Comput. 19, 297–301 (1965)
6. Dunn, J.C.: A fuzzy relative of the ISODATA process and its use in detecting compact well-separated clusters. Journal of Cybernetics 3, 32–57 (1973)
7. Golarai, G., Cavazos, J.E., Sutula, T.P.: Activation of the dentate gyrus by pentylenetetrazol evoked seizures induces mossy fiber synaptic reorganization. Brain Res. 593, 257–264 (1992)
8. Hartigan, J.A., Wong, M.A.: A k-means clustering algorithm. Applied Statistics 28, 100–108 (1979)
9. Hines, M.L., Carnevale, N.T.: Expanding NEURON's repertoire of mechanisms with NMODL. Neural Comput. 12(5), 995–1007 (2000) (review)
10. Hines, M.L., Carnevale, N.T.: The NEURON simulation environment. Neural Comput. 9(6), 1179–1209 (1997) (review)
11. Johnson, S.G., Frigo, M.: A modified split-radix FFT with fewer arithmetic operations. IEEE Transactions on Signal Processing 55, 111–119 (2007)
12. Lytton, W.W.: Neural query system: data-mining from within the neuron simulator. Neuroinformatics 4, 163–176 (2006)
13. Lytton, W.W., Omurtag, A.: Tonic-clonic transitions in computer simulation. Journal of Clinical Neurophysiology 24, 175–181 (2007)
14. MacQueen, J.B.: Some methods for classification and analysis of multivariate observations. In: Proceedings of 5-th Berkeley Symposium on Mathematical Statistics and Probability, vol. 1, pp. 281–297. University of California Press, Berkeley (1967)
15. Sweldens, W.: The lifting scheme: A construction of second generation wavelets. SIAM Journal on Mathematical Analysis 29, 511–546 (1997)
16. Uhlrich, D.J., Manning, K.A., O'Laughlin, M.L., Lytton, W.W.: Photic-Induced Sensitization: Acquisition of an Augmenting Spike-Wave Response in the Adult Rat Through Repeated Strobe Exposure. Journal of Neurophysiology 94, 3925–3937 (2005)

# 6

# Analysis of Spectral Data in Clinical Proteomics by Use of Learning Vector Quantizers

Frank-Michael Schleif[1], Thomas Villmann[1], Barbara Hammer[2],
Martijn van der Werff[3], A. Deelder[3], and R. Tollenaar[3]

[1] University Leipzig, Medical Department
{schleif,villmann}@informatik.uni-leipzig.de
[2] Technical University Clausthal, Computer Science Department
hammer@in.tu-clausthal.de
[3] Leiden University Medical Center, Dept. of Surgery/Biomolecular Mass
Spectrometry Unit
M.P.J.van_der_Werff@lumc.nl

**Summary.** Clinical proteomics based on mass spectrometry has gained tremendous visibility in the scientific and clinical community. Machine learning methods are keys for efficient processing of the complex data. One major class are prototype based algorithms. Prototype based vector quantizers or classifiers are intuitive approaches realizing the principle of characteristic representatives for data subsets or decision regions between them. Examples for such tools are Support Vector Machines (SVM) [1], Kohonens Learning Vector Quantization (LVQ) [2], Self-Organizing Map (SOMs) [2], Supervised Relevance Neural Gas (SRNG) [3] and respective variants. Depending on the task one can distinguish between unsupervised methods for data representation and supervised methods for classification. New developments include the utilization of non-standard metrics (functional norms, scaled Euclidean) and task-dependent automatic metric adaptation (feature selection), fuzzy classification, and similarity based visualization of data. These properties offer new possibilities for analysis of mass spectrometric data. In this contribution we concentrate on recent extensions of SOMs as universal tools in the light of clinical proteomics. We focus on non-standard metrics and biomarker patterns discovery. We consider extensions of the standard SOM and LVQ for handling of more general metrics. In particular, we demonstrate applications of the weighted Euclidean metric and the weighted functional norm (based on weighted $L^p$-norm) or kernelized metrics taking the specific nature of mass-spectra into account. This allows an efficient feature selection, which may be used for biomarker identification. The adaptation of the algorithms to these specific requirements leads to effective tools for knowledge discovery keeping the robustness of the original simple approaches. Further we consider fuzzy classification and regression within the determination of clinical proteomics models. This topic deals with the widely ranged problem of uncertainty of data. Particularly in medicine, the classification of mass spectra may be subject of individual human assessment (based on some expert knowledge), multi-impairment diseases, and incomplete patient/proband information. This leads to the problem of uncertainty of training data in machine learning data bases. We developed a semi-supervised approach based on SOM to process such data. As a result the algorithm

T.G. Smolinski et al. (Eds.): Comp. Intel. in Biomed. & Bioinform., SCI 151, pp. 141–167, 2008.
springerlink.com

provides a fuzzy classification scheme based on prototypes for classification of spectra (Fuzzy Labeled SOM - FLSOM).

We demonstrate the usefulness of the above extensions of the basic prototype based data analysis by SOMs to the analysis of mass spectra in proteomics and related knowledge discovery. In particular, we give application examples for biomarker detection based on feature selection and fuzzy classification of spectra combined with similarity based class visualization.

## 6.1   Introduction

Analysis and visualization of clinical proteomic spectra obtained from mass spectrometric measurements is a complicated issue [4] and has been studied by multiple researchers [5, 6, 7, 8, 9, 10, 11]. One major objective is the search for potential biomarkers in complex body fluids like serum, plasma, urine, saliva, or cerebral spinal fluid [12, 13, 14, 15]. Typically the spectra are given as high-dimensional vectors. Thus, from a mathematical point of view, an efficient analysis and visualization of high-dimensional data sets is required. Moreover, the amount of available data is restricted: usually patient cohorts are small in comparison to data dimension. A further problem is that uncertainty in the data may occur. For example, the clinical diagnosis of a patient may be uncertain (fuzzy). Yet, most of machine learning classification models assume strict (crisp) decisions for training data. All these aspects show that classification learning is a crucial task.

The self-organizing map (SOM) constitutes one of the most popular unsupervised approaches for clustering, visualization and data mining of high-dimensional data [2]. SOMs belong to the prototype based methods of data representation. Due to its inherent regularization abilities SOMs are also applicable

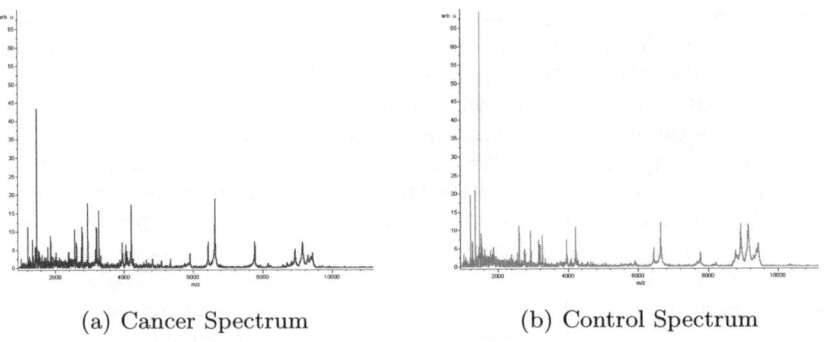

(a) Cancer Spectrum                    (b) Control Spectrum

**Fig. 6.1.** (a) MALDI-TOF spectrum of a colorectal cancer patient and (b) a healthy subject after peptide isolation with C8 magnetic beads. On the Y-axis the relative intensity is shown. The mass to charge ratio (m/z) is demonstrated on the X-axis in Dalton. The spectra are already preprocessed (baseline correction,recalibration) using ClinProTools 2.1.

in case of sparse data sets. Basically, SOMs map the data nonlinearly onto a low-dimensional regular lattice of neurons in a topology-preserving fashion by means of prototype matching, i.e. similar data points are mapped onto nearby or identical neurons under certain conditions [16]. Adaptation takes place as an unsupervised prototype learning. Recently, a semi-supervised counterpart is developed [17]. It allows the determination of a prototype based fuzzy classification model (FLSOM). While the application of (fuzzy) clustering techniques in bioinformatics is not new see e.g. [18] for micoarray analysis or [19] in the field of gene expression an integrated semi-supervised approach like FLSOM has not much been considered in clinical proteomics so far. Especially, in contrast to the widely applied multilayer perceptron [20], prototype based classification allows an easy interpretation, which is of particular interest for many (clinical) applications. Therefore here we focus on *prototype based classifiers* such as the FLSOM, SRNG and SVM whose final models are (much) easier to interpret in the field of clinical proteomics. Beside of a classification model one major objectiv is the identification of relevant features in the original data, to allow the search for potential biomarkers. From a theoretical point of view this problem refers to feature selection for which a wide range of methods have been prosed (see e.g. [21, 22] or [23] for a current overview). Here two types, namely inherent feature selection and some kind of wrapper methods can be identified. The first one tries to identify discriminate features within the model generation process with all data at hand, while the second one tries an indirect identification considering multiple subsets of the features in multiple model generating steps. In this contribution we only consider approaches of the first type with the additional restriction on methods, modifing the metric such that a ranking or weighting of the data features is obtained. This rankings again, allows an easy interpretation of these specific attributes and is sufficiently fast avoiding a large number of model generation step which is a further requirement in the analysis of large data cohorts in the clinical domain.

FLSOM leads to a robust fuzzy classifier where efficient learning of fuzzy labeled or partially contradictory data is possible. Additionally, FLSOM gives the possibility to assess and to visualize *class similarity* by inspection of the generated class map, which represents the label distribution according to the FLSOM lattice structure and the learned class information. However, FLSOM differs from existing extensions of SOM for classification tasks like counterpropagation [24] or Fuzzy SOM [25] fundamentally: In contradiction to these models, for FLSOM the prototype adaptation is also influenced by the class information of the given data such that optimization according to class information is incorporated into the adaptation scheme.

In this contribution, after an introduction of the FLSOM approach and its theory, we apply the algorithm to the problem of classification of mass spectra in case of cancer disease. We show for a data set of colorectal cancer patients and controls, which was also used in a previous study, the successful application of our approach.

## 6.2   Data Analysis by FLSOM

The fuzzy labeled self-organizing map (FLSOM) is a prototype based classification model, which is able to handle fuzzy labeled data (uncertain class decision)
during training and which return fuzzy class decisions during recall. FLSOM is
an extension of the unsupervised self-organizing map (SOM). Therefore, we first
shortly introduce the SOM and thereafter we develop the FLSOM scheme.

### 6.2.1   The Self-organizing Map

As mentioned above, SOMs can be taken as unsupervised learning of topographic
vector quantization with a topological structure (grid) within the set of prototypes (codebook vectors). Roughly speaking, topology preservation thereby
means that similar data points $\mathbf{v} \in V$ are mapped onto identical or neighbored
grid locations which have pointers into the data space (weight vectors). The
principle is depicted in Figure 6.2.

An exact mathematical definition is given in [16]. The weight vectors also are
called prototypes, because they represent parts of the data space.

There exists a wide range of applications in pattern recognition ranging
from spectral image processing to bioinformatics. The mathematics behind the

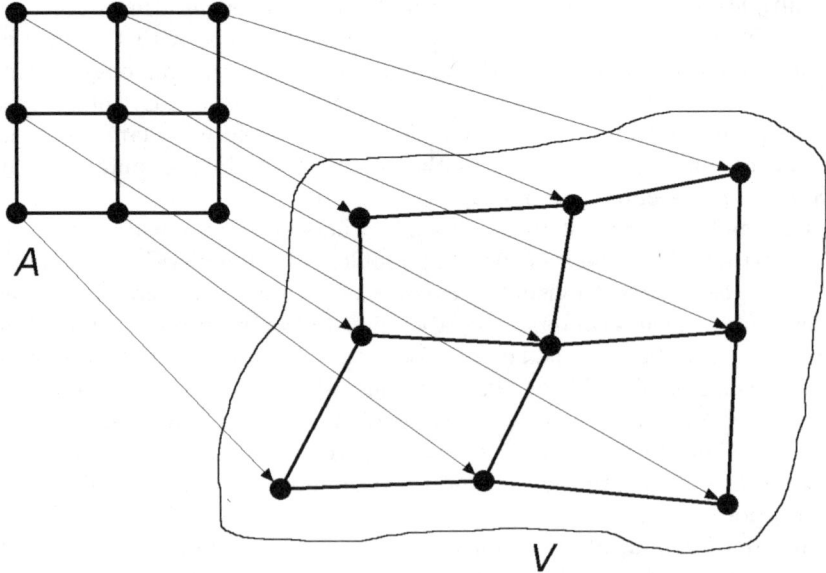

**Fig. 6.2.** The figure shows a mapping of a vector space ($V$) on a rectangular grid
($A$). Prototypes are associated with the rectangular grid by arrows. Multiple points in
the vector space maybe represented by a single prototype which is associated with a
grid position. In case of topological preservation interpretation of the mapping can be
transfered to the potential high dimensional vector space.

original SOM model as proposed by KOHONEN is rather complicated. In particular, the training process does not follow a gradient descent on any cost function
for continuous data distributions [26]. However, HESKES proposed a variant of
the original algorithm which, in practice, leads to at least very similar or identical
results as the original SOM but for which a cost function can be established [27].
We will base our model on this formulation:

Assume that data $\mathbf{v} \in V \subseteq \mathbb{R}^d$ are given distributed according to an underlying distribution $P(V)$. A SOM is determined by a set $A$ of neurons $\mathbf{r}$ equipped
with weight vectors (prototypes) $\mathbf{w_r} \in \mathbb{R}^d$. The neurons are arranged on a lattice structure, which determines the neighborhood relation $N(\mathbf{r}, \mathbf{r}')$ between the
neurons $\mathbf{r}$ and $\mathbf{r}'$. Denote the set of prototypes by $\mathbf{W} = \{\mathbf{w_r}\}_{\mathbf{r} \in A}$. The mapping
description of a trained SOM defines a function

$$\Psi_{V \to A} : \mathbf{v} \mapsto s(\mathbf{v}) = \operatorname*{argmin}_{\mathbf{r} \in A} le(\mathbf{r}) \qquad (6.1)$$

where

$$le(\mathbf{r}) = \sum_{\mathbf{r}' \in A} h_\sigma(\mathbf{r}, \mathbf{r}') \xi(\mathbf{v}, \mathbf{w}_{\mathbf{r}'}) \qquad (6.2)$$

is the local neighborhood weighted error of distances $\xi(\mathbf{v}, \mathbf{w}_{\mathbf{r}'})$. $\xi(\mathbf{v}, \mathbf{w})$ is an
appropriate distance measure, usually the quadratic Euclidean norm $\xi(\mathbf{v}, \mathbf{w_r}) =$
$(\mathbf{v} - \mathbf{w_r})^2$. However, here we only suppose $\xi(\mathbf{v}, \mathbf{w})$ to be arbitrary assuming
differentiability, symmetry and assessing some dissimilarity. The function

$$h_\sigma(\mathbf{r}, \mathbf{r}') = \exp\left(\frac{N(\mathbf{r}, \mathbf{r}')}{2\sigma^2}\right) \qquad (6.3)$$

determines the neighborhood cooperation with range $\sigma > 0$. Large values of $\sigma$
also correspond to high regularization whereas low values ignore this feature. In
this formulation, an input stimulus $\mathbf{v}$ is mapped onto that position $\mathbf{r} \in A$ of the
SOM, the local error $le(\mathbf{r})$ of which is minimum, whereby the average over all
neurons according to the neighborhood is taken. We refer to this neuron $s(\mathbf{v})$ as
the winner.

During the adaptation process a sequence of data points $\mathbf{v} \in V$ is presented to
the map representative for the data distribution $P(\mathcal{V})$. Each time the currently
most proximate neuron $s(\mathbf{v})$ according to (6.1) is determined. All prototypes
are gradually adapted according to the neighborhood degree of the respective
neuron to the winning one by

$$\triangle \mathbf{w_r} = -\epsilon h_\sigma(\mathbf{r}, s(\mathbf{v})) \frac{\partial \xi(\mathbf{v}, \mathbf{w_r})}{\partial \mathbf{w_r}} \qquad (6.4)$$

with a small learning rate $\epsilon > 0$. This adaptation follows a stochastic gradient
descent of the cost function introduced by HESKES [27]:

$$E_{\text{SOM}} = \frac{1}{2C(\sigma)} \int P(\mathbf{v}) \sum_{\mathbf{r}} \delta_{\mathbf{r}}^{s(\mathbf{v})} \sum_{\mathbf{r}'} h_\sigma(\mathbf{r}, \mathbf{r}') \xi(\mathbf{v}, \mathbf{w}_{\mathbf{r}'}) d\mathbf{v} \qquad (6.5)$$

were $C(\sigma)$ is a constant which we will drop in the following, and $\delta_{\mathbf{r}}^{\mathbf{r}'}$ is the usual Kronecker symbol checking the identity of $\mathbf{r}$ and $\mathbf{r}'$.

One main aspect of SOMs is the visualization ability of the resulting map due to its topological structure. Under certain conditions the resulting non-linear projection $\Psi_{V\to A}$ generates a continuous mapping from the data space $V$ onto the grid structure on $A$. This mapping can mathematically be interpreted as an approximation of the principal curve or its higher-dimensional equivalents [28]. Thus, as pointed out above, similar data points are projected on prototypes which are neighbored in the grid space $A$. Further, prototypes neighbored in the lattice space should code similar data properties, i.e. their weight vectors should be close together in the data space according to the dissimilarity measure $\xi$. This property of SOMs is called topology preserving (or topographic) mapping realizing the mathematical concept of continuity. For a detailed and mathematical exact consideration of this topic we refer to [16]. Successful tools for assessing this map property are the topographic function and the topographic product [16], [29].

### 6.2.2   Fuzzy Labeled SOM (FLSOM)

SOM is a well-established model for nonlinear data visualization which, due to its above mentioned topology preserving properties, can also serve as an adequate preprocessing step for data completion, representation or interpolation. The formulation of the adaptation scheme in terms of a gradient descent of a cost function allows an extension to a semi-supervised learning scheme which leads to a classification model. The resulting FLSOM is able to handle uncertainty in class assignments of training data as well as to return fuzzy classification decision in the recall phase. It differs from simple post labeling or separate post-learning of prototype labels as it takes place in counter propagation [24] or Fuzzy-SOM [25] in this way that in FLSOM the prototype adaptation is influenced by the class information. We now explain the model in detail.

Let $N(c)$ be the number of possible data classes. We assume that each training point $\mathbf{v}$ now is equipped with a label vector $\mathbf{x} \in \mathbb{R}^{N(c)}$ whereby each component $x_i \in [0,1]$ determines the soft assignment of $\mathbf{v}$ to class $i$ for $i = 1, \ldots, N(c)$. Hence, we can interpret the label vector as probabilistic or possibilistic fuzzy class memberships. Accordingly, we enlarge each prototype vector $\mathbf{w_r}$ of the map by a label vector $\mathbf{y_r} \in [0,1]^{N(c)}$ which determines the portion of neuron $\mathbf{r}$ assigned to the respective classes. During training, prototype locations $\mathbf{w_r}$ and label vectors $\mathbf{y_r}$ are adapted according to the given labeled training data. For this purpose, we extend the cost function of the SOM as defined in (6.5) to a cost function for fuzzy-labeled SOM (FLSOM) by a term $E_{\mathrm{FL}}$ assessing classification accuracy. Thus the cost function becomes

$$E_{\mathrm{FLSOM}} = (1 - \beta)\, E_{\mathrm{SOM}} + \beta E_{\mathrm{FL}} \tag{6.6}$$

where the factor $\beta \in [0,1]$ is a *balance factor*, which determines the influence of both aspects, the data representation by usual SOM and the classification accuracy. For the classification accuracy term we chose

$$E_{\text{FL}} = \frac{1}{2} \sum_{\mathbf{r}} \int P(\mathbf{v}) \cdot ce(\mathbf{v}, \mathbf{r}) \, dv \qquad (6.7)$$

with *local, weighted classification errors*

$$ce(\mathbf{v}, \mathbf{r}) = g_\gamma(\mathbf{v}, \mathbf{w_r}) \cdot \vartheta(\mathbf{x}(\mathbf{v}), \mathbf{y_r}). \qquad (6.8)$$

$g_\gamma(\mathbf{v}, \mathbf{w_r})$ is a Gaussian kernel defining a neighborhood range in the data space:

$$g_\gamma(\mathbf{v}, \mathbf{w_r}) = \exp\left(-\frac{\xi(\mathbf{v}, \mathbf{w_r})}{2\gamma^2}\right). \qquad (6.9)$$

The value $\vartheta(\mathbf{x}(\mathbf{v}), \mathbf{y_r})$ describes the dissimilarity of the label vectors $\mathbf{x}$ and $\mathbf{y_r}$. Usually, the squared Euclidean distance $\vartheta(\mathbf{x}(\mathbf{v}), \mathbf{y_r}) = (\mathbf{x} - \mathbf{y_r})^2$ is chosen. However, as in the case for the dissimilarity in the data space, other definitions are possible.

This choice of the classification accuracy term $E_{\text{FL}}$ as a sum of weighted data space distances is based on the assumption that data points, close to a prototype $\mathbf{w_r}$, determine the corresponding label, if the underlying class distribution is sufficiently smooth. Note that the kernel $g_\gamma(\mathbf{v}, \mathbf{w_r})$ depends on the prototype locations, such that the classification term $E_{\text{FL}}$ is influenced by both $\mathbf{w_r}$ and $\mathbf{y_r}$. Hence, the gradient of $E_{\text{FL}}$ with respect to $\mathbf{w_r}$ is non-vanishing and yields

$$\frac{\partial E_{\text{FL}}}{\partial \mathbf{w_r}} = -\frac{1}{4\gamma^2} \int P(\mathbf{v}) \cdot g_\gamma(\mathbf{v}, \mathbf{w_r}) \cdot \frac{\partial \xi(\mathbf{v}, \mathbf{w_r})}{\partial \mathbf{w_r}} \cdot \vartheta(\mathbf{x}(\mathbf{v}), \mathbf{y_r}) \, dv \qquad (6.10)$$

which contribute to the overall gradient by

$$\frac{\partial E_{\text{FLSOM}}}{\partial \mathbf{w_r}} = (1 - \beta) \cdot \frac{\partial E_{\text{SOM}}}{\partial \mathbf{w_r}} + \beta \cdot \frac{\partial E_{\text{FL}}}{\partial \mathbf{w_r}} \qquad (6.11)$$

Thus the complete prototype update becomes

$$\triangle \mathbf{w_r} = -\epsilon(1 - \beta) \cdot h_\sigma(\mathbf{r}, s(\mathbf{v})) \frac{\partial \xi(\mathbf{v}, \mathbf{w_r})}{\partial \mathbf{w_r}} \qquad (6.12)$$

$$+ \epsilon\beta \frac{1}{4\gamma^2} \cdot g_\gamma(\mathbf{v}, \mathbf{w_r}) \cdot \frac{\partial \xi(\mathbf{v}, \mathbf{w_r})}{\partial \mathbf{w_r}} \cdot \vartheta(\mathbf{x}(\mathbf{v}), \mathbf{y_r}).$$

The gradient of $E_{\text{FLSOM}}$ with respect to the label determines the adaptation rule for the prototype labels. Because $E_{\text{SOM}}$ is independent on the prototype labels the respective derivative vanishes. We obtain the update rules by taking the derivatives: Labels are only influenced by the second part $E_{\text{FL}}$, which yields

$$\frac{\partial E_{\text{FLSOM}}}{\partial \mathbf{y_r}} = \frac{\partial E_{\text{FL}}}{\partial \mathbf{y_r}} \qquad (6.13)$$

and the corresponding learning rule therefore is

$$\triangle \mathbf{y_r} = \epsilon_l \beta \cdot g_\gamma(\mathbf{v}, \mathbf{w_r})(\mathbf{x} - \mathbf{y_r}) \qquad (6.14)$$

with learning rate $\epsilon_l > 0$. This learning scheme can be seen as a weighted average of the data fuzzy labels of those data $\mathbf{v}$ close to the associated prototype $\mathbf{w_r}$.

### 6.2.3    Topography and Label Distribution in FLSOM

As mentioned above, unsupervised SOMs generate a topographic mapping from
the data space onto the prototype grid under specific conditions. If the classes
are consistently determined with respect to the varying data, one can expect for
supervised topographic FLSOMs that the labels become ordered within the grid
structure of the prototype lattice. In this case the topological order of the prototypes should be transferred to the topological order of prototype labels such that
we have a smooth change of the fuzzy class label vectors between neighbored grid
positions $\mathbf{r}$. This is the consequence of following fact: the neighborhood function
$h_\sigma(\mathbf{r}, \mathbf{s})$ of the usual SOM learning (6.4) forces the topological ordering of the
prototypes. In FLSOM, this ordering is further influenced by the weighted classification error $ce(\mathbf{v}, \mathbf{r})$ (6.8). This classification error term contains the kernel
$g_\gamma(\mathbf{v}, \mathbf{w_r})$, eq. (6.9). Hence, the prototype learning and ordering (6.12) receives
information of both data and class distribution. For high value of the balancing
parameter $\beta$ the latter term becomes dominant. Otherwise, the kernel $g_\gamma(\mathbf{v}, \mathbf{w_r})$
also triggers the label learning (6.14), which is, of course, also dependent on
the underlying learned prototype distribution and ordering. Thus, a consistent
ordering of the labels is obtained in FLSOM.

Hence, the evaluation of the similarities between the prototype label vectors
yields suggestions for the similarity of classes, i.e. similar classes are represented
by prototypes in a local spatial area of the SOM lattice. In case of overlapping
class distributions this topographic class processing leads to prototypes with unclear decision, located between prototypes with clear vote. Further, if classes are
not distinguishable, there will exist prototypes responsive to those data which
have class label vectors containing approximately the same degree of class membership for the respective classes. In this way FLSOM may be used for class
similarity detection.

## 6.3    Data Preprocessing by Wavelet Analysis

The analysis of functional data, is a common task in bioinformatics. Spectral
data as obtained from mass spectrometric measurements in clinical proteomics
are such functional data leading to new challenges for an appropriate analysis.
Here we focus on the determination of classification models for such data. In
general the available approaches for this task initially transform the spectra into
a vector space followed by training a classifier. Hereby the functional nature of
the data is typically lost, which may lead to suboptimal classifier models. Taking
this into account a wavelet encoding is applied onto the spectral data leading
to a compact *functional* representation. Thus, a functional representation of the
data with respect to the used metric and a weighting or pruning of especially
(priory not known) irrelevant function parts of the inputs, would be desirable.
Further feature selection is applied based on a statistical pre-analysis of the
data. Hereby a discriminative data representation is necessary. The extraction
of such discriminant features is crucial for spectral data and typically done by a

parametric peak picking procedure. This peak picking is often focus of criticism because peaks may be insufficiently detected and the functional nature of the data is partially lost. To avoid this difficulties we focus on the approach as given in [30] and apply a wavelet encoding to the spectral data to get discriminative features. The obtained wavelet coefficients are sufficient to reconstruct the signal, still containing all relevant information of the spectra. However this better discriminating set of features is typically more complex and hence a robust approach to determine the desired classification model is needed.

The classification of mass spectra involves in general the two steps peak picking to locate and quantify positions of peaks within the spectrum and feature extraction from the obtained peak list. In the first step a number of procedures as baseline correction, optional denoising, noise estimation and normalization must be applied [31, 32]. Upon these prepared spectra the peaks have to be identified by scanning all local maxima and the associated peak endpoints followed by a S/N thresholding such that one obtains the desired peak list.

The procedure of baseline correction and recalibration (alignment) of multiple spectra is standard, and has been done using ClinProTools in this paper (details in [31])[1]. Here we propose an alternative feature extraction procedure preserving all (potentially small) peaks containing relevant information by use of the discrete wavelet transformation (DWT). The feature extraction has been done by Wavelet analysis using the Matlab Wavelet-Toolbox[2]. Wavelet Analysis is a effective tool in signal processing to encoded and analyse functional signals. The signal is encoded (analysis step) and reconstructed (synthesis step) by a series of parametrized basis functions (daughter wavelets) which are derived from a specific type of function (mother wavelet), with specific mathematical constrains. The principle is described e.g. in [33, 34, 35]. A simple example of a wavelet based signal reconstruction using the Haar wavelet and a more advanced type of wavelet is given in Figure 6.3. As explained in [35] the wavelet analysis allows the respresentation of the signal at multiple resolutions. The signal is encoded on different scalings of the wavelets leading to a fine or more sparse representation of the signal. The sparser the encoding the less information of the original signal is preserved. This can be considered as some kind of compression realized by a low- and high-pass filter approach. Similar like in the classic Fourier analysis the high frequencies encode detail information and the lower frequencies encode the raw structure of the signal.

The usage of wavelet analysis for feature extraction on MS data is a new growing field of research in the bioinformatics domain [36] and offers interesting alternatives to former approaches. Vannucci et al [9] used wavelet analysis to get a discriminant set of features from SELDI-TOF measurements. They used Daubechies wavelets with four vanishing moments (db4) to extract wavelet components from the measurements. In their procedure the large number of features was reduced using probit models with Bayesian methods as a special kind of combined feature selection and classification. They applied the method in the

---

[1]  Biomarker software available at http://www.bdal.de

[2]  The Matlab Wavelet-Toolbox can be obtained from www.mathworks.com

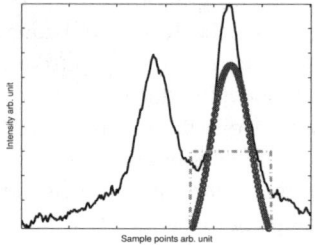
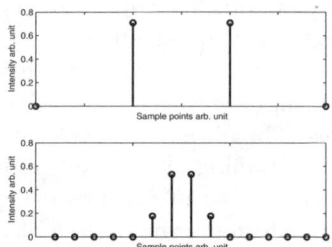

(a) Zoom into spectrum with peak integration. Dashed line: Haar scaling function. Dotted line: biorthogonal bior3.7 scaling function as a weighting function of a generalized quadrature formula.

(b) Reconstruction low-pass filter of Haar(top) and bior3.7(bottom) scaling function.

**Fig. 6.3.**

classification of ovarian cancer data taken from SELDI-TOF measurements and obtained an almost perfect prediction model. Morris et al [8] used the wavelet analysis to denoise spectra taken from MALDI-TOF cancer measurements with a subsequent peak picking on the average spectrum. Thus they used wavelet analysis in a more classical way for denoising and a subsequently feature extraction upon the prepared spectra.

For the second step of feature extraction they applied a peak detection algorithm on the average spectrum of the wavelet-denoised spectra. The final features were defined as the maximum log intensity of the determined peaks and used for further analysis. Another approach using discrete wavelet transformation (DWT) on MS data for two class experiments was recently proposed by Yu et al [10]. They applied a DWT (using a db4 wavelet) on SELDI-TOF ovarian cancer data and used a binning algorithm for an initial reduction of the calculated approximation coefficients (ac). To obtain a further reduction of the coefficients they tuned their classification model incorporating the Kolmogorov-Smirnov-Test with cross-validation by use of a Support Vector Machine classifier (SVM). A similar approach was already taken by Zhu et al [37] where the DWT (with db4) has been applied on MS data followed by a simple feature selection criterion to reduce the number of obtained wavelet coefficients. The statistically motivated criterion considers the between class distance of the obtained wavelet coefficients which gives a rank for each feature and can be used for thresholding. Finally they used a tree based classifier to obtain the final classification model.

Approaches published so far were mainly applied to SELDI-TOF data, are applicable mostly for two classes only and lack a detailed explanation of the underlying processing steps as a whole and the derivation of the process parameters in detail. Especially data pre-treatment is a crucial issue since it may effect all subsequent process steps.

Due to the local analysis property of wavelet analysis the features can still be related back to original mass position in the spectral data which is essential for further biomarker analysis. This effect and the good encoding properties of wavelets makes them preferable in encoding of spectral data. This is especially true with respect to standard alternatives such as the (windowed) Fourier Analysis (FT) which does not or in case of windowed FT not sufficiently allow for a back tracking of the features in the original data. In a first step a feature selection procedure using the Kolmogorov-Smirnoff test (KS-test) was applied. The test was used to identify features which show a significant ($p < 0.01$) discrimination between the two groups (cancer,control). This is done in accordance to [38] were also a generation to a multiclass experiment is given.

### 6.3.1 Feature Extraction and Denoising by Bi-orthogonal Discrete Wavelet Transform

Wavelets have been developed as powerful tools [35, 39] used for noise removal and data compression. The discrete version of the continuous wavelet transform leads to the concept of a multiresolution analysis (MRA). This allows a fast and stable wavelet analysis and synthesis. The analysis becomes more precise if the wavelet shape is adapted to the signal to be analyzed. For this reason one can apply the so called bi-orthogonal wavelet transform [40] which uses two pairs of scaling and wavelet functions. One is for the decomposition/analysis and the other one for reconstruction/synthesis. The advantage of the bi-orthogonal wavelet transform is the higher degree of freedom for the shape of the scaling and wavelet function. In our analysis such a smooth synthesis pair was chosen to avoid artifacts. It can be expected that a signal in the time domain can be represented by a small number of a relatively large set of coefficients from the wavelet domain. The spectra are reconstructed in dependence of a certain approximation level $L$ of the MRA which can be considered as a hard-thresholding. The denoised spectrum looks similar to the reconstruction as depicted in Figure 6.4. The starting point for an argumentation is the simplest example of a MRA which can be defined by the characteristic function $\chi_{[0,1)}$. The corresponding wavelet is the so-called *Haar* wavelet. Assume that the denoised spectrum $f \in L_2(\mathbb{R})$ has a peak with endpoints $2^j k$ and $2^j (k + 1)$, the integral of the peak can be written as

$$\int_{2^j k}^{2^j (k+1)} f(t)dt = \int_{\mathbb{R}} f(t)\chi_{[2^j k, 2^j (k+1))}(t)dt$$

Obviously the right hand side is the Haar DWT scaling coefficient $c_{j,k} = \langle f, \psi_{j,k} \rangle$ at scale $a = 2^j$ and translation $b = 2^j k$. One obtains approximation- and detail-coefficients [40]. The approximation coefficients describe a generalized peak list of the denoised spectrum encoding primal spectral information and depend on the level $L$ which is determined with respect to the measurement procedure. For linear MALDI-TOF spectra a device resolution of $500 - 800 Da$ can be expected. This implies limits to the minimal peak width in the spectrum and hence, the reconstruction level of the Wavelet-Analysis should be able to model corresponding peaks. Another point is the typical mass range used in clinical proteomics

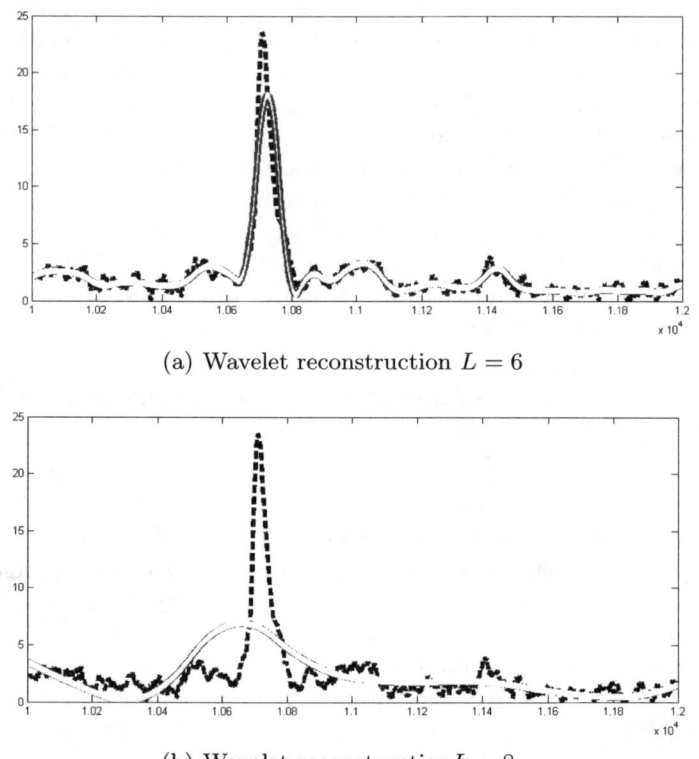

(a) Wavelet reconstruction $L = 6$

(b) Wavelet reconstruction $L = 8$

**Fig. 6.4.** Wavelet reconstruction of the spectra with $L = 6, 8$, $x$ measurement positions, $y$-arbitrary unit. The original signal is plotted with the interrupted line (blue) and the reconstruction with the solid with a white band inside. One observes that a wavelet analysis with $L = 8$ (and 7 as well) is to rough to approximate the sharp peaks.

studies. In general the measurements for linear MALDI-TOF measurements are given in a range of $1 - 10kDa$ such that extremly low or heavy molecules are not present. In case of this mass range the generic change of the peak width, which is increasing with higher masses can be sufficiently modeled by a single wavelet approximation level. A level $L = 4$ is typically sufficient for a linear measured spectrum with $\approx 20000$ measurement points, a level of $L = 6$ has been used for the data with $\approx 65000$ measurement points. (see Figure 6.4). The level $L$ can be automatically determined by considering expected peak width in $Da$ and the reconstruction capabilities of wavelet analysis at a given level. Alternatively multiple levels can be tried and a standard peak picking approach can be applied on both, the original and the reconstructed spectrum. If the obtained peak lists are sufficiently similar, by means, that at least peaks with good S/N values in the original spectrum are sufficiently recovered in the reconstruction the taken level can be considered as acceptable for the experiment. Applying this procedure including the KS-test on the spectra with an initial number of

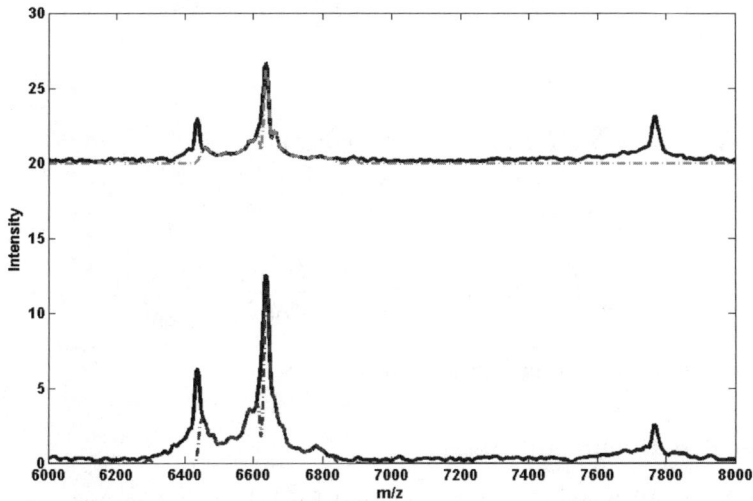

**Fig. 6.5.** Reconstructed region of some spectra of the two classes top control, bottom cancer. The straight lines indicate the reconstruction of the spectra by use of the chosen Wavlet approximation level upon approximation coefficients. The dotted line indicates the same reconstruction but with pruned coefficients which did not pass the statistical test. One observes that regions which are clearly non informative (near to the noise spectrum) are removed but also non-discriminating peaks (by means of the statistical test) are pruned.

$\approx 65000$ measurement points per spectrum one obtains 1036 wavelet coefficients used as representative features per spectrum, still allowing a reliable functional representation of the data. An application of the KS-Test still keeps 199 coefficients for the final analysis. The effect of the KS-Test selection on the wavelet encoded spectra is shown in Figure 6.5.

## 6.4 Classification Dependent Metric Adaptation - Relevance Learning

As mentioned above, the general dissimilarity measure $\xi\left(\mathbf{v}, \mathbf{w_r}\right)$ for the data space $V$ is often chosen as squared Euclidean metric such that the derivative $\frac{\partial \xi(\mathbf{v}, \mathbf{w_r})}{\partial \mathbf{w_r}}$ simply becomes $-2(\mathbf{v} - \mathbf{w_r})$. Yet, other measures also can be applied, for example correlation measures [41]. However, more flexibility is obtained if $\xi\left(\mathbf{v}, \mathbf{w_r}\right)$ is parametrized and the parameters are also subject of optimization according to the given classification task [42], [3].

Generally, we consider a parametrized distance measure $\xi^\lambda(\mathbf{v}, \mathbf{w})$ with a parameter vector $\lambda = (\lambda_1, \dots, \lambda_M)$ with $\lambda_i \geq 0$ and normalization $\sum_i \lambda_i = 1$. Then classification task depending parameter optimization is achieved by gradient descent, i.e. by consideration of $\frac{\partial E_{\mathrm{FLSOM}}}{\partial \lambda_l}$. Formal derivation yields

$$\frac{\partial E_{\text{FLSOM}}}{\partial \lambda_l} = (1 - \beta) \frac{\partial E_{\text{SOM}}}{\partial \lambda_l} + \beta \frac{\partial E_{\text{FL}}}{\partial \lambda_l} \tag{6.15}$$

with

$$\frac{\partial E_{\text{SOM}}}{\partial \lambda_l} = \frac{1}{2} \sum_{\mathbf{r}} \int P(\mathbf{v}) \cdot \delta_{\mathbf{r}}^{s(\mathbf{v})} \sum_{\mathbf{r}'} h_\sigma(\mathbf{r}, \mathbf{r}') \cdot \frac{\partial \xi^\lambda(\mathbf{v}, \mathbf{w_r})}{\partial \lambda_l} d\mathbf{v} \tag{6.16}$$

and

$$\frac{\partial E_{\text{FL}}}{\partial \lambda_l} = -\frac{1}{4\gamma^2} \sum_{\mathbf{r}} \int P(\mathbf{v}) \cdot g_\gamma(\mathbf{v}, \mathbf{w_r}) \cdot \frac{\partial \xi^\lambda(\mathbf{v}, \mathbf{w_r})}{\partial \lambda_l} \cdot \vartheta\left(\mathbf{x}(\mathbf{v}), \mathbf{y_r}\right) d\mathbf{v} \tag{6.17}$$

for the respective parameter adaptation.

### 6.4.1   Scaled Euclidean Metric

In case of $\xi^\lambda(\mathbf{v}, \mathbf{w})$ being the *scaled* squared Euclidean metric

$$\xi^\lambda(\mathbf{v}, \mathbf{w}) = \sum_i \lambda_i (v_i - w_i)^2 \tag{6.18}$$

(with $\lambda_i \geq 0$ and $\sum_i \lambda_i = 1$) the derivative becomes $\frac{\partial \xi(\mathbf{v}, \mathbf{w}_i)}{\partial \mathbf{w}_i} = -2 \cdot \mathbf{\Lambda} \cdot (\mathbf{v} - \mathbf{w}_i)$ with $\mathbf{\Lambda}$ is a diagonal matrix and its $i$-th diagonal entry is $\lambda_i$. The corresponding learning rule for the metric parameter $\lambda_l$ has the form

$$\triangle \lambda_l = -\epsilon_\lambda \frac{1 - \beta}{2} \sum_{\mathbf{r}} h_\sigma(s(\mathbf{v}), \mathbf{r}) \cdot (v_l - (w_{\mathbf{r}})_l)^2 \tag{6.19}$$

$$+\epsilon_\lambda \frac{\beta}{4\gamma^2} \sum_{\mathbf{r}} g_\gamma(\mathbf{v}, \mathbf{w_r}) \cdot (v_l - (w_{\mathbf{r}})_l)^2 \cdot \vartheta\left(\mathbf{x}(\mathbf{v}), \mathbf{y_r}\right) \tag{6.20}$$

(subscript $l$ denoting the component $l$ of a vector) with learning rate $\epsilon_\lambda > 0$. This update is followed by normalization to ensure $\lambda_i \geq 0$ and $\sum_i \lambda_i = 1$.

The parameter optimization of the *scaled* squared Euclidean metric allows a useful interpretation. The parameter $\lambda_i$ is weighting the dimensions of the data space. Hence, optimization of these parameters in dependence on the classification problem leads to a ranking of the input dimensions according to their classification decision relevance. Therefore, metric parameter adaptation of the scaled Euclidean metric is called *relevance learning*. In case of zero-valued $\lambda_i$ this can also be seen as feature selection.

### 6.4.2   Generalized $L^p$-Norm

As pointed out before, the similarity measure $d^\lambda(\mathbf{v}, \mathbf{w})$ is only required to be differentiable with respect to $\lambda$ and $\mathbf{w}$. The triangle inequality has not to be fulfilled necessarily. This leads to a great freedom in the choice of suitable measures

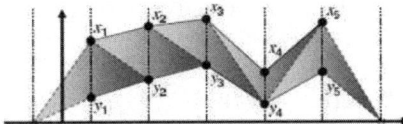 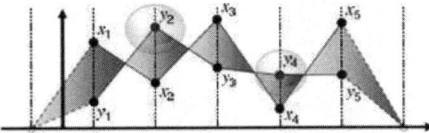

(a) Two functions: Euclidean $= L^p$-norm    (b) Two functions: Euclidean $\neq L^p$-norm

**Fig. 6.6.** Schematical ilustration of the $L^p$-norm. The first plot (a) indicates the case where the distance between two functions is equal considering Euclidean or $L^p$-norm. In the plot (b) parts of the functions are interchanging (crossings) whereby the distances using Euclidean distance is still the same as within plot (a) but for the $L^p$-norm another distance is obtained which indeed gives a more realistic measure of the distances of the two functions.

and allows the usage of non-standard metrics in a natural way. We now review the functional metric as given in [43], the obtained derivations can be plugged into the above equations leading to FLSOM with a functional metric, whereby the data are functions represented by vectors and, hence, the vector dimensions are spatially correlated.

Common vector processing does not take the spatial order of the coordinates into account. As a consequence, the functional aspect of spectral data is lost. For proteome spectra the order of signal features (peaks) is due to the nature of the underlying biological samples and the measurement procedure. The masses of measured chemical compounds are given ascending and peaks encoding chemical structures with a higher mass follows chemical structures with lower masses. In addition multiple peaks with different masses may encode parts of the same chemical structure and hence are correlated.

LEE proposed a distance measure taking the functional structure into account by involving the previous and next values of $x_i$ in the $i$-th term of the sum, instead of $x_i$ alone. Assuming a constant sampling period $\tau$, the proposed norm is:

$$\mathcal{L}_p^{fc}(\mathbf{v}) = \left( \sum_{k=1}^{D} \left( A_{k-1}(\mathbf{v}) + A_{k+1}(\mathbf{v}) \right)^p \right)^{\frac{1}{p}} \tag{6.21}$$

with

$$A_k(\mathbf{v}) = \begin{cases} \frac{\tau}{2}|v_k| & \text{if } 0 \le v_k v_{k-1} \\ \frac{\tau}{2}\frac{v_k^2}{|v_k|+|v_{k-1}|} & \text{if } 0 > v_k v_{k-1} \end{cases} \quad B_k(\mathbf{v}) = \begin{cases} \frac{\tau}{2}|v_k| & \text{if } 0 \le v_k v_{k+1} \\ \frac{\tau}{2}\frac{v_k^2}{|v_k|+|v_{k+1}|} & \text{if } 0 > v_k v_{k+1} \end{cases} \tag{6.22}$$

are respectively of the triangles on the left and right sides of $x_i$. Just as for $L_p$, the value of $p$ is assumed to be a positive integer. At the left and right ends of the sequence, $x_0$ and $x_D$ are assumed to be equal to zero. The concept of the $L^p$-norm is shown in Figure 6.6.

The derivatives for the functional metric taking $p = 2$ are given in [43]. Now we consider the scaled functional norm where each dimension $v_i$ is scaled by a parameter $\lambda_i > 0$ $\lambda_i \in (0,1]$ and $\sum_i \lambda_i = 1$. Then the scaled functional norm is:

$$\mathcal{L}_p^{fc}(\lambda \mathbf{v}) = \left( \sum_{k=1}^{D} (A_{k-1}(\lambda \mathbf{v}) + A_{k+1}(\lambda \mathbf{v}))^p \right)^{\frac{1}{p}} \qquad (6.23)$$

with

$$A_k(\lambda \mathbf{v}) = \begin{cases} \frac{\tau}{2}\lambda_k |v_k| & \text{if } 0 \le v_k v_{k-1} \\ \frac{\tau}{2}\frac{\lambda_k^2 v_k^2}{\lambda_k |v_k| + \lambda_{k-1}|v_{k-1}|} & \text{else} \end{cases} \quad B_k(\lambda \mathbf{v}) = \begin{cases} \frac{\tau}{2}\lambda_k |v_k| & \text{if } 0 \le v_k v_{k+1} \\ \frac{\tau}{2}\frac{\lambda_k^2 v_k^2}{\lambda_k |v_k| + \lambda_{k+1}|v_{k+1}|} & \text{else} \end{cases} \quad (6.24)$$

The prototype update for $p = 2$ changes to:

$$\frac{\partial \delta_2^2(\mathbf{x},\mathbf{y},\lambda)}{\partial x_k} = \frac{\tau^2}{2}(2 - U_{k-1} - U_{k+1})(V_{k-1} + V_{k+1})\triangle_k \qquad (6.25)$$

with

$$U_{k-1} = \begin{cases} 0 & \text{if } 0 \le \triangle_k \triangle_{k-1} \\ \left( \frac{\lambda_{k-1}\triangle_{k-1}}{\lambda_k |\triangle_k| + \lambda_{k-1}|\triangle_{k-1}|} \right)^2 & \text{else} \end{cases} , U_{k+1} = \begin{cases} 0 & \text{if } 0 \le \triangle_k \triangle_{k+1} \\ \left( \frac{\lambda_{k+1}\triangle_{k+1}}{\lambda_k |\triangle_k| + \lambda_{k+1}|\triangle_{k+1}|} \right)^2 & \text{else} \end{cases}$$

$$V_{k-1} = \begin{cases} \lambda_k & \text{if } 0 \le \triangle_k \triangle_{k-1} \\ \frac{\lambda_k |\triangle_k|}{\lambda_k |\triangle_k| + \lambda_{k-1}|\triangle_{k-1}|} & \text{else} \end{cases} , V_{k+1} = \begin{cases} 1\lambda_k & \text{if } 0 \le \triangle_k \triangle_{k+1} \\ \frac{\lambda_k |\triangle_k|}{\lambda_k |\triangle_k| + \lambda_{k+1}|\triangle_{k+1}|} & \text{else} \end{cases}$$

and $\triangle_k = x_k - y_k$ For the $\lambda$-update one observes:

$$\frac{\partial \mathcal{L}_p^{fc}(\lambda \mathbf{v})}{\partial \lambda_k} = \frac{\partial \left( \sum_{k=1}^{D}(A_k(\lambda \mathbf{v}) + B_k(\lambda \mathbf{v}))^p \right)^{\frac{1}{p}}}{\partial \lambda_k}$$

$$= p\left( \sum_{k=1}^{D}(A_{k-1}(\lambda \mathbf{v}) + A_{k+1}(\lambda \mathbf{v}))^p \right)^{\frac{1-p}{p}} \frac{\partial \left[ \sum_{k=1}^{D}(A_k(\lambda \mathbf{v}) + B_k(\lambda \mathbf{v}))^p \right]}{\partial \lambda_k}$$

$$= C_p \frac{\partial \left[ \sum_{k=1}^{D}(A_k(\lambda \mathbf{v}) + B_k(\lambda \mathbf{v}))^p \right]}{\partial \lambda_k}$$

$$= C_p \frac{\sum_{k=1}^{D} \partial \left[ (A_k(\lambda \mathbf{v}) + B_k(\lambda \mathbf{v}))^p \right]}{\partial \lambda_k}$$

$$= C_p \frac{\partial \left[ (A_{k-1}(\lambda \mathbf{v}) + B_{k-1}(\lambda \mathbf{v}))^p + (A_k(\lambda \mathbf{v}) + B_k(\lambda \mathbf{v}))^p + (A_{k+1}(\lambda \mathbf{v}) + B_{k+1}(\lambda \mathbf{v}))^p \right]}{\partial \lambda_k}$$

$$= C_p \left( c_p^{k-1} \frac{\partial [A_{k-1}(\lambda \mathbf{v}) + B_{k-1}(\lambda \mathbf{v})]}{\partial \lambda_k} + c_p^k \frac{\partial [A_k(\lambda \mathbf{v}) + B_k(\lambda \mathbf{v})]}{\partial \lambda_k} + * \right)$$

$$* = c_p^{k+1} \frac{\partial [A_{k+1}(\lambda \mathbf{v}) + B_{k+1}(\lambda \mathbf{v})]}{\partial \lambda_k}$$

with the following expressions

$$c_p^j = p \cdot (A_j(\lambda \mathbf{v}) + B_j(\lambda \mathbf{v}))^{p-1}$$

$$= p \cdot \left( \begin{cases} \frac{\tau}{2}\lambda_j |v_j| & \text{if } 0 \le v_j v_{j-1} \\ \frac{\tau}{2}\frac{\lambda_j^2 v_j^2}{\lambda_j |v_j| + \lambda_{j-1}|v_{j-1}|} & \text{if } 0 > v_j v_{j-1} \end{cases} + \begin{cases} \frac{\tau}{2}\lambda_j |v_j| & \text{if } 0 \le v_j v_{j+1} \\ \frac{\tau}{2}\frac{\lambda_j^2 v_j^2}{\lambda_j |v_j| + \lambda_{j+1}|v_{j+1}|} & \text{if } 0 > v_j v_{j+1} \end{cases} \right)^{p-1}$$

putting all together and with some minor mathematical transformations one obtains:

$$\frac{\partial \mathcal{L}_p^{fc}(\lambda \mathbf{v})}{\partial \lambda_k} = C_p \begin{cases} 0 + c_p^k \left( \frac{\tau}{2} |v_k| \right) & \text{if } 0 \leq v_{k-1} v_k \\ \frac{1}{2} \tau \frac{\lambda_k^2 c_p^k v_k^2 |v_k| - c_p^{k-1} |v_k| v_{k-1}^2 \lambda_{k-1}^2 + 2\lambda_k c_p^k v_k^2 |v_{k-1}| \lambda_{k-1}}{\left( \lambda_k |v_k| + |v_{k-1}| \lambda_{k-1} \right)^2} & \text{if } 0 > v_{k-1} v_k \end{cases}$$

$$+ C_p \begin{cases} c_p^k \left( \frac{\tau}{2} |v_k| \right) + 0 & \text{if } 0 \leq v_{k+1} v_k \\ \frac{1}{2} \tau \frac{\lambda_k^2 c_p^k v_k^2 |v_k| - c_p^{k+1} |v_k| v_{k+1}^2 \lambda_{k+1}^2 + 2\lambda_k c_p^k v_k^2 |v_{k+1}| \lambda_{k+1}}{\left( \lambda_k |v_k| + |v_{k+1}| \lambda_{k+1} \right)^2} & \text{if } 0 > v_{k+1} v_k \end{cases}$$

Using this parametrization one can emphasize/neglect different parts of the function for classification. This distance measure can be put into FLSOM as shown above and has been applied subsequently in the analysis of clinical proteome spectra.

### 6.4.3  Matrix Approach

The metric becomes even more powerful by assigning an individual weight vector $\lambda^j$ to each prototype $\mathbf{w_j}$ (see e.g. [44])

Recently, we have extended the weighted euclidean metric $d_\lambda$ by introducing a full matrix $\Lambda \in \mathbb{R}^{n \times n}$ of relevance factors in the distance measure [45]. The metric has the form

$$d_\Lambda(\mathbf{w}, \mathbf{v}) = (\mathbf{v} - \mathbf{w})^T \Lambda (\mathbf{v} - \mathbf{w})$$

This approach allows to account for correlations between different input features. A set of points equidistant from a prototype can have the shape of a rotated ellipsoidal, whereas the relevance vector $\lambda$ in FLSOM only results in a scaling parallel to the coordinate axis.

For the distance measure $d_\Lambda$ to be well defined, the matrix $\Lambda$ has to be positive (semi-) definite. For this reason, $\Lambda$ is substituted by $\Lambda = \Omega \Omega^T$ with $\Omega \in \mathbb{R}^{n \times n}$. The adaptation formula for $\mathbf{w}$ and $\Omega$ in vector notation is given as:

$$\frac{\partial \xi(\mathbf{v}, \mathbf{w_r})}{\partial \mathbf{w_r}} = -\Lambda(\mathbf{v} - \mathbf{w})$$

$$\frac{\partial \xi(\mathbf{v}, \mathbf{w_r})}{\partial \lambda} = (\Lambda(\mathbf{v} - \mathbf{w})(\mathbf{v} - \mathbf{w})^T)^T + \Lambda(\mathbf{v} - \mathbf{w})(\mathbf{v} - \mathbf{w})^T)$$

Note that we can assume $\Omega^\top = \Omega$ without loss of generality and that the symmetry is preserved under the above update. After each update step $\Lambda$ has to be normalized to prevent the algorithm from degeneration. It is enforced that $\sum_i \Lambda_{ii} = 1$ by dividing all elements of $\Lambda$ by the raw value of $\sum_i \Lambda_{ii}$. In this way the sum of diagonal elements is fixed which coincides with the sum of eigenvalues here. Due to the huge number of parameters it is in general useful to consider a band matrix in the training procedure. This extension of FLSOM is named Matrix FLSOM (MFLSOM).

By attaching local matrices $\Lambda^j$ to the individual prototypes $\mathbf{w_j}$, ellipsoidal isodistances with different widths and orientations can be obtained. The algorithm based on this more general distance measure is called Localized MFLSOM (LFLSOM).

## 6.5   Clinical Data

Serum protein profiling is a promising approach for classification of cancer versus non-cancer samples. The data used in this paper are taken from a colorectal cancer (CRC) study. It contains of measurements taken from cancer patients and healthy individuals [11].

The standardized circumstances for sample collection and the data set are described in detail in [11]. Here it should be mentioned only that for each profile a mass spectrum is obtained within a mass-to-charge-ratio of 1000 to 11000Da. Two sample spectra are depicted in 6.1. The data have been preprocessed as explained before using the approach published in [30]. The spectra are encoded by $\approx 200$ wavelet-coefficients which leads to a data reduction of $\approx 99.9\%$ using the rawdata and is twice the number of peaks as obtained by the standard peak picking approach as proposed in [31]. The preprocessing stage has to be included in the crossvalidation procedure to avoid overfitting, for the considered data set it could be observed that the discriminating wavelet coefficients (with respect to the ks-test) at $p \leq 0.01$ remain the same in a 5−fold cross validation. The wavelet method was used as mentioned in the previous section with $L = 6$.

The data set consist of 123 samples whereby 73 are taken from patients suffering from colorectal cancer and the remaining 50 samples are taken from a matched healthy control group[3]. Colorectal cancer is among the most common malignancies and remains a leading cause of cancer-related morbidity and mortality. It is well recognized that CRC arises from a multistep sequence of genetic alerations that result in the transformation of normal mucosa to aprecursor adenoma and ultimately to carcinoma. Given the natural history of CRC, early diagnosis appears to be the most appropriate tool to reduce disease-related mortality. Currently, there is no early diagnostic test with sufficient diagnostic quality, which can be used as a routine screening tool. Therefore, there is a need for new biomarkers for colorectal cancer that can improve early diagnosis, monitoring of disease progression and therapeutic response and detect disease recurrence. Furthermore, these markers may give indications for targets for novel therapeutic strategies. In addition to potential markers validated by further post analysis on identified masses, generic classification models with high validity maybe of value as well.

## 6.6   Experiments

The available data set for investigation consists of overall 123 proteomic expression profiles generated by MALDI-TOF mass spectrometry (MS) labeled into two classes. We consider a data set which is measured in the correct context of MALDI-MS and clinical proteomics and which is generated in accordance to proven best clinical practices [11]. This data set takes also recent research results on data collection and standardization in clinical proteomics [46, 12] into

---

[3] In the article of [11] some additional selections with respect to the cancer group has been done - here we work on the whole data set.

account, which stengthens the validity of the analyzed data in contrast to some other public available data sets[4].

Single parts of the priorly presented processing e.g. the FLSOM method on its own, have been applied on other data sets already [41]. This has been done also in case of multiple classes [41, 48], which are currently not easily available in clinical proteomics, although multiple class experiments are more and more evolving.

In [11] an experimental setting was shown focusing on Fishers Linear Discriminant Analysis (LDA) combined with a principal component (PCA) approach to reduce the dimensionality of the underlying problem with promising results. Here, the peak picking was avoided by a simple binning approach and the PCA was used to get a sufficient reduction of the dimensionality of the feature space. PCA is focusing on maximal explained variance in the data [49], this however is typically not a good criteria in the analysis of clinical proteome spectra because the group separations are in general not indicated by large variations in the intensities [32]. Hence a PCA approach will in general fail to give sufficient results. Although the PCA got sufficient results in [11] a more generic approach for the analysis of clinical proteome spectra taken form MALDI-MS is desirable.

Such an approach is to determine the decision plane with respect to the known class label information which is pointed out by multiple authors e.g. [50, 51]. Taking these into account we focus on a supervised data analysis and reduce the dimensionality of the data by use of a problem specific wavelet analysis combined with a statistical selection criterium. we avoid statistical assumptions with respect to the underlying data sets, but take only measurement specific knowledge into account.

Hence we have a 199-dimensional space of wavelet coefficients and we use multiple algorithms and metrics to determine classification models. We focus of the presented FLSOM algorithm which beside a classification model leads to a (under some constraints) topological preserving visualization of these high dimensional data.

To be comparable with the study in [11], we trained in a first investigation a FLSOM with data only of the groups A and B. We used a $7 \times 3$ FLSOM lattice, the size of which is determined by a growing SOM (GSOM) [52]. GSOM generates optimum hyper-cubical SOM lattice structures obtained by evaluation of the local receptive fields of the prototypes to achieve a topologic preserving SOM mapping [53]. GSOMs seems to be more robust than the alternative of e.g. ART networks [54] which sensitively depends on a proper choice of the vigilance parameter to be estimated by the user. The balancing parameter was declared as $\beta = 0.85$, which emphasizes the classification term in (6.12) but prevents instabilities for higher values [55]. To be generalizing and regularizing we used the inherent regularization abilities of SOMs by non-vanishing neighborhood range $\sigma$ in the neighborhood function $h_\sigma$ in (6.3). To do so and to prevent violations in topology preservation the remaining value was chosen as $\sigma = 0.5$ [56]. In case

---

[4] There is an ongoing discussion on that topic and on former obtained results and data sets see e.g. [7, 47].

of a topogical preserving map clear class separations can be identified by a clear labeling and potential empty fields between different groups on the map. Due to the class similarity attribute of the FLSOM sub groups of the data can be identified in this way and data with unclear class assignment are made visible by a similar fuzzy class labeling and a corresponding identifiable region on the map.

## 6.7   Results

A typical FLSOM obtained from multiple runs is depicted in Figure 6.9. One observes a clear separation between cancer and control data. The overlapping region between the classes is rather small which is also supported by the relative good crossvalidation results for the linear classification models. For this data set the obtained FLSOM using different metrics are topological preserving. The FLSOM approaches obtained $\approx 86\%$ cross validation accuracy in a 5-fold cross-validation, using scaled Euclidean metric which is a similar good accuracy as in [11]. In addition to the good classification accuracy a ranking of individual features as well as a planar visualization of the high dimensional data is obtained. The latter one allows for interpretation of similarities between sub groups of patients (see Figure 6.9)

The relevance parameters $\lambda_i$ of the scaled Euclidean metric are adapted parallely. This leads to a ranking of the input dimensions according to their importance for classification. A typical relevance profile using scaled Euclidean metric is depicted in Figure 6.7. The most important frequencies are indicated by straight arrows in Fig. 6.8, dashed arrows refer to further highly relevant frequencies. The depicted frequencies contribute substantially to classification accuracy and, therefore, are important for distinction of the classes.

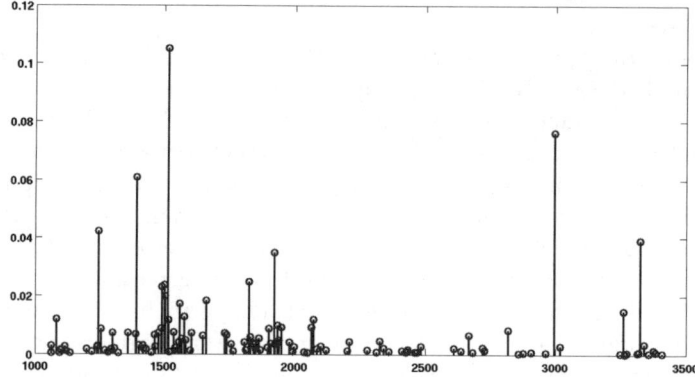

**Fig. 6.7.** Visualization of a typical relevance profile obtained by FLSOM using scaled Euclidean metric. Peaks with larger values indicate higher relevance with respect to the classification task. The x-axis indicates the relative mass position of the corresponding wavelet coefficient in the original spectrum. The y-axis is a relevance measure $\in [0, 1]$.

A comparison of the FLSOM results using the different metrics and alternative algorithms is given in Table 6.1. It should be noted, that in [11] a part of the cancer class spectra has been removed from the model generation due to quality constraints, while in our analysis all spectra have been used. The lower three rows of the table contain results obtained on alternative data preparations, namely peaklists (CPT results) and the preparation as given in [11]. In [11] a leave one out (LOO) cross validation has been used to determine the generalization ability of the approach, LOO is a restriction which is typical for small data sets. LOO however has some drawbacks as pointed out in [57, 58, 59]. We used a 5-fold CV in accordance to the suggestions in [57] because the number of sample is not so small and they are reliable homogeneous per group as depicted in Figure 6.8.

One observes that the results are competitive with respect to other classifiers but it should be mentioned again that FLSOM is not focusing on classification but equally on visualization and interpretability of the given high dimensional data sets. In that way classification accuracy as well as a modeling of the data distribution is optimized. In average the different methods obtain a cross validation accuracy of ≈ 89% using the presented generic preprocessing approach. The wavelet prepared data perform similar than a standardized peak picking approach with other parameters fixed. The approach in [11] obtained slightly better results in the LOO cross validation but is too much focused on explained variance which can not be generalized to other clinical proteomics problems in general. Considering the cross validation results in Table 6.1 it can be observed, that similar results were obtained using the different metrics. However the metrics itself show different properties. The relevance profile of the scaled Euclidean metric indicates most important data features in a univariate interpretation whereas the generalized $L^p$ norm takes local neighborhoods or correlations in the data space into account while keeping the functional nature of the MS spectra. Therefore also descents in the function and not just peaks as well as correlative

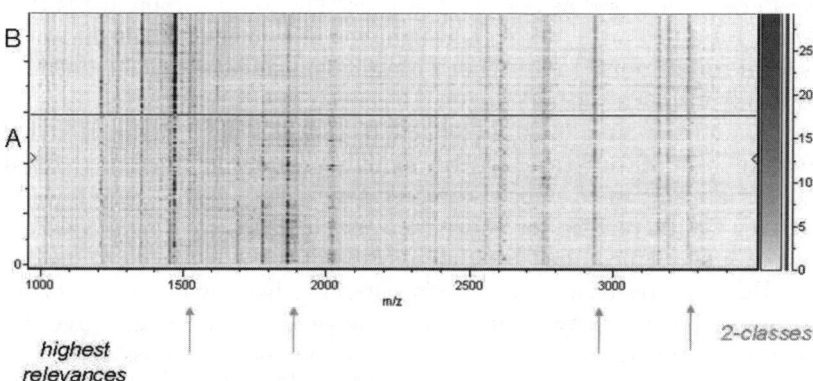

**Fig. 6.8.** A gel view of the two classes with the cancer class (region A) and control class (region B). The relevant mass positions are indicated by arrows (bottom) using the relevance profile of FLSOM with scaled Euclidean metric.

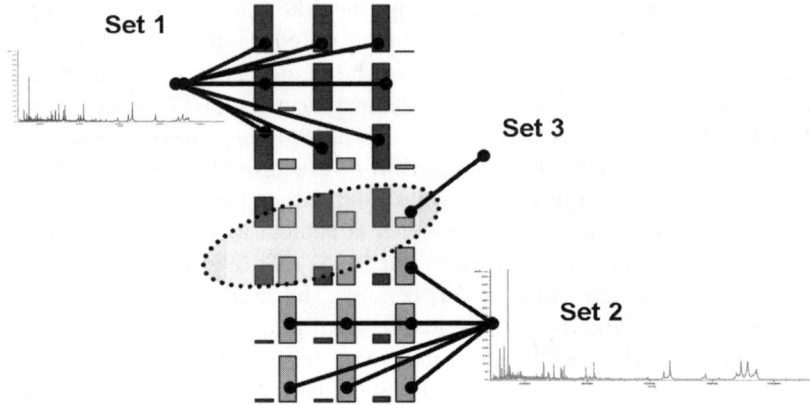

**Fig. 6.9.** Visualization of the FLSOM using the $L^p$ metric. The FLSOM consists of $7 \times 3$ cells with each cell containing two bars indicating a fuzzy labeling. The first bar is responsible for the cancer class, while the second for control. A high bar for cancer indicates that spectra which are mapped to the considered cell are more likely to belong to cancer than to control. A clear separation of the two classes with a small overlap region can be observed. For each spectrum in the data set an associated cell on the grid by the SOM mapping can be identified. A raw analysis shows three sets of spectra. Set $1, 2$ contains quite homogenous spectra of the corresponding classes while the spectra in set 3 show multiple inhomogenities e.g. some of the cancer spectra show a bad S/N ratio for peaks and are in overall more noisy. There are also some spectra which show strong fluctuation in the intensities. Considering the mapping as well as the fuzzy label of the corresponding map a specific clustering of the high dimensional data is obtained. In case of multiple classes this further leads to a similarity highlighting of the different classes.

effects can be interpreted as relevant features. The Matrix approach is the most generic used metric and is able to highlight the relevance of single dimensions as well as local correlations. In 6.1 results are shown using the matrix metric with a limited bandwidth. The primal diagonal has been weighted by 1, the direct neighbors by 0.5 and the remaining diagonals are pruned out. The functional metric has indicated alternative regions with similar separation capability. Relatively small peaks are identified which, combined with the neighborhood are indeed informative. Characteristic for those regions identified in the considered data is, that not a single peak has been identified but a trace of a local biochemical pattern. Here the pattern typically consists of a peak with moderate intensities and small but not perfect differences between the two classes and a valley close to the peak with a quite clear (but also not perfect) missing of mass information for one class. This valley could not be identified as a peak by a peak picking procedure because the region has no peak characteristic, nevertheless it could be observed that for one class at this valley mass intensities has been measured whereas for the other class the intensities are zero or very low. This trace of information can be further analysed by e.g. MS/MS techniques to test if a

**Table 6.1.** Recognition and cross validation accuracies for FLSOM using different similarity measures in comparison to alternative standard approaches. The results for LDA/PCA are taken from the article [11]. It should be noted, that in [11] a part of the cancer class spectra has been removed from the model generation due to quality constraints, while in our analysis all spectra have been used. The lower three rows of the table contain results obtained on alternative data preparations, namely peaklists (CPT results) and the preparation as given in [11]. The approach available in CPT with SVM+kNN first determines a ranking of the peaks by interpretation of the weight vector of a linear SVM. In a second step a kNN classifier is trained on the best peaks.

| Method | Rec. | CV - 5 fold |
|---|---|---|
| FLSOM-EUC | 89.62% | 86.12% |
| FLSOM-$L^p$ | 89.23% | 86.17% |
| FLSOM-M | 83.74% | 87.94% |
| SRNG-EUC | 100% | 90.24% |
| SVM-Linear | 96.75% | 89.43% |
| SVM-kNN (CPT)-LOO | 96.58% | 92.52% |
| SVM-kNN (CPT)-5CV | 96.58% | 87.84% |
| LDA+PCA -LOO | 92.9% | 92.6% |

potential useful pattern can be observed which in the current linear measurement has not been sufficiently resolved so far.

The respective learned data distribution using FLSOM with the $L^p$-norm is depicted in Fig.6.9 Each square represents a label vector $\mathbf{y_r}$ of a prototype $\mathbf{w_r}$. The position is according localization $\mathbf{r}$ in the grid. The height of the bars reflects the fuzzy amount for the respective class as indicated above. These findings are in agreement with clinical expectations. We observe the fine conformity of the detected class similarities with the clinical expectations. Hence, FLSOM successfully discovered the underlying class structure. It should be noted, that the FLSOM gives a similar but a bit worse predicition rate in comparison to the other algorithms using a 5-fold CV, but in addition to a prediction model a topology of the data has been obtained and could be interpreted. This is not directly possible by use of SRNG or SVM. Another point is that in case of SVM the model consists of extreme cases or data points which describe the class boundaries whereas for SRNG and FLSOM the model is given by prototypes which forms local classifier models considering there receptive field. This allows a fine granular interpretation of patient data with respect to the model and, in case of FLSOM, additionally with respect to the map.

## 6.8  Conclusions

We presented an extension of the SOM for supervised classification tasks, which takes classification task explicitly into account for prototype adaptation during the gradient descent based adaptation process. Each prototype dynamically adapt its assigned class label depending on the balancing between clustering and classification in the FLSOM model. In this way the statistical as well as label properties of the data influence prototype positions and fuzzy label learning. The visualization abilities of SOMs based on the topology preservation property of unsupervised SOMs then can be used for visual inspection of the class labels of the prototypes which may allow a better understanding of the underlying classification decision scheme. Further, the FLSOM is able to detect class or sub group similarities as shown in the experimental section.

The FLSOM has been applied to the classification of mass spectrometric data (profiles) of cancer disease and controls. Beside a comparable classification accuracy the model automatically discovered the class similarities in good agreement to clinical expectations. Thereby it could be recovered that the data labeled as cancer and control do not form dense sets for each group but are overlapping. The overlapping region forms in fact a new sub group of the data. Samples belonging to this group has been manually reanalyzed with respect to the original spectra and found to be of either bad quality or somehow specific with respect to the anamnesis data.

Hence, the FLSOM allows a more specific interpretation of the classification models, by interpreting clinical or patient specific findings with respect to its representing prototype of the FLSOM. Thus, FLSOM can be used not only for classification and visualization but also for detection of class dependencies. This effect becomes even more apparent for multi class data sets as already shown in [41]. Further, if only partially labeled data are available, FLSOM can be taken as a semi-supervised learning approach.

## Acknowledgment

The authors are grateful to M. Kostrzewa and T. Elssner for useful discussions and support in interpretation of the results (both Bruker Daltonik Leipzig, Germany. Further, we would like to give a special acknowledge to the valuable comments of reviewer one.).

## References

1. Vapnik, V.: Statistical Learning Theory. Wiley, New York (1998)
2. Kohonen, T. (ed.): Self-Organizing Maps, Springer Series in Information Sciences, vol. 30. Springer, Berlin (1995) (2nd Ext. Ed. 1997)
3. Hammer, B., Strickert, M., Villmann, T.: Supervised neural gas with general similarity measure. Neural Proc. Letters 21(1), 21–44 (2005)
4. Pusch, W., Flocco, M., Leung, S., Thiele, H., Kostrzewa, M.: Mass spectrometry-based clinical proteomics. Pharmacogenomics 4, 463–476 (2003)
5. Petricoin, E., Ardekani, A., Hitt, B., et al.: Use of proteomic patterns in serum to identify ovarian cancer. Lancet 359, 572–577 (2002)
6. Wulfkuhle, J., Petricoin, E., Liotta, L.: Proteomic applications for the early detection of cancer. Nat. Rev. Cancer 3, 267–275 (2003)
7. Ransohoff, D.: Lessons from controversy: ovarian cancer screening and serum proteomics, J. Natl. Cancer Inst. 97, 315–319 (2005)

8. Morris, J.S., Coombes, K.R., Koomen, J., Baggerly, K.A., Kobayashi, R.: Feature extraction and quantification for mass spectrometry in biomedical applications using the mean spectrum. Bioinformatics 21(9), 1764–1775 (2005)
9. Vannucci, M., Sha, N., Brown, P.J.: Nir and mass spectra classification: Bayesian methods for wavelet-based feature selection. Chem. and Int. Lab Systems 77, 139–148 (2005)
10. Yu, J.S., Ongarello, S., Fiedler, R., et al.: Ovarian cancer identification based on dimensionality reduction for high-throughput mass spectrometry data. Bioinformatics 21(10), 2200–2209 (2005)
11. de Noo, M., Deelder, A., van der Werff, M., zalp, A., Martens, B.: MALDI-TOF serum protein profiling for detection of breast cancer. Onkologie 29, 501–506 (2006)
12. Fiedler, G., Baumann, S., Leichtle, A., Oltmann, A., Kase, J., Thiery, J., Ceglarek, U.: Standardized peptidome profiling of human urine by magnetic bead separation and matrix-assisted laser desorption/ionization time-of-flight mass spectrometry. Clinical Chemistry 53(3), 421–428 (2007)
13. Schäffeler, E., Zanger, U., Schwab, M., et al.: Magnetic bead based human plasma profiling discriminate acute lymphatic leukaemia from non-diseased samples. In: 52nd ASMS Conference. TPV 420 (2004)
14. Schipper, R., Loof, A., de Groot, J., Harthoorn, L., van Heerde, W., Dransfield, E.: Salivary protein/peptide profiling with seldi-tof-ms. Annals of the New York Academy of Science 1098, 498–503 (2007)
15. Guerreiro, N., Gomez-Mancilla, B., Charmont, S.: Optimization and evaluation of seldi-tof mass spectrometry for protein profiling of cerebrospinal fluid. Proteome science 4, 7 (2006)
16. Villmann, T., Der, R., Herrmann, M., Martinetz, T.: Topology Preservation in Self–Organizing Feature Maps: Exact Definition and Measurement. IEEE Transactions on Neural Networks 8(2), 256–266 (1997)
17. Schleif, F.M., Elssner, T., Kostrzewa, M., Villmann, T., Hammer, B.: Analysis and Visualization of Proteomic Data by Fuzzy labeled Self Organizing Maps. In: Proc. of CBMS 2006, pp. 919–924 (2006)
18. Wang, J., Bo, T.H., Jonassen, I., Myklebost, O., Hovig, E.: Tumor classification and marker gene prediction by feature selection and fuzzy c-means clustering using microarray data. BMC Bioinformatics 4, 60 (2003)
19. Arima, C., Hanai, T., Okamoto, M.: Gene expression analysis using fuzzy k-means clustering. Genome Informatics 14, 334–335 (2003)
20. Bishop, C.: Pattern Recognition and Machine Learning. Springer, Science+Business Media, LLC, New York (2006)
21. Pudil, P., Novovicova, J.: Floating search methods in feature selection. Pattern Recognition Letters 15, 1119–1125 (1994)
22. Somol, P., Pudil, P.: Adaptive floating search methods in feature selection. Pattern Recognition Letters 20, 1157–1163 (1999)
23. Guyon, I., Gunn, S., Nikravesh, M., Zahed, L.A.: Feature Extraction - Foundations and Applications. Springer, Heidelberg (2006)
24. Hecht-Nielsen, R.: Counterprogagation networks. Appl. Opt. 26(23), 4979–4984 (1987)
25. Vuorimaa, P.: Fuzzy self-organizing map. Fuzzy Sets and Systems 66(2), 223–231 (1994)
26. Erwin, E., Obermayer, K., Schulten, K.: Self-organizing maps: Ordering, convergence properties and energy functions. Biol. Cyb. 67(1), 47–55 (1992)
27. Heskes, T.: In: Oja, E., Kaski, S. (eds.) Kohonen Maps, pp. 303–316. Elsevier, Amsterdam (1999)

28. Hastie, T., Stuetzle, W.: Principal curves. J. Am. Stat. Assn. 84, 502–516 (1989)
29. Bauer, H.U., Pawelzik, K.R.: Quantifying the neighborhood preservation of Self-Organizing Feature Maps. IEEE Trans on Neural Networks 3(4), 570–579 (1992)
30. Schleif, F.M., Hammer, B., Villmann, T.: Supervised Neural Gas for Functional Data and its Application to the Analysis of Clinical Proteom Spectra. In: Sandoval, F., Prieto, A.G., Cabestany, J., Graña, M. (eds.) IWANN 2007. LNCS, vol. 4507, pp. 1036–1044. Springer, Heidelberg (2007)
31. Ketterlinus, R., Hsieh, S.Y., Teng, S.H., Lee, H., Pusch, W.: Fishing for biomarkers: analyzing mass spectrometry data with the new clinprotools software. Bio techniques 38(6), 37–40 (2005)
32. Schleif, F.M.: Prototype based Machine Learning for Clinical Proteomics. Ph.D. Thesis, Technical University Clausthal, Technical University Clausthal, Clausthal-Zellerfeld, Germany (2006)
33. Daubechies, I.: Ten lectures on wavelets. In: CBMS-NSF Regional Conference Series in Applied Mathematics, Philadelphia, PA. Society for Industrial and Applied Mathematics (SIAM), vol. 61 (1992)
34. Mallat, S.: A wavelet tour of signal processing. Academic Press, San Diego (1998)
35. Louis, A.K., Maaß, A.P.: Wavelets: Theory and Applications. Wiley, Chichester (1998)
36. Lio, P.: Wavelets in bioinformatics and computational biology: state of art and perspectives. Bioinformatics 19(1), 2–9 (2003)
37. Zhu, H., Yu, C.Y., Zhang, H.: Tree-based disease classification using protein data. Proteomics 3, 1673–1677 (2003)
38. Waagen, D., Cassabaum, M., Scott, C., Schmitt, H.: Exploring alternative wavelet base selection techniques with application to high resolution radar classification. In: Proc. of the 6th Int. Conf. on Inf. Fusion (ISIF 2003), pp. 1078–1085. IEEE Press, Los Alamitos (2003)
39. Leung, A., Chau, F., Gao, J.: A review on applications of wavelet transform techniques in chemical analysis: 1989-1997. Chem. and Int. Lab. Sys. 43(1), 165–184(20) (1998)
40. Cohen, A., Daubechies, I., Feauveau, J.C.: Biorthogonal bases of compactly supported wavelets. Comm. Pure Appl. Math. 45(5), 485–560 (1992)
41. Villmann, T., Strickert, M., Brüß, C., Schleif, F.M., Seiffert, U.: Visualization of fuzzy information in fuzzy-classification for image sagmentation using MDS. In: Proc. of ESANN 2007, pp. 103–108 (2007)
42. Hammer, B., Villmann, T.: Generalized relevance learning vector quantization. Neural Netw 15(8-9), 1059–1068 (2002)
43. Lee, J., Verleysen, M.: Generalizations of the Lp Norm for time series and its application to Self-Organizing Maps. In: Cottrell, M. (ed.) 5th Workshop on Self-Organizing Maps, vol. 1, pp. 733–740 (2005)
44. Hammer, B., Schleif, F.M., Villmann, T.: On the generalization ability of prototype-based classifiers with local relevance determination, Technical Reports University of Clausthal IfI-05-14, p. 18 (2005)
45. Schneider, P., Biehl, M., Hammer, B.: Relevance Matrices in LVQ. In: Proc. of ESANN 2007, pp. 37–42 (2007)
46. Baumann, S., Ceglarek, U., Fiedler, G., Lembcke, J., Leichtle, A., Thiery, J.: Standardized approach to proteomic profiling of human serum based magnetic bead separation and matrix-assisted laser esorption/ionization time-of flight mass spectrometry. Clinical Chemistry 51, 973–980 (2005)
47. Check, E.: Proteomics and cancer: Running before we can walk? Nature 429, 496–497 (2004)

48. Villmann, T., Schleif, F.M., Merenyi, E., Hammer, B.: Fuzzy Labeled Self Organizing Map for Classification of Spectra. In: Sandoval, F., Prieto, A.G., Cabestany, J., Graña, M. (eds.) IWANN 2007. LNCS, vol. 4507, pp. 556–563. Springer, Heidelberg (2007)
49. Hastie, T., Tibshirani, R., Friedman, J.: The Elements of Statistical Learning. Springer, New York (2001)
50. Zhang, Z., Page, G., Zhang, H.: Fishing Expedition - A supervised approach to extract patterns from a compendium of expression profiles. In: Lin, S.M., Johnson, K.F. (eds.) Methods of Microarray Data Analysis II. Kluwer Academic Publishers, Dordrecht (papers from CAMDA 2001 (2002)
51. Lee, Y., Lee, C.K.: Classification of multiple cancer types by multicategory support vector machines using gene expression data. Bioinformatics 19(9), 1132–1139 (2003)
52. Villmann, T., Bauer, H.U., Villmann, T.: Proceedings of WSOM 1997, Workshop on Self-Organizing Maps, Helsinki University of Technology Neural Networks Research Centre, June 4-6, pp. 286–291 (1997)
53. Bauer, H.U., Villmann, T.: Growing a Hypercubical Output Space in a Self–Organizing Feature Map. IEEE Transactions on Neural Networks 8(2), 218–226 (1997)
54. Carpenter, G., Grossberg, S.: The Handbook of Brain Theory and Neural Networks, 2nd edn., pp. 87–90. MIT Press, Cambridge (2003)
55. Villmann, T., Hammer, B., Schleif, F.M., Geweniger, T.: Fuzzy classification by fuzzy labeled neural gas. Neural Networks 19(6-7), 772–779 (2006)
56. Der, R., Herrmann, M.: Instabilities in Self-Organized Feature Maps with Short Neighborhood Range. In: Verleysen, M. (ed.) Proc. ESANN 1994, European Symp. on Artificial Neural Networks, pp. 271–276. D facto conference services, Brussels, Belgium (1994)
57. Molinaro, A., Simon, R., Pfeiffer, R.: Prediction error estimation: A comparison of resampling methods. Bioinformatics 21(15), 3301–3307 (2005)
58. Kearns, M.J., Mansur, Y., Ng, A., Ron, D.: An experimental and theoretical comparison of model selection methods. Machine Learning 27, 7–50 (1997)
59. Bartlett, P.L., Boucheron, S., Lugosi, G.: Model selection and error estimation. Machine Learning 48, 85–113 (2002)

# 7

## Computational Intelligence Techniques in Image Segmentation for Cytopathology

Andrzej Obuchowicz, Maciej Hrebień, Tomasz Nieczkowski,
and Andrzej Marciniak

Institute of Control and Computation Engineering
University of Zielona Góra
ul. Podgórna 50, 65-246 Zielona Góra, Poland
{A.Obuchowicz,M.Hrebien,T.Nieczkowski,A.Marciniak}@issi.uz.zgora.pl

**Summary.** A variety of computational intelligence approaches to nuclei segmentation in the microscope images of fine needle biopsy material is presented in this chapter. The segmentation is one of the most important steps of the automatic medical diagnosis based on the analysis of the microscopic images, and is crucial to making a correct diagnostic decision. Due to complex nature of biological images, standard segmentation methods are not effective enough. In this chapter we present and discuss some modified versions of watershed algorithm, active contours, cellular automata, Grow-Cut technique, as well as new approaches like fuzzy sets of I and II type, and the sonar-like method.

## 7.1 Introduction

Segmentation of the object of interest is one of the most critical tasks in image analysis and therefore it has been the subject of considerable research activity over the last four decades. During all this time we have witnessed a tremendous development of new, powerful instruments for detecting, storing, transmitting, and displaying images but automatic segmentation still remained a challenging problem.

This fact is easy to notice in medical applications, where image segmentation is particularly difficult due to restrictions imposed by image acquisition, pathology and biological variation. Biomedical image segmentation is a sufficiently complex problem that no single strategy has proven to be completely effective. Due to a complex nature of biomedical images, it is practically impossible to select or develop automatic segmentation methods of generic nature, that could be applied for any type of these images, e.g. for either micro- and macroscopic images, cytological and histological ones, MRI and X-ray, and so on.

In this chapter we are focused on the microscopic images of the Fine Needle Biopsy (FNB) material taken from the breast cancer. In the last decade we have been observing a dynamic growth in the number of research works conducted in the area of breast cancer diagnosis. Many university centers and commercial institutions [18] are focused on this issue because of the fact that breast cancer is becoming the most common form of cancer disease of today's female population. The attention covers not only curing the external effects of the disease [2, 43]

T.G. Smolinski et al. (Eds.): Comp. Intel. in Biomed. & Bioinform., SCI 151, pp. 169–199, 2008.
springerlink.com

but also its fast detection in the early stadium. Thus, the construction of a fully automatic cancer diagnosis system supporting a human expert has became a challenging task.

Many nowadays camera-based automatic breast cancer diagnosis systems have to face the problem of cells and their nuclei separation form the rest of the image content [19, 35, 38, 48]. This process is very important because the nucleus of the cell is the place where breast cancer malignancy can be observed. Thus, much attention in the construction of the expert supporting diagnosis system have to be paid to the segmentation stage.

The main difficulty of the segmentation process is due to incompleteness and uncertainty of the information contained in the image. Imperfection of the data acquisition process in the form of noise, chromatic distortion and deformity of the cytological material caused by its preparation additionally increases the complicity of the problem. The nature of the image acquisition (3D to 2D transformation) and the method of scene illumination also affects the image's luminance and sharpness. In many cases one must also deal with a low-cost CCD sensor whose quality and resolution capabilities are rather small.

Until now many segmentation methods have been proposed [3, 4, 16, 33, 47, 42] but, unfortunately, each of them introduces different kinds of additional problems and usually works in practice under given assumptions and/or needs the end-user's interaction/co-operation [19, 41, 48, 49]. Since many nowadays cytological projects assume full automation and real-time operation with a high degree of efficacy, a method free of drawbacks of the already known approaches has to be constructed.

The aim of this chapter is presentation a variety of computational intelligence approaches to nuclei segmentation in the microscope images of fine needle biopsy material. Some of them are modified versions of cytological image segmentation methods adopted for fine needle biopsy images, that is the watershed algorithm [10], active contours [11], cellular automata [27], GrowCut technique [9], and decision three technique [29]. Some of them exemplify quite new proposals of segmentation techniques: sets of I and II type approach [7], the sonar-like method [28]. One can also find here a description of the denoising and contrast enhancement techniques, pre-segmentation and a fully automatic nuclei localization mechanism used in our approaches. The quality and applicability of described segmentation methods are still investigated by our team. The final judgement can be precised after finishing the next steps of automatical breast cancer diagnosis, i.e. morphometric parameters calculating and classification. The quality of classification will be testimony of the applied segmentation.

## 7.2   Image Segmentation of Cytological FNB Microscope Images

### 7.2.1   Segmentation within Image Analysis

Image segmentation consists in subdividing an image into its constituent regions that hopefully correspond to structural units in the scene or distinguish objects

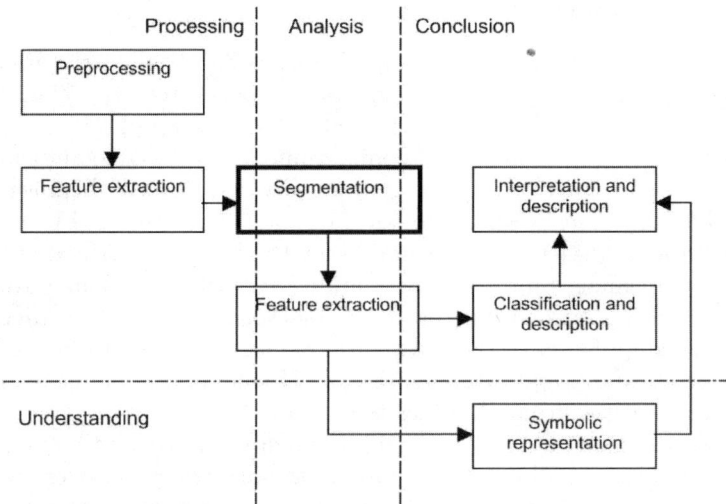

**Fig. 7.1.** An automatic vision system

of interest. The level to which this subdivision is carried depends on the problem being solved and applied approach. In the Fig.7.1 one can notice that segmentation is a crucial and a central step in whole image analysis system. The dashed lines in the picture illustrate vagueness of the definitions adopted in literature (see e.g. [26, 36]). Some of them are deliberately restrictive and assume that no contextual information is utilized in the segmentation [36]. In this approach segmentation does not involve identifying and isolating segments. The process consists in subdividing an image according to some homogeneity criteria defined for individual pixels; it does not attempt to recognize the individual segments nor their relationships to one another. In the most opposite approach, segmentation is defined as a process of isolation of components that correspond to the physical objects in the scene. In this case, a feedback from the subsequent steps is taken into account, i.e. analysis outcomes and relations between isolated image regions are important factors constituting the segmentation criteria. In life-crucial applications (e.g. medical diagnosis) even the intervention of a human operator is required. Different patterns of such interaction can be found in [14, 26, 32].

As a consequence of application-driven approach in defining the process of segmentation, no single standard methodology has emerged. In principle, there is no general theory of image segmentation, though some attempts to build a functional model have been made (see for example [50]). Moreover, the problem of segmentation is ill-defined and can be perceived as one of psychophysical perception and therefore not susceptible to a purely analytical solution [8]. As a consequence, no single standard method of image segmentation has been elaborated, although there are methods that have received some degree of popularity.

## 7.2.2   Testing Database of Images

What we have on input is the cytological images database of the Fine Needle Biopsy material gained in a cooperation with experts from the Zielona Góra's *Onkomed* medical center [21, 22]. The database consists of 750 images of 75 clinical cases, including 25 images of benign tumour, 25 of malignant tumour and 25 of fibroadenoma. There are 10 images for every case - one of them is magnified by 100 and 9 magnified by 400. Photographs x400 where selected in such a way as to contain at least 10 nucleus suitable for the morphometric analysis. As a result, the set of images for one case contains at least 90 nucleuses to analyze. The image itself is coded using the RGB colorspace and is not subject to any kind of lossy compression (a raw color bitmap format), and with a resolution of $704 \times 578$ pixels and 24-bit color depth (16.7M colors). The number of distinct colors in images varies from about 10 to 60 thousands.

What we expect on output is a binary segmentation mask with one pixel separation rule which will allow more robust morphometric parameters estimation in our future work. Additionally, the proposed segmentation algorithm should be insensitive to colors of contrasting pigments used for preparation of the cytological material (see an example in Fig. 7.2).

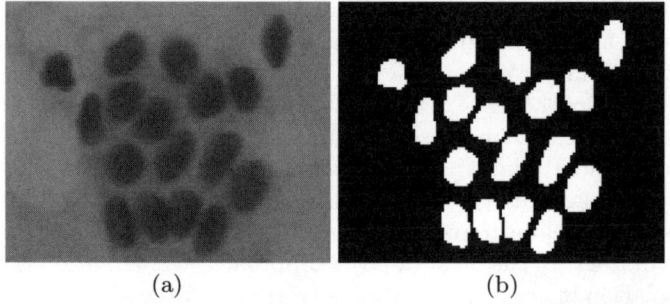

(a)                                 (b)

**Fig. 7.2.** Exemplary fragment of: (a) cytological image, (b) appropriate segmentation mask

## 7.2.3   Image Filtering and Preparation

The quantity of information contained in a color image is surplus at the early stage of image processing. The color components do not carry as important information as luminosity so they can be removed to reduce processing complexity. An RGB color image can be converted to greyscale by calculating a luminance value in the same way as it is calculated for YCbCr color space [37].

Since a great deal of images have a low contrast, an enhancement technique is needed to improve theirs quality. In our research we use a simple histogram processing with the linear transform of images levels of intensities, namely a cumulated sum approach [36].

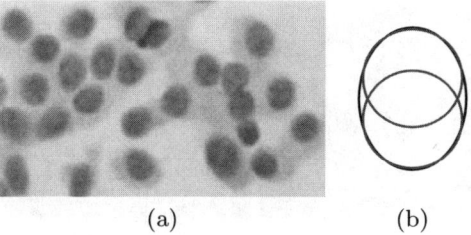

**Fig. 7.3.** Examplary fragment of cytological image with circular nuclei

| 1 | 1 | 1 |
|---|---|---|
|   |   |   |
| -1 | -1 | -1 |

| 1 | 2 | 1 |
|---|---|---|
|   |   |   |
| -1 | -2 | -1 |

| 3 | 2 | 1 |
|---|---|---|
| 2 |   | -2 |
| -1 | -2 | -3 |

| 1 |   |   |
|---|---|---|
|   |   |   |
|   |   | -1 |

**Fig. 7.4.** Gradient masks used in our experiments

If we look closely at the nuclei we have to segment, we notice that they all have an elliptical shape (Fig. 7.3). Most of them resemble an ellipse but, unfortunately, detection of the ellipse is computationally expensive. The shape of the ellipse can be approximated by a given number of circles (as shown in Fig. 7.3b). The detection of circles is much simpler in the sense of the required computations because we have only one parameter, which is the radius $R$. These observations and simplifications constitute grounding for a nucleus pre-segmentation algorithm – in our approach we try to find such circles with different radii in a given feature space.

The Hough transform [1, 6, 51] can be easily adopted for the purpose of circle detection. The transform in a discrete space can be defined by:

$$HT_{discr}(R, i_0, j_0) = \sum_{i=i_0-R}^{i_0+R} \sum_{j=j_0-R}^{j_0+R} g(i,j)\delta\Big((i-i_0)^2 + (j-j_0)^2 - R^2\Big), \quad (7.1)$$

where $g$ is a two-dimensional feature image and $\delta$ is Kronecker's delta (equal to unity at zero). $HT_{discr}$ plays the role of an accumulator which accumulates the levels of feature image $g$ similarity to the circle placed at the $(i_0, j_0)$ position and defined by the radius $R$.

The feature space $g$ can be created by many different ways. In our approach we use a gradient image as the feature indicating the occurrence or absence of the nucleus in a given fragment of the cytological image. The gradient image is a saturated sum of gradients estimated in eight directions on the greyscale image prepared in the pre-processing stage. The base gradients can be calculated using, e.g., Prewitt's, Sobel's mask methods [12, 44] or their heavy or light versions (Fig. 7.4).

Thresholding the values in the accumulator by a given $\theta$ value we can obtain a very good pre-segmentation mechanism with a lower threshold strategy (see, for

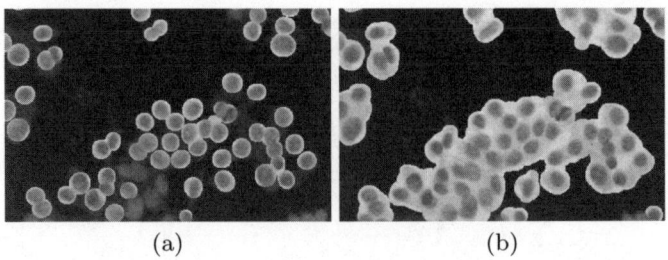

(a)                                    (b)

**Fig. 7.5.** Exemplary results of the pre-segmentation stage for two different $\theta$ threshold strategies: (a) high and (b) low

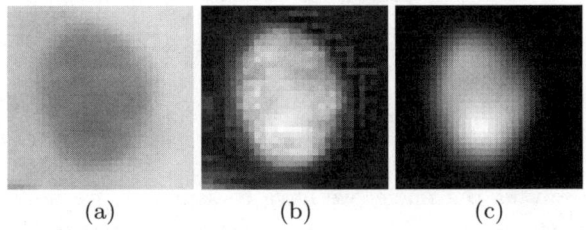

(a)                 (b)                 (c)

**Fig. 7.6.** Exemplary fragment of: (a) cytological image, (b) Euclidian distance to the mean background color, (c) smoothed out version of (b)

instance, Fig. 7.5). Since the threshold value strongly depends on the database and the feature image $g$, the method can only be used as a pre-segmentation stage. A smaller value of the threshold causes fast removal of unimportant information from the background, and what we achieve is a mask, which approximately defines the places where are the objects we have to segment (nuclei in this case) and where is the background. Such a mask can constitute a base for more sophisticated and detail-oriented algorithms.

The results obtained from the pre-segmentation stage can lead us to the estimation of an average background color. Such information can be used to model the nuclei as a color distance between the background and the objects, which fulfils the requirements of the lack of any color dependency in the imaged material (the color of contrasting pigments may change in the future). In our research we tried few distance metrics: Manhattan's, Chebyshev's, the absolute Hue value from the HSV colorspace, but the Euclidian one gives us visually the best results (Fig. 7.6ab):

$$D_{euclid} = \sqrt{(I_R - B_R)^2 + (I_G - B_G)^2 + (I_B - B_B)^2}, \qquad (7.2)$$

where $B$ is the average background color estimated for the $I$ input image.

Since the modeling distance can vary in the local neighborhood (see Fig. 7.6b), mostly because of camera sensor simplifications, a smoothing technique is needed to reconstruct the nuclei shape. The smoothing operation in our approach relies on the fact that this sort of 2D signal can be modeled as a sum of sinusoids [20] with defined amplitudes, phase shifts and frequencies. Cutting all low amplitude frequencies off (leaving only a few significant ones with the highest amplitude) will result in a signal deprived of our problematic local noise effect (Fig. 7.6c). What we finally achieve is a three-dimensional modeled terrain where hills correspond to nuclei.

The localization of objects on a modeled map of nuclei can be performed locally using various methods. In our approach we have chosen an evolutionary (1+1) search strategy [25, 30, 31] mostly because it is simple, quite fast despite appearances, can be easily parallelized due to its nature and settles very well in local extrema, which is very important in our case.

The search in our approach can be conducted in two versions: single-point and multi-point. In the single-point version it is only allowed to have only one marker pointing a nucleus while in the multi-point one it is allowed to have more than one marker pointing the same nucleus.

The used watershed algorithm as a final segmentation method forced us to create two population of individuals. The first population localizes the background. Specimens are moved through the mutation stage $Y_i^t = X_i^t + r^t N_i(0, 1)$ with a constant movement step ($r^t = 1$) preferably to places with a smaller density of population to maximize background coverage. The second population localizes the nuclei. Specimens are moved with a decreasing movement step ($r^t = R_{max}(1/R_{max})^{t/t_{max}}$) to group very fast the population near local extremum in the first few epochs and to finally work on details in the ending ones. The movement of individuals is preferred to be directed towards places with a higher population density to create the effect of nuclei localization.

The fitness function $\phi$ calculates the average *height* of the terrain in a given position including the nearest neighborhood defined by the smallest radius detected by the Hough transform in the pre-segmentation stage. Such a definition of the fitness function avoids a possible split of the population, localized near a nucleus with multimodal character of its shape, giving only one marker for a nucleus (Fig. 7.7b).

Finally, the nucleus is localized in the place where the density of the population searching for hilltops in the modeled terrain is locally maximal. As we have

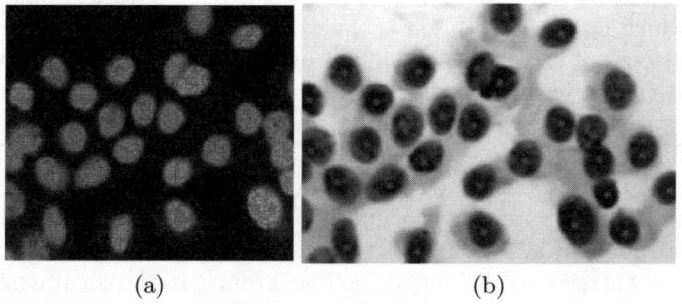

(a)                              (b)

**Fig. 7.7.** Exemplary single-point localization: (a) screenshot after 8 epochs, (b) final result (localization points are marked with red asterisks)

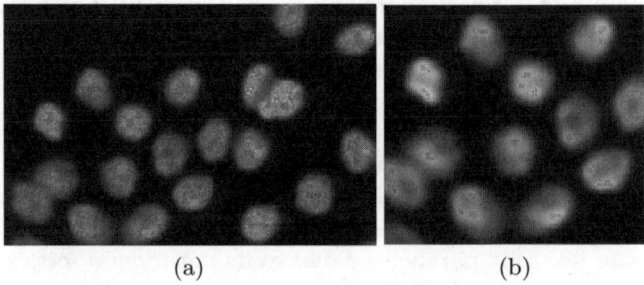

(a) (b)

**Fig. 7.8.** Exemplary multi-point localization: (a) screenshot after few epochs, (b) final result

mentioned earlier, the method is quite fast and just a few epochs are needed to observe a visible progress in nuclei localization and background coverage (Fig. 7.7a).

The algorithms that do not have such tight requirements concerning only one single marker per nucleus, that is they allow multiple markers pointing the same one, not optimal or even false localization points and can take information about the background location from the pre-segmentation mask [9], can use much simplified version of the above presented $(1 + 1)$ search strategy. In such cases we can use only one population, that is the one searching for nuclei and the fitness function is simply the terrain *height* at an individuals position. The number of iterations of the algorithm can also be reduced, because we need only an approximate localization of nuclei (Fig. 7.8). Thus, the algorithm is the same and the only difference with the one described above is the fitness function $\phi$ and reduced number of epochs.

### 7.2.4   Watershed Segmentation

**Method description**

The watershed segmentation algorithm is inspired by natural observations, such as a rainy day in the mountains [12, 36, 37]. A given image can be defined as a terrain on which nuclei correspond to valleys (upside down the terrain modeled in previous steps). The terrain is flooded by rainwater and arising puddles are starting to turn into basins. When the water from one basin begins to pour away to another, a separating watershed is created.

The flooding operation has to be stopped when the water level reaches a given $\theta$ threshold. The threshold should preferably be placed somewhere in the middle between the background and a nucleus localization point. In our approach the nuclei are flooded to the half of the altitude between the nucleus localization point and the average height of the background in the local neighborhood. Since the images we have to deal with are spot illuminated during the imaging

```
// For each basin
∀p ∈ P assign a label (i + 1)

for θ ∈ 0 : Δ : 1
    for ∀p ∈ P
        color p to the level ⩽ θ * Ψ(p)/2
    end
end
```

**Algorithm 1.** The simplified version of the watershed algorithm

operation (resulting in a modeled terrain higher in the center of the image and much lower in the corners), this mechanism protects the basins from being over-flooded and, in consequence, nuclei from being undersegmented.

The simplified version of the watershed algorithm is given in Alg. 1. The coloring to the level of $\theta$ implements the flooding operation. It also considers a possible situation of watershed building when there is a neighbor nearby with another label. $\Delta$ defines water level increase in each iteration of the algorithm and $\Psi$ defines the difference between the $p$ valley's depth and the background's height in its local neighborhood.

## Typical results for cytological images

Exemplary results of the presented watershed segmentation method and common errors observed in our hand-prepared benchmark database can be divided into four classes:

- *class 1*: good quality images with only small irregularities and rarely generated subbasins (a basin in another basin) (Fig. 7.9ab),
- *class 2*: errors caused by fake circles created by spots of fat (Fig. 7.9cd),
- *class 3*: mixed nucleus types: red and purple in this case and those reds which are more purple than yellow (background) are also segmented, which is erroneous (Fig. 7.9ef),
- *class 4*: poor quality image with a bunch of nuclei glued together, which causes basin overflooding and, in consequence, undersegmentation (Fig. 7.9gh).

The conducted experiments show that the watershed algorithm gives a 68.74%, on average, agreement with the hand-prepared templates using a simple XOR metric. Most errors are located at boundaries (see, for instance, Fig. 7.13a) of nuclei where the average distance between the edges of segmented and reference objects is about 3.28 pixels. The XOR metric is underestimated as a consequence of not very high level of water flooding the modeled terrain, but the shape of the nuclei seems to be preserved, which is important in our future work – the estimation of morphometric parameters of segmented nuclei.

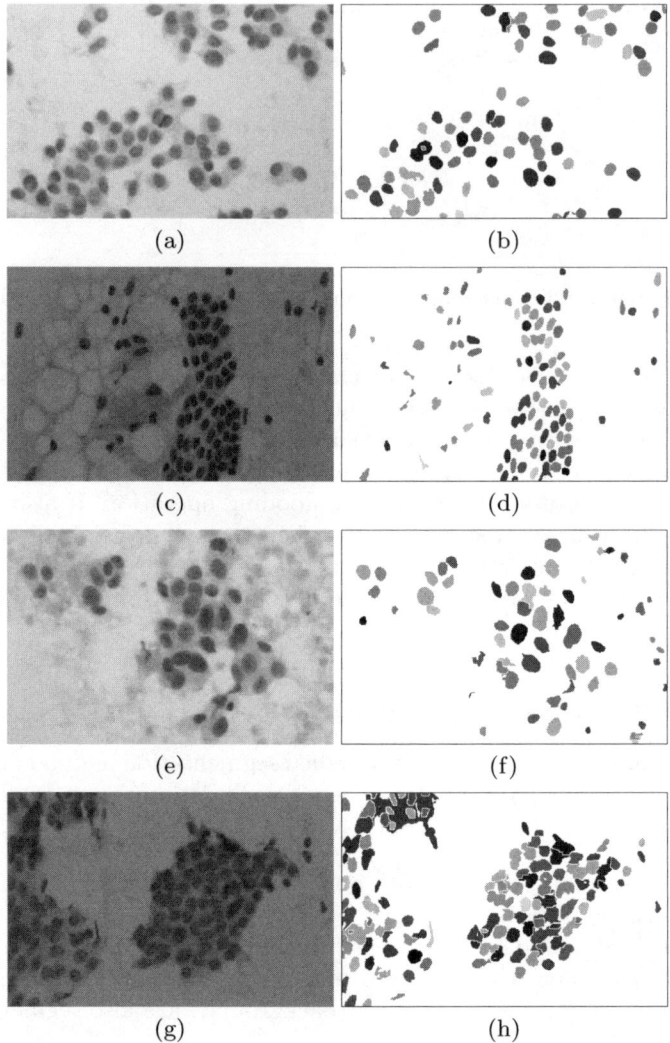

(a)  (b)

(c)  (d)

(e)  (f)

(g)  (h)

**Fig. 7.9.** Exemplary results of the watershed segmentation

## 7.2.5  Active Contour Technique

### Method description

The active contouring technique can be considered as a more advanced region growing method [44]. The algorithm groups neighboring pixels when a given homogeneity and similarity criteria is met. All joined pixels create a segment, which boundary spreads in all directions until another segment is met or the new candidates for joining introduce not acceptable error. The algorithm is stopped

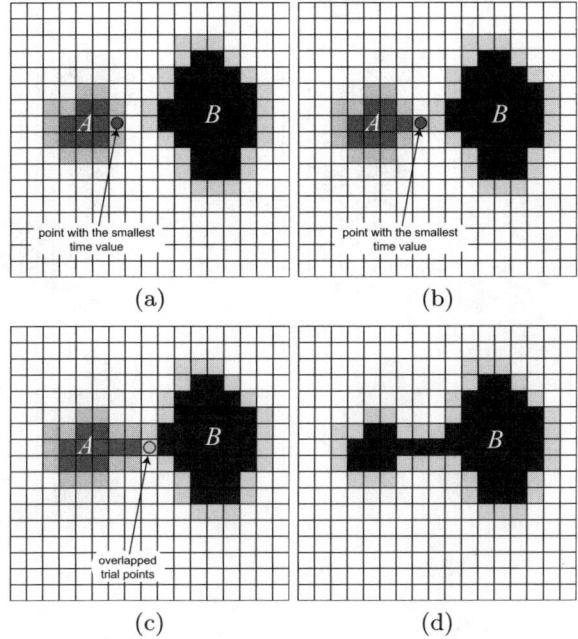

(a)                                    (b)

(c)                                    (d)

**Fig. 7.10.** Illustration of the contour merging operation

when all pixels get a label, that is the object in the image is separated from the background.

The images we are dealing with can contain more than a single object per image. Additionally, the assumption of the project is that the segmentation process have to be fully automatic (there is no human operator which manually initializes the method). This two factors forces us to modify the algorithm to meet the stated requirements. Thus, the algorithm, which in our case is based on fast marching method (FMM) [39], must have multilabel extension [40] and the seeding process has to be done without end-user's interaction.

In the proposed approach the multilabel FMM is initialized with a pre-segmentation mask and the results obtained from the multi-point nuclei localization stage. The background-object boundary from the pre-segmentation mask is the initial seed for the background segment. The nuclei localization points, on the other hand, are initial seeds for the object segments. The most important in this method is that the initialization mask and the nuclei localization points do not have to be perfect – all fake initial markers are fully acceptable and they do not have any influence on the final segmentation result and its quality.

The contour expansion speed of the multilabel FMM is governed globally by the function [11]:

$$F = \frac{1}{|g(x,y) - \bar{g}(i)|^3 + 1},$$ (7.3)

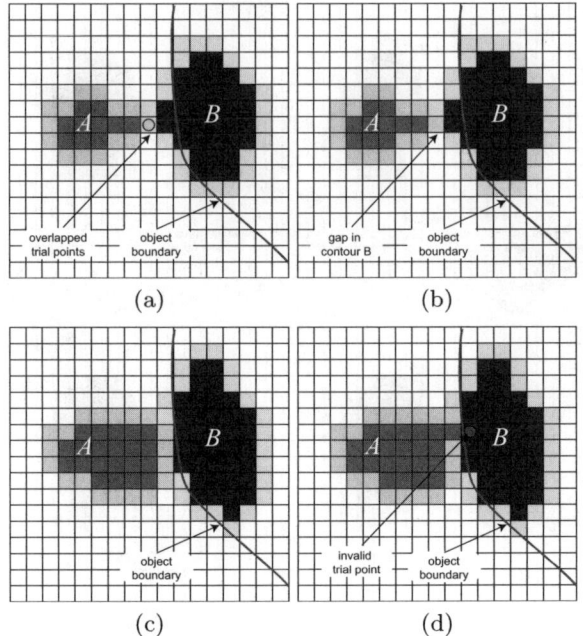

(a)                    (b)

(c)                    (d)

**Fig. 7.11.** Illustration of the contour pushing operation

where $g(x, y)$ is the color under the contour and $\bar{g}(i)$ is the mean color under the $i$-th segment. Such a definition of the expansion speed slows down the contour near object (nucleus) boundary. Two very close to each other spreading segments can meet while the algorithm execution. The two meeting segments can be merged (the smaller one into the bigger) when their mean color difference is below certain threshold (Fig. 7.10). The segments not classified to be merged can bush back the segment with the lower difference between considered pixel and mean color of each segment (Fig. 7.11). The pushing operation can be performed only once to reduce contour oscillation known from the classical approach and the pushed back segment can not move father at this place.

**Typical results for cytological images**

The conducted experiments show that the modified multilabel FMM algorithm is very stable and robust to initialization errors. Visually, segmentation quality is promising and gives good detection of even small objects (Fig. 7.12). Unfortunately, the algorithm has problems with connected nuclei and detect them as a one single object, which is erroneous. The average XOR metric score with the hand-prepared templates is only 22.32% and the average distance between the edges of segmented and reference objects is about 4.1 pixels.

Despite the mentioned problems the shape of segmented nuclei seems to be represented accurately and most errors are located at the boundaries of the

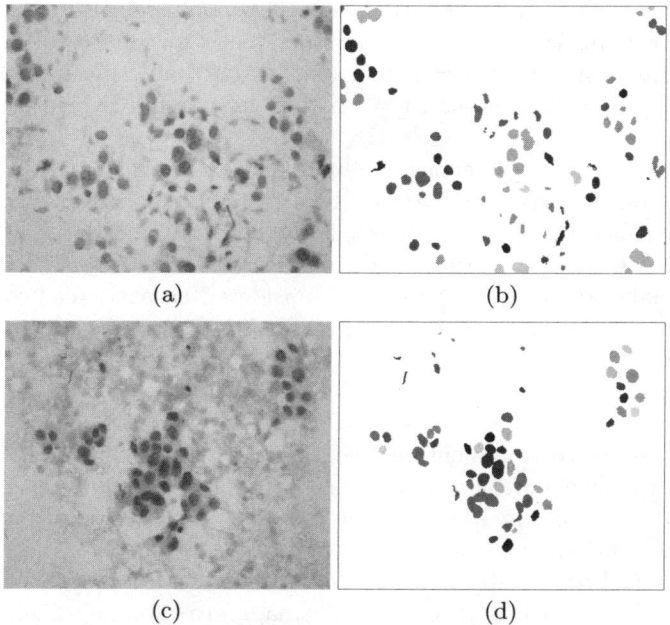

Fig. 7.12. Exemplary results of the active contouring segmentation

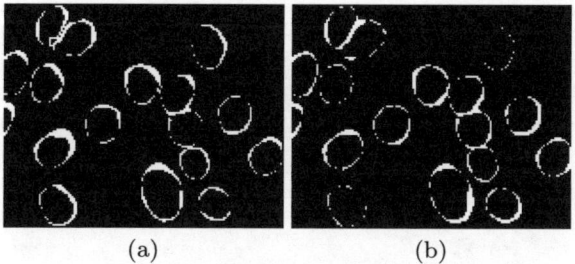

Fig. 7.13. Exemplary XOR results with a fragment of the hand-prepared segmentation mask for: (a) watershed algorithm, (b) active contouring technique

segmented objects (see, for instance, Fig. 7.13b). This illustrates that the proper selection of merging threshold and detection of overlapping nuclei is still a challenge and has to be improved in the future works.

### 7.2.6  GrowCut Cellular Automata Segmentation

#### Method description

The next technique inspired by natural observations is the GrowCut cellular automata segmentation algorithm [46]. It imitates growth and struggle for domination of rivalry bacteria colonies. Each type of bacteria represents a single type of objects used in segmentation. The GrowCut algorithm was originally developed for multi-label intelligent scissors tasks for photo-editing tools. It requires manual initialization of the seed pixels, but concatenated with a proper pre-segmentation method gives a fully automated hybrid segmentation technique.

The GrowCut algorithm defines a cellular space $P$ as $k \times m$ array, where $k$ and $m$ are dimensions of the image. Each of the array cells is an automaton described by the state triplet $(l_p, \theta_p, \mathbf{C}_p)$, where $l_p$ is the label of the cell, $\theta_p$ is the *strength* of the cell and $\mathbf{C}_p$ is the feature vector of the cell defined by associated image pixel. An unlabeled image may be then considered as particular configuration state of cellular automata, where initial states for $\forall p \in P$ are set to:

$$l_p = 0, \quad \theta_p = 0, \quad \mathbf{C}_p = RGB_p, \tag{7.4}$$

where $RGB_p$ is the three dimensional vector of pixel $p$ color in RGB space. The final goal of the segmentation is to assign each pixel to one of the $K$ possible labels. As stated before, we use two labels in segmentation of cytological images – the *nuclei* and the *background*.

In a single evolution step each cell (the bacteria) tries to attack all its neighbors. The evolution goal is to occupy all image area starting from a group of previously initialized pixels. Cell neighbors are defined by neighborhood system. In our approach the Moore neighborhood system was used:

$$N(p) = \left\{ q \in Z^n : \| p - q \|_\infty := \max_{i=1,\ldots,n} |p_i - q_i| = 1 \right\}. \tag{7.5}$$

The attack power is defined as a function of attacker $q$ and defender $p$ strengths and the distance between their feature vectors: $\mathbf{C}_q$ and $\mathbf{C}_p$. The basic rule of

```
// For each cell
for ∀p ∈ P
    // copy previous state
    l_p^{t+1} = l_p^t
    θ_p^{t+1} = θ_p^t
    // neighbors try to attack
    // current cell
    for ∀q ∈ N(p)
        if g(|| C_p − C_q ||_2) · θ_q^t > θ_p^t
            l_p^{t+1} = l_q^t
            θ_p^{t+1} = g(|| C_p − C_q ||_2) · θ_q^t
        end
    end
end
```

**Algorithm 2.** The GrowCut algorithm

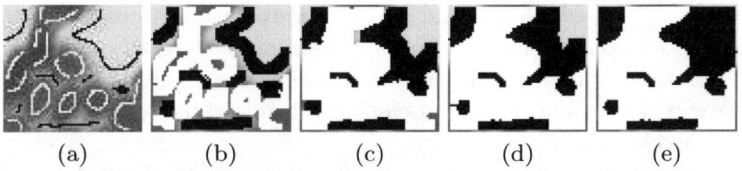

(a)            (b)            (c)            (d)            (e)

**Fig. 7.14.** Exemplary segmentation of a FNB image with the GrowCut algorithm. White – *nucleus* labeled cells, black – *background* labeled cells: (a) the seed, (b) step 2, (c) step 4, (d) step 6, (e) final 19-th step.

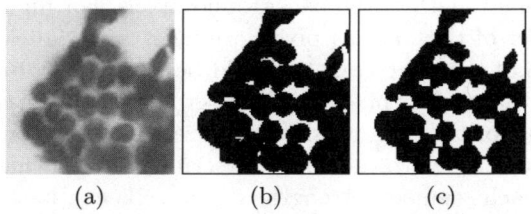

(a)                    (b)                    (c)

**Fig. 7.15.** Exemplary segmentation with the GrowCut algorithm initialized with thresholding result: (a) exemplary image, (b) thresholding result, (c) GrowCut result

automaton state change at time $t + 1$ is shown in Alg. 2. The $g$ function is monotonous, decreasing and bounded to $[0, 1]$. For the purpose of this work, simple $g$ function was used, as proposed in [46]:

$$g(x) = 1 - \frac{x}{\max \parallel \mathbf{C} \parallel_2}, \qquad (7.6)$$

where $\max \parallel \mathbf{C} \parallel_2$ is calculated as a feature vector length for white pixel (RGB $= [255, 255, 255]$). As the strength of each cell is increasing and bounded, so the method is guaranteed to converge. Thus for any seed configuration of the image, after finite number of evolution steps, all cells are labeled and their states seize to change. Fig. 7.14 shows subsequent steps of the GrowCut segmentation for a manually initialized cytological image.

The GrowCut algorithm requires initialization of a number of cells with proper labels for each separate, consistent group of pixels (segment seed). To allow for application of the algorithm to the automated diagnostic system we employ the information from the pre-segmentation and the nuclei localization stage to initialize the seed pixels. At this point almost any rough segmentation technique (e.g. thresholding) can be also applied as the pre-segmentation (Fig. 7.15), however our research shows that initialization which leaves unclassified pixels at objects boundaries performs better. One of the techniques, which results can be utilized at the GrowCut algorithm initialization stage is the pre-segmentation mask obtained using the Hough transform. The transform result is a set of circles covering regions of the image, where nuclei are located. Pixels enclosed inside these regions are initially labeled as the *nucleus* pixels. Remaining pixels of the image are labeled as the *background*. For this type of initialization, all the image pixels are classified before the first GrowCut evolution step. The goal of the algorithm application is only to adjust the segments edges to real boundaries of objects. Therefore, to enforce the proper direction and the range of label changes within following evolution steps, associating suitable values of initial strength for both of the pixel classes is necessary.

The appropriate direction of label changes depends on the $\theta$ threshold value, used at pre-segmentation stage. For lower values of threshold, Hough transform results in a number of background pixels located in boundaries of regions labeled as *nucleus*. These pixels should change their labels do *background* in process of actual segmentation. Thus, the initial strength value of *nucleus* labeled pixels has to be less than strength of the background pixels. For higher values of the $\theta$ threshold a number of the *nucleus* pixels are incorrectly labeled as *background*. In this case labels of boundary pixels should be changed to *nucleus*. Therefore, initial strength of the *nucleus* pixels has to be greater than the *background* pixels.

The GrowCut algorithm can be also initialized with the result obtained from the multi-point nuclei localization stage described above. Due to only few initialized pixels of each segment, strengths of the cells can be set to equal values for both classes. It allows for automation of the segmentation process. However, more uninitialized pixels results in more evolution steps and so greater computational cost.

### Typical results for cytological images

Typical results of the GrowCut cellular automata, initialized with the result obtained from the multi-point nuclei localization stage, can be seen in Fig. 7.16. For the exemplary image the proportion of incorrectly labeled pixels was about 6%. However, the shape of the identified nuclei segments is too ragged (due to camera sensor interlace), so additional smoothing post-segmentation stage is needed for this combination of techniques.

The problem with the Hough transform and the GrowCut cellular automata hybrid is that the optimum proportion of the initial *nucleus* pixels strength should be estimated to achieve good segmentation quality. The proportion strongly depends on the analyzed image contrast and the pigment used, so a

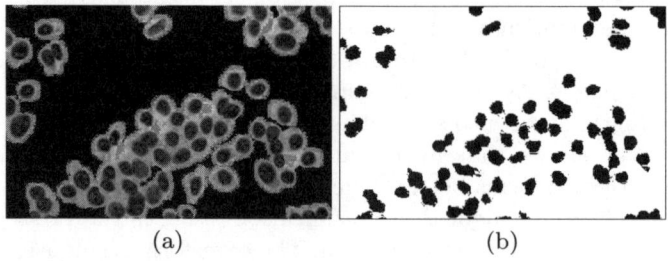

(a)                                        (b)

**Fig. 7.16.** Exemplary result of the segmentation (b) with the GrowCut cellular automata, initialized with multi-point nuclei localization points (a)

potential automated diagnostic system should be learned beforehand. The second hybrid – the GrowCut cellular automata initialized with multi-point nuclei searching algorithm, can be applied with fixed initial strength values for the *nucleus* and the *background* seed pixels. However it is much more computationally expensive due to more cellular automata iterations required.

### 7.2.7 Fuzzy Sets of Type I and II in Thresholding

The technique used in this subsection belongs to object attribute-based methods and is based on a type-1 fuzzy thresholding technique. However type-1 fuzzy sets still have some inherent uncertainties. There are (at least) four different sources of uncertainties in type-1 fuzzy logic systems [23]: uncertainty about meanings of the words that are used (words mean different things to different people), uncertainty about consequents, uncertainty about measurements, which may be noisy and uncertainty about the data that is used to tune parameters. Thus, to address this problem type-2 fuzzy sets (T2FSs) have been formulated, which let us model and minimize the effects of this uncertainties. Such sets are fuzzy sets whose membership grades themselves are T1FSs; they are very useful in circumstances where it is difficult to determine an exact membership function for a fuzzy set; therefore, they are useful for incorporating uncertainties [15].

Nevertheless a general T2FSs computational complexity is severe and it is very difficult to justify the use of any other kind of secondary membership functions (e.g. right now there is no best choice for a T1FS, therefore secondary membership functions only complicate the matter). Thus interval T2FSs were introduced — when the T2FSs are interval T2FSs, all secondary grades equal one [23].

Another drawback of the thresholding techniques is that they are, in general, monochrome techniques. Compared to gray scale, color provides information in addition to the intensity. Color is useful or even necessary for pattern recognition and computer vision. Thus the other part of this paper is concerned with the adaptation of this monochrome technique to use extra information of color images.

**Interval type-2 fuzzy sets**

An interval type-2 fuzzy set (IT2 FS) $\tilde{A}$ is characterized as [24]

$$\tilde{A} = \int_{x \in X} \int_{u \in J_x \subseteq [0,1]} \frac{du\, dx}{ux} = \int_{x \in X} \left( \int_{u \in J_x \subseteq [0,1]} \frac{du}{u} \right) dx \Big/ x \qquad (7.7)$$

where $x$, the *primary variable*, has domain $X$; $u \in U$, the *secondary variable*, has domain $J_x$ at each $x \in X$; $J_x$ is called the *primary membership* of $x$ and is defined in (7.11); and, the *secondary grades* of all $\tilde{A}$ equal 1. Note that (7.7) means: $\tilde{A} : X \rightarrow \{[a, b] : 0 \leqslant a \leqslant b \leqslant 1\}$. Uncertainty about $\tilde{A}$ is conveyed by the union of all the primary memberships, which is called the *footprint of uncertainty* (FOU) of $\tilde{A}$ (see Fig. 7.17), i.e.

$$FOU(\tilde{A}) = \bigcup_{\forall x \in X} J_x = \{(x, u) : u \in J_x \subseteq [0,1]\} \qquad (7.8)$$

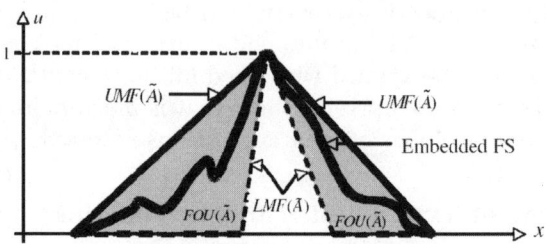

**Fig. 7.17.** FOU (shaded), LMF (dashed), UMF (solid) and an embedded FS (wavy line) for IT2 FS $\tilde{A}$

The upper membership function (UMF) and *lower membership function* (LMF) of $\tilde{A}$ are two type-1 membership functions that bound the FOU (Fig. 7.17). The UMF is associated with the upper bound of $FOU(\tilde{A})$ and is denoted $\overline{\mu}_{\tilde{A}}(x)$, $\forall x \in X$, and the LMF is associated with the lower bound of $FOU(\tilde{A})$ and is denoted $\underline{\mu}_{\tilde{A}}(x)$, $\forall x \in X$, i.e.

$$\overline{\mu}_{\tilde{A}}(x) \equiv \overline{FOU(\tilde{A})} \qquad \forall x \in X \tag{7.9}$$

$$\underline{\mu}_{\tilde{A}}(x) \equiv \underline{FOU(\tilde{A})} \qquad \forall x \in X \tag{7.10}$$

Note that $J_x$ is an *interval set*, i.e.

$$J_x = \left\{ (x, u) : u \in \left[ \underline{\mu}_{\tilde{A}}(x), \overline{\mu}_{\tilde{A}}(x) \right] \right\} \tag{7.11}$$

so that $FOU(\tilde{A})$ in (7.8) can also be expressed as

$$FOU(\tilde{A}) = \bigcup_{\forall x \in X} \left[ \underline{\mu}_{\tilde{A}}(x), \overline{\mu}_{\tilde{A}}(x) \right] \tag{7.12}$$

The upper and lower membership degrees $\overline{\mu}_{\tilde{A}}$ and $\underline{\mu}_{\tilde{A}}$ can also be defined by means of linguistic hedges like *dilation* and *concentration* on a principle membership function $\mu_A$. Because hedges are usually available as pairs, that represent diagonally different modifications of the basic term, so it seems practical to use linguistic hedge and its reciprocal value to draw the FOU. Thus, upper and lower membership values can be defined as follows [45]:

$$\overline{\mu}_{\tilde{A}}(x) = [\mu_A(x)]^{1/\alpha}, \tag{7.13}$$

$$\underline{\mu}_{\tilde{A}}(x) = [\mu_A(x)]^{\alpha} \ . \tag{7.14}$$

where $\alpha \in (1, \infty)$. However, according to [45] $\alpha \gg 2$ is usually not meaningful for image data.

## Image thresholding with type-2 fuzzy sets

Measures of the fuzziness estimate the average vagueness in fuzzy sets. Intuitively, one should expect that if the set is maximally ambiguous then the fuzziness measure should be maximum. On the other hand, the fuzziness of the crisp set using any measure should be zero, as there is no ambiguity about whether an element belongs to the set or not. When the membership value approaches either 0 or 1, vagueness in the set decreases. Thus a fuzzy set is the most vague when $\mu_A(x) = 0.5 \; \forall x$ [34].

The most common measure of fuzziness, introduced by [17], is the *linear index of fuzziness*. For an $M \times N$ image subset $A \subseteq X$ with $L$ gray levels $g \in [0, L-1]$, the histogram $h(g)$ and the membership function $\mu_A(g)$, the linear index of fuzziness $\gamma_l$ can be defined as follows:

$$\gamma_l(A) = \frac{2}{MN} \sum_{g=0}^{L-1} h(g) \times \min[\mu_A(g), 1 - \mu_A(g)]. \qquad (7.15)$$

But if images or thresholds are to be interpreted as T2FSs then there is a need for a new measurement. In this case one can ask how *ultrafuzzy* is a fuzzy set? If the degrees of the membership can be defined without any uncertainty (T1FSs), then clearly the ultrafuzziness should be minimum (=0). For the case that individual membership values can only be indicated as an interval, the amount of the ultrafuzziness should increase. And while absolutely nothing is known about the nature of membership degrees of the problem at hand, then the ultrafuzziness should be maximal (=1). With respect to these thoughts and the way a T2FS is defined, a measure of ultrafuzziness $\tilde{\gamma}$ for an $M \times N$ image subset $\tilde{A} \subseteq X$ with $L$ gray levels $g \in [0, L-1]$, histogram $h(g)$ and membership function $\tilde{\mu}_A(g)$ can be defined as follows [45]:

$$\tilde{\gamma}(\tilde{A}) = \frac{1}{MN} \sum_{g=0}^{L-1} h(g) \times [\overline{\mu}_{\tilde{A}}(g) - \underline{\mu}_{\tilde{A}}(g)], \qquad (7.16)$$

where $\overline{\mu}_{\tilde{A}}(g) = [\mu_A(g)]^{1/\alpha}$ and $\underline{\mu}_{\tilde{A}}(g) = [\mu_A(g)]^{\alpha}$, $\alpha \in (1, 2]$. This basic definition relies on the assumption that singletons sitting on the FOU are all equal in height (which is the reason why the IT2 FS is used). Thus, only the variation in the length of the FOU can be measured.

The general algorithm for the image thresholding based on type II fuzzy sets and measures of the ultrafuzziness can be formulated as follows: 1) Select the shape of the principle (skeleton) membership function $\mu_A(g)$ and initialize $\alpha$; 2) Calculate the image histogram; 3) Initialize the position of the membership function; 4) Shift the membership function along the gray-level range; 5) Calculate in each position upper and lower membership values $\overline{\mu}_{\tilde{A}}(g)$ and $\underline{\mu}_{\tilde{A}}(g)$; 6) Calculate in each position the amount of the ultrafuzziness, using Eq. 7.16; 7) Locate the position $g_{opt}$ with the maximum ultrafuzziness; 8) Threshold the image with $T = g_{opt}$.

For the thresholding algorithm to be complete, there is a need to define a suitable principle membership function. In this paper we are using following membership functions.

Function defined by [13] as follows:

$$\mu_A(g) = \begin{cases} \frac{1}{1+|g-\mu_0|/C}, & g \leqslant T, \\ \\ \frac{1}{1+|g-\mu_1|/C}, & g > T, \end{cases} \tag{7.17}$$

where $C$ is a constant value such that $0.5 \leqslant \mu_A(g) \leqslant 1$, e.g. $C = g_{max} - g_{min}$; the average gray levels of the background $\mu_0 = \sum_{g=0}^{T} gh(g) / \sum_{g=0}^{T} h(g)$ and the object $\mu_1 = \sum_{g=T}^{L-1} gh(g) / \sum_{g=T}^{L-1} h(g)$, for a certain threshold $T$.

## Color quantization

In the case of a widespread 24 bit color image representation, the number of possible colors is over 16 millon and exhaustive searching is computationally expensive or even unfeasible. However, in most cases images do not occupy the entire color gamut, and the number of colors used is much lower. Thus quantization of the color space is a viable choice.

A quantized image $M \times N$ may be regarded as a mapping defined by

$$q : M \times N \to R \subseteq \Psi \tag{7.18}$$

where $\Psi = (r, g, b) | 0 \leqslant r, g, b \leqslant 255$ is the RGB color space, $R = \{\bar{r}_1, \bar{r}_2, \ldots, \bar{r}_k\}$ is a set of representative colors used in the quantized image [5].

Hence, the color quantization can be divided into two parts: color palette design, in which a desirable number of colors (usually 8–256) is specified, and pixel mapping, in which each pixel is assigned to one of the colors in the designed palette. The goal is to achieve the lowest perceivable difference between the quantized image and the original one.

The color palette design can be obtained by simply dividing a color cube into a smaller cube, but the result is usually poor. Better results are achieved by means of clustering algorithms such as K-means or fuzzy c-means. However, a major drawback of these algorithms is a high computational time. On the other hand, there are still fast quantization algorithms characterized by high quality performance and time efficiency.

When a palette has been designed, what remains is to assign the original color of each pixel in the input image to their best match in the color palette. The simplest way is to compute the distances between the original color vectors and all color vectors of the new palette, then choose the one with the minimum distance. However, faster methods can be used, such as binary tree search or k-d tree search.

Of course, quantization of the color space alone is not enough to be appropriate for the image segmentation. One of the reasons is that the new color palette is disordered, and therefore the histogram of the image is also chaotic and not useful for thresholding. One of the way to deal with that problem is to sort the new color palette. When sorted, the palette, and therefore the histogram, should consist of the organized data, that visually resembles the original image. In this approach the color information (palette) is discarded and the image is treated like a monochrome one, by using only the frequency of the color occurrence.

This technique requires a definition of the distance measure, such as the Euclidean distance, in order to correctly sort color vectors. However, the RGB color space is non-uniform and thus it is hard to measure color differences. Hence, other color space, like the CIE $L^*a^*b^*$ which is perceptually uniform and efficient in measuring a small color difference, can be used to sort the color palette, or even as a substitute of the RGB space for the entire color quantization process. This approach can be considered as a nonlinear projection of 3-D space onto a lower dimensional 1-D space. Also the big advantage of this method is that it can be used for any type of monochrome segmentation techniques.

### Exemplary results

RGB images were converted to gray levels, for the non-fuzzy reference the Otsu technique was employed. Exemplary results obtained for techniques can be seen on Fig. 7.18. For the recursive thresholding the best results were achieved by F1 with HW MF followed closely by F2 and Otsu techniques.

Afterwards, the color quantization approach, described in Sec. 7.2.7 was utilized in order to quantize color cell images. Color images were quantized using

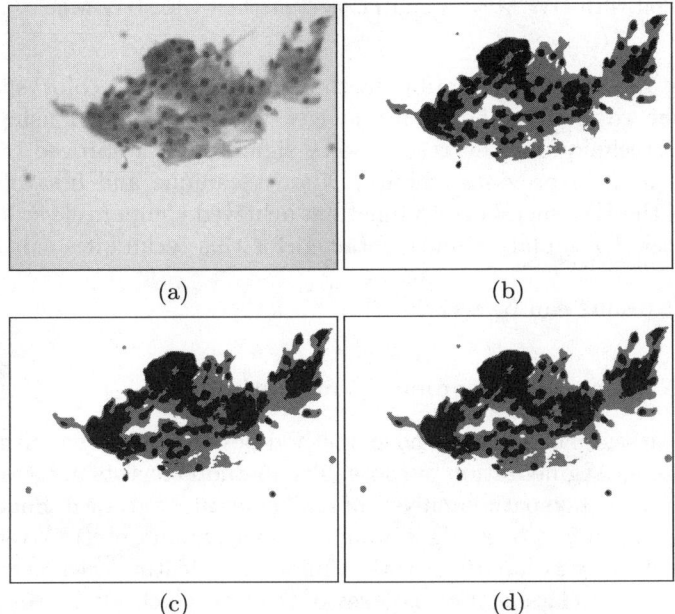

(a)    (b)

(c)    (d)

**Fig. 7.18.** The comparison of different methods of conversion from RGB images to gray levels: original image (a), Otsu method (b), Fuzzy-1 method with HW MF (c) and Fuzzy-2 method with HW MF

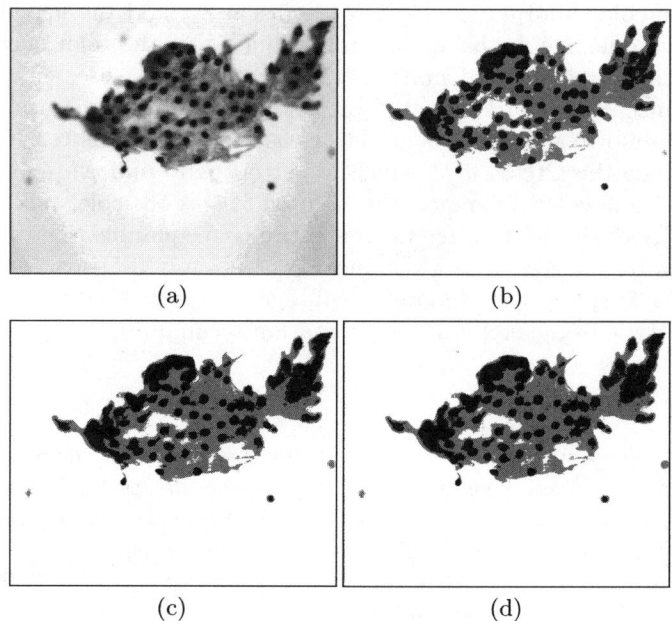

(a)    (b)

(c)    (d)

**Fig. 7.19.** Color quantization approach: $L^*a^*b^*$, $L^*a^*b^*$ sorted (a), Otsu method (b), Fuzzy-1 method with HW MF (c) and Fuzzy-2 method with HW MF

a minimum variance quantization method, using $L^*a^*b^*$ color spaces. Subsequently, the resulted monochrome images were thresholded using recursive thresholding techniques. The results were significantly improved over normal gray images used in previous example. Otsu technique and fuzzy based techniques, with the HW membership function, achieved comparable score with minor differences. Exemplary results obtained for this techniques can be seen on Fig. 7.19.

The exact results can be seen in [7].

### 7.2.8    The Sonar-Like Segmentation Method

The sonar-like segmentation is a novel method developed by the authors for the cytological image segmentation purpose. The method consists in classification of image pixels based on spatial analysis of a pixel feature variance. Each class represents a visual artefact (e.g. edges, uniform inner regions, etc.). Artefact classes are distinguished by comparison with a number of feature variance templates. The result of pixel classification is a set of regions which after proper merging allows for identification of actual objects. The name of the method originates from a similarity of the feature variance analysis mechanism to the physical phenomenon of a sonar sound wave speed alternation between water regions of different physical condition (temperature, pressure, etc.).

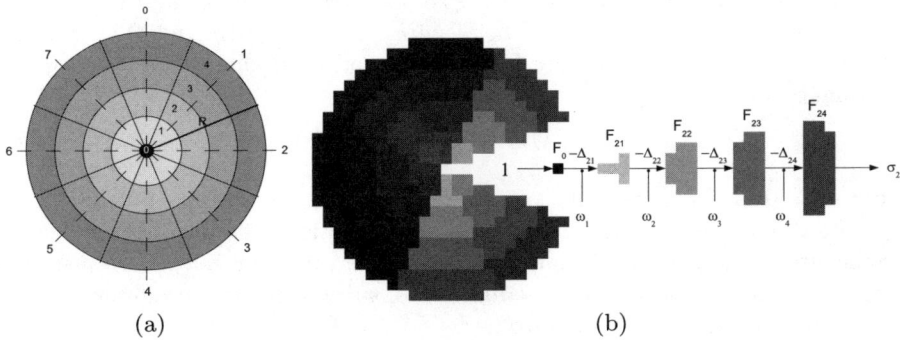

**Fig. 7.20.** Sonar neighborhood model for $s = 8$, $r = 4$ and $\theta = 0$ (a) and its discrete realization for range $R = 8$ (b)

The base concept for the feature variance analysis is the sonar pixel neighborhood. It is defined as a set of concentrically situated rings of pixels. All of the rings have the same width and together they cover a circle, which centre is in the examined pixel. The circle is then partitioned into a number of slices along concentrically located bearings. Intersections of the rings and the slices are the neighborhood sectors. The number of sectors in each ring is the same, and it equals to the number of the neighborhood bearings.

Formally, the sonar neighborhood $N_{sonar}$ can be defined as quadruplet:

$$N_{sonar} = (R, r, s, \theta) \tag{7.19}$$

where $R$ is the neighborhood range, $r$ is the number of rings, $s$ is the number of bearings (sectors in each ring) and $\theta$ is the angle between the zero-sector axis and the 'north' bearing. The neighborhood bearings are numbered from 0 and rings are numbered from 1. The zero-ring is the examined pixel itself. Figure 7.20 shows a visual model of the sonar neighborhood for $s = 8$, $r = 4$ and $\theta = 0$ (a) and its discrete realization for $R = 8$ (b).

The sonar neighborhood applied to an actual pixel allows for calculation of the sonar vector $\boldsymbol{\sigma}$. The length of the vector equals to the number of bearings, so each of the vector values represents the variance of the examined feature along single bearing. The values bounded to $[0, 1]$ are calculated on the basis of a set of differences of the feature statistics between subsequent sectors along the bearing. The statistics used in the presented application of the method was arithmetic mean of the luma component of the YCbCr color model.

Figure 7.20(b) shows a scheme of a sonar vector value calculation for a single bearing. The value can be seen as the energy of the sound wave signal after passing from the central point through subsequent sectors along bearing. Passage through each boundary between two adjacent sectors decreases the signal energy by $\Delta_{si}$ value, which is calculated as a weighted function of the sectors feature statistics difference:

$$\Delta_{si} = \delta(|f_{s,i} - f_{s,i-1}|)w_i \tag{7.20}$$

where $f_{s,i}$ is the feature statistics of the sector at $s$-bearing of the $i$-th ring, and $w_i$ is the weight for the boundary between $i-1$-th and $i$-th ring. The function $\delta$ is a monotonous nondecreasing function, bounded to $[0, 1]$. The function can be defined as follows:

$$\delta(t) = \begin{cases} \frac{t}{d} & if \ t \leq d, \\ 1 & if \ t > d, \end{cases} \tag{7.21}$$

where $d > 0$ is the sensitivity parameter defining the threshold value of the statistics difference over which the value of $\delta$ function remains 1. If the feature variance in the neighborhood is high enough, the value can reach zero just after subtracting only few of the delta values. If so, subsequent feature differences are ignored and the value remains zero.

The weights for subsequent boundaries are inversely proportional to the distance of the boundary to the central point, so differences between sectors closest to the neighborhood centre have the greatest impact on the sonar vector value. The weights sequence can be arbitrarily chosen or defined as a function of the ring number, for example:

$$w_i = \frac{1}{2^i w} \tag{7.22}$$

where $w$ is the arbitrary weighting factor.

The calculated sonar vector can be visualized with a radar plot. A shape of the plot expresses the feature variance in the neighborhood of the examined pixel. Later in this section, the sonar vectors visualizations are stated as sonar views.

The sonar vector matrix representing the image, prepared as described above, is used for a classification of the image pixels. The classification is performed by the comparison of the pixel sonar vector with a number of template vectors. The template vectors can be prepared manually or calculated automatically by the evaluation of average sonar vectors for a set of images classified with reference segmentation masks.

Some of the searched artifacts, like edges, have a number of same-shaped, but rotated sonar vectors. To reduce computational cost regarding introduction of a number of sub-classes for single artefact, a normalization of sonar vectors can be performed beforehand. The normalization process is a cyclic rotation of the vector values, until the lowest value is at the zero-bearing. After such an operation, each of the artefact sub-class sonar vectors equals the vector of the unrotated artefact with lowest value at the zero-bearing. So, a single class template can be produced. Figure 7.21 shows sonar views and normalized sonar views for pixels located at differently aligned nucleus edges.

Actual classification of the image pixels is preformed by finding the minimal mean squared error between the pixel sonar vector and each of the template vectors. The mean squared error $MSE$ of the sonar vector compared to the $C$-class template vector is defined as follows:

$$MSE_C(\boldsymbol{\sigma}) = E((\boldsymbol{\sigma} - \boldsymbol{\sigma_C})^2) = \frac{1}{s} \sum_{i=0}^{s} (\sigma_i - \sigma_{Ci})^2 \tag{7.23}$$

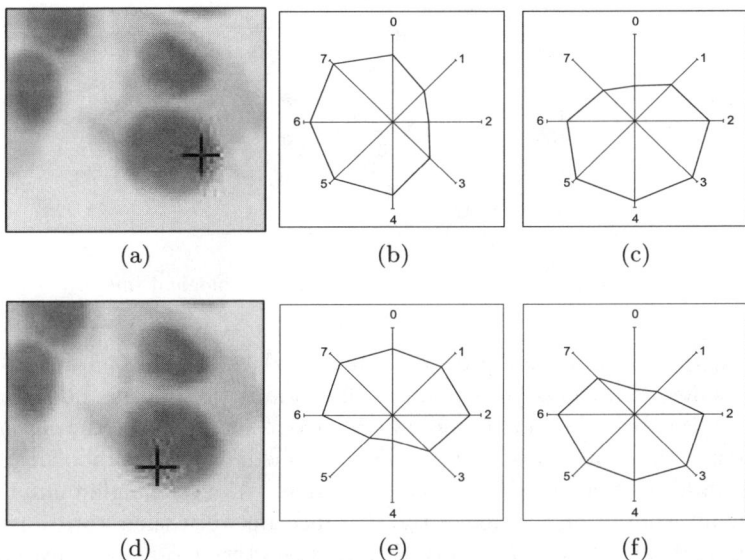

(a)  (b)  (c)

(d)  (e)  (f)

**Fig. 7.21.** Sonar views and normalized sonar views for pixels located at differently aligned nucleus edges: (a), (d) - location of pixels; (b), (e) - sonar views; (c), (f) - normalized sonar views

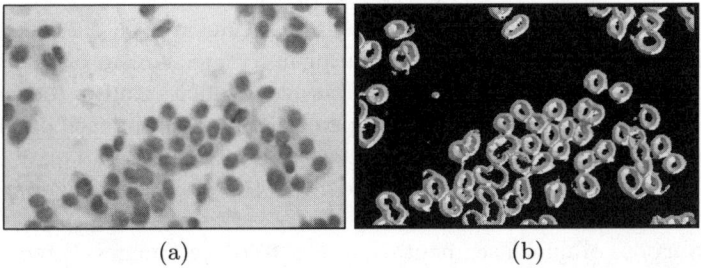

(a)  (b)

**Fig. 7.22.** Cytological image (a) and the result of the pixel classification with Sonar (b)

where $\sigma$ is the sonar vector, $\sigma_C$ is the sonar vector template of the class $C$, and $s$ is the number of bearings. The pixel is labeled with the label of the class with the minimal mean squared error. Figure 7.22 shows a sample cytological image and the result of the pixel classification with Sonar. For the 'edge' class the saved number of the rotation steps is marked with the greyscale, from white color for the bearing 0 to dark grey for the bearing 7.

At this stage all artifacts are identified and located. For cytological FNB images the expected result is an identification and location of nuclei. Each of the nuclei consists of a single interior and a number of adjacent edge-artifacts. To achieve the proper segmentation, an additional step of artifacts merging needs to be performed.

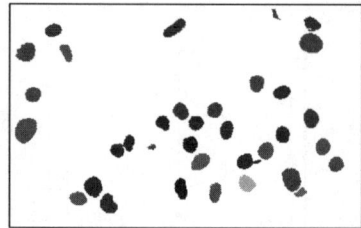

**Fig. 7.23.** Result segmentation of the cytological image

As can be observed in the Figure 7.22, actual edges of objects are surrounded from both sides with the edge segments. One side is the nucleus boundary, and the other is the background boundary. Adjacent segments of the actual edge have opposite rotation, so it can be distinguished which neighboring interior segment should be merged with the edge segment. Along a single boundary, the edge segments should be merged if the distance between their rotations equals 1. The segments should be also merged with the interior segment located below the 'north' edge. The result of the segment merging process is the segmentation of the image along actual edges of objects (Fig. 7.23).

### 7.2.9    Decision Tree Method

Another algorithm of the pixel-based segmentation area is the decision tree-based method of pixel classification. The primary mechanism of the method is classification of image pixels with a decision tree, which input is the pixel color, and the output is the probability of the pixel membership in each of the applied object-classes. It is assumed that the numerical information on a pixel color (e.g. RGB color components) is sufficient to identify the pixel as a member of one of a number of object-classes.

For the purpose of nuclei segmentation in cytological images, three classes are introduced: the nucleus (N), the background (B) and the inter-nucleus (I) class. The last of the three classes represents all objects which cannot be unambiguously labeled as a member of one of the former classes. These pixels are elements of erythrocytes and cytoplasm.

The decision tree training process is based on a set of manually prepared three-color image masks along with regarding training images. All image masks classify a set of pixels to the three classes. Classified pixels of all the training images together are used as the information for the decision tree training process. The result of the process is a decision tree, which assigns probabilities of the pixel being a member of each class. Due to no spatial information is used in the training process there is no requirement for the masks of images to be completed. Some of the mask pixel can be left unlabeled. It allows excluding ambiguous pixels from the decision tree training process. Figure 7.24 shows a sample training image (a) with regarding mask (b).

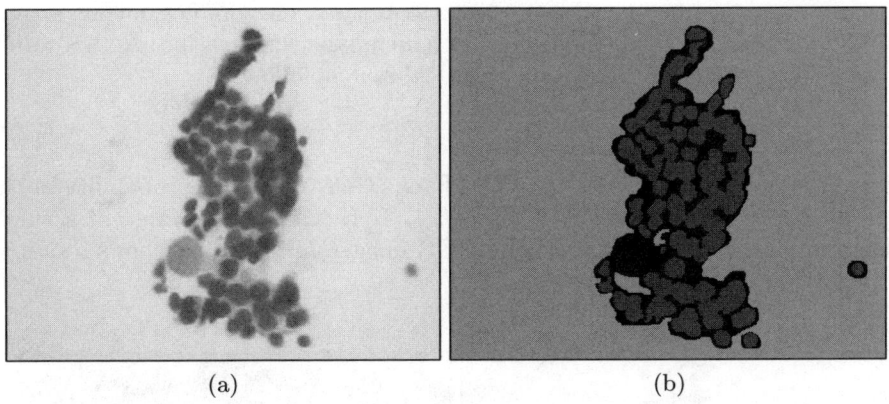

(a)                            (b)

**Fig. 7.24.** Sample training image (a) with regarding mask (b). Red - the nucleus pixels; green - the background pixels; blue - inter-nucleus pixels; black - unclassified (ambiguous) pixels

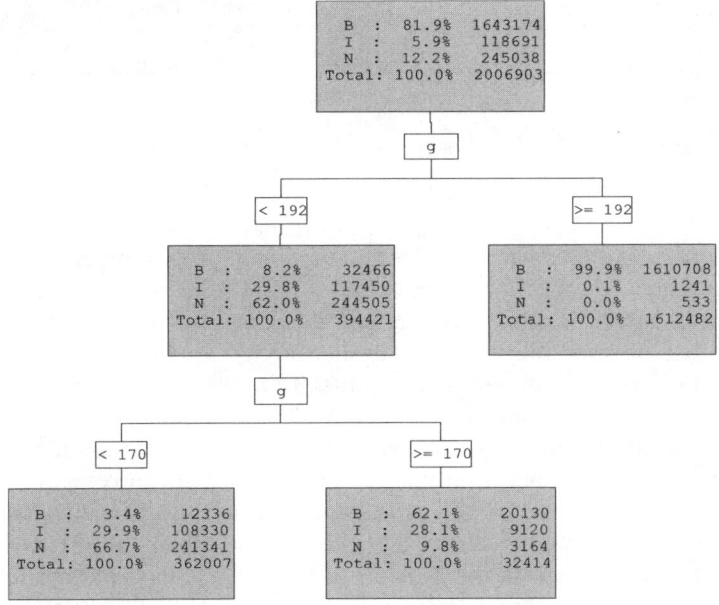

**Fig. 7.25.** Sample decision tree with 3 leaves

As stated before, the output of the decision tree for a pixel is a set of probabilities of the pixel membership in each class. Figure 7.25 presents a sample decision tree with 3 leaves.

The output of the decision tree can be visualized by rendering the image with pixel color components proportional to the respective probabilities. Using the same components for classes as in mask images, one can get output images similar to the masks, however there will be no unlabeled pixels left. Actual values of color components for pixels can be calculated as follows:

$$C = P(O)C_{max} \tag{7.24}$$

where $C$ is the color component, $C \in \{R, G, B\}$, $P(O)$ is the probability of the class $O$ membership, $O \in \{N, I, B\}$, $C_{max}$ is the maximum value of a single color component (255 for 24-bit color RGB images). Figure 7.26 shows a sample output image.

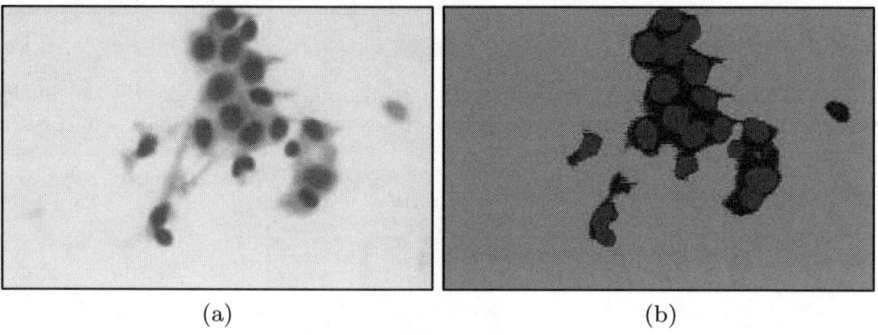

(a)                                          (b)

**Fig. 7.26.** Sample result of the decision tree application: input image (a), output image (b)

The number of combinations of class-membership probabilities in the decision tree output is much lower than the number of RGB color components in the processed image. Output images have the maximum number of distinct colors equal to the number of probability combinations. So the application of the decision tree can be perceived as the problem complexity reduction technique for another segmentation algorithm. Due to the output of the method are rendered images along with probability matrices, almost any segmentation technique can be applied on the actual segmentation stage. For example the thresholding with 3D homograms can be applied for the actual segmentation. In the authors' research the decision tree prepared with the SAS Enterprize Miner had 21 to 56 leaves. The number of leaves equals the number of bins of a homogram needed to perform the segmentation, so the decision tree can play the role of color quantizer based on the information on the image color characteristics.

## 7.3 Conclusions

In this chapter we bring together the latest results from researchers involved in state-of-the-art work in cytological image segmentation, providing both a survey on segmentation well-known techniques supporting such processes as measurement, visualization, registration and reconstruction of image and a collection of new approaches elaborated for the case of cytopathologic scans. A wide variety of new methods is presented, including solutions based on fuzzy sets of types I and II, clustering, decision trees, shape detection, active contours and many others as well as they hybrids. Issues of automated segmentation of cell nuclei are broadly described on the examples of microscopic cytological images obtained via fine needle biopsy technique. Although some of the predictions would probably be shared by many people working in this field, this presentation still will be subjective and personal. In our opinion, perspectives for further development of cytological image segmentation are closely connected with computational intelligence, closer interaction between system and a human operator as well as semantic image interpretation.

# References

1. Ballard, D.: Generalizing the Hough transform to Detect Arbitrary Shapes. Pattern Recogn 13(2), 111–122 (1981)
2. Boldrini, J., Costa, M.: An Application of Optimal Control Theory to the Design of Theoretical Schedules of Anticancer Drugs. Int. J. Appl. Math. and Comput. Sci. 9(2), 387–399 (1999)
3. Carlotto, M.: Histogram analysis using a scale space approach. IEEE Trans. Pattern Analysis and Machine Intelligence 9(1), 121–129 (1987)
4. Chen, C., Luo, J., Parker, K.: Image segmentation via adaptive K-mean clustering and knowledge-based morphological operations with biomedical applications. IEEE Trans. Image Processing 7(12), 1673–1683 (1998)
5. Cheng, H.D., Chen, C.H., Chiu, H.H., Xu, H.: Fuzzy homogeneity approach to multilevel thresholding. IEEE Trans. Image Processing 7(7), 1084–1088 (1998)
6. Duda, R., Hart, P.: Use of the Hough Transformation to Detect Lines and Curves in Picture. Comm. ACM 15, 11–15 (1972)
7. Dziekan, L., Marciniak, A., Obuchowicz, A.: Segmentation of color cytological images usin type II fuzzy sets. In: Korbicz, J., Patan, K., Kowal, M. (eds.) Fault Diagnosis and Fault Tolerant Control, pp. 263–270. Academic Publishing House EXIT, Warszawa (2007)
8. Fu, K.S., Mui, J.K.: A survey on image segmentation. Pattern Recogn 13, 3–16 (1981)
9. Hrebień, M., Nieczkowski, T., Korbicz, J., Obuchowicz, A.: The Hough transform and the GrowCut method in segmentation of cytological images. In: Proc. Int. Conf. Signals and Electronic Systems ICSES 2006, Łódź, Poland, vol. 1, pp. 367–370 (2006)
10. Hrebień, M., Korbicz, J., Obuchowicz, A.: Hough transform (1+1) search strategy and watershed algorithm in segmentation of cytological images. In: Proc. 5th Int. Conf. Comp. Recogn. Systems CORES 2007. Adv. in Soft Computing, vol. 45, pp. 550–557. Springer, Heidelberg (2007)
11. Hrebień, M., Steć, P.: Fine Needle Biopsy Material Segmentation with Hough Transform and Active Contouring Technique. Journal of Medical Inform. & Techn. 10, 25–34 (2007)
12. Gonzalez, R., Woods, R.: Digital Image Processing. Prentice-Hall, Englewood Cliffs (2002)
13. Huang, L.K., Wang, M.J.: Image thresholding by minimizing the measure of fuzziness. Pattern Recogn. 28, 41–51 (1995)

14. Jain, A.K.: Fundamentals of Digital Image Processing. Prentice-Hall, Englewood Cliffs (1989)
15. Karnik, N.N., Mendel, J.M., Liang, Q.: Type-2 fuzzy logic systems. IEEE Trans. Fuzzy Syst. 7, 643–658 (1999)
16. Kass, M., Witkin, A., Terauzopoulos, D.: Snakes: active contour models. In: Proc. 1st Int. Conf. on Computer Vision, pp. 259–263 (1987)
17. Kaufmann, A.: Introduction to the Theory of Fuzzy Subsets—Fundamental Theoretical Elements. Academic Press, New York (1975)
18. Kimmel, M., Lachowicz, M., Świerniak, A. (eds.): Cancer Growth and Progression, Mathematical Problems and Computer Simulations. Int. J. of Appl. Math. and Comput. Sci. 13(3) (2003) (special Issue)
19. Lee, M., Street, W.: Dynamic learning of shapes for automatic object recognition. In: Proc. 17th Workshop Machine Learning of Spatial Knowledge, pp. 44–49 (2000)
20. Madisetti, V., Williams, D.: The Digital Signal Processing Handbook. CRC Press, Boca Raton (1997)
21. Marciniak, A., Monczak, R., Kołodziński, M., Prętki, O.A.: Test base for the breast cancer diagnosis using FNB method. In: Proc. Nat. Conf. Artificial Intelligence in Biomedical Engineering SIIB 2004, Kraków, Poland, [4] CD-ROM (2004) (in Polish)
22. Marciniak, A., Obuchowicz, A., Monczak, R., Kołodziński, M.: Cytomorphometry of Fine Needle Biopsy Material from the Breast Cancer. In: Proc. 4th Int. Conf. Comp. Recogn. Systems CORES 2005. Adv. in Soft Computing, pp. 603–609. Springer, Heidelberg (2005)
23. Mendel, J.M., John, R.I.: Type-2 Fuzzy Sets Made Simple. IEEE Trans. Fuzzy Systems 10(2), 117–127 (2002)
24. Mendel, J.M.: An architecture for making judgments using computing with words. Int. J. Appl. Math. Comput. Sci. 12(3), 325–335 (2002)
25. Michalewicz, Z.: Genetic Algorithms + Data Structures = Evolution Programs. Springer, Heidelberg (1996)
26. Nevatia, R.: Image Segmentation. In: Young, T.Y., Fu, K.S. (eds.) Handbook of Pattern Recognition and Image Processing. Academic Press, NY (1986)
27. Nieczkowski, T., Obuchowicz, A.: Application of cellular automaton for enhancing segmentation results of breast cancer fine needle biopsy microscope images. In: Kłopotek, M., Tchórzewski, J. (eds.) Proceedings of Artificial Intelligence Studies, vol. 3, pp. 71–78. University of Podlasie Press, Siedlce (2006)
28. Nieczkowski, T., Obuchowicz, A.: 'Sonar' - Region of Interest Identification and Segmentation Method for Cytological Breast Cancer Images. In: Proc. 5th Int. Conf. Comp. Recogn. Systems CORES 2007. Adv. in Soft Computing, vol. 45, pp. 566–573. Springer, Heidelberg (2007)
29. Nieczkowski, T., Obuchowicz, A.: Application of decision trees to filtering and segmentation of breast cancer fine needle biopsy microscope images. Biocybernetics and Biomedical Engineering 27(4), 59–70 (2007)
30. Obuchowicz, A.: Evolutionary Algorithms for Global Optimization and Dynamic System Diagnosis. Lubuskie Scientific Society Press, Zielona Góra (2003)
31. Obuchowicz, A., Korbicz, J.: Evolutionary methods in designing diagnostic systems. In: Korbicz, J., Kościelny, J.M., Kowalczuk, Z., Cholewa, W. (eds.) Fault Diagnosis: Models, Artificial Intelligence, Applications, pp. 301–331. Springer, Heidelberg (2004)
32. Olabarriaga, S.D., Smeulders, A.W.M.: Interaction in the segmentation of medical images: A survey. Medical Image Analysis 5, 127–142 (2001)

33. Otsu, N.: A threshold selection method from grey-level histograms. IEEE Trans. Systems, Man and Cybernetics 9(1), 62–66 (1979)
34. Pal, N.R., Bezdek, J.C.: Measures of fuzziness: a review and several new classes. In: Yager, R.R., Zadeh, L.A. (eds.) Fuzzy Sets, Neural Networks and Soft Computing, Van Nostrand Reinhold, New York (1994)
35. Pena-Reyes, C., Sipper, M.: Envolving fuzzy rules for breast cancer diagnosis. In: Proc. Int. Symp. on Nonlinear Theory and Application, vol. 2, pp. 369–372. Polytechniques et Universitaires Romandes Press (1998)
36. Pratt, W.K.: Digital image processing. Wiley, Chichester (2001)
37. Russ, J.: The Image Processing Handbook. CRC Press, Boca Raton (1999)
38. Setiono, R.: Extracting rules from pruned neural networks for breast cancer diagnosis. Artificial Intelligence in Medicine, 37–51 (1996)
39. Sethian, J.: Fast marching methods. SIAM Review 41(2) (1999)
40. Steć, P., Domański, M.: Video Frame Segmentation Using Competitive Contours. In: Proc. 13th European Signal Processing Conference, Antalya, Turkey (2005)
41. Street, W.: Xcyt: A system for remote cytological diagnosis and prognosis of breast cancer. In: Jain, L. (ed.) Soft Computing Techniques in Breast Cancer Prognosis and Diagnosis, pp. 297–322. World Scientific Publishing, Singapore (2000)
42. Su, M., Chou, C.: A modified version of the K-means algorithm with a distance based on cluster symmetry. IEEE Trans. Pattern Analysis and Machine Intelligence 23(6), 674–680 (2001)
43. Świerniak, A., Ledzewicz, U., Schättler, H.: Optimal Control for a Class of Compartmental Models in Cancer Chemotherapy. Int. J. of Appl. Math. and Comput. Sci. 13(3), 357–368 (2003)
44. Tadeusiewicz, R.: Vision Systems of Industrial Robots. WNT, Warszawa (in Polish) (1992)
45. Tizhoosh, H.R.: Image thresholding using type II fuzzy sets. Pattern Recognition 38, 2363–2372 (2005)
46. Vezhnevets, V., Konouchine, V.: "GrowCut" – Interactive Multi-Label N-D Image Segmentation by Cellular Automata. In: Proc. 15th Int. Conf. on Comp. Graphics and Appl. GraphiCon 2005, Novosibirsk, Russia, pp. 150–156 (2005)
47. Vincent, L., Soille, P.: Watersheds in digital spaces: an efficient algorithm based on immersion simulations. IEEE Trans. Pattern Analysis and Machine Intelligence 13(6), 583–598 (1991)
48. Wolberg, W., Street, W., Mangasarian, O.: Breast cytology diagnosis via digital image analysis. Analytical and Quantitative Cytology and Histology 15(6), 396–404 (1993)
49. Zhou, P., Pycock, D.: Robust statistical models for cell image interpretation. Image and Vision Computing 15(4), 307–316 (1997)
50. Zouagui, T., Benoit-Cattin, H., Odet, C.: Image segmentation functional model. Pattern Recognition 37(9), 1785–1795 (2002)
51. Żorski, W.: Image Segmentation Methods Based on the Hough Transform. Studio GiZ, Warszawa (in Polish) (2000)

# 8

# Curvature Flow Based 3D Surface Evolution Model for Polyp Detection and Visualization in CT Colonography

Dongqing Chen[1], Aly A. Farag[1], M. Sabry Hassouna[1], Robert L. Falk[2], and Gerald W. Dryden[3]

[1] Computer Vision & Image Processing (CVIP) Laboratory
Department of Electrical & Computer Engineering, University of Louisville
Louisville, Kentucky 40292, USA
{dqchen,farag}@cvip.louisville.edu
[2] Department of Medical Imaging, Jewish Hospital & St. Mary's Healthcare
Louisville, Kentucky 40202, USA
[3] Division of Gastroenterology/Hepatology, Department of Medicine,
University of Louisville
Louisville, Kentucky 40292, USA

**Summary.** Computerized Tomography (CT) colonography is an emerging noninvasive technique for screening and diagnosing colon cancers. Since colonic polyps grow outward from the colon wall, they are modeled as protrusion shapes. In this chapter, we propose a novel anisotropic 3D surface evolution model for detecting protrusion shape based colonic polyp on the curved surface. The important feature of the proposed model is that it can detect protrusions with both convex and concave shapes. Protrusion shapes are defined as the extension beyond the usual limits or above a plane surface. Based on Gaussian and mean curvature flows, the approach works by locally deforming the convex or concave surface until the second principal curvature goes to zero. The diffusion directions are changed to prevent convex surfaces from converting into concave shapes, and vice versa. The deformation field quantitatively measures the amount of protrudeness. We also designed a new color coding scheme for better visualization of the detected polyps. The proposed method has been evaluated by using synthetic phantoms and real colon datasets.

## 8.1 Introduction

Colorectal cancer is the second leading cause of death caused by cancers and the third most common form of cancer in the United States [1]. Most colorectal cancers begin as a polyp, which is a small, harmless growth on the colon wall. As a polyp gets larger, it can develop into a cancer.

Since colorectal cancer is largely preventable, several screening tests such as digital rectal exam, fecal occult blood test, flexible sigmoidoscopy, double-contrast barium enema, and colonoscopy are recommended for all people age 50 and above. Although the optical colonoscopy is currently the gold standard for colorectal cancer screening, it is invasive, uncomfortable and inconvenient.

T.G. Smolinski et al. (Eds.): Comp. Intel. in Biomed. & Bioinform., SCI 151, pp. 201–222, 2008.
springerlink.com

Computer tomographic colonography (CTC), also known as virtual colonoscopy (VC), is a minimally invasive technique and rapidly evolving diagnostic tool for the location, detection and identification of benign polyps in the early stage before their malignant transformation.

Since colonic polyps generally grow from the colon wall into the lumen like dome structures, they are normally modeled as protrusion shapes in the literature [20]- [28]. Different differential geometry-based algorithms were proposed for protrusion shaped-based polyp detection. Yoshida and Näppi [20] used the curvature analysis to characterize polyps, folds and colon walls in the extracted colon. Summers et al. [21] investigated the feasibility of geometric features-based shape analysis for polyp detection. Huang et al. [22] investigated three surface patch fitting methods for curvature estimation, which include Cubic B-spline, paraboloid, and quadratic polynomials. Napel et al. [24] attempted to develop an algorithm to classify the output of CTC. They aimed at eliminating false positives (FPs) and increasing specificity without sacrificing sensitivity. Yao et al. [25] presented an automatic method to segment colonic polyps, which was based on a combination of knowledge-based intensity adjustment, fuzzy c-mean (FCM) clustering and deformable models. In [26], two major improvements were made to extend to a 3D polyp segmentation. The two improvements were summarized as follows: 1) knowledge-guided intensity adjustment was extended to 3D, and 2) active contour models for 2D cases were replaced with 3D dynamic deformable surfaces. As a result, Yao et. al. [27] proposed their entire framework for colonic polyp segmentation based on fuzzy clustering and deformable models. Once the 2D polyp segmentation finished one slice, the procedure was propagated to the neighboring slices. Finally, all 2D segmentations were stacked up to generate a 3D segmentation for the whole volumetric dataset.

Although most of the current methods modeled the polyp as an approximately spherical or elliptical polypoid shape, real polyps have irregular shapes.

Curvature flow-based curve and surface evolution methods have been widely used in computer vision, pattern recognition and medical image analysis [2]. A famous application of curve or surface evolution is the deformable model which mainly includes the explicit deformable models (e.g. snake [10], balloon force model [12] and gradient vector flow (GVF) model [11]) and the implicit deformable models (e.g. level sets [14, 13, 3], geodesic active contour [17], and prior shapes based model [15, 16]).

To our knowledge, there are few models which apply curve or surface evolution methods for protrusion shape detection.

In [18], a 2D object border was divided into a set of local and global indentation and protrusion segments by extending the classic curvature scale-space filtering method. Then, the object shape was represented by arranging the resultant segments in hierarchical structure.

Recently, Wijk et al. [29] proposed an idea on surface evolution for protrusion detection. The points on convex parts of the protrusion iteratively moved inward and finally the protrusion was flattened. Protrudeness was quantitatively measured by the displacement amount, and protrusion candidates were detected by thresholding the displacement field. The method can locate the convex protrusion shapes with constraint $\kappa_2 > 0$, where $\kappa_2$ is the minimum principal curvature. However, only protrusions with convex shapes were detected by the constraint.

In [19], change detection was used in shape representation. Compared with convex changes with equal magnitude, concave shape had a dramatic advantage. Concavity introduced or removed from an object contour were more easily observed.

In this chapter, we present a comprehensive framework for colonic polyp detection and visualization. Based on Gaussian and mean curvature flows, we propose a novel anisotropic algorithm for 3D surface evolution to detect protrusion shape based colonic polyps. We have validated the proposed method using simulated phantom containing convex and concave synthetic polyps with different shapes and sizes. We also present the results using real colon datasets to demonstrate the effectiveness of the new surface evolution algorithm.

The rest of the chapter is organized as follows. Section 8.2 presents the surface parametric formulation and the mathematical fundamental of surface evolution and protrusion shape detection by this method. Section 8.3 introduces the proposed new surface evolution function based on the Gaussian and mean curvature flows. Section 8.4 discusses a new color coding scheme to highlight the detected protrusion shape based colonic polyps. The proposed method is validated by two synthetic phantoms and real colon datasets in Section 8.5. Section 8.6 concludes the paper.

## 8.2  Mathematical Background

### 8.2.1  Surface Parametric Formulation

Accurate protrusion shape representation and detection on the curved surface are important in computer vision, psychology and medical image analysis. The protrusion is defined as follows:

**Definition 1.** *Extension beyond the usual limits, or above a plane surface [7].*

A more formal presentation follows from the description of the surface shape using the principal curvatures. The definitions of convex and concave protrusions are given in **Definition** 2 and **Definition** 3, respectively.

**Definition 2.** *Convex protrusions are those regions on the surface where the minimum principal curvatures are larger than zero, which implies that the maximum principal curvatures are definitely larger than zero [29].*

**Definition 3.** *Concave protrusions are those regions on the surface where the minimum principal curvatures are less than zero, which implies that the maximum principal curvatures are less than zero as well.*

In this chapter, we use $S$ to denote a compact 3D surface which is regular, orientable in $R^3$. The definitions of compact, regular and orientable are defined as follows [4, 5, 6].

**Definition 4.** *Compact: The set K is compact if and only if it is bounded and closed.*

A compact surface has a triangulation with a finite number of triangles.

**Definition 5.** *Regular: A subset $M \subset R^3$ is called a regular surface if for each point $p \in M$, there exists a neighborhood $V$ of $p$ in $R^3$, and a map $X : U \to R^2$ onto $V \cap R^2$, such that*

1. *X is differentiable;*
2. *$X : U \to V \cap M$ is homeomorphic; and*
3. *Each map $X : U \to M$ is a regular patch.*

*where, U is an open set.*

**Definition 6.** *Orientable: A regular surface $M \subset R^3$ is called orientable, if each tangent space $M_p$ has a complex structure $J_p : M_p \to M_p$, such that $p \to J_p$ is a continuous function.*

### 8.2.2  Surface and Three Dimensional Differential Geometry

Considering a regular surface represented as $S(u, v) : R^2 \to R^3$, where $(u, v) \in [0, 1] \times [0, 1]$ and it can be expressed in the vector form as $S(u, v) = [x(u, v), y(u, v), z(u, v)]^T$ as shown in Figure 8.1.

The surface is assumed to have the derivatives as follows:

$$\frac{\partial}{\partial \mathbf{P}} S = \begin{pmatrix} x_u & x_v \\ y_u & y_v \\ z_u & z_v \end{pmatrix} \tag{8.1}$$

where, $\mathbf{P} = [u, v]^T$.

Let us define $S_u = (x_u, y_u, z_u)$ and $S_v = (x_v, y_v, z_v)$ as the first partial derivatives, $S_{uu} = (x_{uu}, y_{uu}, z_{uu})$, $S_{uv} = (x_{uv}, y_{uv}, z_{uv})$ and $S_{vv} = (x_{vv}, y_{vv}, z_{vv})$ as the second partial derivatives on the tangent plane of $S$, such that the first fundamental forms $E$, $F$ and $G$ can be expressed as follows.

$$E = S_u \bullet S_u$$

$$F = S_u \bullet S_v$$

$$G = S_v \bullet S_v$$

where, $\bullet$ denotes the dot product.

The unit surface normal $\hat{\mathbf{N}}$ can be calculated as:

$$\hat{\mathbf{N}} = \frac{S_u \times S_v}{\|S_u \times S_v\|} \tag{8.2}$$

where, $\times$ denotes the cross product, and $\|\bullet\|$ is the norm operator.

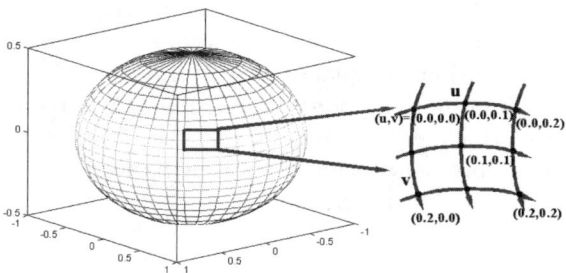

**Fig. 8.1.** 3D surface example parameterized by $(u, v) \in [0, 1] \times [0, 1]$ and its surface patch for demo

The second fundamental forms $L$, $M$ and $N$ of $S$ are expressed as follows.

$$L = S_{uu} \bullet \hat{\mathbf{N}}$$

$$M = S_{uv} \bullet \hat{\mathbf{N}}$$

$$N = S_{vv} \bullet \hat{\mathbf{N}}$$

### 8.2.3  Curvature Formulation

Given the first fundamental forms $E$, $F$ and $G$, and the second fundamental forms $L$, $M$ and $N$ of surface $S$, the maximum principal curvature $\kappa_1$ and the minimum principal curvature $\kappa_2$ are defined as the two roots $\lambda_1$ and $\lambda_2$, which satisfy the following identity:

$$\begin{vmatrix} E\lambda - L & F\lambda - M \\ F\lambda - M & G\lambda - N \end{vmatrix} = 0 \tag{8.3}$$

Simplifying Equation 8.3, we can get

$$(EG - F^2)\lambda^2 + (2FM - EN - GL)\lambda + (LN - M^2) = 0 \tag{8.4}$$

As a result, Gaussian curvature $K$ and mean curvature $H$ can be defined as

$$K = \kappa_1 \kappa_2 = \lambda_1 \lambda_2 = \frac{LN - M^2}{EG - F^2} \tag{8.5}$$

$$H = \frac{1}{2}(\kappa_1 + \kappa_2) = \frac{1}{2}(\lambda_1 + \lambda_2) = \frac{EN - 2FM + GL}{2(EG - F^2)} \tag{8.6}$$

Then the two principal curvatures can be computed as

$$\kappa_1 = H + \sqrt{H^2 - K} \tag{8.7}$$

$$\kappa_2 = H - \sqrt{H^2 - K} \tag{8.8}$$

Unfortunately, some problems exist for computing the principal curvatures as follows: 1) under discrete cases, $H^2 - K$ in Equations (8.7) and (8.8) is not always guaranteed to be greater than or equal to zero; 2) $\kappa_1$ and $\kappa_2$ computed in this way can not provide any information about direction; and 3) large neighborhood for the high accuracy of principal curvature increases the computational complexity.

We solve the above problems using accurate curvature estimation addressed by Taubin. Please see [33] for details.

### 8.2.4  Surface Representation and Evolution

In terms of the two principal curvatures, 3D surface shapes are represented in Table 8.1. Figure 8.2 shows some examples of 3D shape mesh surfaces.

**Table 8.1.** Surface shape representation by curvature analysis

|  | $\kappa_1 < 0$ | $\kappa_1 = 0$ | $\kappa_1 > 0$ |
|---|---|---|---|
| $\kappa_2 < 0$ | elliptic concave $(H < 0)$ | parabolic surface | hyperbolic surface $(H \neq 0)$ |
| $\kappa_2 = 0$ | parabolic surface | plane $(H = 0)$ | parabolic surface |
| $\kappa_2 > 0$ | hyperbolic surface $(H \neq 0)$ | parabolic surface | elliptic convex $(H > 0)$ |

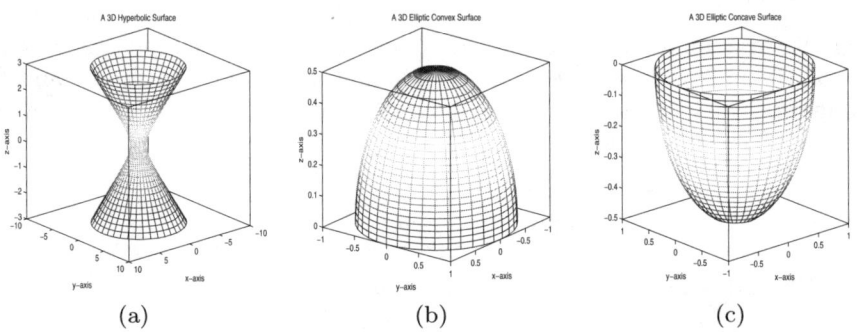

(a)    (b)    (c)

**Fig. 8.2.** Examples of 3D surface shape by curvature analysis, (a) Hyperbolic surface (K< 0 & H≠ 0), (b) Elliptic convex surface (K> 0 & H> 0) and (c) Elliptic concave surface (K> 0 & H< 0)

### 8.2.5  Mathematical Background on Surface Evolution

Let $S(X,t) : R^2 \times [0,T) \rightarrow R^3$ denote a family of closed surfaces, where t parameterizes the family and $X = (x, y, z)$ parameterizes the surface. Assume that this family of surfaces complies with the following partial different equation (PDE).

$$\frac{\partial S(X,t)}{\partial t} = \alpha(X,t)\overrightarrow{\mathbf{T}}(X,t) + \beta(X,t)\overrightarrow{\mathbf{N}}(X,t) \qquad (8.9)$$

with $S(X,0)$ as the initial condition, and **N** representing the inward unit normal.

**Lemma 1.** *The geometry of the deformation of $S(X, t)$ is only dependent of the normal component of the velocity field.*

*Proof.* Let us use the implicit format $\Phi(X, t)$ of the family of close surfaces $S(X, t)$: $\Phi(X, t) = 0$.

Taking derivative of $\Phi(X, t) = 0$ , we can get

$$\frac{\partial \Phi(X, t)}{\partial t} = 0 \tag{8.10}$$

In terms of chain rule, the above equation can be written as follows.

$$\frac{\partial \Phi}{\partial t} + \frac{\partial \Phi}{\partial X} \bullet \frac{\partial X}{\partial t} = 0 \tag{8.11}$$

$$\Phi_t + \bigtriangledown \Phi \bullet \vec{\mathbf{V}} = \vec{0} \tag{8.12}$$

where, $\bigtriangledown \Phi$ and $\vec{\mathbf{V}}$ represent gradient of $\Phi$ and velocity field, respectively.

If $\vec{\mathbf{V}}$ is denoted by $\vec{\mathbf{V}} = \alpha(v)\vec{\mathbf{T}} + \beta(v)\vec{\mathbf{N}}$, where $\alpha(v)$ and $\beta(v)$ are the velocity components in the tangent and normal direction, respectively, then

$$\Phi_t = - \bigtriangledown \Phi \bullet (\alpha(v)\vec{\mathbf{T}} + \beta(v)\vec{\mathbf{N}}) \tag{8.13}$$

$$\Phi_t = -\alpha(v) \bigtriangledown \Phi \bullet \vec{\mathbf{T}} - \beta(v) \bigtriangledown \Phi \bullet \vec{\mathbf{N}} \tag{8.14}$$

Since $\bigtriangledown \Phi \bullet \vec{\mathbf{T}} = \vec{0}$,

$$\Phi_t = -\beta(v) \bigtriangledown \Phi \bullet \vec{\mathbf{N}} = \beta\vec{\mathbf{N}}. \tag{8.15}$$

Following Lemma 1, the most general geometric deformation for a family of surface $S \colon R^2 \to R^3$ is given by

$$\frac{\partial S}{\partial t} = \beta\vec{\mathbf{N}} \tag{8.16}$$

where the geometric velocity is in the direction of the 3D normal $\vec{\mathbf{N}}$.

### 8.2.6  Laplacian Method Based Surface Evolution

In [34], the diffusion equation controls the evolution process of a 3D surface

$$\frac{\partial S}{\partial t} = \bigtriangledown^2 S \tag{8.17}$$

where, $\bigtriangledown^2$ is the Laplacian operator and $S$ represents a given 3D surface.

Equation 8.17 could be solved by the following finite difference approach.

$$S^{t+1} = S^t + \lambda \bigtriangledown^2 S^t \tag{8.18}$$

where, $\lambda$ is the regularization parameter governing the iterative process, and $t$ is the iteration time.

Under discrete case, the diffusion equation at mesh vertex $i$, is expressed

$$\frac{\partial X_i}{\partial t} = \lambda L(X_i) \tag{8.19}$$

$$with \quad L(X_i) = \left(\frac{1}{N_1} \sum_{j \in 1-ring} X_j\right) - X_i$$

where, $\lambda$ is the regularization parameter. $X_i$ is the position of the mesh point i, $N_1$ is the total number of vertices inside the 1-ring neighborhood of $X_i$, which includes all the neighboring vertices directly connecting with $i$ on the mesh surface.

Equation 8.19 could be solved by the bi-conjugate gradient method.

$$(\mathbf{I} - \lambda dt \mathbf{L}) X_i^{t+1} = X_i^t \tag{8.20}$$

where, $\mathbf{I}$ is the identity matrix.

## 8.2.7   Curvature Flow Based Surface Evolution

Desbrun *et al.* [35] proposed to use mean curvature flow to replace the Laplacian diffusion, thus the surface evolution becomes:

$$\frac{\partial X_i}{\partial t} = -H_i \overrightarrow{n}_i \tag{8.21}$$

where, $H_i$ and $\overrightarrow{n}_i$ are the mean curvature and the unit normal vector at $X_i$, respectively.

In [35], the diffusion was applied to all mesh vertices. The surface moved along the normal vector direction at a speed proportional to the mean curvature, and finally achieved the desired smooth result with respect to the shape. The surface evolution did not depend on the choice of external coordinate system, thus it was dependent on the intrinsic properties of mesh surface.

Wijk *et al.* [29] used Gaussian curvature flow for colonic polyp detection. Their method only diffused on a limited number of mesh points with $\kappa_2 > 0$, instead of the entire mesh surface, to reduce the computational complexity.

They introduced a '$force$' term by minimizing the second principal curvature $\kappa_2$. The resulting equation becomes

$$L(X_i) = F_i(\kappa_2) \tag{8.22}$$

The '$force$' field initially balanced the displacement prescribed by the Laplacian and was updated by solving Equation 8.23:

$$F_i^{t+1} = F_i^t - \kappa_2^t \frac{A_{1-ring}}{2\pi} \overrightarrow{n}_i \quad with \quad F_i^{t=0} = L(X_i) \tag{8.23}$$

where, $A_{1-ring}$ is the surface area of the 1-ring neighborhood around $i$.

Since the minimum principal curvature $\kappa_2 > 0$, was considered, the method only worked for convex protrusion shape polyps. We change the main constraint $\kappa_2 > 0$ to the Gaussian curvature $K > 0$ for both convex and concave protrusion shapes.

## 8.3   A New 3D Surface Evolution Model

In this section, we propose an anisotropic formula for surface evolution to detect the general protrusions with elliptic convex, concave and irregular shapes.

We consider constraints mainly on elliptic and hyperbolic points by simplifying the idea proposed in [36]. It aims to determine the appropriate moving direction of the velocity vector depending on two principal curvatures $\kappa_1$ and $\kappa_2$. The velocity diffusion directions are summarized in Table 8.2. The proposed algorithm deforms the local colon wall until the protrusions are flattened and the diffusion directions are changed to prevent convex surface from converting into concave shape, and vice versa. This idea is illustrated in Figure 8.3.

To achieve the above goal, we introduce the following diffusion equation:

$$\frac{\partial S}{\partial t} = \begin{cases} sgn(H)K\overrightarrow{n} & \text{if } K > 0 \quad and \quad H \neq 0 \\ \alpha K\overrightarrow{n} & \text{if } K < 0 \quad and \quad H \neq 0 \\ 0 & \text{if } K = 0 \quad or \quad H = 0 \end{cases} \quad (8.24)$$

where, $sgn$ is the sign function with

$$sgn(H) = \begin{cases} 1 \ \text{if } H \geq 0 \\ \text{-}1 \ \text{if } H < 0 \end{cases}$$

where, $K$ and $H$ are the Gaussian and mean curvatures, respectively.

**Table 8.2.** Gaussian and mean curvature flows based velocity diffusion direction

| Surface Shape Representation | Evolution Direction |
|---|---|
| Elliptic Convex $(K > 0, H > 0)$ | Moving Inward: $-\overrightarrow{n}$ |
| Elliptic Concave $(K > 0, H < 0)$ | Moving Outward: $\overrightarrow{n}$ |
| Hyperbolic Surface $(K < 0, H \neq 0)$ | Moving Inward(or Outward): $-\overrightarrow{n}$ (or $\overrightarrow{n}$) |
| Parabolic $(K = 0)$ | Not Moving |
| Others $(K \neq 0, H = 0)$ | Not Moving |

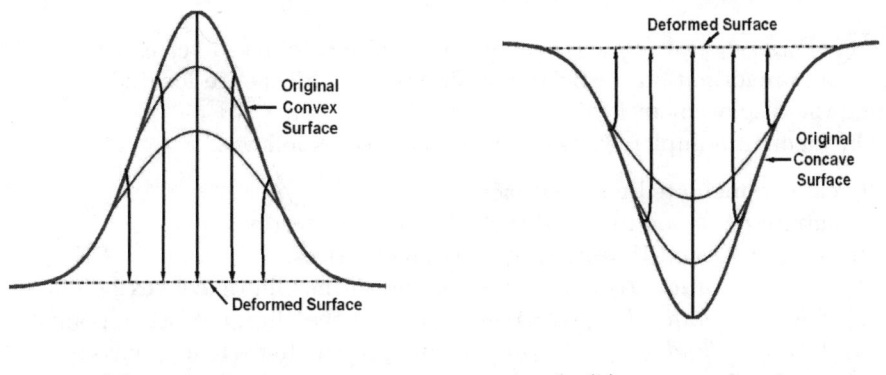

(a)convex shape                    (b) concave shape

**Fig. 8.3.** Typical polypoid shape surface evolution

As we know, $H$ could be zero under the discrete case even if $H \neq 0$ under the continuous case. In order to avoid this instability, the Equation 8.24 takes the following format by adding a strict constraint.

$$\frac{\partial S}{\partial t} = \begin{cases} sgn(H)K\overrightarrow{n} & \text{if } K > 0 \quad and \quad |H| \geq \epsilon \\ \alpha K\overrightarrow{n} & \text{if } K < 0 \quad and \quad |H| \geq \epsilon \\ 0 & \text{if } else \end{cases} \tag{8.25}$$

where, $|\alpha| < 1$ and $\epsilon > 0$.

Considering Equation 8.19 and Equation 8.25, at each $i$, we propose a new anisotropic diffusion function for surface evolution in term of $\kappa_1$ and $\kappa_2$.

$$\frac{\partial X_i}{\partial t} = -\beta(\kappa_1, \kappa_2)\overrightarrow{n}_i \tag{8.26}$$

with,

$$\beta_i(\kappa_1, \kappa_2) = \begin{cases} e^{-(1+K_i^{\gamma})}sgn(H)K_i & \text{if } K_i > 0, |H_i| \geq \epsilon \\ e^{-(1+K_i^{\gamma})}\alpha K_i & \text{if } K_i < 0, |H_i| \geq \epsilon \\ 0 & else \end{cases} \tag{8.27}$$

where, $1 \leq \gamma < \infty$.

After we introduce the new $'force'$ term, Equation 8.22 becomes

$$L(X_i) = \overline{F}_i(\kappa_1, \kappa_2) \tag{8.28}$$

By substituting $\kappa_2$ in Equation 8.23 by $\beta_i(\kappa_1, \kappa_2)$, we can get the resulting equation.

$$\overline{F}_i^{t+1} = \overline{F}_i^t - \frac{A_{1-ring}}{2\pi}\beta_i(\kappa_1, \kappa_2)\overrightarrow{n}_i \quad with \quad \overline{F}_i^{t=0} = L(X_i) \tag{8.29}$$

After mesh surface deformation, the displacement value is estimated by the following equation.

$$disp_i = |(P_{final})_i - (P_{initial})_i| \tag{8.30}$$

where, $(P_{initial})_i$ and $(P_{final})_i$ denote the positions of mesh vertex $i$ before and after mesh deformation, respectively. Protrusion objects are located by thresholding the displacement field.

The algorithm implementation is summarized as follows.

1. Create triangulated mesh surface.
2. Compute $\kappa_1$, $\kappa_2$ and normal vector $\overrightarrow{n}$ at each vertex $i$.
3. In term of K, and H, search the satisfied vertices:
   1) if $K > 0$ $and$ $H > 0$, corresponding to the elliptic convex points.
   2) if $K > 0$ $and$ $H < 0$, corresponding to the elliptic concave points.
   3) if $K < 0$ $and$ $H < 0$, corresponding to the hyperbolic points.
4. Compute $\beta(\kappa_1, \kappa_2)$ using Equation 8.27.
5. Compute $\overline{F}_i^{t=0}$ using $\overline{F}_i^{t=0} = L(X_i)$.
6. Update $\overline{F}_i^{t+1}$ iteratively using Equation 8.29.

7. Recompute $\kappa_1$, $\kappa_2$ and $\overrightarrow{n}$ at each vertex $i$, until obtain new positions of all satisfied points.
8. Detect polyp candidates by thresholding the displacement field by using Equation 8.30.

## 8.4  Color Coding Scheme for Visualization

### 8.4.1  Background

In virtual colonoscopy examination, image display formats can change the visibility and accuracy of the detected colonic polyps with protrusion shapes.

Index color coding is one of the important methods for improving the visualization effect of data. The main idea behind index color can be summarized as follows. First, 3D datasets are fed to the geometric features or statistical features criteria for protrusion shape based polyp detection. Second, protrusion candidates are typically assigned a color using their values indicated in a lookup table. Finally, the protrusion candidates are highlighted and distinguished from other colonic structures by the assigned colors. Summers *et al.* [21] used elliptic curvature as the primary shape criterion and three more strict shape criteria to detect colonic polyps. The results of synthetic polyp detection were visualized by coloring the colon using curvature analysis of the colon inner surface.

However, the lookup table is sensitive to the proper values of hue, saturation, value and alpha opacity (HSVA). Since it is not easy to balance the HSVA values, it is hard to understand and manipulate all the associated parameters.

Näppi *et al.* [37] presented a shape-scale color mapping approach for colon coloring. Characteristic signatures of the shape-scale signature were determined for different type lesions. Once the characteristic signature was determined, a unique color was assigned to a given type lesion. Since the RGB color space of the characteristic signatures may generate false colors, the $YC_rC_b$ was used to perform the color interpolation. This scheme was somewhat complicated and difficult to implement.

### 8.4.2  A New Color Coding Scheme

In this section, we propose a new color coding scheme. It is mainly based on creating an isosurface of 3D colon object, generating two polygonal datasets and assigning different colors to the two datasets to distinguish polyps from other tissues. This color coding denomination is analogous to placing two different things in two separated rooms and assigning different colors in order to visualize one from the other easily. The algorithm implementation is summarized as shown in Figure 8.4.

**Algorithm (An Easy Color Coding Scheme)**

1. Reconstructing 3D colon object;
2. Computing continuous curve skeleton of the 3D volumetric colon object and performing fly-through navigation using [38, 39];

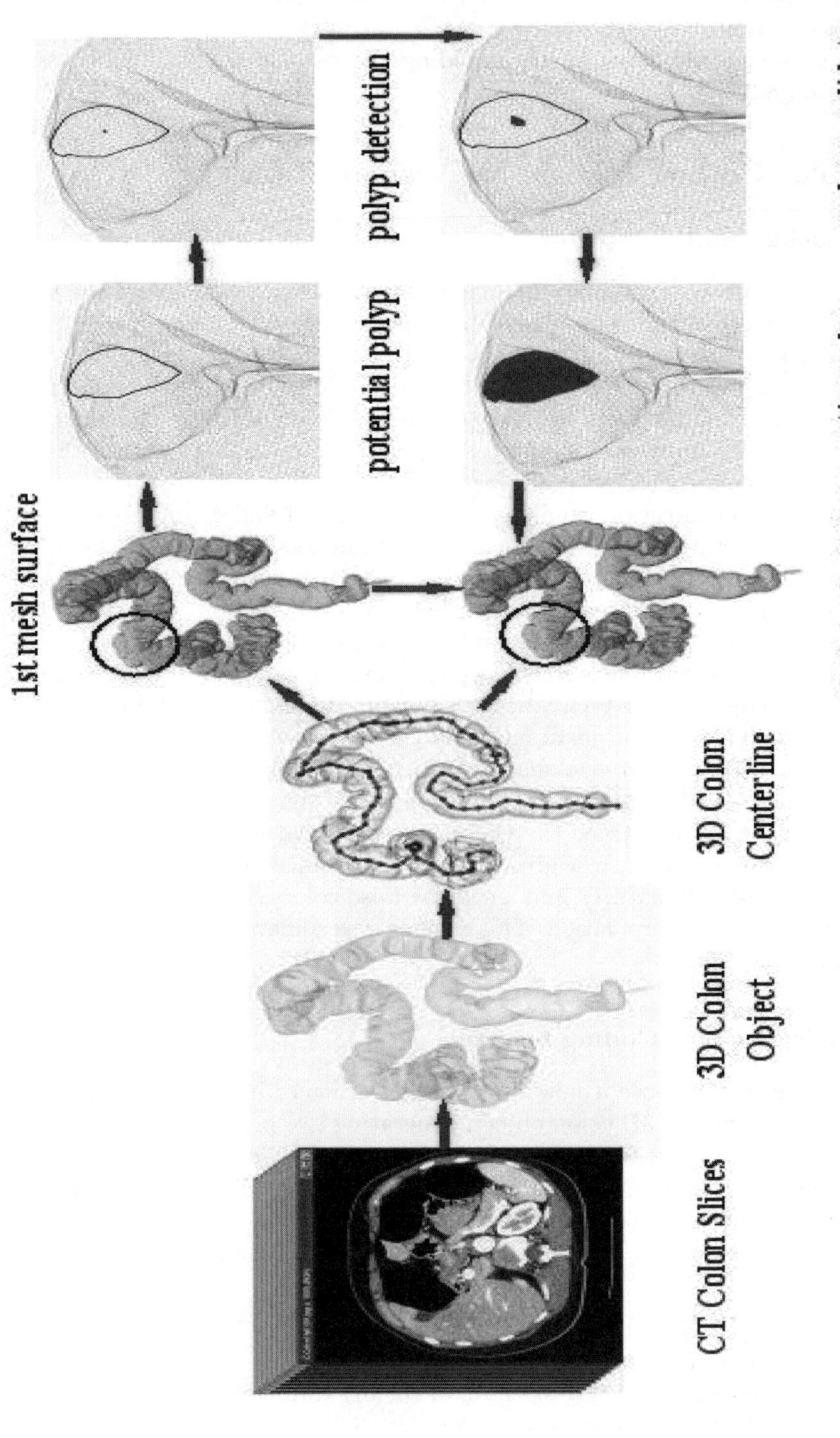

**Fig. 8.4.** The proposed color coding scheme

3. Creating isosurface of the 3D colon object using the standard *marching cubes* (MC) algorithm proposed in [40];
4. Generating the first polygonal dataset of triangle mesh on the isosurface;
5. Creating the second polygonal dataset with the same topology and geometric properties as the one in Step 4, but void at each point address. (Here, the topology is the set of properties invariant under certain geometric transformations, and geometry is the instantiation of the topology, the specification of position in 3D space. In this chapter, the topology of the polygonal dataset is triangle mesh, and we specify their geometric properties as providing voxel coordinates).
6. LOOP: Checking each triangle vertex of the first dataset:
   1) if it is a polyp candidate, insert the scalar values of the vertex and its neighboring vertices at the same location in the second polygonal dataset;
   2) if not, keep the vertex at the same location in the second polygonal dataset void;
   3) go back to LOOP, until all points in the first dataset are finished;
7. Assigning background color to the first colon polygonal dataset (containing colon inner wall and haustral folds) and foreground color to the second polygonal dataset (only containing polyp candidates).

Compared with other existing color coding methods, our algorithm differs in Step 5 and Step 6. It creates a second polygonal dataset with the same geometry and topology structures of the original 3D CT colon object, then associates polyp and non-polyp candidates with different datasets, finally assigns foreground color to the polyp candidates, and background color to the non-polyp tissues.

## 8.5 Validation, Result and Discussion

### 8.5.1 Validation

In this section, we validate the proposed 3D anisotropic surface evolution algorithm using different synthetic cylindrical phantoms with voxel size $1.0 \times 1.0 \times 1.0mm^3$. In this chapter, we assume that the surface normal is pointing into the cylinder phantoms and real colons. If the orientation of the entire isosurface is consistently defined, we can find three convex and one concave protrusions inserted at different locations as shown in Figure 8.5. The second convex protrusion shape is created as a ellipse-like protrusion with size $10 \times 12 \times 8mm^3$, while the other three spherical shapes are of sizes $20mm$, $30mm$ and $10mm$, respectively. The proposed algorithm runs iteratively until all the protrusion shapes become nearly flat, which means all Gaussian curvatures of those the vertices go to zeros.

Figure 8.6 shows that the proposed 3D surface evolution algorithm works on convex protrusion shape. The results of the first convex protrusion shape shown in Figure 8.5 after 20, 50 and 100 iterations are shown in (b) through (d). The results of the concave protrusion shape are shown in (a) through (d) in Figure 8.7.

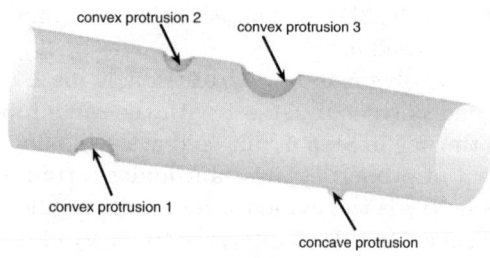

(a)

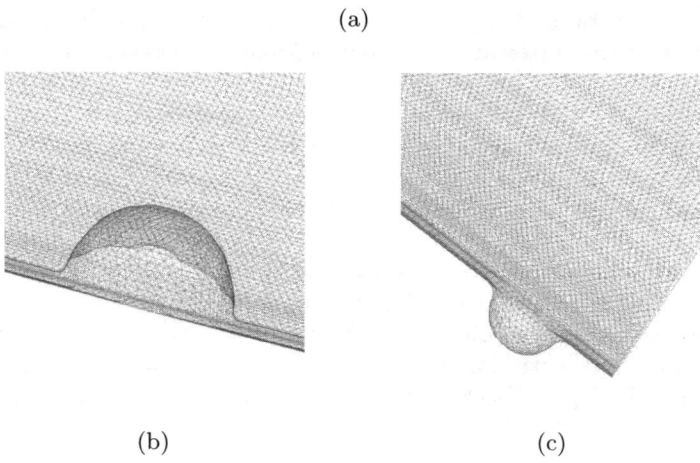

(b)                                    (c)

**Fig. 8.5.** The synthetic phantom I (a) consisting of three convex and one concave protrusion shapes, (b) local mesh surface of protrusion shape 1, and (c) local mesh surface of the concave protrusion shape

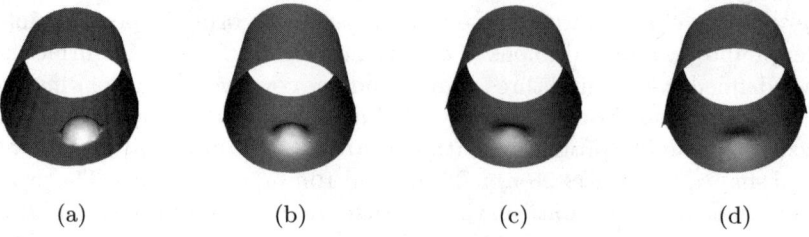

(a)              (b)              (c)              (d)

**Fig. 8.6.** Deformity of the first convex protrusion shape shown in Figure 8.5, (a) original protrusion shape, (b) after 20 iterations, (c) after 50 iterations and (d) after 100 iterations

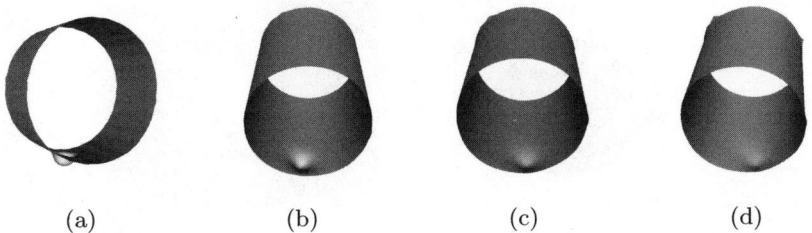

(a)                    (b)                    (c)                    (d)

**Fig. 8.7.** Deformity of the concave protrusion shape shown in Figure 8.5, (a) original protrusion shape, after (b) 20, (c) 50 and (d) 100 iterations

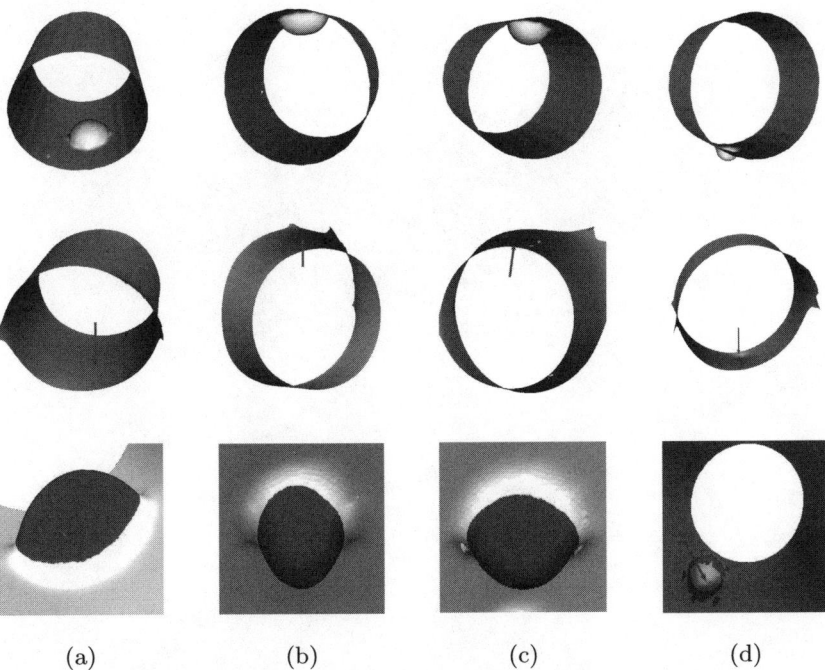

(a)                    (b)                    (c)                    (d)

**Fig. 8.8.** Four synthetic polyps (1, 2, 3 and 4) as shown from (a) to (d) on the first row. The second row shows the 'flattened' protrusion shapes, while the third row shows the results detected by the method proposed in this paper.

The final results for the whole phantom are illustrated in Figure 8.8. The four original synthetic protrusions are shown individually on the first row. After the iteration completes, the protrusion shape is flattened, which is shown on the second rows. The detection results are shown on the third row. Since the proposed method considers $K > 0$ (equivalently, $|\kappa_1| > |\kappa_2| > 0$) instead of $\kappa_2 > 0$ only, it can work for concave and convex protrusions unlike the method proposed in [29] (see Fig. 8.9). Since the fourth concave protrusion, with small

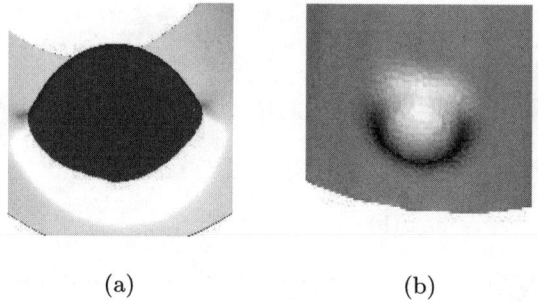

(a)                              (b)

**Fig. 8.9.** (a) One example detected and (b) the one missed by he method proposed in [29]

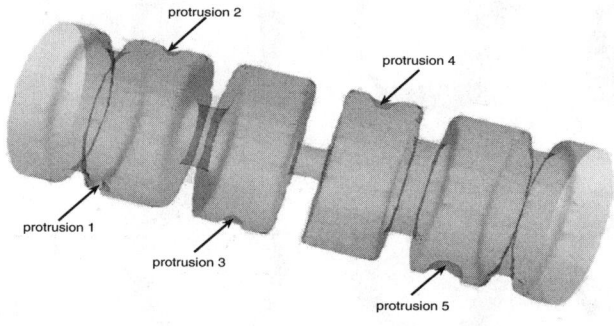

(a)

(b)

**Fig. 8.10.** A more complicated synthetic phantom II (a) consisting five protrusion shapes either on or between the folds, and (b) local mesh surface of protrusion 5

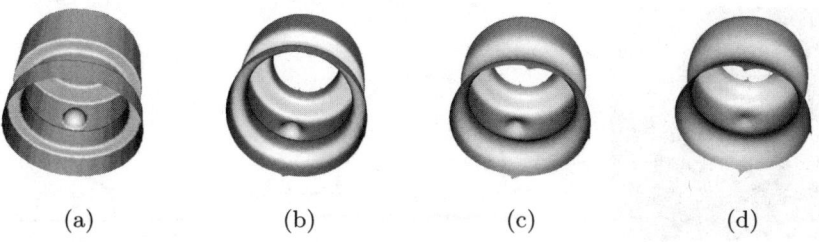

**Fig. 8.11.** Deform of the fifth convex protrusion shape shown in Figure 8.10, (a) original protrusion shape, after (b) 20, (c) 50 and (d) 100 iterations

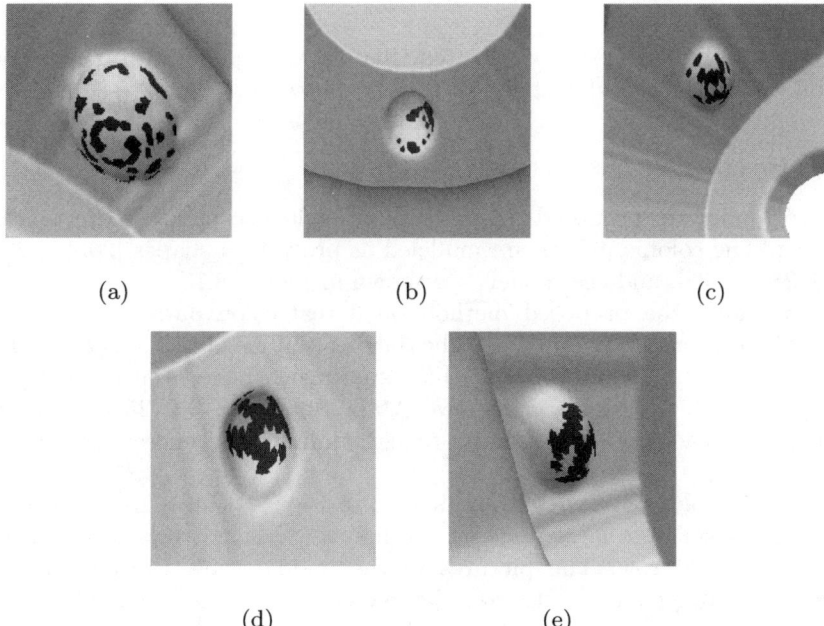

**Fig. 8.12.** Detection results of five protrusion shapes by the proposed method, and they are shown from (a) to (e) corresponding to the protrusion 1 to 5 in phantom II

size, is created to simulate the true concavity as a part of polyp surface, we find that the concavity is detected and highlighted.

A more complicated phantom is generated to test the performance of the proposed algorithm. Five thick folds with diameters ranging from $15mm$ to $60mm$ are simulated in the phantom as shown in Figure 8.10. Five protrusion shapes are created and inserted as shown on the folds or between two folds.

Figure 8.11 shows the proposed algorithm iteratively deforms the protrusion 5 in phantom II. Even though the protrusion is located between two folds and the

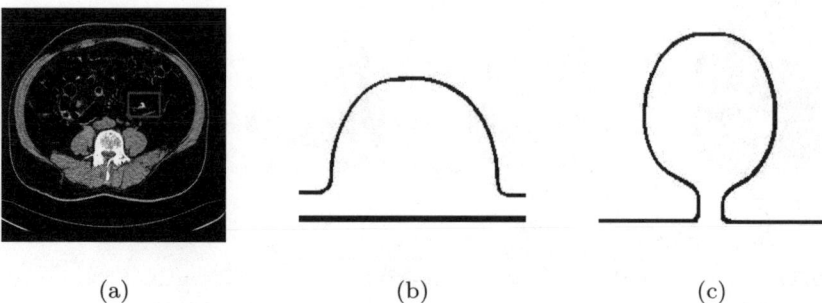

(a)                          (b)                          (c)

**Fig. 8.13.** Polyp modeled as protrusion shapes, (a) a colon CT slice, (b) sessile polyp model, and (c) pedunculated polyp model

geometry property is more complicated than those in phantom I, the proposed method works well and detects all of them as shown in Figure 8.12.

### 8.5.2   Result and Discussion

We have applied the proposed algorithm in CT colonoscopy for colorectal cancer screening. The colonic polyps are modeled as protrusion shapes [20, 21, 22, 23, 24, 30, 28, 31, 32], and the models are shown in Figure 8.13.

We validated the proposed method on 3 real colon datasets acquired by Siemens Sensation CT scanner [8]. The dataset volume is $512 \times 512 \times 580$ with voxel size $0.74 \times 0.74 \times 0.75mm^3$. Our experiments have been carried out on a computer with 8G Memory and two AMD Opteron (TM) 252 CPUs at 2.6 GHz each. For each dataset, it always takes 20 minutes to generate the detection results.

Three real polyps are circled and shown in Figure 8.14(a) from top to bottom. They range from large sizes on the first and second rows to the medium size on the third row. The pictures in the middle show the deformed surface after the polyps are flattened. The detected results are highlighted in red and generated by thresholding the displacement fields. They are illustrated in Figure 8.14(c).

The proposed model is mainly controlled by three parameters $\alpha$, $\gamma$ and $\epsilon$ in Equation 8.27. Theoretically, $\alpha$ is used to control the hyperbolic surface. However, there are only a few colonic polyps having such shapes, hence, a very small value for $\alpha$ between $(0, 0.0005)$ is selected based on the real experiment results.

$\gamma$ controls the shape of anisotropic filter, which is important for smoothing the local shape to avoid the abrupt changes. Normally, when $\gamma$ is larger than 4, the size of the designed anisotropic filter is small, which is not recommended. Experimental results showed that $\gamma = 3$ provides the reasonable smooth shape.

As we know that under the discrete case, $H$ could be zero even if $H \neq 0$ under the continuous case, which is controlled by $\epsilon$ introduced in this paper. If $\epsilon$ is too small, such as $\epsilon < 0.01$, this instability could not be avoided.

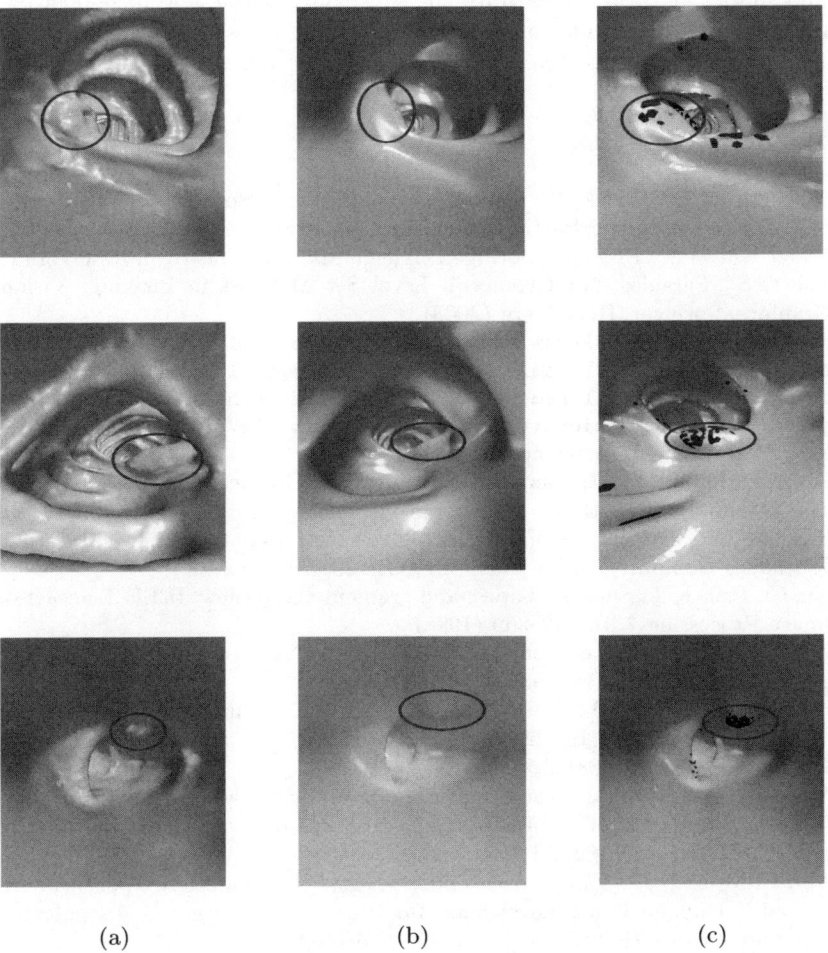

(a)               (b)               (c)

**Fig. 8.14.** (a) Three real colonic polyps as shown on three rows, (b) deformed mesh rendering, which shows that how the colon looks like after the polyp is *'removed'*, and (c) detection results using the proposed method in this paper

## 8.6  Conclusion

In this chapter, we have presented a novel anisotropic 3D surface evolution formula for protrusion shape-based colonic polyp detection and a color coding scheme for visualization. The surface evolution algorithm incorporates Gaussian and mean curvature flows. The proposed 3D surface evolution model works on detecting the protrusion shapes with convex or concave shapes.

The polyp detection and visualization framework has been validated by the synthetic polyps and real colon datasets. For the synthetic polyps with different size and shapes at various locations, the sensitivity is 100, and false positive is 0. Future work mainly includes using more real colon datasets to test its robustness and evaluate its false positive.

# References

1. Abbruzzese, J., Pollock, R.: Gastrointestinal Cancer. Springer, Heidelberg (2004)
2. Sapiro, G.: Geometric Partial Differential Equations and Image Analysis. Cambridge University Press, Cambridge Oakleigh Madrid Cape Town New York (2001)
3. Osher, S., Paragios, N.: Geometric Level Set Methods in Imaging, Vision and Graphics. Springer, Heidelberg (2003)
4. http://mathworld.wolfram.com/CompactSurface.html
5. http://mathworld.wolfram.com/RegularSurface.html
6. http://mathworld.wolfram.com/OrientableSurface.html
7. http://medical-dictionary.thefreedictionary.com/protrusion
8. http://www.usa.siemens.com/CTScanning
9. Olver, P., Sapiro, G., Tannenbaum, A.: Invariant Goemetric Evolutions of Surfaces and Volumetric Smoothing. SIAM Journal Applied Math 57(1), 176–194 (1997)
10. Kass, M., Witkin, A., Terzopolos, D.: Snakes: active contour models. International Journal of Computer Vision 1(1), 321–331 (1987)
11. Xu, C., Prince, J.: Snakes, shapes, and gradient vector flow. IEEE Transactions on Image Processing 7(3), 359–369 (1998)
12. Cohen, L.: On active contour models and balloons. Computer Vision, Graphics, and Image Understanding 53(2), 211–218 (1991)
13. Osher, S., Sethian, J.: Fronts Propagating with Curvature-Dependent Speed: Algorithms Based on Hamilton-Jacobi Formulations. Journal of Computational Physics 79, 12–49 (1988)
14. Malladi, R., Sethian, J., Vemuri, B.: Shape Modeling with Front Propagation: A Level Set Approach. IEEE Transactions on Pattern Analysis and Machine Intelligence 17(2), 158–175 (1995)
15. Abd-El-Munim, H., Farag, A.: A Shape-Based Segmentation Approach: An Improved Technique Using Level Sets. In: Proc. 10th International Conference on Computer Vision (ICCV 2005), pp. 930–935 (2005)
16. Abd-El-Munim, H., Farag, A.: Curve/Surface Representation and Evolution using Vector Level Sets with Application to the Shape-based Segmentation Problem. IEEE Transactions on Pattern Analysis and Machine Intelligence 29(6), 945–958 (2007)
17. Caselles, V., Kimmel, R., Sapiro, G.: Geodesic Active Contours. In: Proc. 5th International Conference on Computer Vision (ICCV 1995), pp. 694–699 (1995)
18. Lee, T., Atkins, M., Li, Z.: Indentation and Protrusion Detection and Its Applications. In: Kerckhove, M. (ed.) Scale-Space 2001. LNCS, vol. 2106, pp. 335–343. Springer, Heidelberg (2001)
19. Barenholtz, E., Cohen, E., Feldman, J., Singh, M.: Detection of change in shape: an advantage for concavities. Cognition 89, 1–9 (2003)
20. Yoshida, H., Näppi, J.: Three-Dimensional Computer-Aided Diagnosis Scheme for Detection of Colonic Polyps. IEEE Transactions on Medical Imaging 20(12), 1261–1274 (2001)
21. Summers, R., Beaulieu, C., Pusanik, L., Malley, J., Jeffrey, R., Glazer, D., Napel, S.: Automated Polyp Detector for CT Colonography: Feasibility Study. Radiology 12(216), 284–290 (2000)

22. Huang, A., Summers, R., Hara, A.: Surface Curvature Estimation for Automatic Colonic Polyp Detection. In: Proc. SPIE Medical Imaging, vol. 5746, pp. 392–402 (2005)
23. Paik, D., Beaulieu, C., Rubin, G., Acar, B., Jeffery, R., Yee, J., Dey, J., Napel, S.: Surface Normal Overlap: A Computer-Aided Detection Algorithm With Application to Colonic Polyps and Lung Nodules in Helical CT. IEEE Transaction on Medical Imaging 6(23), 661–675 (2004)
24. Acar, B., Beaulieu, C., Göktürk, S., Tomasi, C., Paik, D., Jeffery, R., Yee, J., Napel, S.: Edge Displacement Field-based Classification for Improved Detection of Polyps in CT Colonography. IEEE Transaction on Medical Imaging 12(21), 1461–1467 (2002)
25. Yao, J., Miller, M., Summers, R.: Automatic segmentation and detection of colonic polyps in CT colonography based on knowledge-guided deformable models. In: Proc. SPIE Medical Imaging, vol. 5031, pp. 370–380 (2003)
26. Yao, J., Summers, R.: Three-dimensional colonic polyp segmentation using dynamic deformable surfaces. In: Proc. SPIE Medical Imaging, vol. 5369, pp. 280–289 (2004)
27. Yao, J., Miller, M., Franazek, M., Summers, R.: Colonic Polyp Segmetattion in CT Colonography-Based on Fuzzy Clustering and Deformable Models. IEEE Transactions on Medical Imaging 23(11), 1344–1352 (2004)
28. Chen, D., Hassouna, M., Farag, A., Falk, R.: Principal Curvature Based Coloinc Polyp Detection. International Journal of Computer Assisted Radiology and Surgery 2(supp. 1), 6–8 (2007)
29. Wijk, C., Ravesteijn, V., Vos, F., Truyen, R., Vries, A., Stoker, J., Vliet, L.: Detection of Protrusions in Curved Folded Surfaces Applied to Automated Polyp Detection in CT Colonography. In: Larsen, R., Nielsen, M., Sporring, J. (eds.) MICCAI 2006. LNCS, vol. 4191, pp. 471–478. Springer, Heidelberg (2006)
30. Chen, D., Hassouna, M., Farag, A., Falk, R.: An Improved 2D Colonic Polyp Segmentation Framework Based on Gradient Vector Flow Deformable Model. In: Yang, G.-Z., Jiang, T., Shen, D., Gu, L., Yang, J. (eds.) MIAR 2006. LNCS, vol. 4091, pp. 372–379. Springer, Heidelberg (2006)
31. Chen, D., Hassouna, M., Farag, A., Falk, R.: A New Color Coding Scheme For Easy Polyp Visualization in CT-Based Virtual Colonoscopy. In: Proc. SPIE Symposium on Medical Imaging (SPIE Medical Imaging 2007). vol. 6514, pp. 21–32 (2007)
32. Chen, D., Hassouna, M., Farag, A., Falk, R., Dryden, G.: Geometric Features Based Framework for Colonic Polyp. In: Proc. IEEE International Conference on Image Processing (ICIP 2007), vol. 5, pp. 17–20 (2007)
33. Taubin, G.: Estiamting the Tensor of Curvature of a Surface from a Polyhedral Approximation. In: Proc. Fifth International Conference on Computer Vision (ICCV 1995), pp. 902–907 (1995)
34. Kobbelt, L., Campagna, S., Vorsatz, J., Seidel, H.: Interactive Multi-Resolution Modeling on Arbitrary Meshes. In: Proc. SIGGRAPH 1998, pp. 105–144 (1998)
35. Desbrun, M., Meyer, M., Schröder, P., Barr, A.: Implicit Fairing of Irregular Meshes Using Diffusion and Curvature Flow. In: Proc. SIGGRAPH 1999, pp. 317–324 (1999)
36. Zhao, H., Xu, G.: Triangular Surface Mesh Fairing via Gaussian Curvature Flow. Journal of Computational and Applied Mathematics 195(1), 300–311 (2006)
37. Näppi, J., Frimmel, H., Yoshida, H.: Virtual Endoscopic Visualization of the Colon by Shape-Scale Signatures. IEEE Transaction on Information Technology in Biomedicine 9(1), 120–131 (2005)

38. Hassouna, M., Farag, A.: PDE-Based Three Dimensional Path Planning For Virtual Endoscopy. In: Christensen, G.E., Sonka, M. (eds.) IPMI 2005. LNCS, vol. 3565, pp. 529–540. Springer, Heidelberg (2005)
39. Hassouna, M., Farag, A., Falk, R.: Differential Fly-Throughs (DFT): A General Framework For Computing Flight Paths. In: Duncan, J.S., Gerig, G. (eds.) MICCAI 2005. LNCS, vol. 3750, pp. 654–661. Springer, Heidelberg (2005)
40. Lorensen, W., Cline, H.: Marching Cubes: A High Resolution 3D Surface Construction Algorithm. Computer Graphics 21(4), 163–170 (1997)

# 9

# Assisting Cancer Diagnosis with Fuzzy Neural Networks

Feng Chu[1], Wei Xie[2], Farideh Fazayeli[1], and Lipo Wang[1]

[1] School of Electrical and Electronic Engineering, Nanyang Technological University,
Singapore 639798
elpwang@ntu.edu.sg
[2] Institute for Infocomm Research, Singapore 119613

**Summary.** Cancer diagnosis from huge microarray gene expression data is an important and challenging bioinformatics research topic. We used a fuzzy neural network (FNN) proposed earlier for cancer classification. This FNN contains three valuable aspects i.e., automatically generating fuzzy membership functions, parameter optimization, and rule-base simplification. One major obstacle in microarray data set classifier is that the number of features (genes) is much larger than the number of objects. We therefore used a feature selection method based on t-test to select more significant genes before applying the FNN. In this work we used three well-known microarray databases, i.e., the lymphoma data set, the small round blue cell tumor (SRBCT) data set, and the ovarian cancer data set. In all cases we obtained 100% accuracy with fewer genes in comparison with previously published results. Our result shows the FNN classifier not only improves the accuracy of cancer classification problem but also helps biologists to find a better relationship between important genes and development of cancers.

## 9.1 Introduction

The advent of DNA microarray datasets has opened a new research area for biologists and data mining experts. An experiment with DNA microarray chips can measure the expression level of thousands of genes in different tissues and samples simultaneously. It gives a global understanding into the molecular level of the living organism [1, 2]. This new technology provides biologists with a snapshot of the whole genome. Although it has improved traditional diagnostic problems, there still are many issues that should be addressed. However, the discovery of genes which are involved in a particular disease, including cancer, is a challenging task. Accurate and precise diagnosis is critical for future treatment.

The abundance of data has led to an unalterable demand for data mining in order to perform classification. Hence, analyses of this kind require hybrid knowledge of biology and data mining.

The difference in the structure of this kind of datasets, i.e., a huge number of genes (or features) and the scarcity of samples, gives rise to the need for new classification methods specifically designed to suit this particular application. There are small number of relevant and non-redundant genes from thousands which actually play a part in differentiating the samples. Feature selection

T.G. Smolinski et al. (Eds.): Comp. Intel. in Biomed. & Bioinform., SCI 151, pp. 223–235, 2008.
springerlink.com        © Springer-Verlag Berlin Heidelberg 2008

becomes crucial in this field for both computational efficiency and accuracy issues. Here we have chosen the t-test statistic for feature ranking.

Machine learning methods are used for classification of huge volumes of data. Many machine learning approaches have been proposed in the literature which give good results, including neural networks [3], support vector machines [4], nearest shrunken centroids [5], and so on.

In this chapter, we discuss a hybrid of feature ranking and neural network to obtain higher accuracy for cancer diagnosis. At first, we obtain the most significant genes by applying a t-test. Next, we classify samples using a fuzzy neural network (FNN) proposed earlier by Frayman and Wang [12, 13, 5].

### 9.1.1    Structure of the Chapter

The chapter is organized as follows. Section 9.2 summarizes the basics of t-test based gene importance ranking method. In Section 9.3, the basic principles behind the structure and algorithm of the FNN proposed by Frayman and Wang [12, 13] are outlined. Section 9.4 shows the potential of the FNN on classifying three important microarray data sets, namely, the lymphoma data set, the SRBCT data set, and the ovarian cancer data set. We discuss our results and draw some conclusions in the last section. Some of the results in this chapter have been presented at conferences [14, 15].

## 9.2    Gene Importance Ranking

One major problem with microarray dataset classification is that the number of features (genes) is much larger than the number of data instances. Also most genes may be irrelevant or redundant for cancer identification. One can use feature selection algorithms to get the most significant genes related to the diagnosis. Here we have used t-test feature selection which is outperformed Principle Component Analysis(PCA). In the lymphoma data set the best testing accuracy using PCA as a feature selector was 92.31 % when 6 genes were input to the classifier comparing to 100 % accuracy achieve with t-test.

At first, we ranked the genes according to their T-score (TS). Then, we selected the genes with the highest ranking. We used a threshold to find a group of the genes with the largest TS values. This threshold can be experimentally decided. For example, we first select TS=0.3. If we found the result looks OK and not much noise is included, we enlarge the threshold a little bit to include more genes, e.g., TS=0.4. If obviously a lot of noise included, then stop. These most important genes are fed to the FNN classifier as input, as described in the next section.

The TS of a gene $i$ is defined as follows [16, 17]:

$$TS_i = max\{|\frac{\overline{x}_{ik} - \overline{x}_i}{m_k s_i}|, k = 1, 2, ...K\} \qquad (9.1)$$

$$\overline{x}_{ik} = \sum_{j \in C_k} \overline{x}_{ij}/n_k \tag{9.2}$$

$$\overline{x}_i = \sum_{j=1}^{n} x_{ij}/n \tag{9.3}$$

$$s_i^2 = \frac{1}{n-K} \sum_{k} \sum_{j \in C_k} (x_{ij} - \overline{x}_{ik})^2 \tag{9.4}$$

$$m_k = \sqrt{1/n_k + 1/n} \tag{9.5}$$

where $K$ is the number of classes. $C_k$ refers to class $k$ that includes $n_k$ samples. $x_{ij}$ is the expression value of gene $i$ in sample $j$. $\overline{x}_{ik}$ is the mean expression value in class $k$ for gene $i$. $n$ is the total number of samples. $\overline{x}_i$ is the general mean expression value for gene $i$. $s_i$ is the pooled within-class standard deviation for gene $i$. In fact, the TS used here is a t-statistic between a specific class and the overall centroid of all classes [16].

## 9.3  The FNN

In this section, we review the basic principles behind the structure and algorithm of the FNN proposed by Frayman and Wang [12, 13].

Fig. 9.1 represents the structure of generation and the learning algorithm of the FNN. In the following subsections, we provide more details of the FNN and its mechanism. First the basic structure of the FNN will be explained, and then the training algorithm of the FNN will be represented. Finally, we will present a rule base modification method for improving the accuracy of our network.

### 9.3.1  FNN Structure

The FNN structure is designed as shown in Fig. 9.2 which is composed of four layers.

Let $x_j$ ($j = 1, 2, .., n$), and $y_l$ ($l = 1, 2, .., m$) be the input and output vectors respectively. The first layer of the FNN represents input variables, meaning that there are $n$ nodes in the first layer. The second layer represents the input membership function. Here we use two equally spaced triangular membership functions along the operating range of each input variable. In a such way these membership functions will satisfy $\epsilon - completeness$, which means that for a given value of $x$ of one of the inputs in the operating range we can always find a linguistic label $A$ such that $\mu_A(x) \geq \epsilon$. If $\epsilon - completeness$ is not satisfied, there may be no rule applicable for a new data input. Piecewise-linear triangular membership function is chosen for computational efficiency [18] as shown in Fig. 9.3.

$$\mu(x) = \begin{cases} \frac{x-l}{c-l}, & l \leq x \leq c \\ \frac{x-r}{c-r}, & c \leq x \leq r \\ 0, & \text{otherwise} \end{cases} \tag{9.6}$$

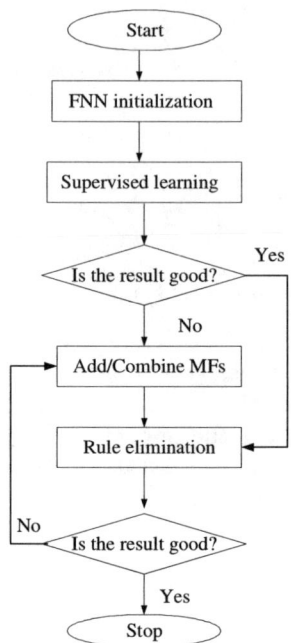

**Fig. 9.1.** The structure generation and the learning algorithm of the FNN

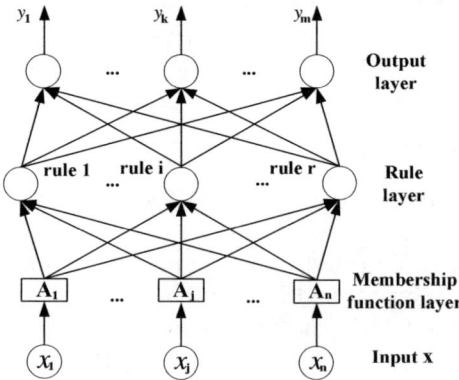

**Fig. 9.2.** The structure of the FNN

The leftmost and rightmost membership functions are threshold to cover for the whole operating range of each input.

The third layer is the rule layer which is initially empty since at the beginning, there are no rules in the rule set. The last layer of the FNN represents target variables (class labels), meaning that there are $m$ nodes in the output layer.

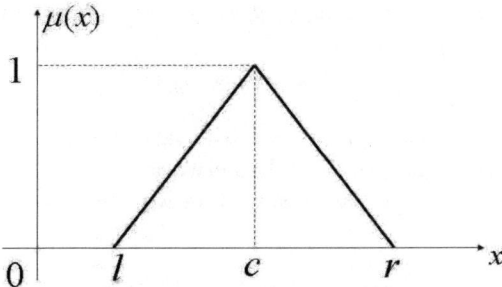

**Fig. 9.3.** The triangular Membership Function

Each rule node is connected to each input membership function node and each output node. The input membership functions are encoded as fuzzy connection weights between the input layer and the rule layer.

The initial rule $i$ is expressed in the input terms, such as,

Rule $i$: if $x_1$ is $A_{x1}^i, \cdots$, and $x_n$ is $A_{xn}^i$, then $y_1 = \omega_l^i, \cdots, y_m = \omega_m^i$,

where $\omega_l^j$ is a real number and $A_q^i$ $(q = x_1, x_2, \cdots, x_n)$ is the membership function of the antecedent part of rule $i$ for node $q$ in the input layer [19].

Each rule performs a product of its inputs. Membership value $\mu_i$ of the premise of the *i-th* rule is calculated as a fuzzy AND using the product operator as follows,

$$\mu_i = A_{x1}^i(x_1) \times A_{x2}^i(x_2) \times \cdots \times A_{xn}^i(x_n) \tag{9.7}$$

Finally, output $y_l$ of the FNN is obtained using the weighted average [20]:

$$y_l = \frac{\sum_i \mu_i \times \omega_l^i}{\sum_i \mu_i} \tag{9.8}$$

### 9.3.2 FNN Training

The FNN uses general learning rule for training the network [21]:

$$y_l^i(k+1) = y_l^i(k) - \eta \frac{\partial \varepsilon_l}{\partial y_l^i} \tag{9.9}$$

The learning rules for $\omega_l^i$ and $A_q^i$ are:

$$\omega_l^i(k+1) = \omega_l^i(k) - \eta \frac{\partial \varepsilon_l}{\partial \omega_l^i} \tag{9.10}$$

$$A_q^i(k+1) = A_q^i(k) - \eta \frac{\partial \varepsilon_l}{\partial A_q^i} \tag{9.11}$$

where $\eta$ is the learning rate. The objective is to minimize the error function:

$$\varepsilon_l = \frac{1}{2} \times (y_l - y_{dl})^2 \qquad (9.12)$$

where $y_l$ is the current output, and $y_{dl}$ is the target output.

An adaptive learning rate provides the network with higher convergence speed and learning performance, i.e., accuracy. Here we use a heuristic for tuning the learning rate $\eta$:

1. If the error undergoes five consecutive reductions, increase $\eta$ by 5%.
2. If the error undergoes three consecutive combinations of one increase and one reduction, decrease $\eta$ by 5%.
3. Otherwise, keep $\eta$ unchanged.

Since we dynamically update the learning rate, initial value for $\eta$ becomes insignificant as long as it is not too large.

The learning error $\varepsilon_l$ is reduced toward zero or a pre-specified small value $\varepsilon_{def} > 0$ as the iteration number $k$ increases.

### 9.3.3    Rule Base Modification

In this part we present some post processing routines to obtain a better model. Initially, we add a new membership function to each input at the point of the maximum output error, following Higgins and Goodman [18]. Each triangular membership function is represented by 3 points. One point with unity membership function is placed at the point of the maximum output error. Other two points with zero membership value are placed at the centers of the two neighboring regions, respectively.

As the output of the network is not a binary 0 or 1, but a number ranging from 0 to 1, we can speed up the convergence of the network substantially by eliminating the error whose deviation from the target value is the greatest.

Next, If the degree of overlapping of membership functions is greater than a pre-specified threshold, we combine those membership functions. We used the following fuzzy similarity measure [23]

$$Degree(A_1 = A_2) = E(A_1, A_2) = \frac{M(A_1 \cap A_2)}{M(A_1 \cup A_2)}, \qquad (9.13)$$

where $\cap$ and $\cup$ denote the intersection and the union of two fuzzy sets $A_1$ and $A_2$, respectively. $M(.)$ is the size of fuzzy set, and $0 \leq E(A_1, A_2) \leq 1$. We can delete irrelevant inputs if an input variable ends up with only one membership function. We can thus reduce the size of the rule base.

Finally, we evaluate the rules on the basis of accuracy and simplicity. For handling the tradeoff between accuracy and simplicity, we use a weighting parameter, namely, the compatibility grade (CG) of each fuzzy rule. CG for the rule $j$ is calculated by the product operator as follows,

$$\mu_j(x) = \mu_{j1}(x_1) \times \mu_{j2}(x_2) \times \cdots \times \mu_{jn}(x_n) \qquad (9.14)$$

when the system provides correct classification results.

We eliminate those rules which have a CG less than a threshold and when a rule node is eliminated, all connection links and its input membership nodes are deleted as well. By varying CG threshold a user is able to specify the degree of rule compactness. The size of the rule base can thus be kept minimal. If eliminating rules causes less accuracy than required, we add another rule into the rule set as described above. Otherwise we stop the procedure of generating classification rules.

The FNN combines three valuable aspects, namely, automatically generating fuzzy membership functions [19], parameter optimization [21], and rule-base simplification to achieve good performance. We have applied the FNN to some well-known microarray datasets. The results are presented in the next section.

## 9.4  Experimental Results

### 9.4.1  Lymphoma Data

The lymphoma data set (http://llmpp.nih.gov/lymphoma) [7] contains 42 samples derived from diffuse large B-cell lymphoma (DLBCL), 9 samples from follicular lymphoma (FL), and 11 samples from chronic lymphocytic leukaemia (CLL). The entire data set includes the expression data of 4026 genes. In this data set, a small part of data is missing. For handling missing attribute values, we used a k-nearest neighbor algorithm [22]. We randomly divided the data set into 2 parts with 31 samples as the training set and 31 samples as the test set.

At first, we chose 174 most significant genes with highest TSs from all 4026 genes using our training set (Table 9.1). Then, we chose the best subset of genes from this 174 gene collection which yield the highest accuracy of classification applying the FNN.

We applied the FNN into all possible subsets of genes as follows. We used an expanding window to choose different subsets of genes. Let $g_i$ be a gene with the TS ranking $i$, and $w$ be the size of the window. Here the window size is

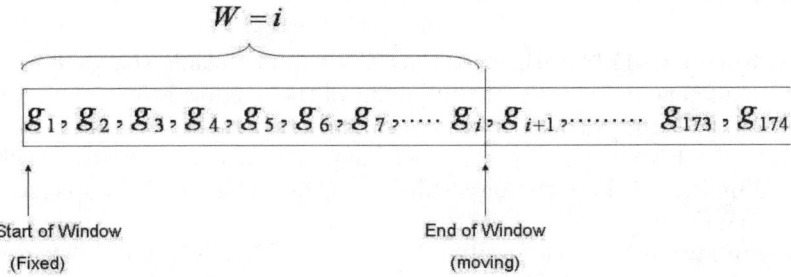

**Fig. 9.4.** Subset features selection as input for the FNN using expanding window

**Table 9.1.** Lymphoma gene importance ranking : 174 genes with the highest TSs, in the order of decreasing TSs (Gene ID is defined in [7])

| Rank | Gene ID | Gene Description |
|---|---|---|
| 1 | GENE2307X | (CD23A=low affinity II receptor for Fc fragment of IgE; Clone=1352822) |
| 2 | GENE3320X | (Similar to HuEMAP=homolog of echinoderm microtubule associated protein EMAP; Clone=1354294) |
| 3 | GENE708X | *Ki67 (long type); Clone=100 |
| 4 | GENE2393X | *MDA-7=melanoma differentiation-associated 7=anti-proliferative; Clone=267158 |
| 5 | GENE1622X | *CD63 antigen (melanoma 1 antigen); Clone=769861 |
| 6 | GENE1641X | *Fibronectin 1; Clone=139009 |
| 7 | GENE2391X | (Unknown; Clone=1340277) |
| 8 | GENE1636X | *Fibronectin 1; Clone=139009 |
| 9 | GENE1644X | (cathepsin L; Clone=345538) |
| 10 | GENE1610X | *Mig=Humig=chemokine targeting T cells; Clone=8 |
| 11 | GENE707X | (Topoisomerase II alpha (170kD); Clone=195630) |
| 12 | GENE689X | *lamin B1; Clone=1357243 |
| 13 | GENE695X | *mitotic feedback control protein Madp2 homolog; Clone=814701 |
| 14 | GENE1647X | *cathepsin B; Clone=261517 |
| 15 | GENE537X | (B-actin,1099-1372; Clone=143) |
| ... | ... | ... |
| 165 | GENE1539X | *lysophospholipase homolog (HU-K5); Clone=347403 |
| 166 | GENE2385X | *Unknown UG Hs.124382 ESTs; Clone=1356466 |
| 167 | GENE719X | (Myt1 kinase; Clone=739511) |
| 168 | GENE2415X | (Unknown; Clone=1289937) |
| 169 | GENE527X | *glutathione-S-transferase homolog; Clone=1355339 |
| 170 | GENE1598X | *Similar to ferritin H chain; Clone=1306027 |
| 171 | GENE1192X | *Interferon-induced guanylate-binding protein 2; Clone=545038 |
| 172 | GENE731X | *Chromatin assembly factor-I p150; Clone=1334875 |
| 173 | GENE769X | *14-3-3 epsilon; Clone=266106 |
| 174 | GENE724X | (Hyaluronan-mediated motility receptor (RHAMM); Clone=756037) |

changed from 1 to 174. With expanding the window inside the genes we get a new subset of genes where the starting point of the window is fixed, i.e., it always starts from the gene with the highest ranking (Fig. 9.4). For each window size we trained the FNN with all genes inside the window and evaluated it with the test set. Finally, we chose the best subset of genes which had the lowest testing and training error.

The training and testing results are shown in Fig. 9.5. The FNN performs very well. Training and testing error becomes 0 with applying only the first 9 genes in Table 9.1 as the input to the FNN.

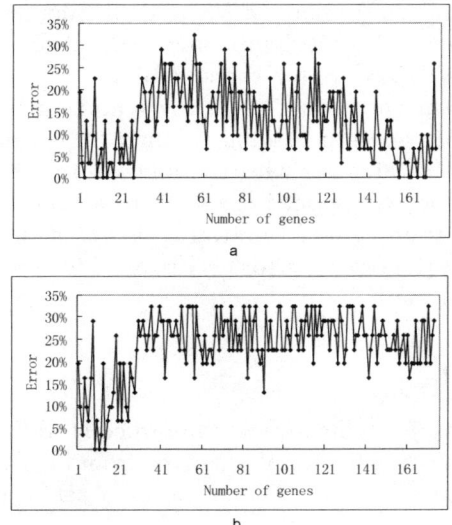

**Fig. 9.5.** The training (a) and the testing (b) results for the lymphoma data

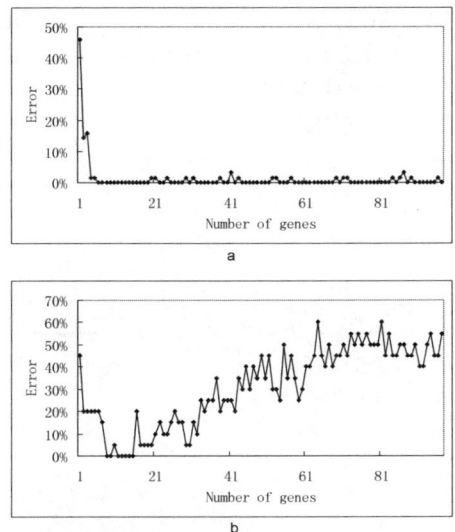

**Fig. 9.6.** The training (a) and the testing (b)results for the SRBCT data

## 9.4.2    SRBCT Data

The SRBCT data (http://research.nhgri.nih.gov/microarray/Supplement/) [3] contains the expression data of 2308 genes. There are total of 63 training samples and 25 testing samples preprovided i.e., the splie has been done as a part of the problem, 5 of the testing samples are not SRBCTs. The 63 training samples contain 23 Ewing family of tumors (EWS), 20 rhabdomyosarcoma (RMS), 12 neuroblastoma (NB) and 8 Burkitt lymphomas (BL). And the 20 SRBCT testing samples contain 6 EWS, 5 RMS, 6 NB and 3 BL.

We followed the same procedure as in the lymphoma dataset. We initially ranked the entire 2308 genes according to their TSs using training samples. Then we picked out the 96 genes with the highest TSs. Here the window size is changed from 1 to 96. We applied the FNN to all possible subsets of genes. Fig. 9.6 presents the training and the testing errors of the classification. Both the training error and the testing error decreases to 0 when the top 8 genes are chosen as the input of the FNN.

### 9.4.3    Ovarian Data

The ovarian data (http://genome-www.stanford.edu/ovarian_cancer/) [11] contains 125 samples, including 68 samples derived from breast cancer and 57 samples derived from ovarian cancer. The entire data set includes the expression data of 3363 genes.

Similarly, we randomly divided the data into 2 parts at the beginning, with 75 samples as the training set, and 50 samples as the test set. We picked out 100 genes with the highest TS from the training sample. Here the window size is changed from 1 to 100. By applying the FNN to all possible subsets of genes, training and testing errors decreased to 0 with only 3 genes. The training and the testing results are presented in Fig.9.7.

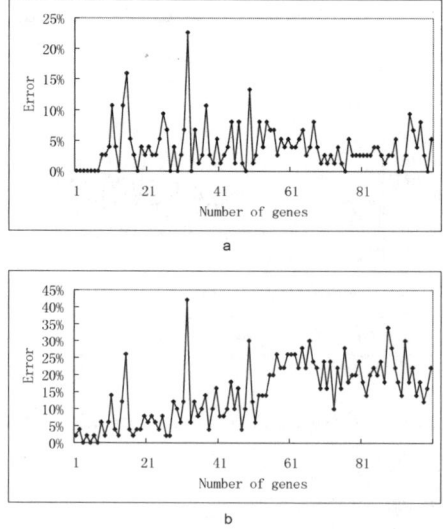

**Fig. 9.7.** The training (a) and the testing (b) results for the Ovarian data

## 9.5   Conclusion

We compared our method with 3 well-known methods proposed earlier, i.e., MLP neural network [3], Nearest shrunken centroids [5] and SVM [9].

Tables 9.2, 9.3, and 9.4 present the comparison between methods. We obtained 9 genes with 100% accuracy for the lymphoma dataset using the FNN compared to the 48 genes required by nearest shrunken centroids.

For the SRBCT data, the FNN needs only 8 genes to obtain the same accuracy, i.e., 100%, while the well-known evolutionary algorithm reported by Deutsch [8] used 12 genes.

Schaner et al. [11] used at least 61 genes to classify the ovarian cancer and the breast cancer data sets to obtain 100% accuracy while the FNN needed only 4 genes to obtain the same accuracy.

Comparing our results with results from other existing methods, our proposed method noticeably improves the accuracy of microarray classification with much smaller numbers of genes. It can help biologists to classify huge volumes of data, difficult to classify with traditional clinical methods. The smaller number of genes required by the FNN along with high accuracy helps biologists in focusing on a small number of significant genes to discover the relationships between those genes which are involved in the development of cancers.

**Table 9.2.** Comparisons of results all with 100% accuracy for the SRBCT data

| Method | Number of genes required |
|---|---|
| MLP neural network [3] | 96 |
| Nearest shrunken centroids [5] | 43 |
| SVM [9] | 20 |
| Evolutionary algorithm [8] | 12 |
| Our FNN | 8 |

**Table 9.3.** Comparisons of results all with 100% accuracy for the Ovarian data

| Method | Number of genes required |
|---|---|
| Hierarchical clustering [11] | 61 |
| Our FNN | 4 |

**Table 9.4.** Comparisons of results all with 100% accuracy for the Lymphoma data

| Method | Number of genes required |
|---|---|
| Nearest shrunken centroids [5] | 48 |
| Our FNN | 9 |

# References

1. Schena, M., Shalon, D., Davis, R.W., Brown, P.O.: Quantitative monitoring of gene expression patterns with a complementary DNA microarray. Science 270, 467–470 (1995)
2. Golub, T.R., Slonim, D.K., Tamayo, P., Huard, C., Gaasenbeek, M., Mesirov, J.P., Coller, H., Loh, M.L., Downing, J.R., Caligiuri, M.A., et al.: Molecular classification of cancer: class discovery and class prediction by gene expression monitoring. Science 286, 531–537 (1999)
3. Khan, J.M., Wei, J.S., Ringner, M., Saal, L.H., Ladanyi, M., Westermann, F., Berthold, F., Schwab, M., Antonescu, C.R., Peterson, C., et al.: Classification and diagnostic prediction of cancers using gene expression profiling and artificial neural networks. Nature Medicine 7, 673–679 (2001)
4. Brown, M.P., Grundy, W.N., Lin, D., Cristianini, N., Sugnet, C.W., Furey, T.S., Jr. M., Haussler, D.: Knowledge-based analysis of microarray gene expression data by using support vector machines. In: Proc. Natl Acad. Sci., USA, vol. 97, pp. 262–267 (2000)
5. Tibshirani, R., Hastie, T., Narashiman, B., Chu, G.: Diagnosis of multiple cancer types by shrunken centroids of gene expression. In: Proc. Natl. Acad. Sci., USA, vol. 99, pp. 6567–6572 (2002)
6. Tibshirani, R., Hastie, T., Narasimhan, B., Chu, G.: Class predicition by nearest shrunken centroids with applications to DNA microarrays. Statistical Science 18, 104–117 (2003)
7. Alizadeh, A.A., Eisen, M.B., Davis, R.E., Ma, C., Lossos, I.S., Rosenwald, A., Boldrick, J.C., Sabet, H., Tran, T., Yu, X., et al.: Distinct types of diffuse large B-cell lymphoma identified by gene expression profiling. Nature 403, 503–511 (2000)
8. Deutsch, J.M.: Evolutionary algorithms for finding optimal gene sets in microarray prediction. Bioinformatics 19, 45–52 (2003)
9. Lee, Y., Lee, C.K.: Classification of multiple cancer types by mulitcategory support vector machines using gene expression data. Bioinformatics 19, 1132–1139 (2003)
10. Dudoit, S., Fridlyand, J., Speed, T.P.: Comparison of discrimination methods for the classification of tumors using gene expression data. Journal of the American Statistical Association 97, 77–87 (2002)
11. Schaner, M.E., Ross, D.T., Ciaravino, G., Sorlie, T., Troyanskaya, O., Diehn, M., Wang, Y.C., Duran, G.E., Sikic, T.L., Caldeira, S.: Gene expression patterns in ovarian carcinomas. Molecular Biology of the Cell 14, 4376–4386 (2003)
12. Frayman, Y., Wang, L.: Data mining using dynamically constructed fuzzy neural networks. In: Wu, X., Kotagiri, R., Korb, K.B. (eds.) PAKDD 1998. LNCS, vol. 1394, pp. 122–131. Springer, Heidelberg (1998)
13. Frayman, Y., Wang, L.: A Dynamically-constructed fuzzy neural controller for direct model reference adaptive control of multi-input-multi-output nonlinear processes. Soft Computing 6, 244–253 (2002)
14. Chu, F., Xie, W., Wang, L.: Gene selection and cancer classification using a fuzzy neural network. In: Proceedings of the North-American Fuzzy Information Processing Conference (NAFIPS 2004), vol. 2, pp. 555–559 (2004)
15. Xie, W., Chu, F., Wang, L., Lim, E.T.: A fuzzy neural network for intelligent data processing. In: Proceedings of SPIE, Data Mining, Intrusion Detection, Information Assurance, and Data Networks Security 2005, vol. 5812, pp. 283–290 (2005)
16. Devore, J., Peck, R.: Statistics: the Exploration and Analysis of Data, 3rd edn. Duxbury Press, Pacific Grove, CA (1997)

17. Tusher, V.G., Tibshirani, R., Chu, G.: Significance analysis of microarrays applied to the ionizing radiation response. In: Proc. Natl. Acad. Sci., USA, vol. 98, pp. 5116–5121 (2001)
18. Higgins, C.M., Goodman, R.M.: Fuzzy rule-based networks for control. IEEE Transactions on Fuzzy Systems 2, 82–88 (1994)
19. Wang, L.X., Mendel, J.M.: Generating fuzzy rules by learning from examples. IEEE Trans. Syst., Man, Cybern 22, 1414–1427 (1992)
20. Sugeno, M., Kang, G.T.: Structure identification of fuzzy model. Fuzzy Sets Syst. 28, 15–33 (1988)
21. Werbos, P.: Beyond regression: new tools for prediction and analysis in the behavioral sciences. Ph.D. thesis, Harvard University, Cambridge, MA (1974)
22. Troyanskaya, O., Cantor, M., Sherlock, G., Brown, P., Hastie, T., Tibshirani, R., Botstein, D., Altman, R.B.: Missing value estimation methods for DNA microarrays. Bioinformatics 17, 520–525 (2001)
23. Dubios, D., Prade, H.: A unifying view of comparison indices in a fuzzy set theoretic framework. In: Yager, R.R. (ed.) Fuzzy sets and possibility theory: Recent Developments, Pergamon, NY (1982)

# 10

## Computational Intelligence in Clinical Oncology: Lessons Learned from an Analysis of a Clinical Study

B. Haibe-Kains[1,2], C. Desmedt[2], S. Loi[3], M. Delorenzi[4], C. Sotiriou[2], and G. Bontempi[1]

[1] Machine Learning Group, Université Libre de Bruxelles, Brussels, Belgium
[2] Functional Genomics Unit, Institut Jules Bordet, Brussels, Belgium
[3] Peter MacCallum Cancer Center, East Melbourne, Victoria, Australia
[4] Bioinformatics Core Facility, Institut Suisse de Recherche Expérimentale sur le Cancer, Lausanne, Switzerland

**Summary.** In this chapter, we present a retrospective clinical study where the adoption of computational intelligence approaches for performing knowledge extraction from gene expression data enabled an improved oncological clinical analysis. This study focuses on a survival analysis of estrogen receptor (ER) positive breast cancer patients treated with tamoxifen. The chapter describes each step of the gene expression data analysis procedure, from the quality control of data to the final validation going through normalization, feature transformation, feature selection, and model building. Each section proposes a set of guidelines and motivates the specific choice made for this particular study. Finally, the main guidelines that emerged from this study are the use of simple and effective techniques rather than complex non-linear models, the use of interpretable methods and the use of scalable computational solutions able to deal with multiplatform and multisource data.

## 10.1 Introduction

Recent advances in biomedical measurement technologies, such as gene expression profiling, expose clinicians to an exponential increase of complex data. Major challenges on the computational side arise from the huge dimensionality of the data, the relatively low number of samples, the high redundancy of input variables, the heterogeneity of the data sources and the high level of noise. This is why traditional clinical analysis needs the help of computational intelligence approaches to manage the complexity of the analysis task, without being overwhelmed by the massive amount of data or mislead by spurious patterns [1].

In this chapter, we present a retrospective clinical study in which the adoption of computational intelligence approach for performing knowledge extraction from gene expression data enabled an improved oncological clinical analysis. We focus on a survival analysis of estrogen receptor (ER) positive breast cancer (BC) patients treated with tamoxifen, a well-known treatment in BC therapy. The patients included in the study are heterogeneous with respect to their clinical

T.G. Smolinski et al. (Eds.): Comp. Intel. in Biomed. & Bioinform., SCI 151, pp. 237–268, 2008.
springerlink.com                                          © Springer-Verlag Berlin Heidelberg 2008

behavior and response to tamoxifen therapy. It is known that current biomarkers (e.g. expression levels of ER) give little insight into tumor biology and potential response to treatment. Indeed 30-40% of women with ER-positive disease develop distant metastases and die despite tamoxifen treatment, illustrating the urgent clinical need for new biomarkers that can predict which women with ER-positive BC are at high risk of relapse despite the use of tamoxifen. These patients would benefit from new therapies such as the aromatase inhibitors [2, 3].

Gene expression profiling of tumors appears to be a promising new strategy for predicting clinical outcome in BC patients. According to recent studies [4, 5, 6, 7] the heterogeneity of clinical response can be correlated with different molecular "portraits". Additionally, gene classifiers have been developed that can distinguish subgroups of patients with different prognoses or responses to therapies [8]. Due to the pressing clinical need, several other investigators have also developed gene classifiers for ER-positive BC patients treated with tamoxifen [9, 10, 11, 12]. After an initial period of enthusiasm about the potentiality of computational techniques, problematic issues about the design of the analysis and the performance assessment of gene classifiers were raised [13, 14]. Moreover, skepticism about the robustness of the approach arose in the medical community because of the small overlap of genes in *signatures* (i.e. set of highly discriminating genes) derived from different clinical studies on similar cohorts of BC patients [15]. In this chapter, we address these open questions and describe the entire computational procedure leading to the development of a model which predicts the risk of recurrence in tamoxifen treated patients. Thereby we also investigate the potential of this model to reveal new biological processes associated with the clinical outcome of these patients.

The chapter is organized as a set of sections, each serving as illustration of a single step of the clinical study (entirely represented in Figure 10.1). Each section introduces the issue, refers to existing reviews on the topic and discusses the most relevant methodological aspects. A subsection called *Tamoxifen Study* details the methodological choices made during the study, the motivations behind them and the related results. The section terminates with a subsection called *Lessons Learned* where we summarize the main lessons to be retained by the reader.

In particular, Section 10.1 introduces the clinical problem and emphasizes the need of computational intelligence techniques. Section 10.2 illustrates the properties of gene expression data collected in large clinical studies. Section 10.3 serves as an introduction to survival data and the related statistical methods. Sections 10.4 and 10.5 discuss the preprocessing of gene expression data, in particular quality controls and normalization. Sections 10.6 and 10.7 emphasize the importance of dimension reduction and the issue of stability in feature selection. Section 10.8 details the model building procedure. Sections 10.9 and 10.10 summarizes the accuracy assessment of the model in a cross-validation framework and an independent validation set respectively. Section 10.11 gives biological interpretation of the model. Conclusions are drawn in Section 10.12.

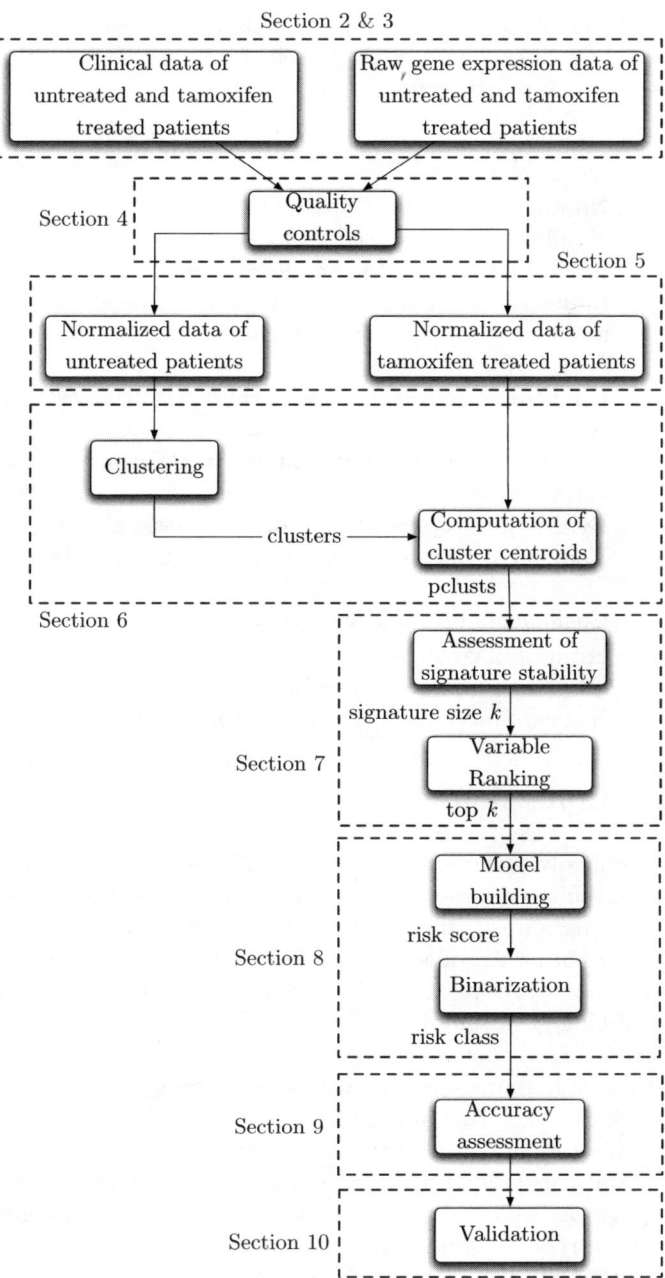

**Fig. 10.1.** Design of the survival analysis of tamoxifen treated BC patients from gene expression data. Each step is delimited by a dashed red box with the corresponding section number in this chapter.

It is worth to mention that the present chapter complies with the *research reproducibility* guidelines proposed in [16] in terms of availability of the code and reproducibility of results and figures[1].

**Notations**

| | |
|---|---|
| $N$ | Number of samples or patients. |
| $n$ | Number of input variables. |
| $X$ | Set of input variables, i.e. gene expressions. |
| $G$ | Indicator variable for class ($G = 0$ for the low-risk class and $G = 1$ for the high-risk class). |
| $S$ | Scoring function. |
| $X, Y, \ldots$ | Upper case letters represent random variables (except $N$). |
| $x, y, \ldots$ | Lower case letters represent the realization of random variables (except $n$). |
| $\mathbf{x}, \mathbf{X}, \boldsymbol{\beta}, \ldots$ | Bold letters represents vectors or matrices. |
| $T_i$ | Time of occurrence/censoring for the sample $i$, $i \in \{1, \ldots, N\}$. |
| $\beta$ | Coefficients of a linear regression model. |
| $\hat{\beta}$ | Estimated coefficients. |
| $S(t)$ | Survivor function depending on time $t$. |
| $\lambda(t)$ | Hazard function depending on time $t$. |

## 10.2   Data Collection

Clinical studies involving gene expression profiling include usually very few patients due to the limited number of clinical records and the high costs of biological experiments. In this context investigators need to develop partnerships with several institutions in order to collect clinical information and biological samples for sufficiently large cohorts of patients. A basic requirement is that protocols for the management of biological samples and clinical information must be well-defined to reduce heterogeneity between data sources. In spite of these precautions, some heterogeneity remains and this will have to be taken into account during the analysis (see Sections 10.3 and 10.5).

A typical dataset is composed of two types of data, clinical and gene expression data. Clinical data include traditional clinical variables (e.g. age, tumor size, nodal status, ...) as well as survival data (see Section 10.3). Gene expression data represent expression values for a large number of genes.

There exist several gene expression profiling technologies to simultaneously measure the expression of a large number of genes. They differ by two main characteristics, the gene expression measurement and the manufacturing. Some

---

[1] Raw gene expression and clinical data are publicly available in the GEO public database [17] and the Sweave version of the chapter including the standalone R code [18] is available at http://www.ulb.ac.be/di/map/bhaibeka/cichapter/.

technologies use two channels to measure *relative* gene expressions, i.e. the ratio between the gene expressions from a control (also called reference) and a target biological sample. Other technologies use only one single channel and therefore measure *absolute* gene expressions. The manufacturing of the chips dedicated to gene expression profiling differs, some technology using cDNA sequences while others synthesizing directly sequences of oligonucleotides on the chips. An in-depth description of different gene expression technologies is given in [19].

The choice of technology is not obvious since all technologies have advantages and limitations. The single-channel technology allows for exchange of raw data between different laboratories whereas the dual-channel technologies requires the laboratories to use the same reference. At the same time, systematic biases (see Section 10.5) are typically lower in dual-channel technology. In terms of manufacturing process, the use of short oligonucelotide sequences improves the robustness at the cost of a reduced specificity of the probes. For a full comparison we refer the reader to [19].

### 10.2.1   Tamoxifen Study

The dataset collected for our retrospective clinical study consists of 414 early-stage BC patients, diagnosed between 1980 and 1995. Samples came from three different institutions, Guy Hospital (London, United Kingdom), Uppsala University Hospital (Uppsala, Sweden) and John Radcliffe Hospital (London, United Kingdom). These institutions are henceforth referred to as GUY, KI and OXF respectively. Among the patients, 137 were untreated (labeled by U) and 277 were treated by tamoxifen (labeled by T). For the sake of simplicity, in the following we will use an abridged cohort notation where the first letters denotes the hospital and the last one the absence/presence of treatment (e.g. GUYT stands for the cohort of tamoxifen treated patients from the Guy hospital). Note that all the patients from GUY were treated (no GUYU cohort).

We used the Affymetrix technology for the gene expression profiling of our tumor samples. This technology measures absolute gene expressions (single-channel) and uses short sequences of oligonucletides. The choice was motivated by the renown of the Affymetrix technology and the need of sharing experiments between different laboratories.

Samples from OXFU, OXFT and GUYT were shipped to the Institut Jules Bordet (Brussels, Belgium) where RNA was extracted and samples were hybridized. In KIU and KIT samples RNA was extracted at the Karolinska Institute (Stockholm, Sweden) and samples hybridized at the Genome Institute of Singapore (Singapore). In OXF and KI samples gene expression profiling was performed with Affymetrix HG-U133A and HG-U133B Genechips. HG-U133PLUS2 Genechips were used for GUY samples. Note that there are $\approx 45000$ common probes between the HG-U133A/B and HG-U133PLUS2 chips. Raw data are publicly available at the Gene Expression Omnibus (GEO) database[2] [17] with accession number GSE6532.

---

[2] http://www.ncbi.nlm.nih.gov/geo/

Since data came from different institutions and experiments were made in different laboratories, special care need to be taken to detect and to deal with heterogeneity in clinical data (see Section 10.3) and gene expression data (see Section 10.5).

---

**Lessons Learned**

In large gene expression profiling studies, the collected data are often heterogeneous since they are usually generated by different laboratories using different technologies. Methods to detect and to deal with this heterogeneity are essential in order to allow the samples to be considered in the same analysis.

---

## 10.3  Survival Data

Unlike conventional computational intelligence tasks, most clinical studies rely on censored survival data characterized by event or discontinued observations [20]. Consider for instance a five-years follow-up of a group of BC patients after the surgical operation. In this case what is relevant is the dependence between occurrence and timing of the first metastasis and a set of explanatory variables (aka features). If we narrow our focus to a binary dependent variable (i.e. presence or absence of metastasis) conventional classification methods (e.g. logistic regression, linear discriminant analysis, support vector machines) [21] would serve our purpose. However, the analysis would ignore the information related to the timing of the event. For instance, it is intuitive to suppose that the aggressiveness of the tumor is related to the time of metastasis appearance during the follow-up. A possible solution to this problem could come from the adoption of a conventional regression strategy where the time to metastasis plays the role of dependent variable. In this case the problem would be how to deal with patients where no metastasis appeared during the five years follow-up. Such cases are referred to as *censored*. Note that by simply discarding these cases or setting them to a constant value we would introduce a large bias in our analysis.

Survival analysis combines the information of censored and uncensored data by statistical modeling. The occurring times of events are assumed to be realizations of some random variable $T$. Two functions are widely used to describe the probability distribution of $T$ :

- The survivor function $S(t) = \Pr\{T > t\}$ ($S(0) = 1$) measuring the probability for an individual to survive until time $t$.

- The hazard function $\lambda(t) = \lim\limits_{\Delta t \to 0} \dfrac{\Pr\{t \leq T < t + \Delta t \mid T \geq t\}}{\Delta t}$ measuring the instantaneous risk[3] that an event will occur in the interval $[t, t + \Delta t]$. Note that the probability term is conditional to a survival up to time $t$.

---

[3] Although it may be helpful to think of the hazard as the instantaneous probability of an event at time $t$, this quantity is not a probability and may be greater than 1. This is due to the division by $\Delta t$. Although the hazard has no upper bound, it cannot be smaller than 0.

The survivor function and the hazard function are equivalent ways of describing a continuous probability distribution [20]. The Kaplan-Meier (KM) estimator (also known as the product-limit estimator) returns a non-parametric maximum likelihood estimation of the survivor function [22]. Semi-parametric regression models [23] use the hazard function to model the relationship between the survival outcome and a set of explanatory variables $x$. A well known example is the Cox regression model [24] which represents the hazard of a patient $i$ as

$$\lambda_i(t) = \lambda_0(t) \exp\left(\beta_1 x_{1i} + \cdots + \beta_n x_{ni}\right) \tag{10.1}$$

The hazard for individual $i$ at time $t$ is the product of two factors :

- A baseline hazard function $\lambda_0(t) \geq 0$ which represents the hazard function of an individual where all the explanatory variables are set to zero.
- An exponential function where the argument is a linear function of the explanatory variables. This linear function is also denoted as the *linear predictor* or the *risk score* in literature.

Note that while the estimation of the $\beta$ term is obtained by maximizing a partial likelihood function [24], the estimation of the baseline term is not required. In fact, model 10.1 is often used in the proportional hazard form to compute the ratio

$$\frac{\lambda_i(t)}{\lambda_j(t)} = \exp\left\{\beta_1(x_{1i} - x_{1j}) + \cdots + \beta_n(x_{ni} - x_{nj})\right\}$$

of the hazards between an individual $i$ and an individual $j$ for $i, j \in \{1, \ldots, N\}$.

An extension of the Cox model allows for multiple strata where patients are classed into disjoint groups, each sharing the estimation of the coefficient $\beta$ but with a distinct baseline hazard function. This is useful in multicenter clinical studies where it is is reasonable to assume that different cohorts of patients are characterized by different baseline survival curves.

### 10.3.1   Tamoxifen Study

In our retrospective clinical study the event was defined as the appearance of the first distant metastasis. In the following we will use the acronym DMFS (which stands for Distant Metastasis Free Survival) to denote the time to event occurrence.

The survival dataset was collected during a follow-up period longer than 10 years. Note that 21 patients were not considered due to missing survival data and that only few events were observed (27% of the patients). Also, 82% (70%) of the patients did not experience any event till the first 5 (10) years.

In spite of well defined criteria for patient inclusion, we observed differences between cohorts in terms of demographics (e.g. age) and survival. This is well illustrated by Figure 10.2 which shows the KM survival curves for each single cohort. We observe here that patients from KIT tend to have distant metastasis earlier than patients from other institutions. This fact suggested the use of cohort labels as strata indicator in the Cox model (see Sections 10.7 and 10.8).

> **Lessons Learned**
>
> In large retrospective clinical studies, we usually observe heterogeneity in population sampling (pool of several cohorts of patients). In survival analysis, stratification in Cox model addresses efficiently this issue.

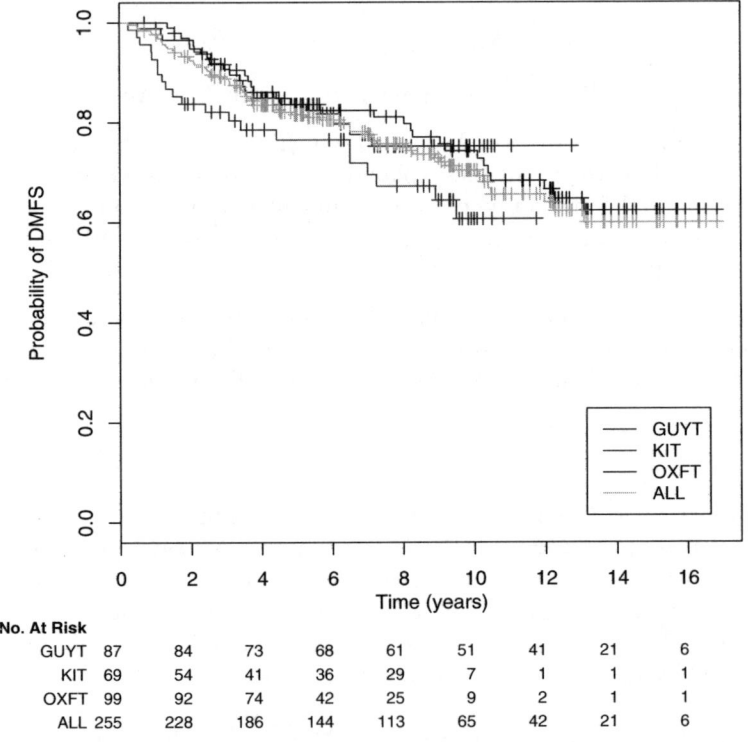

**Fig. 10.2.** Survival curves stratified by patient cohort. ALL stands for the global cohort including GUYT, KIT and OXFT. The "+" symbol represents the censoring.

## 10.4  Quality Controls

Profiling gene expression in biological samples is an expensive, time consuming and highly noisy process. As a consequence, it is essential to make the best use of the information contained in the gene expression data and to ascertain its quality. Before starting the data analysis, preliminary checks are suggested in order to raise evidence of quality problems. In some cases, chips could appear beyond correction and the only recommended solution would be to discard them. For a review on existing methods for quality control we refer the reader to [25]. Here we will focus on the quality guidelines issued for the Affymetrix technology by [26].

Two types of quality controls for Affymetrix chips are adopted :

- Single-chip quality controls : These controls concern one chip at a time. An example is the use of raw image analysis to detect hybridization artifacts like large areas of low intensity due to air bubbles.
- Multi-chip quality controls : These controls [25] target a set of quantities whose "values should be comparable over all chips of a dataset" [26], like scale factors, background intensities and percentage of present calls. Scale factors is a robust measure of the mean level of intensities on a chip. Background intensity is the intensity measured in an empty area (with no hybridization) and returns a measure of the background level. Percentage of present calls measures the proportion of genes being expressed (intensity significantly higher than background) on the chip. Once these quality controls have been carried out, the identification and the consequent discard of the anomalous chips is done.

Note that single-chip controls are well standardized and can be easily performed by the technicians in charge for the hybridizations. This is not the case for multi-chip quality controls which are more complex and would benefit from the availability of standardized support tools. Unfortunately such tools are currently missing though initiatives like the MicroArray Quality Control project[4] (MAQC) are expected to provide solutions in the coming years.

### 10.4.1  Tamoxifen Study

Once single-chip quality controls were performed by the laboratory technicians, we focused on multi-chip quality controls. We defined a simple acceptance criterion which relies on three conditions: the scale factor must lie within 3-fold to a median chip, the average background intensity must lie within 3-fold to a median chip, and the percentage of present calls must lie within 1.5-fold to a median chip. In order to account for the technology variability (e.g. probes on HG-U133B are related to genes that are typically less expressed than those on HG-U133A), we separately applied the criterion to each Affymetrix platform. Moreover, the untreated and the tamoxifen treated datasets were treated separately given their different role in the following analysis.

We obtained the following results. While no untreated chips were discarded, 25 treated chips were removed. For instance, the HG-U133B chip of patient OXFT_1173 was not retained in the study because of an average background intensity equal to 234 (admitted range was [36, 111]) and a percentage of present calls equal to 8% (admitted range was [27%, 43%]).

---

**Lessons Learned**

Quality controls for gene expression profiling are complex and need automatic procedures to help the detection of poor quality chips in large datasets. Although single-chip quality controls are well standardized, multi-chip quality controls still require an ad-hoc procedure.

---

[4] http://edkb.fda.gov/MAQC/

## 10.5  Normalization

Normalization deals with systematic variations between experimental conditions (technical variation) which are not related to effective biological differences. Normalization methods aim to compensate for systematic technical differences between chips in order to enhance the analysis of biological differences between samples. Plenty of normalization methods specific to existing gene expression profiling technologies have been proposed in literature. Similarly to quality controls, they can be grouped in two main classes:

- Single-chip normalization methods : These are low complexity methods which use only one single chip to define the normalization transformation (e.g. mean scaling). A widely used single-chip normalization for Affymetrix technology is the Microarray Suite 5 (MAS5) [26].
- Multi-chip normalization methods : These methods use a set of chips to fit a (possibly) complex normalization transformation. This class of methods is sometimes referred to as *model-based* normalization methods. Widely used multi-chip normalization methods for Affymetrix technology are the Robust Multichip Average (RMA) [27], RMA using sequence information (GCRMA) [28], DNA-Chip Analyzer (dChip) [29] and Variance Stabilization Normalization (VSN) [30].

An overview of these normalization methods is given in [25]. Several studies addressed the question about the impact of normalization methods on gene expression analysis [31, 32, 33], but no gold-standard exists for normalization of Affymetrix data.

Two aspects of clinical studies should drive the choice of a normalization method :

- Data sources are usually heterogeneous (see Section 10.2) which could reduce the efficacy of normalization methods. It follows that special care needs to be taken in assessing the quality of a multi-source gene expression dataset. In the next subsection, we will sketch a computational intelligence method we adopted to detect heterogeneity in our study.
- The integration of a data-driven predictive tool (e.g. a gene classifier built from a gene expression dataset) in daily clinical routine is not a trivial task. Since the tool is expected to be used with patients who did not take part in the original study, methods to transform the new patient expression data into the normalized data space used for training the tool should be taken into account. Until now, most multi-chip normalization methods do not provide the possibility to save the fitted model for normalization transformation in order to apply it to new patients. This can be detrimental to the effectiveness of the approach in a real setting (e.g. need for a re-training, bias related to the laboratory, . . . ). For these reasons, nowadays single-chip normalization methods are still largely preferred in clinical studies.

## 10.5.1   Tamoxifen Study

We first normalized raw probe intensities with the MAS5 algorithm, a single-chip normalization method. The choice was justified by the reduced complexity and memory requirements of the method. Then, in order to account for the heterogeneity of sources, we developed a computational intelligence method to detect systematic variations in chips. The method relies on a preliminary clustering analysis to detect hidden structure in gene expression data and a subsequent contigency analysis to test the presence of a systematic effect related to the data sources.

**Table 10.1.** Contingency table between the cohorts of patients in columns and the five main clusters of patients in rows. Hierarchical clustering was performed on gene expression data after MAS5 normalization.

|            | GUYT | KIT | KIU | OXFT | OXFU |
|------------|------|-----|-----|------|------|
| cluster.1  | 0    | 0   | 0   | 0    | 68   |
| cluster.2  | 0    | 61  | 52  | 0    | 1    |
| cluster.3  | 0    | 18  | 16  | 0    | 0    |
| cluster.4  | 0    | 0   | 0   | 87   | 0    |
| cluster.5  | 86   | 0   | 0   | 0    | 0    |

**Table 10.2.** Contingency table between the cohorts of patients (in columns) and the five main clusters of patients (in rows). Hierarchical clustering was performed on gene expression data after MAS5 normalization and after robust standardization for probes.

|            | GUYT | KIT | KIU | OXFT | OXFU |
|------------|------|-----|-----|------|------|
| cluster.1  | 14   | 15  | 13  | 9    | 9    |
| cluster.2  | 21   | 18  | 20  | 25   | 25   |
| cluster.3  | 22   | 20  | 16  | 22   | 14   |
| cluster.4  | 18   | 16  | 14  | 17   | 16   |
| cluster.5  | 11   | 10  | 5   | 14   | 5    |

The clustering step was performed by a hierarchical clustering algorithm [34, 35]. The algorithm allows an easy partition of the dataset in a number of clusters equal to the number of cohorts (five). Two-way contingency tables reporting the observed frequencies of patients within cohorts (in columns) and within clusters (in rows), were used to test the association between cohorts and clusters ($\chi^2$ test [36]). We used the Cramer's V statistic [37] to quantify the strength of association. The values range from 0 to 1, with 0 indicating independence and 1 indicating a perfect association. Traditionally, values of 0.36 to 0.49 indicate a substantial association, and values of 0.50 or more indicate a strong association.

The obtained results (see Table 10.1) revealed a strong association between clusters and cohorts (Cramer's V of 0.86, $\chi^2$ test p-value of 0) and supported the fact that the use of single-chip techniques alone may not be sufficient to remove systematic variation related to heterogenous data sources.

In order to correct the bias, we went through a second round of normalization by performing a robust standardization ($\frac{x-\text{median}(x)}{\text{iqr}(x)}$ where iqr is the interquartile range of $x$) for probes in each cohort separately. The resulting clustering exhibited a low association with cohorts, as shown in Table 10.2 (Cramer's V of 0.08, $\chi^2$ test p-value of 8.30E-01).

---

**Lessons Learned**

Normalization methods may suffer from high heterogeneity of data sources. Single-chip normalization methods are less effective but easier to apply in real clinical studies as each new patient can be normalized separately. Unsupervised methods allow for the detection of systematic associations between gene expressions and sources.

---

## 10.6   Feature Transformation

Feature transformation is the data analysis step which aims to transform the original input space into a lower dimension space (called *feature* space) preserving most of the information available within the data. Dimensionality reduction is particularly important in gene expression analysis because of the following data characteristics :

- High feature-to-sample ratio (several thousands of genes and only few hundreds of samples).
- High correlation (co-expressed genes may be involved in common metabolic pathways).
- High noise due to the complex technology of gene expression profiling.

Feature transformation is expected to be beneficial in terms of better visualization, understanding and representation of the data. At the same time it is worth to mention that feature transformation in gene expression datasets exposes the procedure to a risk of overfitting and selection bias [38].

This section will focus on the use of unsupervised techniques for feature transformation. Three main methods are available for unsupervised feature transformation : compression, kernel and clustering methods. Compression and kernel methods transform the original input space into a new one whose dimensions are a linear combination of the original variables. These new variables are difficult to interpret from a biological and medical point of view. However, recent advances [39] were made to mitigate this problem in computing the gene expression deviations from normal tissues in order to build new features that are disease-specific. Unfortunately, such a database of normal breast tissues is very rare and did not exist for Affymetrix technology.. Overview of compression and kernel methods are given in [40]. Clustering methods rely on the fact that many genes are co-expressed and that their expression is highly correlated. The approach consists in finding clusters of highly correlated probes and in summarizing each set of clustered genes by the centroid (or prototype) of the cluster [41]. The transformed features are expected to have lower variance than the original ones, yet remaining easy to interpret from a biological and medical point of view. Another advantage of the approach is that the procedure is robust to missing genes, making then easier to map the transformed variables to different microarray platforms.

### 10.6.1 Tamoxifen Study

We performed an unsupervised clustering on the untreated dataset (OXFU and KIU cohorts) in order to avoid any correlation between the transformation step

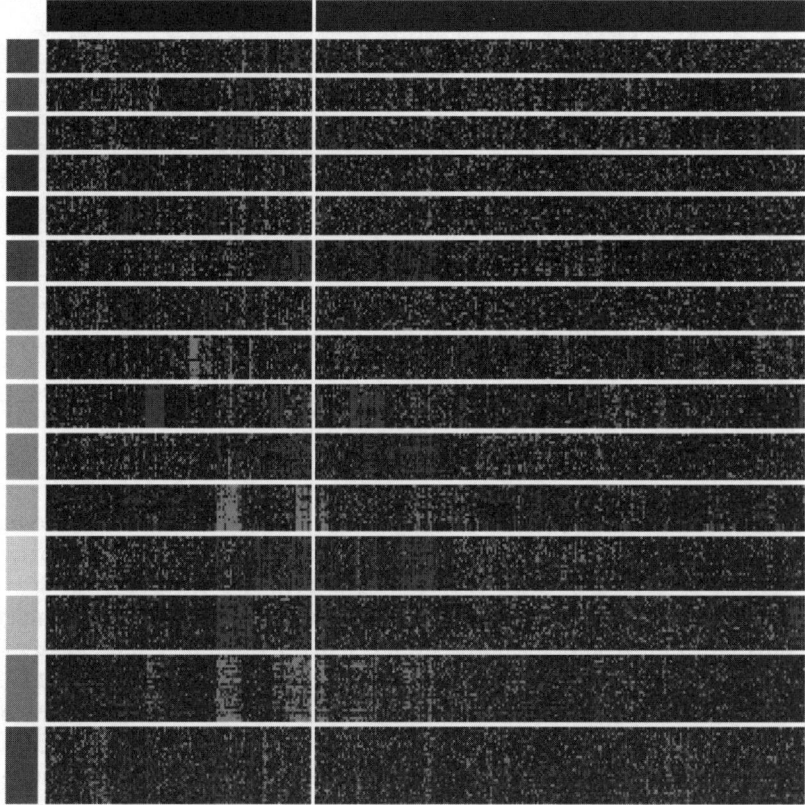

**Fig. 10.3.** Heatmap of probe clusters identified using OXFU and KIU cohorts. Color bars above the heatmap represent the untreated patients (OXFU and KIU cohorts, in blue) and the tamoxifen treated patients (OXFT, KIT and GUYT cohorts, in red). Color bars at left represent the different clusters. For clarity, only the 15 largest clusters are represented.

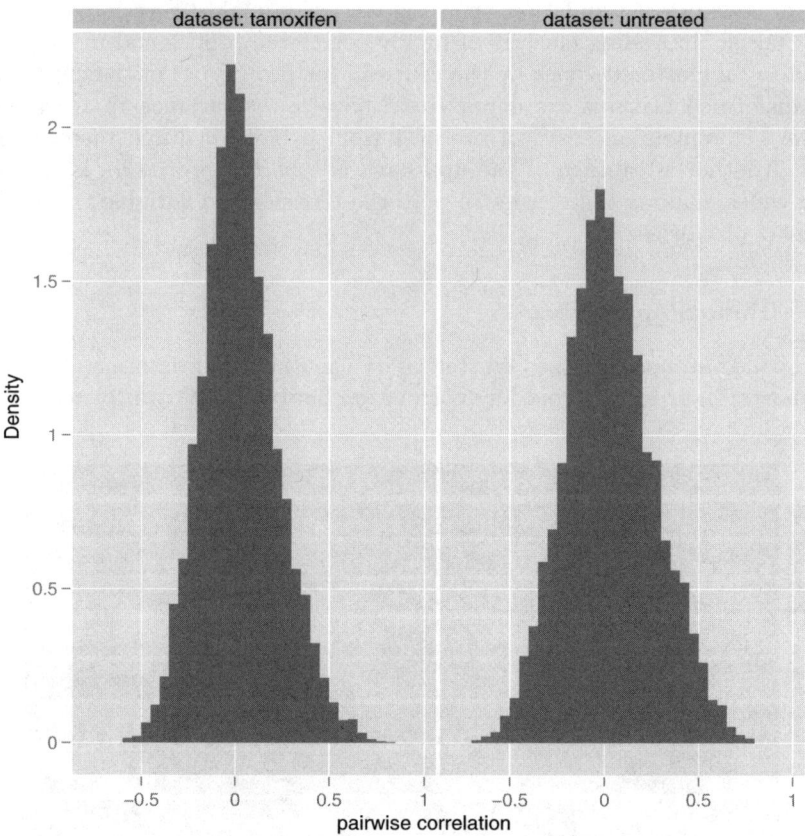

**Fig. 10.4.** Histogram of pairwise Spearman's correlations for the tamoxifen treated and untreated datasets

and the survival model fitting. This allowed to obtain dimensionality reduction without increasing the risk of potential overfitting. After removing a large proportion (80%) of low variance probes, we performed a hierarchical clustering using Spearman's correlation similarity metric and complete linkage. The generated dendrogram was then cut at the upper quartile of all heights. Clusters were discarded if there were less than 5 known Entrez gene ids[5] per cluster, 117 clusters remaining for further analysis. We represented the gene expressions of the largest clusters using a heatmap, i.e. a graphical display of the gene expressions (green to red for low to high level of gene expression) in two dimensions, where the rows denote the probes and the columns denote the patients. Figure 10.3 shows the heatmap of the largest clusters in all the cohorts.

The clustering computed on the untreated dataset was then used to compute the centroids in the tamoxifen treated dataset, then obtaining the set of

---

[5] http://www.ncbi.nlm.nih.gov/Entrez/index.html

transformed variables (called *pclust*). Note that unlike components in Principal Components Analysis (PCA), the *pclust* features are not orthogonal by construction, as shown by the histogram of pairwise Spearman's correlations in Figure 10.4. In particular it is possible to observe that some of them are highly correlated (maximum absolute pairwise correlation of 0.83 for untreated and tamoxifen treated datasets). Note that we were unable to perform a disease-specific genomic analysis (DSGA [39]) as we did not have any dataset of normal breast tissues.

---

**Lessons Learned**

Dimension reduction by feature transformation is an important step in order to reduce the noise and the complexity of the data. At the same time, keeping the new features interpretable is essential since the aim of clinical studies is not only to build efficient tools but also to bring new insights into biology and medicine. A combination of hierarchical clustering and simple average is well suited for this type of analysis.

---

## 10.7   Feature Selection

While the role of feature transformation is to reduce the dimensionality of the data, feature selection [41, 42] seeks which features, among the available ones, provide the largest amount of information about the survival of the patients. There are three main categories of feature selection methods : filter, wrapper and embedded methods. Filter methods assess the relevance of features ignoring the effects of the selected feature subset on the accuracy of the model. Wrapper methods assess subsets of features according to their relevance for a given model. The method conducts a search for a good subset using the model itself as part of the evaluation function (e.g. forward, backward and stepwise feature selections). Embedded methods perform feature selection as part of the model fitting and are usually specific to given models (e.g. classification trees and regularization techniques).

This section will focus on a specific filter method : variable ranking. First, variable ranking assesses the relevance of each individual feature $x_j, j = \{1, 2, \ldots, n\}$ according to a univariate scoring function $\mathcal{S}(j)$ supposed to be proportional to the relevance of the variable $x_j$ with respect to the prediction tasks. Second, it sorts all the features in a decreasing order according to the value of $\mathcal{S}(j)$. Eventually it selects the number of features to be selected according to some specific criterion.

Variable ranking is an intuitive technique which enjoys two interesting properties :

- Computational scalability : It is computationally efficient since it requires only the computation of the $n$ scores and the consequent sorting.
- Statistical scalability : variable ranking, like many filter methods, avoids the estimation of multivariate models to account for the relevance of a set of

variables. If on one hand, this exposes the technique to some redundancy (large bias), on the other hand it preserves the approach from overfitting risks (low variance) [43]. This property is particularly appealing in a gene expression study context where the noise is high and the number of features is huge, even after feature transformation.

Any variable ranking technique requires the definition of a criterion to select the optimal number of features. This is particularly important in a clinical study involving gene expression profiling since this determines the size of the gene signature which is distinctive of the phenomenon under examination. At the same time the task is extremely difficult given the reduced number of samples and the need of using the same dataset for both feature selection and model building. Cross-validation criteria have been proposed in literature [44] to select the number of variables. Although a cross-validation strategy relies on a multiple fold training and test strategy, it is important to remark that it is still prone to over-fitting if it is not kept independent with respect to the model building procedure. For instance re-using a dataset already employed to select a feature set (e.g. by cross-validation) in order to assess the quality of a predictive model (e.g. again by cross-validation) would return over-optimistic results about the quality of the modeling procedure. A way to minimize the covariance between the selection and the model building procedure would be not to rely on supervised criteria but rather on unsupervised measures. Another limitation of cross-validation criteria is due to the fact that, like other sampling frameworks (e.g. bootstrap), it generates different subsets of features for each fold or repetition. This is particularly annoying in a clinical setting where the variability of the selection reduces the confidence of the doctors in the computational intelligence tool and casts doubts about the efficacy of the approaches.

As an alternative to sampling approaches, recent studies [45, 46] introduced criteria for feature selection stability in gene expression data. Although stability reduces only the variance component of the prediction error (expressed conventionally as a bias/variance sum) the large amount of noise and the high dimensionality of the input space suggest that this term could be the most important to address in the bias-variance trade-off. The second advantage deriving from the use of stability measures would be a reinforced confidence of doctors in the gene signature outcomes of clinical studies.

### 10.7.1  Tamoxifen Study

We used a univariate variable ranking where the scoring function $S$ is based on the likelihood ratio (LR) statistic of the univariate Cox regression. The LR statistic is

$$LR = 2(l(\hat{\beta}) - l(0))$$

where $l(\hat{\beta})$ is the log partial likelihood of the model with coefficient $\hat{\beta}$. The null hypothesis distribution of the LR statistic is a $\chi^2$ with one degree of freedom.

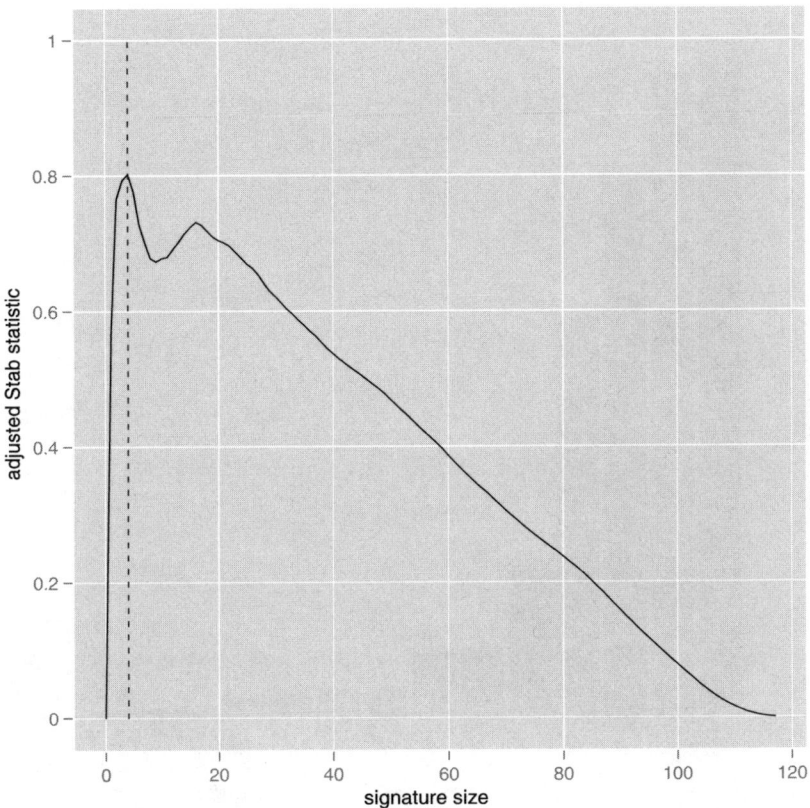

**Fig. 10.5.** Evolution of $Stab_{adj}$ with respect to the signature size $k$ in M10FOLDCV. The vertical dashed line (in green) represents the signature size $k = 4$.

The selection of the signature size was performed according to a stability criterion inspired to [46]. This criterion assesses the stability of the ranking for different signature size and selects the most stable size. Let $X$ be the set of features and $freq(x_j)$ be the number of sampling steps in which a feature $x_j \in X$ has been selected out of $m$ sampling steps. The set $X$ is sorted by frequency into the set $x_{(1)}, x_{(2)}, \ldots, x_{(n)}$ where $freq(x_{(i)}) \geq freq(x_{(j)})$ if $i < j$ where $i, j \in \{1, 2, \ldots, n\}$. A first measure of stability for a given signature size $k$ is returned by

$$Stab(k) = \frac{\sum_{i=1}^{k} freq(x_{(i)})}{km}$$

This statistic is equal to 1 if the same signature is always selected over sampling steps. In the case of no overlap, $Stab$ is equal to $\dfrac{1}{m}$ if $k > 0$ and 0 otherwise.

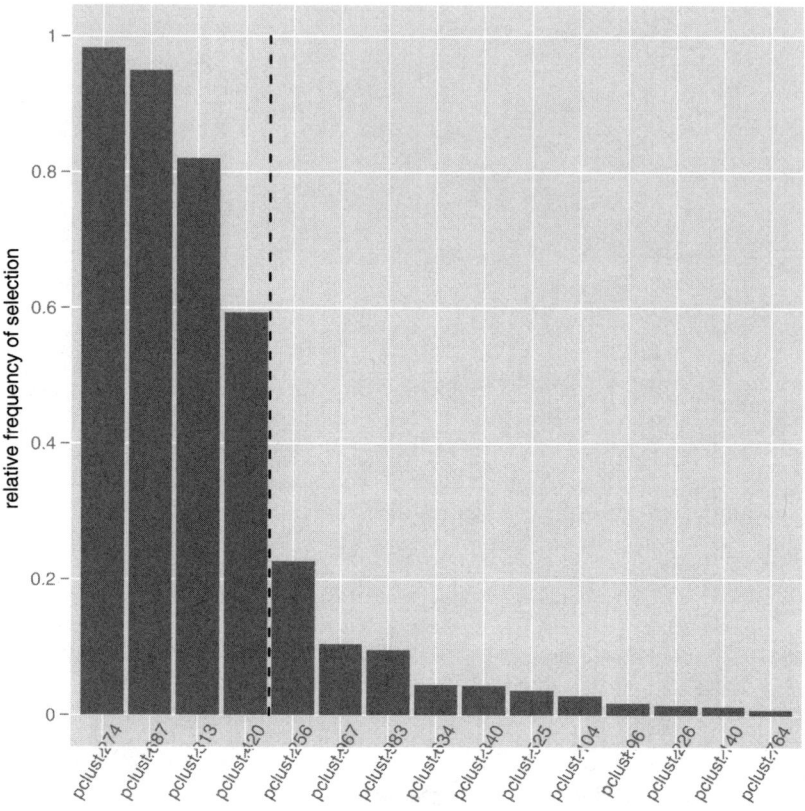

**Fig. 10.6.** Frequency of each feature during feature selection in M10FOLDCV. The vertical dashed line (in green) represents the limit for the first 4 *pclusts*. For clarity, only the 15 most selected features are shown.

However, since the *Stab* statistic can be made artificially high by simply increasing $k$, we formulated an adjusted statistic

$$Stab_{adj}(k) = \max\left\{0, Stab(k) - \alpha\frac{k}{n}\right\}$$

where $\alpha$ is a penalty factor depending on the number of selected features. In our study the penalty factor $\alpha$ was fixed to 1 in order to facilitate the selection of the trade-off between signature size and stability. Indeed, the $Stab_{adj}$ criterion is equal to 0 for the two extreme cases, i.e. when either no feature or all ones are selected.

We used two types of cross-validation to assess the stability of the feature selection : leave-one-out (LOOCV) and multiple ten fold cross-validation (M10FOLDCV). The M10FOLDCV consists in randomizing the order of patients each time a new ten fold cross-validation is performed.

Figure 10.5 sketches the evolution of $Stab_{adj}$ with respect to the signature size $k$. We observed a global maximum at $k = 4$ and a local maximum at $k = 16$. We selected the size $k = 4$ according to the stability criterion. Interestingly enough, this size appeared at the end of the study to be sufficiently large to allow for a biological interpretation of the results.

Figure 10.6 sketches the frequency of selection over all sampling steps for each *pclust* in the signature of size $k = 4$. Over the 19 different *pclusts* that were selected at least once, the 4 first *pclusts* were selected more than half of the times.

---

**Lessons Learned**

Feature selection is computationally intensive and prone to overfitting. A low complexity and high stability filter method, like variable ranking may have a positive effect on final accuracy by reducing the variance of the feature selection step.

---

## 10.8  Model Building

Building a survival model from expression data is a complex task which put the data analyst in front of several alternatives. Here we enumerate some of the hardest dilemmas to be solved :

- Linear v.s. non-linear : Linear models are simpler, more stable than non-linear models but not capable of dealing with complex dependencies. On the other hand, the higher complexity of non-linear models reduces the prediction bias at the cost of an increased variance.
- Univariate v.s. multivariate : Multivariate models deal more effectively with redundancies than univariate ones but demand ill-conditioned and computationally intensive estimation procedures.

The nature of gene expression data (very large dimensionality, few samples and high noise) evokes the potential risks of a non-linear and multivariate approach. At the same time, a simple univariate model would not be able to account for the multiple interactions underlying the cancer phenomenon.

The demand for a multivariate model which should be able at the same time to return accurate prediction and to avoid instability may have an effective answer in the adoption of additive modeling schemes [47]. The interest of an additive approach lies in the fact that the linear combination of several univariate models returns a model which is simple, yet able to address multivariate tasks. What is less attractive in a survival context is the need of iterative algorithm to fit the additive weights. A simple method to fit additive weights is provided by combination schemes, commonly used in computational intelligence literature [48], to combine several models in an effective manner. The following section presents the additive solution adopted in the context of our study.

## 10.8.1   Tamoxifen Study

The model building step started by comparing a multivariate Cox approach with an additive combination of univariate Cox models. The multivariate approach appeared to be more computationally intensive and less accurate. We decided then to develop an additive predictor which combines a set of univariate Cox models. In this model, the contribution $\log \lambda_j(t)$ of each single univariate Cox model is weighted according to its accuracy, measured by the likelihood ratio statistic (see Section 10.7). The outcome of the model for a given patient is given by the risk score

$$RS = \sum_{j=1}^{k} w_j p_j$$

where the signature size $k = 4$, $p_j = \beta_j X_j$ is the output of the $j$th linear predictor and $w_j$ is the normalized likelihood ratio statistic of the $j$th univariate Cox model such that $\sum_{j=1}^{k} w_j = 1$.

Although the risk score is a quantity that sheds light on the degree of risk of a patient, doctors find often more convenient (and more intelligible) to be provided with a crisp attribution of a patient to a low or high-risk class. In order to transform the risk score into a Boolean class we relied on existing medical expertise according to which there is an a priori 30% probability for a treated patient to belong to the high risk class. On the basis of a 70:30 partition we transformed the continuous model in the classifier

$$RG = I(RS)$$

where $I$ is an indicator function which takes the value 1 if $RS$ is included in the upper 30% of the risk scores and 0 otherwise.

---

**Lessons Learned**

Because of low variance, additive combination of univariate models guarantees robustness and accuracy in gene expression survival analysis.

---

## 10.9   Accuracy Assessment

Once a survival model is built, it is time to assess its accuracy. This section will discuss three accuracy criteria commonly used in survival analysis : the concordance index ($C$-index), the logrank test for KM survival curves and the hazard ratio (HR).

## 10.9.1   Concordance Index

The $C$-index [49] estimates the probability that, for a pair of randomly chosen comparable patients, the patient with the higher score will experience an event within a shorter time than the other patient.

The $C$-index takes the form

$$C\text{-index} = \frac{\sum_{i,j \in \Omega} I\left(RS_i, RS_j\right)}{|\Omega|}$$

where

$$I\left(RS_i, RS_j\right) = \begin{cases} 1 \text{ if } RS_i > RS_j \\ 0 \text{ if } RS_i \leq RS_j \end{cases}.$$

and $RS_i$ and $RS_j$ are the risk scores for the $i$th and the $j$th patient, respectively. Here $\Omega$ consists of all the pairs of patients $\{i, j\}$ who meet one of the following conditions :

1. Both patients $i$ and $j$ experienced event and time $t_i$ of patient $i$ is shorter than time $t_j$ of patient $j$.
2. Only patient $i$ experienced event and time $t_i$ is shorter than time $t_j$ of patient $j$.

Confidence intervals and p-values for this statistic are computed making an assumption of asymptotic normality [50].

### 10.9.2   Logrank Test

The logrank test, also known as the Mantel-Haenzel test, is widely used to test the null hypothesis that the survivor functions (see Section 10.3) are the same in a group 0 and a group 1, i.e. $H_0 : S_0(t) = S_1(t) \; \forall t > 0$.

This statistic quantifies how the observed timing of events diverges from the expected one according to the null hypothesis. Note that the logrank test is purely a test of significance and as such it does return neither estimate of the size of the difference between the groups nor a confidence interval.

### 10.9.3   Hazard Ratio

The hazard ratio is a summary of the risk difference between two survival curves. Proportional hazards regression model assumes that the relative risk of event between two groups is constant at each interval of time (see Section 10.3).

Let $G$ be an indicator variable, which takes the value zero if an individual is in the first group and unity if an individual is in the second group. If $g_i$ is the value of $G$ for the $i$th individual in the study, the hazard function for this individual can be written as

$$\lambda_i(t) = \lambda_0(t) \exp(\beta g_i)$$

where $g_i = 1$ if the $i$th individual is in the second group or zero otherwise. Because of the type of the indicator variable $G$, $\lambda_0(t)$ is the hazard function for an individual in the first group. Moreover, the hazard function for any individual in the second group is $\psi \lambda_0(t)$ (proportional hazards), so $\psi$ is the hazard ratio with $\psi = \exp(\beta)$.

The above mentioned criteria return measures of accuracy for a given survival model once a suitable test set is provided. However, the availability of only a limited amount of samples together with the complexity of the analysis raise a concern about how to obtain these measures without any excess of optimism about the accuracy of the modeling. This is particularly dangerous if undetected correlation between the model choices and the test set are established [13]. In order to avoid over-optimistic assessment the validation procedure (e.g. cross-validation) should include the entire modeling process and specifically the feature selection [51].

### 10.9.4  Tamoxifen Study

We used the same cross-validation procedures than in Section 10.7 to assess the performance of the entire modeling procedure : leave-one-out (LOOCV) and multiple ten fold cross-validation (M10FOLDCV).

Figure 10.7 sketches the survival curves for the low and high-risk groups of patients computed in LOOCV. The two curves were significantly different according to the logrank test (p-value of 1.64E-03). The proportions of patients

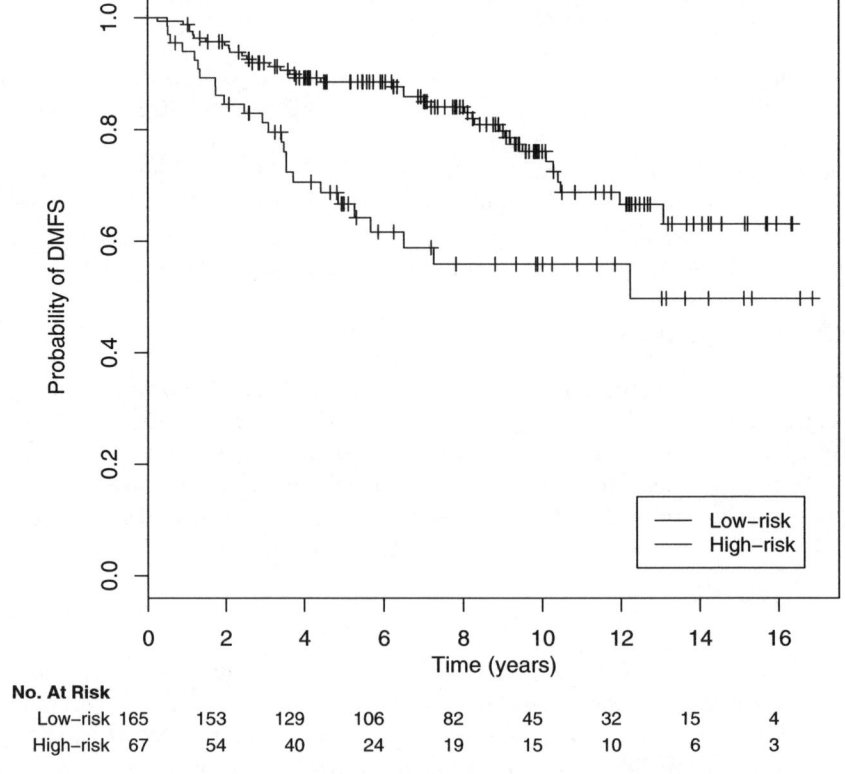

| No. At Risk | | | | | | | | |
|---|---|---|---|---|---|---|---|---|
| Low–risk 165 | 153 | 129 | 106 | 82 | 45 | 32 | 15 | 4 |
| High–risk 67 | 54 | 40 | 24 | 19 | 15 | 10 | 6 | 3 |

**Fig. 10.7.** Survival curves of the low and high-risk classes as defined by the method using a signature size $k = 4$ in LOOCV. The "+" symbol represents the censoring.

**Table 10.3.** Proportions of patients who did not experience any event at 3, 5 and 10 years

|  | 3.years | 5.years | 10.years |
|---|---|---|---|
| low.risk | 0.91 | 0.89 | 0.76 |
| high.risk | 0.80 | 0.67 | 0.56 |

who did not experience any event at 3, 5 and 10 years for the low and high-risk groups are given in Table 10.3.

The hazard ratio between these two groups of patients was estimated by Cox regression stratified by cohorts (see "ALL" symbol in the forestplot of Figure 10.8). It was equal to 2.26 with 95% CI [1.35,3.80] and was significant according to the LR test (p-value of 1.96E-03). In order to visualize the variability of hazard ratios between the different cohorts, a forestplot is drawn in Figure 10.8. The method performed better (higher hazard ratio) in the OXFT cohort than in the KIT and the GUYT cohorts. According to a test of heterogeneity [52] (p-value of 3.29E-01), there was not enough evidence in the data to show that variability in hazard ratio estimation between cohorts was due to other factors than sampling error alone.

In summary all the accuracy assessments supported the effectiveness of the approach. In order to complete our analysis we performed also a specific study about the effectiveness of the stability criterion adopted to choose the signature size equal to $k = 4$.

Figure 10.9 shows the evolution of the LOOCV and M10FOLDCV performance ($\log_2$ hazard ratio with 95% CI) with respect to the signature size. We observed, as for stability criterion, that signature sizes $k$ from 2 to 5 and $k \geq 16$ have reasonable performance. Additionally, we observed a relationship between feature selection stability and performance (see Figures 10.5 and 10.9 respectively) : models with stable signatures exhibited higher performance than models with unstable signatures in agreement with the bias-variance trade-off.

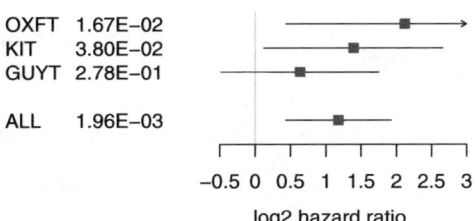

**Fig. 10.8.** Forestplot of $\log_2$ hazard ratios between low and high-risk classes as defined by the method using a signature size $k = 4$ in LOOCV. Each line represents a cohort of patients with the last one, labeled "ALL", representing all the cohorts together. Horizontal blue lines represent the 95% CI and the squares represent the point estimate of HR.

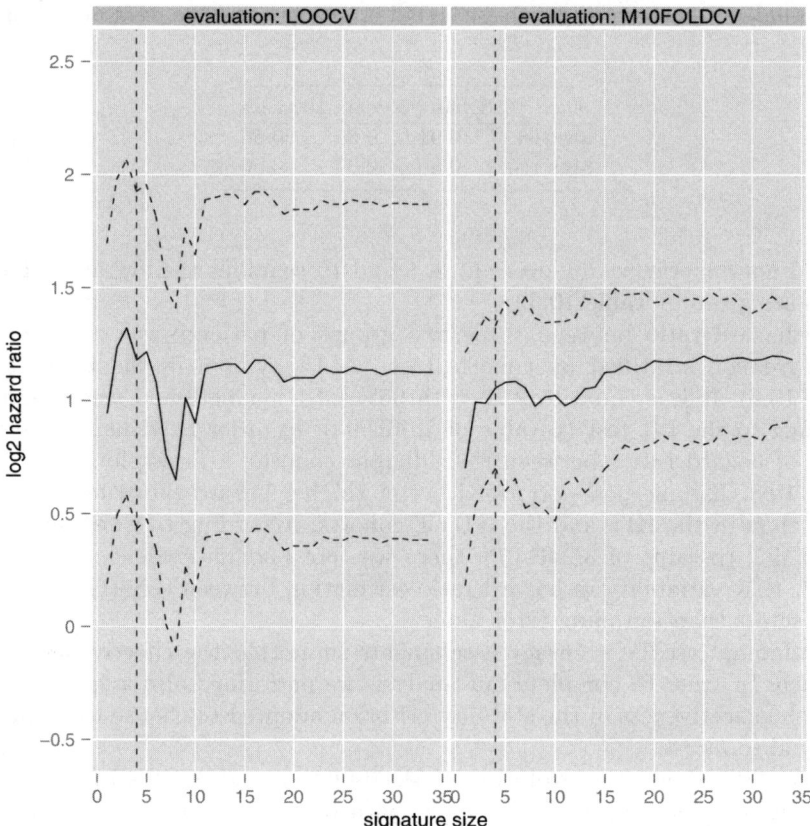

**Fig. 10.9.** Evolution of the $\log_2$ hazard ratio and its 95% confidence interval with respect to the signature size $k$ in LOOCV and M10FOLDCV. For clarity, only the signature sizes $k \leq 35$ are shown, the performance being stable for larger $k$.

**Lessons Learned**

The weighted combination of univariate linear models allows the implementation of a multivariate predictor which has low sensitivity to noise in the data and enables model interpretation. There is a relationship between feature selection stability and accuracy : stable signatures exhibit higher performance than unstable ones.

## 10.10   Validation

Once the modeling procedure has been positively assessed, it comes the time of testing the obtained model in real settings and for new patients. It should be evident that only this phase may give a reliable and thorough answer to the expectations of the doctors. Unfortunately a new and independent dataset is not always affordable in a clinical study because of the difficulty and cost of adding more patients. As a result most bioinformatics studies do not go beyond some cross-validated study, as the one presented in the previous section. Now, as said before, the complexity of the analysis, the noise in the data, the scarcity of samples and the large number of variables cast doubts on the reliability of cross-validation results, mainly if the data analyst has not been sufficiently cautious in avoiding (indirect) correlation between the training and the test set. It is therefore recommended to claim the success of a clinical genomic study only when a final validation step has been accomplished.

### 10.10.1 Tamoxifen Study

In order to provide a more convincing assessment of the quality of the proposed approach, our clinical study relied on an independent validation dataset provided by 77 patients diagnosed between 1980 and 1995 at the Guy Hospital (London, United Kingdom). This new cohort of samples, called GUYT2, were hybridized one year after the first three cohorts of tamoxifen treated patients, i.e. GUYT,

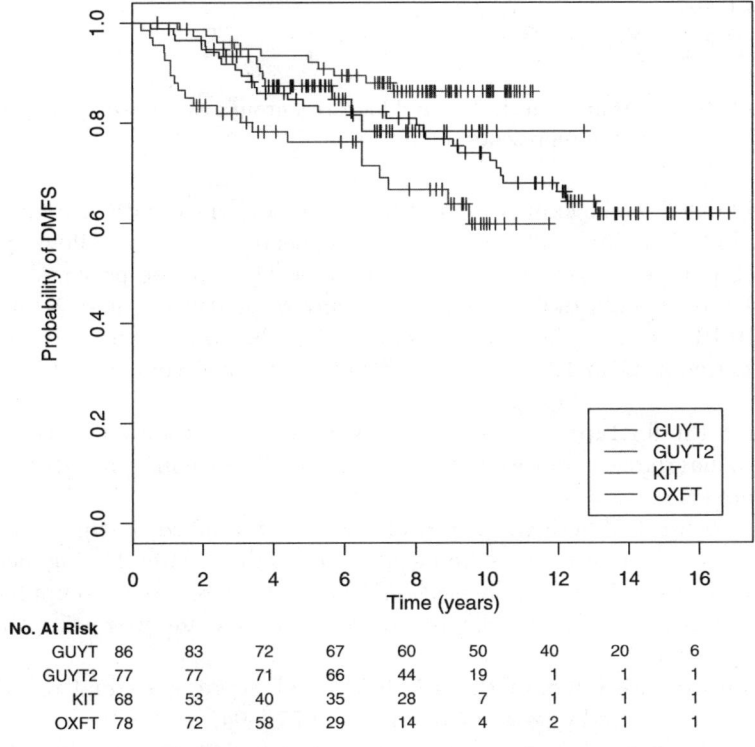

| No. At Risk | | | | | | | | |
|---|---|---|---|---|---|---|---|---|
| GUYT | 86 | 83 | 72 | 67 | 60 | 50 | 40 | 20 | 6 |
| GUYT2 | 77 | 77 | 71 | 66 | 44 | 19 | 1 | 1 | 1 |
| KIT | 68 | 53 | 40 | 35 | 28 | 7 | 1 | 1 | 1 |
| OXFT | 78 | 72 | 58 | 29 | 14 | 4 | 2 | 1 | 1 |

**Fig. 10.10.** Survival curves stratified by patient cohort, GUYT2 being the validation cohort. The "+" symbol represents the censoring.

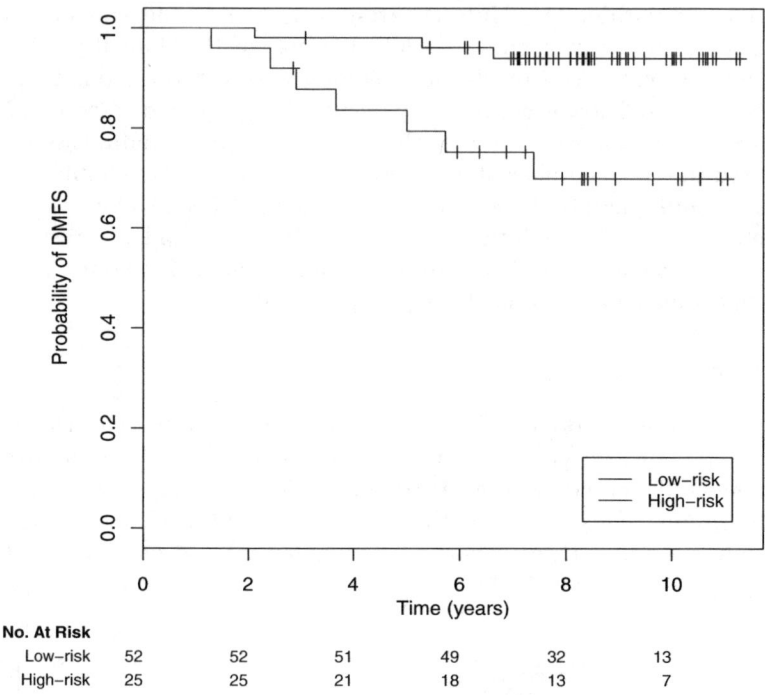

**Fig. 10.11.** Survival curves of the low and high-risk groups as defined by the TAMCOX model. The "+" symbol represents the censoring.

KIT and OXFT. Gene expression profiling was performed with Affymetrix HG-U133PLUS2 Genechips. This dataset was collected during a follow-up period longer than 10 years. Few events were observed (13% of the patients) and 92% (86%) of the patients did not experience any event till the first 5 (10) years. Figure 10.10 sketches the survival curves for the four cohorts of tamoxifen treated patients. GUYT2 cohort had a better survival compared to the other cohorts.

To build the final model (referred to as TAMCOX), we used all the patients of the original dataset and we repeated the whole modeling procedure in using a signature size of $k = 4$.

Quality controls described in Section 10.4, were applied on this new dataset and no chip was identified as being of poor quality. After having performed the normalization described in Section 10.5, risk scores were computed using the TAMCOX model. Binary classification was then computed using the cutoff previously defined.

The concordance index of the TAMCOX risk score was equal to 0.74 with 95% CI [0.60,0.89] and one-tailed p-value of 5.57E-04.

TAMCOX classified 52 patients in the low-risk class and 25 patients in the high-risk class. Figure 10.11 sketches the survival curves of the low and high-risk groups of patients. The two curves were significantly different according to

**Table 10.4.** Proportions of patients who did not experience any event at 3, 5 and 10 years

|          | 3.years | 5.years | 10.years |
|----------|---------|---------|----------|
| low.risk | 0.98    | 0.96    | 0.94     |
| high.risk| 0.84    | 0.75    | 0.70     |

the logrank test (p-value of 4.78E-03). The proportions of patients who did not experience any event at 3, 5 and 10 years for the low and high-risk classes are given in Table 10.4.

The hazard ratio between these two classes of patients was equal to 5.63 with 95% CI [1.45,21.79] and was significant according to the LR test (p-value of 8.02E-03).

Since the validation cohort was different in terms of survival, we could have expected lower performance of the model compared to previous results. Notwithstanding, TAMCOX model classified with success this set of new patients.

The success of the validation step would suggest the adoption of the TAMCOX model in clinical routine. However, this is not still a reality because of the complexity and the price of gene expression profiling technology for daily clinic use. A more affordable usage of the model might consist in replacing each *pclust* by a specific probe and design a low-cost test relying on a cheaper technology like the real-time polymerase chain reaction (RT-PCR) [53].

> **Lessons Learned**
>
> Validation set is mandatory to honestly assess the accuracy of the outcome of a survival analysis. The passage from research prototype to daily clinic tool is still made difficult by the large cost of the gene expression profiling technology.

## 10.11   Biological Interpretation

The biological interpretation of gene expression analysis results is made difficult by the possibly large number of genes in the signature. However, bioinformatics tools based gene ontologies [54], i.e. a structured vocabulary for the annotation of genes, can help biologists and doctors to retrieve useful statistical and biological information from the gene signatures.

### 10.11.1   Tamoxifen Study

The biological functions of each of the 4 clusters were analyzed in the context of a curated list of published molecular interactions by Ingenuity Pathway Analysis[6]. Table 10.5 lists the high level functions with statistically significant enrichment

---
[6] http://www.ingenuity.com/products/pathways_analysis.html

**Table 10.5.** Biological interpretation for each cluster of probes used in the final model

| Probes cluster | Nr of probes | Top high-level function overall | Top network functions |
|---|---|---|---|
| pclust.274 | 9 | Cancer, Cell cycle, Cellular growth and proliferation, DNA replication, recombination and repair | DNA replication, recombination and repair, Cell cycle |
| pclust.313 | 13 | Cancer, Cell cycle | Cancer, Tumor Morphology, Cell cycle |
| pclust.420 | 9 | Amino acid metabolism, Small molecule biochemistry | Cellular development, Cellular growth |
| pclust.687 | 6 | Cellular growth and proliferation, Cancer, Cell Morphology, Small molecule | Cancer, Cell cycle |

for each cluster. Three out of the four clusters are related to cell cycle function, supporting our previously reported finding that cell cycle and proliferation genes may help to define two subgroups of ER-positive BC tumors associated with statistically distinct clinical outcome [7]. For the fourth cluster, no definite conclusion can be drawn since only 2 genes could be used for functional and network analysis. However, when looking individually at these genes one can notice that CDK4 is also involved in the cell cycle. However, the presence of genes in the predictor that allude to other biological pathways apart from cell cycle function, such as DNA replication, recombination and repair or small molecule biochemistry, may facilitate further understanding of the upstream mechanisms behind tamoxifen resistance.

---

**Lessons Learned**

An interpretable model makes possible a biological characterization of the gene signature. The related biological insights adds value to the clinical genomic study.

---

## 10.12   Conclusion

Modeling the relationship between gene expression and clinical data, such as survival data, on the basis of a small number of samples is a big challenge for computational intelligence techniques. In this chapter, we discussed the main issues that arise in a retrospective clinical study aiming to transform raw data into useful medical information. In particular we focused on the automatic quality control, the normalization in presence of data heterogeneity, the feature transformation and the feature selection, the model building, the accuracy assessment and the external validation. We could summarize the guidelines that emerged from our real experience in the following points :

- Heterogeneity, noise, large dimensionality and scarcity of samples suggest that simple and effective techniques should be preferred to complex nonlinear models. The rationale is that the gain in robustness and stability would largely compensate for the lack of complexity.

- Interpretability matters in clinical study and should be always be taken into consideration when choosing a methodology.
- Large clinical studies ask for computational solutions able to deal with multiplatform and multisource configurations.

The importance of these recommendations is still more evident when we think that the near future of bioinformatics and biomedicine will be characterized by the need of integrating more and more sources of high dimensional data (integrative bioinformatics). High dimensional data will be generated by new gene expression profiling technologies, single nucleotide polymorphism (SNP), and comparative genomic hybridization (CGH) at a continuously growing rate. In front of this overwhelming amount the data, the doctors will be more and more demanding of effective, interpretable and robust computational techniques able to return exploitable information from genomic data.

# References

1. Zizka, J., Hudik, T.: Machine Learning - Based Knowledge Extraction from Complex Clinical Oncological Data. In: Knowledge Extraction and Modelling Conference (2006)
2. Pritchard, K.I.: Aromatase inhibitors in adjuvant therapy of breast cancer: Before, instead of, or beyond tamoxifen. Journal of Clinical Oncology 23(22), 4852–4858 (2005)
3. Loi, S., Piccart, M., Haibe-Kains, B., Desmedt, C., Harris, A., Bergh, J., Ellis, P., Miller, L., Liu, E., Sotiriou, C.: Prediction of early distant relapses on tamoxifen in early-stage breast cancer (BC): a potential tool for adjuvant aromatase inhibitor (AI) tailoring. In: Proceedings of the American Society of Clinical Oncology Meeting, Orlando, abstract 509 (2005)
4. Perou, C.M., Sorlie, T., Eisen, M.B., van de Rijn, M., Jeffrey, S.S., Rees, C.A., Pollack, J.R., Ross, D.T., Jonhsen, H., Aklslen, L.A., Fluge, O., Pergamenschikov, A., Williams, C., Zhu, S.X., Loning, P.E., Borresen-Dale, A.L., Brown, P.O., Botstein, D.: Molecular portraits of human breast tumours. Nature 406(6797), 747–752 (2000)
5. Sorlie, T., Tibshirani, R., Parker, J., Hastie, T., Marron, J.S., Nobel, A., Deng, S., Johnsen, H., Pesich, R., Geister, S., Demeter, J., Perou, C., Lonning, P.E., Brown, P.O., Borresen-Dale, A.L., Botstein, D.: Repeated observation of breast tumor subtypes in indepedent gene expression data sets. Proc. Natl. Acad. Sci. USA 1(14), 8418–8423 (2003)
6. Sotiriou, C., Neo, S.Y., McShane, L.M., Korn, E.L., Long, P.M., Jazaeri, A., Martiat, P., Fox, S., Harris, A.L., Liu, E.T.: Breast cancer classification and prognosis based on gene expression profiles from a population-based study. Proc. Natl. Acad. Sci. 100(18), 10393–10398 (2003)
7. Loi, S., Haibe-Kains, B., Desmedt, C., Lallemand, F., Tutt, A., Gillett, C., Ellis, P., Harris, A., Bergh, J., Foekens, J.A., Klijn, J., Larsimont, D., Buyse, M., Bontempi, G., Delorenzi, M., Piccart, M., Sotiriou, C.: Definition of clinically distinct molecular subtypes in estrogen receptor positive breast carcinomas through use of genomic grade. Journal of Clinical Oncology 25(10), 1239–1246 (2007)

8. Sotiriou, C., Piccart, M.J.: Taking gene-expression profiling to the clinic: when will molecular signatures become relevant to patient care? Nature Cancer Review 7, 545–553 (2007)
9. Ma, X.J., Wang, Z., Ryan, P.D., Isakoff, S.J., Barmettler, A., Fuller, A., Muir, B., Mohapatra, G., Salunga, R., Tuggle, J.T., Tran, Y., Tran, D., Tassin, A., Amon, P., Wang, W., Wang, W., Enright, E., Stecker, K., Estepa-Sabal, E., Smith, B., Younger, J., Balis, U., Michaelson, J., Bhan, A., Habion, K., Baer, T.M., Brugge, J., Haber, D.A., Erlander, M.G., Sgroi, D.S.: A two-gene expression ratio predicts clinical outcome in breast cancer patients treated with tamoxifen. Cancer Cell 5, 607–616 (2004)
10. Paik, S., Shak, S., Tang, G., Kim, C., Bakker, J., Cronin, M., Baehner, F.L., Walker, M.G., Watson, D., Park, T., Hiller, W., Fisher, E.R., Wickerham, D.L., Bryant, J., Wolmark, N.: A multigene assay to predict recurrence of tamoxifen-treated, node-negative breast cancer. New England Journal of Medicine (351), 2817–2826 (2004)
11. Jansen, M., Foekens, J.A., van Staveren, I.L., Dirkzwager-Kiel, M.M., Ritstier, K., Look, M.P., van Gelder, M.E.M., Sieuwerts, A.M., Portengen, H., Dorssers, L.C., Jlijn, J., Berns, M.: Molecular clasification of tamoxifen-resistant breast carcinomas by gene expression profiling. Journal of Clinical Oncology 23(4), 732–740 (2005)
12. Oh, D.S., Troester, M.A., Usary, J., Hu, Z., He, X., Fan, C., Wu, J., Carey, L.A., Perou, C.M.: Estrogen-regulated genes predict survival in hormone receptor-positive breast cancers. Journal of Clinical Oncology 24(11) (2006)
13. Ransohoff, D.F.: Rules of evidence for cancer molecular marker discovery and validation. Nature Cancer Review 4, 309–314 (2004)
14. Michiels, S., Koscielny, S., Hill, C.: Prediction of cancer outcome with microarrays: a multiple random validation strategy. Lancet 365, 488–492 (2005)
15. Ein-Dor, L., Kela, I., Getz, G., Domany, E.: Outcome signature genes in breast cancer: Is there a unique set? Bioinformatics 21, 171–178 (2005)
16. Gentleman, R.: Reproducible research: A bioinformatics case study. Statistical Applications in Genetics and Molecular Biology 4(1) (2005)
17. Barrett, T., Suzek, T.O., Troup, D.B., Wilhite, S.E., Ngau, W.C., Rudnev, P.D., Lash, A.E., Fujibuchi, W., Edgar, R.: NCBI GEO: mining millions of expression profiles - database and tool. Nucleic Acids Research 33, D562 (2005)
18. R Development Core Team, R: A language and environment for statistical computing. R Foundation for Statistical Computing, Vienna, Austria. ISBN 3-900051-07-0 (2007)
19. Parmigiani, G., Garett, E.S., Irizarry, R.A., Zeger, S.L.: The Analysis of Gene Expression Data. Springer, Heidelberg (2003)
20. Allison, P.D.: Survival Analysis Using SAS: A Practical Guide. SAS Institute Inc. (1995)
21. Duda, R.O., Hart, P.R., Stork, D.G.: Pattern classification. John Wiley and Sons, Chichester (2001)
22. Kaplan, E.L., Meier, P.: Nonparametric estimation from incomplete observations. Journal of American Statistical Asscoiation 53, 451–457 (1958)
23. Therneau, T.M., Grambsch, P.M.: Modeling Survival Data: Extending the Cox Model. Springer, Heidelberg (2000)
24. Cox, D.R.: Regression models and life tables. Journal of the Royal Statistical Society Series B 34, 187–220 (1972)
25. Gentleman, R., Huber, W., Carey, V.J., Irizarry, R.A., Dudoit, S.: Bioinformatics and Computational Biology Solutions Using R and Bioconductor. Springer, Heidelberg (2005)

26. Affymetrix, GeneChip Expression Analysis (2002)
27. Irizarry, R.A., Boldstad, B.M., Collin, F., Cope, L.M., Hobbs, B., Speed, T.R.: Summaries of affymetrix genechip probe level data. Nucleic Acids Research 31(4) (2003)
28. Wu, Z., Irizarry, R.A.: Preprocessing of oligonucleotide array data. Nature Biotechnology 22, 656–658 (2004)
29. Li, C., Wong, W.H.: Model-based analysis of oligonucleotide arrays: model validation, design issues and standard error application. Genome Biology 2(8), 1–11 (2001)
30. Huber, W., von Heydebreck, A., Sultman, H., Poustka, A., Vingron, M.: Variance stabilization applied to microarray data calibration and to the quantification of differential expression. Bioinformatics 18(1), S96–S104 (2002)
31. Ploner, A., Miller, L.D., Hall, P., Bergh, J., Pawitan, Y.: Correlation test to assess low-level processing of high-density oligonucletide microarray data. BMC Bioinformatics 6(80), 1–20 (2005)
32. Bolstad, B.M., Irizarry, R.A., Astrand, M., Speed, T.P.: A comparison of normalization methods for high density oligonucleotide array data based on variance and bias. Bioinformatics 19(2), 185–193 (2003)
33. Harr, B., Schlotterer, C.: Comparison of algorithms for the analysis of affymetrix microarray data as evaluated by co-expression of genes in known operons. Nucleic Acids Research 34(2), 8 (2006)
34. Hartigan, J.A.: Clustering Algorithms. Wiley, Chichester (1975)
35. Eisen, M., Spellman, P., Brown, P., Botstein, D.: Cluster analysis and display of genome-wide expression patterns. PNAS 95, 14863–14868 (1998)
36. Chernoff, H., Lehmann, E.L.: The use of maximum likelihood estimates in chi-square tests for goodness-of-fit. The Annals of Mathematical Statistics 25, 579–586 (1954)
37. Cramer, H.: Mathematical Methods of Statistics. Princeton University Press, Princeton (1999)
38. Ambroise, C., McLachlan, G.: Selection bias in gene extraction on the basis of microarray gene-expression data. Proc. Natl. Acad. Sci. USA 99, 6562–6566 (2002)
39. Nicolau, M., Tibshirani, R., Borresen-Dale, A.L., Jeffrey, S.S.: Disease-specific genomic analysis: identifying the signature of pathologic biology. Bioinformatics 23(8), 957–965 (2007)
40. Cristianini, N., Shawe-Taylor, J.: An Introduction to Support Vector Machines and Other Kernel-Based Learning Methods. Cambridge University Press, Cambridge (2000)
41. Guyon, I., Elisseeff, A.: An introduction to variable and feature selection. Journal of Machine Learning Research 3, 1157–1182 (2003)
42. Kohavi, R., John, G.H.: Wrappers for feature subset selection. Artificial Intelligence 97(1-2), 273–324 (1997)
43. Hastie, T., Tibshirani, R., Friedman, J.: The elements of statistical learning. Springer, Heidelberg (2001)
44. Weiss, S.M., Kulikowski, C.A.: Computer Systems that learn. Morgan Kaufmann, San Mateo (1991)
45. Pang, S., Havukkala, I., Hu, Y., Kasabov, N.: Classification consistency analysis for bootstrapping gene selection. Neural Computing and Applications 18(6), 527–539 (2007)
46. Davis, C.A., Gerick, F., Hintermair, V., Friedel, C.C., Fundel, K., Kuffner, R., Zimmer, R.: Reliable gene signatures for microarray classification: assessment of stability and performance. Bioinformatics 22(19), 2356–2363 (2006)

47. Hastie, T., Tibshirani, R.: Generalized Additive Models. Chapman and Hall, London (1990)
48. Kittler, J., Hatef, M., Duin, R., Matas, J.: On combining classifiers. IEEE Transactions on Pattern Analysis and Machine Intelligence 20(3), 226–238 (1998)
49. Harrell, F.E.: Tutorial in biostatistics: multivariable prognostic models: issues in developing models, evaluating assumptions and adequacy, and measuring and reducing errors. Statistics in Medicine 15, 361–387 (1996)
50. Pencina, M.J., D'Agostinno, R.B.: Overall C as a measure of discrimination in survival analysis: model specic population value and condence interval estimation. Statistics in Medicine 23, 2109–2123 (2004)
51. Varma, S., Simon, R.: Bias in error estimation when using cross-validation for model selection. BMC Bioinformatics 7(91), 1471–2105 (2006)
52. Cochrane, W.G.: Problems arising in the analysis of a series of similar experiments. Journal of the Royal Statistical Society 4, 102–118 (1937)
53. Freeman, W.M., Walker, S.J., Vrana, K.E.: Quantitative RT-PCR: pitfalls and potential. Biotechniques 26(1), 124–125 (1999)
54. Ashburner, M., Ball, C.A., Blake, J.A., Botstein, D., Butler, H., Cherry, J.M., Davis, A.P., Dolinski, K., Dwoght, S.S., Eppig, J.T., Harris, M.A., Hill, D.P., Issel-Tarver, L., Kasarskis, A., Lewis, S., Matese, J.C., Richardson, J.E., Ringwald, M., Rubin, G.M., Sherlock, G.: Gene ontology: tool for the unfication of biology. the gene ontology consortium. Nature Genetics 25, 25–29 (2000)

Computational Intelligence in Bioinformatics

# 11

# Artificial Immune Systems in Bioinformatics

Vitoantonio Bevilacqua[1], Filippo Menolascina[12], Roberto T. Alves[3],
Stefania Tommasi[2], Giuseppe Mastronardi[1], Myriam Delgado[3],
Angelo Paradiso[2], Giuseppe Nicosia[4], and Alex A. Freitas[5]

[1] Polytechnic of Bari, Via E. Orabona 4, 70125 Bari, Italy
`bevilacqua@poliba.it`
[2] National Cancer Institute 'Giovanni Paolo II', Via F. Hahnemann 10,
70126 Bari, Italy
`f.menolascina@ieee.org`
[3] Federal Technological University of Paraná, Av. 7 de setembro, 3165 Curitba, Brazil
`r.t.alves@gmail.com`
[4] University of Catania, Viale A. Doria 6, 95125 Catania, Italy
`nicosia@dmi.unict.it`
[5] Computing Laboratory University of Kent, Canterbury, CT2 7NF, UK
`a.a.freitas@kent.ac.uk`

**Summary.** Artificial Immune Systems (AIS) represent one of the most recent and promising approaches in the branch of bio-inspired techniques. Although this open field of research is still in its infancy, several relevant results have been achieved by using the AIS paradigm in demanding tasks such as the ones coming from computational biology and biochemistry. The chapter will show how AIS have been successfully used in computational biology problems and will give readers further hints about possible implementations in unexplored fields. The main goal of the contribution lays in providing both theoretical foundations and hands-on experience that allow researchers to figure out novel applications of AIS in bioinformatics and, at the same time, providing researchers with necessary insights for implementation in daily research. The contribution will be organised in 5 sections.

## 11.1 Introduction

Artificial Immune Systems (AIS) represent one of the most recent and promising approaches in the branch of bio-inspired techniques. Although this open field of research is still in its infancy, several relevant results have been achieved by using the AIS paradigm in demanding tasks such as the ones coming from computational biology and biochemistry. Artificial immune systems (AIS) can be defined as computational systems inspired by theoretical immunology, observed immune functions, principles and mechanisms in order to solve problems. Their development and application domains follow those of soft computing paradigms such as artificial neural networks (ANN), evolutionary algorithms (EA) and fuzzy systems (FS). Soft computing was the term coined to address a new trend of co-existence and integration that reflects a high degree of interaction among several computational intelligence approaches like artificial neural network, evolutionary

T.G. Smolinski et al. (Eds.): Comp. Intel. in Biomed. & Bioinform., SCI 151, pp. 271–295, 2008.
springerlink.com

algorithms and fuzzy systems. The idea of integrating different computational intelligence paradigms in order to create hybrids combining the strengths of different approaches is not new. Following the previous concepts when in 2002 de Castro and Timmis introduced AIS as a new soft computing paradigm they gave birth to a new challenge to a have a great potential to interact the new born technique with the other previously existing. Strictly speaking evolution and immune system are biologically very correlated to each other in fact the process of natural selection can be seen to act the immune system at two levels. First recall that lymphocytes multiply based on their affinity with a pathogen. The higher affinity lymphocytes are selected to reproduce, a process usually named immune microevolution. The mechanism of immune microevolution is very important. The clonal selection principle presupposes that a very large number of *B-cells* containing antigenic receptors is constantly circulating throughout the organism. The great diversity of this repertoire is a result of the random genetic recombination of gene fragments from different libraries plus the random insertion of gene sequences during cell development. This availability of different solutions guarantees that at least one cell will produce an antibody capable of recognizing, thus binding with, any antigen that invades the organism. The antigen-antibody binding stimulate the production of clones of the selected cells, where successive generations result in exponential growth of the selected antibody type. Some of these antibodies remain in circulation even after the immune response ceases, constituting a sort of immune memory. Other cells differentiate in plasma cells, producing antibodies in high rates. Finally during reproduction, some clones suffer an affinity maturation process, where somatic mutations are inserted with high rates (hypermutation) and, combined with a strong selective mechanism, improve the capability (*Ag-Ab* affinity and clone size) of these antibodies to recognize and respond to the selective antigens. Secondly, there is surely an immune contribution to natural selection, which acts by allowing the multiplication of those people carrying genes that are most able to provide maximal defense against infectious diseases coupled with minimal risk of autoimmune diseases. At this time the majority of the immune algorithms currently developed have an evolutionary like type of learning of embodied process and several techniques from one strategy have been used to enhance another. I-PAES presented and discussed in the section 11.3.1 is an example of hybridization between a particular class of evolutionary algorithms called multi-objective and immune inspired operators namely cloning and hypermutaion.

The success of the AIS paradigm is based on two key properties of its theoretical foundations: recognition and adaptation/optimisation. When an animal is exposed to an antigen, some subpopulation of its bone marrow derived cells (*B lymphocytes*) respond by producing antibodies (*Ab*). Each cell secretes a single type of antibody, which is relatively specific for the antigen. By binding to these antibodies (*cell receptors*), and with a second signal from accessory cells, such as the T-helper cell, the antigen stimulates the *B cell* to proliferate (divide) and mature into terminal (non-dividing) antibody secreting cells, called plasma cells. The process of cell division (mitosis) generates a clone, i.e., a cell or set of cells that are the progenies of a single cell. While plasma cells are the most active antibody secretors, large B lymphocytes, which divide rapidly, also secrete antibodies, albeit at a lower rate. On the other hand, T cells play a central role in the regulation of the *B cell* response and are preeminent in cell mediated immune responses, but will not be explicitly accounted for the development of our model. Lymphocytes, in addition to proliferating and/or differentiating into plasma cells, can differentiate into long-lived B memory cells. Memory cells circulate through the blood, lymph and tissues, and when exposed to a second antigenic stimulus commence to differentiate into large lymphocytes capable of producing high affinity antibodies, pre-selected for the specific antigen that had stimulated the primary response. Fig 11.1 depicts the clonal selection principle.

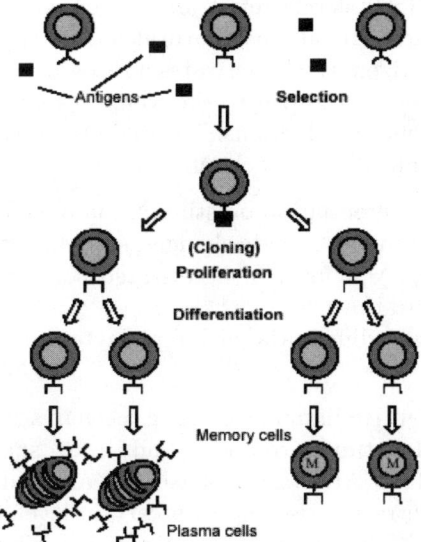

**Fig. 11.1.** Clonal selection principle in natural immune systems

The clonal selection and affinity maturation principles are used to explain how the immune system reacts to pathogens and how it improves its capability of recognizing and eliminating pathogens [1]. In a simple form, clonal selection states that when a pathogen invades the organism, a number of immune cells that recognize these pathogens will proliferate; some of them will become effector cells, while others will be maintained as memory cells. The effector cells secrete antibodies in large numbers, and the memory cells have long life spans so as to act faster and more effectively in future exposures to the same or a similar pathogen. During the cellular reproduction, the cells suffer somatic mutations with high rates and, together with a selective force, the higher affinity cells in relation to the invading pathogen differentiate into memory cells. This whole process of somatic mutation plus selection is known as affinity maturation. To a reader familiar with evolutionary biology, these two processes of clonal selection and affinity maturation are much akin to the (macro-)evolution of species. There are a few basic differences however, between these immune processes and the evolution of species. Within the immune system, somatic cells reproduce in an asexual form (there is no crossover of genetic material during cell mitosis), the mutation suffered by an immune cell is proportional to its affinity with the selective pathogen (the higher the affinity, the smaller the mutation rate), and the number of progenies of each cell is also proportional to its affinity with the selective pathogen (the higher the affinity, the higher the number of progenies). Evolution in the immune system occurs within the organism and, thus it can be viewed as a micro-evolutionary process. As we know, in fact, immunology suggests that the natural Immune System (IS) has to assure recognition of each potentially dangerous molecule or substance, generically called antigen ($Ag$), by antibodies ($Ab$). The IS first recognises an antigen as "dangerous" or external invaders and then adapts (by affinity maturation) its response to eliminate the threat. To detect an antigen, the IS activates a recognition process. In vertebrate organisms, this task is accomplished by the complex machinery made by cellular interactions and molecular productions. The main features of the clonal selection theory that will be explored in this chapter are [1]]:

- Proliferation and differentiation on stimulation of cells with antigens;
- Generation of new random genetic changes, subsequently expressed as diverse antibody patterns, by a form of accelerated somatic mutation (a process called affinity maturation);
- Elimination of newly differentiated lymphocytes carrying low affinity antigenic receptors.

To illustrate the adaptive immune learning mechanism, consider that an antigen $Ag1$ is introduced at time zero and it finds a few specific antibodies within the animal (see Fig. 11.2. After a lag phase, the antibody against antigen $Ag1$ appears and its concentration rises up to a certain level, and then starts to decline (*primary response*). When another antigen $Ag2$ is introduced, no antibody is present, showing the specificity of the antibody response [1]. On the other hand, one important characteristic of the immune memory is that it is associative: *B cells* adapted to a certain type of antigen $Ag1$ presents a faster and more efficient secondary response not only to $Ag1$, but also to any structurally related antigen $Ag1 + Ag2$. This phenomenon is called immunological cross-reaction, or cross-reactive response. This associative memory is contained in the process of vaccination and is called *generalization capability*, or simply generalization, in other artificial intelligence fields, like neural networks [1].

Receptor editing offers the ability to escape from local optima on an affinity landscape. Fig 11.3 illustrates this idea by considering all possible antigen-binding sites depicted in the x-axis, with the most similar ones adjacent to each other. The Ag-Ab affinity is shown on the y-axis. If it is taken a particular antibody ($Ab1$) selected during a primary response, then point mutations allow the immune system to explore local areas around Ab1 by making small steps towards an antibody with higher affinity, leading to a local optima ($Ab1$ *). Because mutations with lower affinity are lost, the antibodies can not go down

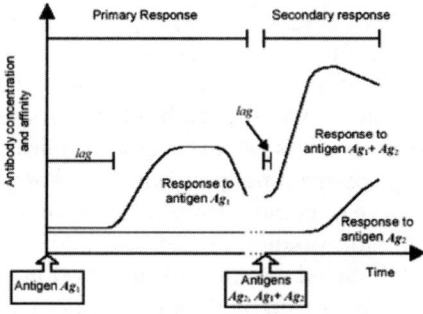

**Fig. 11.2.** Immune response plotted as antibody concentration over time

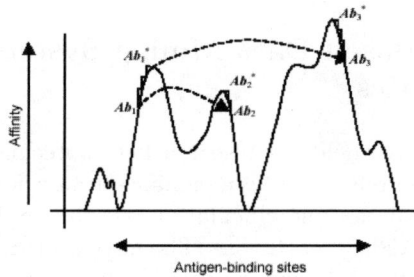

**Fig. 11.3.** Antibody affinity as function of the specific antigen binding site

the hill. Receptor editing allows an antibody to take large steps through the landscape, landing in a locale where the affinity might be lower (*Ab2*). However, occasionally the leap will lead to an antibody on the side of a hill where the climbing region is more promising (*Ab3*), reaching the global optimum. From this locale, point mutations can drive the antibody to the top of the hill (*Ab3 \**). In conclusion, point mutations are good for exploring local regions, while editing may rescue immune responses stuck on unsatisfactory local optima.

Computational immunology is the research field that attempts to reproduce in silico the behavior of the natural IS. From this approach, the new field of Artificial Immune Systems (AIS) attempts to use theories, principles, and concepts of modern immunology to design immunity-based system applications in science and engineering [1]. AIS are adaptive systems in which learning takes place by evolutionary mechanisms similar to biological evolution. These different research areas are tied together: the more we learn from in silico modelling of natural systems, the better we are able to exploit ideas for computer science and engineering applications. Thus one wants, first, to understand the dynamics of such complex behavior when they face antigenic attack, and second, one wishes to develop new algorithms that mimic the natural IS under study. Thus the final system may have a good ability to solve computational problems otherwise difficult to be solved by conventional specialised algorithms. The computational and predictive power of AIS offers researchers a promising approach for trying to solve well known and challenging problems like knowledge discovery from huge biological databases (e.g. coming from high throughput platforms) as well as protein folding or function prediction and multiple sequence alignment. The chapter will show how AIS have been successfully used in computational biology problems and will give readers further hints about possible implementations in unexplored fields. The main goal of the contribution lays in providing both theoretical foundations and hands-on experience that allow researchers to figure out novel applications of AIS in bioinformatics and, at the same time, providing researchers with necessary insights for implementation in daily research.

## 11.2     Immunity-Based Data Mining Systems in Bioinformatics

Recent advances in active fields of research like biotechnology and electronics allowed biomedical research to make a significant step forward in the acquisition of fundamental tools for the elucidation of complex bio-processes like the ones behind cancer or Alzheimer disease. The advent of High-Throughput (HT) platforms has revolutionized the way researchers working in life sciences thought at their role in experiments. HT devices allowed researchers to concentrate on higher tasks like experimental design and results interpretation at the same time avoiding him minding of hundreds when not thousands of repeats of the same protocols for the different patients or mRNA sequences for instance. Microarrays are, probably, one of the most evident examples of this change of perspectives: gene expression evaluation for a panel of even only a few tens of genes took several days to be completed before their introduction, now we are able to obtain gene expression level for thousands of genes in the time of an overnight hybridization. Together with expression microarrays we can mention copy number monitoring microarrays (commonly referred to as aCGH technique), High-Throughput Sequencers, and Mass Spectrometers. In the next sections we will go through a brief analysis of the main open problems in bioinformatics and will discuss about how they can be addressed using immunity based data mining algorithms. A short introduction on data mining principles and potentialities is given in order to help unexperienced readers understanding concepts behind statements.

### 11.2.1     Data Bases and Information Retrieval in Biology

Devices coming from the integration of experiences gained in diverse fields like physics, chemistry, biology and engineering, in this way helped researchers in boosting their work and in quickly obtaining results of their experiments. The capabilities of these different kinds of approach pushed the interest for the establishment of data repositories for newly generated results. Data-bases entered the world of biology. Larger and larger amounts of data started to fill public databases (leaving apart literature databases which, of course, need a separated analysis) giving rise to what we can rename "Moore's law in biology" [2] (that just like the original Moore's law in electronics, models future progress in biotechnology [3]). However the main advantages provided by novel devices soon revealed to be their main weak point. The availability of large amount of data as results didn not yield of information drawn from these data; this phenomenon characterized both early and more recent years in life sciences research bringing to the so-called "gap". Roughly speaking, researchers indicate, with this term, an estimate of the difference between the amount of available data and the amount of these data that have been sufficiently interpreted [4]. In the recent years we have observed a worrying widening in this gap: this means that we are making quite large investments with a ROI (return on investments) that still keeps low. In order to maximize the information yield of each experiment several alternative solutions have been proposed being probably data warehousing the most successful. Data warehouses are the natural evolution of data bases; described for the first time by William Immon [5] they are integrated, subject-oriented, time-variant and non-volatile data collection processes implemented with the precise aim to build a unique decision support system. The distinction between data bases and data warehouses is clear: as advanced data bases, data warehouse provide data analysis functionalities that ease the process of knowledge extraction from highly dense data repository. In this context grew significant experiences like the GEO (Gene Expression for Omnibus, [7]), SMD (Stanford Microarray Database, [8]) and ArrayExpress [9]. This is the evident that data warehouse can greatly help researchers in reducing the gap providing a valuable aid in filling the last real hole in experimental processes automation: results interpretation.

## 11.2.2  Mining the Data: Converting Data to Knowledge

Data mining, also known as Knowledge Discovery in Data-bases (KDD), has been defined as *"The nontrivial extraction of implicit, previously unknown, and potentially useful information from data"* [6] (a more practical definition of data mining will be given in the following section); it uses machine learning, statistical and visualization techniques to discover and present knowledge in a form easily comprehensible to humans. Data mining grew at the border line among statistics, computer science and artificial intelligence and soon became a golden tool to solve problems spacing from Customer Relationship Management (CRM, [10]) to Decision Making Support in medicine [15]. Data mining in bioinformatics, then, can be considered as a useful tool for modelling complex processes allowing researchers speeding the pace towards treatments for diseases like cancer: for instance several works have successfully tried to exploit the potentialities of rule induction systems in breast cancer associated survival [56, 57] and cancer evolution modelling [58]. It can be argued that data mining was born from several diverse disciplines, in the effort of overcoming intrinsic limitations of the single approaches. It is particularly evident if we compare the expressive power of typical statistical inference approaches and propositional or first order logic on the other hand. Huge efforts have been spent, in the recent past, in order to speed up one of the central tasks in current research in bioinformatics, that is the transformation process that converts *data* in *knowledge* passing through *information* [16]. Data mining software, then, became more and more common: researchers soon realized the valuable aid algorithms could have given to their researchers and the amount of paper describing algorithms for information extraction grew faster and faster [40, 41, 45]. Comprehensive software suites for data mining purposes are currently largely used in bioinformatics and include both open-source and proprietary solutions. Among commercial packages we can list SPSS, SAS, Clementine and E-Miner. Open source suites are well represented by:

- Weka [18]
- Rapid Miner (formerly YALE) [19]
- Orange [20]

In particular Weka has gained a relevant success in the field of data mining due to its flexibility and versatility. Thanks to these characteristics Weka has been customized and redistributed in several different flavours (BioWeka [21] devoted to biological sequences mining and Weka4WS [22], the GRID-enable Weka implementation). Due to a simple but efficient modular organization Weka allowed third-party developers to add functionalities to the core package. It is the case of "Weka Classification Algorithms" project managed by Jason Brownlee who has implemented several bioinspired [11, 12, 13] data mining algorithm in a customized version of Weka [14]. One of the most interesting aspects of this implementation consists in the presence of a wide variety of Artificial Immune System based data mining algorithms. Both the *black* and *white box* flavours are represented in the set of proposed algorithms. The distinction between black and white box algorithms will be described in the following paragraph, however it can be argued that white box approaches provide the user with tools to easily interpret the way it reached a certain results, on the contrary to what happens with black box algorithms (think at how complex is the interpretation of neural network predictions and how simple is interpreting rules induced from a dataset). Among black box Immunity based algorithm we can mention:

*Clonalg.* The Clonal Selection Algorithm, originally called CSA in [55], and renamed to CLONALG in [61] is said to be inspired by the following elements of the clonal selection theory:

- Maintenance of a specific memory set
- Selection and cloning of most stimulated antibodies
- Death of non-stimulated antibodies
- Affinity maturation (mutation)
- Re-selection of clones proportional to affinity with antigen
- Generation and maintenance of diversity

The goal of the algorithm is to develop a memory pool of antibodies that represents a solution to an engineering problem. In this case, an antibody represents an element of a solution or a single solution to the problem, and an antigen represents an element or evaluation of the problem space.

*CSCA.* The Clonal Selection Classifier Algorithm is an evolution of the concept behind Clonalg since it tries to maximise classification accuracy and minimise misclassification accuracy still using clonal selection paradigms.

*Immunos.* The Immunos [54] algorithm has been mentioned a number of times in AIS literature [37, 38, 39]. It is claimed as being one of the first immune-inspired classification systems. Immunos tries to mimic in a very precise way the mechanisms underlying immune response to antigen attacks and this has led to a quite complex classification system still under discussion.

*AIRS.* The Artificial Immune Recognition System [42] algorithm was one of the first AIS technique designed specifically and applied to classification problems. After an initialisation phase the algorithm cycles through each antigen (record in the dataset) in order to select best fitting memory cells through a powerful resource competition stage.

On the other hand white box AIS based paradigms can be found in:

- IFRAIS
- AIS based rule induction with boosting

These approaches will be deeply discussed in the next section.

### 11.2.3 Algorithmic Approaches to Data-Mining in Biology

As previously stated data mining is an interdisciplinary research field, involving areas such as machine learning, statistics, databases, expert systems and data visualization, whose main goal is to extract knowledge (or patterns) from real-world data sets [17, 18]. This section focuses on the classification (supervised learning) task of data mining. In essence, the goal of the classification task is to assign each example (data instance or record) to a class, out of a predefined set of classes, based on the values of attributes describing that example. In the context of bioinformatics an example could be, for instance, a protein; the classes could be protein functions; and the attributes describing the protein could be, say, physico-chemical properties of the amino acids composing the protein. It is important that the attributes describing an example are relevant for predicting its class. Hence, it would be a mistake to use a clearly irrelevant attribute, say the name of the patient, as an attribute to predict whether or not a patient will get a certain disease. In bioinformatics, ideally, the classification model should satisfy two requirements. First, it should have a high predictive accuracy, or generalization ability, correctly predicting the class of new examples unseen during the training of the system. Second, it should be comprehensible to users (biologists), so that it can be interpreted in the context of existing biological knowledge and potentially further validated through new biological experiments. Concerning the issue of comprehensibility of the classification model discovered from the data, it should be noted that some classification algorithms are designed to maximize only predictive accuracy, representing the classification model in a way that cannot be understood by the user - therefore ignoring the comprehensibility requirement. Typical examples of algorithms in this category are support vector machines [24] and neural networks [25]. In this case the classification model is a "black box", which does not give the user any insight about the data or explanations about the classification of new examples. By contrast, some classification algorithms use a representation which is comprehensible to the user, therefore returning "knowledge" to the user. In this section we focus on one popular kind of comprehensible representation, namely IF-THEN classification rules, and algorithms that use this kind of representation are called rule induction algorithms [23]. In rule induction algorithms the classification model is represented by a set of classification rules. These rules are of the form: "IF antecedent THEN consequent", where the antecedent represents a conjunction of conditions and the consequent represents the class predicted for all examples (data instances, records) that satisfy the antecedent. Each condition in the antecedent typically specifies a value or a range of values for a given attribute of the data being mined - e.g., "gender = female", "age < 21".

The first AIS for rule induction in the classification task of data mining was proposed in [27], and named IFRAIS (Induction of Fuzzy Rules with an Artificial Immune System). IFRAIS will be discussed in the next section. In this section we just highlight that this system discovers fuzzy classification rules. Fuzzy rules are in general more natural and more comprehensible to human beings than crisp rules, and the fuzzy rule representation also has the ability of coping well with the uncertainties frequently associated with data in biological databases [28]. Other algorithms based on AIS for rule induction are discussed in detail in [66, 67].

### Current Models

*Artificial Immune Systems in Bio-medical Data Mining: IFRAIS Study Case* As mentioned earlier, IFRAIS is an AIS that discovers fuzzy classification rules from data. Recall that the rule antecedent is formed by a conjunction of conditions. Each attribute can be either continuous (real-valued, e.g. the molecular weight of a protein) or categorical (nominal, e.g. the name of a species), as usual in data mining. Categorical attributes are inherently crisp, but continuous attributes are fuzzified by using a set of three linguistic terms (low, medium, high). Hence, in the case of conti-nuous attributes, IFRAIS discovers fuzzy rules having conditions such as: "molecular weight is large". IFRAIS discovers fuzzy classification rules by using the sequential covering approach for rule induction algorithms [18]. This is an iterative process which starts with an empty set of rules and the full training set (containing all training examples). At each iteration, IFRAIS is run to discover the best possible classification rule for the current training set, which is then added to the set of discovered rules. Then the examples correctly covered by the discovered rule (i.e. the examples satisfying the antecedent of that rule and having the class predicted by the rule) are removed from the training set, so that a smaller training set is available for the next iteration. This process is repeated until all (or a large part of the) training examples have been covered by the discovered rules. In order to discover classification rules, IFRAIS uses essentially clonal selection and hypermutation procedures. The basic ideas are as follows. Each antibody corresponds to a candidate fuzzy classification rule. During an IFRAIS run, the better the classification accuracy of an antibody, the more likely it is to be selected for cloning. In addition, once an antibody is cloned, the rate of mutation of a clone is inversely proportional to the classification accuracy of the antibody. Hence, the principles of clonal selection and hypermutation drive the evolution of the population of antibody towards better and better classification rules. In [34] [35] IFRAIS was successfully employed to discover fuzzy classification rules for female breast cancer familiarity profiling. IFRAIS' results were validated using statistical driven approaches using Gene Ontology through GO Miner [40]. Competitive results obtained by IFRAIS seem to encourage new efforts in this field. A biological interpretation of the results carried out using Gene Ontology is currently under investigation.

### 11.2.4  Application of AIS based Data Mining in Bioinformatics

As we previously stated several examples of application of AIS based data mining systems in bioinformatics can be retrieved in literature. Artificial Immune Systems-derived algorithms have been employed in familiarity profiling [34], prognosis prediction [58] and estrogen receptor modelling [59] in breast cancer. For a brief comparative overview of the performances of these kinds of systems in the context of aCGH data analysis the reader is referred to [60]. Previously de Castro and colleagues focused on the use of Hierarchical Artificial Immune Network paradigm for the problem of gene expression clustering [63, 64] and for rearrangement study of gene expression [62]. AIS/K-NNK-NN hybrid data mining algorithm have been tested for cancer classification in [43]. Tsanakova and colleagues, instead, focused on the problem of gene signature finding in the context of diffuse large B-Cell lymphoma [44]. A similar perspective has been reported by Ando and colleagues in [65] for the problem of acute leukemia classification. PCA-AIRS hybrid systems have been employed in the diagnosis of lung cancer [46] and [47]. A hybrid system based on fuzzy weighting pre-processing and AIRS has been described and employed in the diagnosis of heart, hepatitis and thyroid diseases in [48, 49, 50] respectively. Research currently being carried out by Alves and colleagues is mainly focused on the application of a multi-label AIS based data mining system to the problem of protein function prediction [36].

## 11.3  Immune Algorithms in Structural Bioinformatics and Proteomics

### 11.3.1  The Multi-objective Immunological Algorithm

Central to the field of protein structural biology is a set of observations, hypothesis and so-called paradoxes. The *Thermodynamic hypothesis* postulates that the native state of a protein is the state of lowest free energy of the protein system under physiological conditions.

The free energy of a protein can be modelled as function of the different interactions within the protein. These interactions (local, non-local, hydrophobic, entropic effects, hydrogen bonding) depend on the positions of the atoms of the protein. The set of atomic coordinates providing the minimum possible value of the free energy corresponds to the native conformation of the protein. Since the interactions comprising the energy function are highly non-convex, the protein structure prediction (PSP) problem must be tackled as a global optimization problem.

For the past fifty years, the PSP problem has been defined as a *large single-objective optimization problem*, with researchers employing Molecular Dynamics, Monte Carlo methods and Evolutionary Algorithms [71, 69, 72, 73, 70]. In this section, we reason by computational experiments that it would be more suitable to model the PSP problem as a *multi-objective optimization problem*. The goal of the research is to find a set of *equivalent* three-dimensional folded conformations, relying on the observation that the folded state is one of only a small *ensemble* of all possible conformations [74]. We adopt a multi-objective approach in order to obtain "good" non-dominated compact solutions near or inside the folded state.

PAES is a multi-objective optimizer which uses a simple $(1+1)$ local search evolution strategy. Nonetheless, it is capable of finding diverse solutions in the Pareto optimal set because it maintains an archive of non-dominated solutions which it exploits to accurately estimate the quality of new candidate solutions. At each iteration $t$, a candidate solution $c_t$ and a mutated solution $m_t$ must be compared for dominance. Acceptance is simple if one solution dominates the other. If neither solution dominates the other, the new candidate solution is compared with the reference population of previously archived non-dominated solutions. If the comparison fails to favor one solution over the other, the chosen solution is the one which resides in the least crowded region of the space. A maximum size of the archive is always maintained. The crowding procedure is based on recursively dividing up the $M$-dimensional objective space in $2^d$ equal-sized hypercubes, where $d$ is a user defined depth parameter. The algorithm continues until a given, fixed number of *iterations* is reached.

PAES by itself has proved to be a very useful MOEA with successful application in many different fields. However, when applied to the PSP problem, we have observed poor performance both in terms of energy function and final structure obtained. The complexity of the funnel landscape of the PSP problem, which is characterized by a huge number of local minima, coupled with the goal of producing a "good" conformation from a structural point of view ($RMSD$ and $DME$), clearly poses many problems (e.g., premature convergence, trapping in local minima, etc).

*I-PAES* [76] is a modified version of PAES with a different solution representation (polypeptide chain) and immune inspired (*cloning* and *hypermutation*) operators. The algorithm starts by initializing a random conformation. The torsion angles $(\phi, \psi, \chi_i)$ are generated randomly from the constraint regions. Next,

```
I-PAES(depth, archive_size, objectives)
1. t := 0;
2. Initialize(c); /*Generate initial random solution*/
3. Evaluate(c); /*Evaluation of initial solution*/
4. AddToArchive(c); /*Add c to archive*/
5. while(not(Termination()))
            /*Start Immune phase*/
6.      (c₁ᶜˡᵒ, c₂ᶜˡᵒ) := Cloning(c); /*Clonal expansion phase*/
7.      (c₁ʰʸᵖ, c₂ʰʸᵖ) := Hypermutation(c₁ᶜˡᵒ, c₂ᶜˡᵒ); /*Affinity maturation phase*/
8.      Evaluate(c₁ʰʸᵖ, c₂ʰʸᵖ); /*Evaluation phase*/
10.     if(c₁ʰʸᵖ dominates c₂ʰʸᵖ) m := c₁ʰʸᵖ;
10.     else if(c₂ʰʸᵖ dominates c₁ʰʸᵖ) m := c₂ʰʸᵖ;
10.     else m := Best(c₁ʰʸᵖ, c₂ʰʸᵖ); /*min E_charmm selection*/
12.         AddToArchive(Worst(c₁ʰʸᵖ, c₂ʰʸᵖ)); /*max E_charmm selection*/
            /*End Immune phase*/
            /*Start (1+1)-PAES*/
10.     if(c dominates m) discard m;
11.     else if(m dominates c)
12.         AddToArchive(m);
13.         c := m;
14.     else if(m is dominated by any member of the archive) discard m;
15.     else test(c, m, archive_size, depth);
16.     t := t + 1;
17. endwhile
```

The pseudo-code uses the following equations in the listing:

$$(c_1^{clo}, c_2^{clo}) := \text{Cloning}(c);$$
$$(c_1^{hyp}, c_2^{hyp}) := \text{Hypermutation}(c_1^{clo}, c_2^{clo});$$
$$\text{Evaluate}(c_1^{hyp}, c_2^{hyp});$$

**Fig. 11.4.** Pseudo-code of I-PAES

the energy of the conformation (a point in the landscape) is evaluated. The protein structure in internal coordinates (torsion angles) is transformed in Cartesian coordinates. The CHARMM energy potential of the structure is then computed using routines from TINKER Molecular Modeling Package[1].

Figure 11.4 shows the pseudo-code of the algorithm.

## 11.3.2 Open Questions in Proteomics

Given a protein with unknown biological function, its function(s) can be determined in a biological laboratory or via theoretical/computational methods. In a biological laboratory, the determination of protein functions is usually performed by experimental methods such as X-ray, crystallography or nuclear magnetic resonance. Theoretical/computational methods include homology modelling (based on previous knowledge) or ab-initio methods [29]. The problem of protein function prediction can be naturally cast as a classification problem. In this context, a protein is considered as an example (record) to be classified, and a list of predefined protein functions that can be assigned to each protein are the classes

---

[1] http://dasher.wustl.edu/tinker/

to be predicted by the classification algorithm. The ultimate goal is to predict the functions of proteins whose function is not yet known, based on attributes describing characteristics of the proteins. Protein function prediction is a very active research area for several reasons, such as the urgent and crucial need for a better understanding of proteins related to diseases, developing of more effective medical drugs, preventive medicine, etc. In any case, the very large volume of data stored in biological databases makes it infeasible to manually determine the function of each protein in those databases. Hence, several bioinformatics studies have been performed with the aim of developing computational methods for predicting protein function [26]. At present, the biological functions that can be performed by proteins are defined in a structured, standardized dictionary of terms called the Gene Ontology [30]. The GO consists of a dictionary that defines gene products independent from species. GO actually consists of 3 separate "domains" (very different types of GO terms): molecular function, biological process and cellular component. The GO is structurally organized in the form of a direct acyclic graph (DAG), where each GO term represents a node of the hierarchical structure. The inter-node relationships are of the type "is a" or "part of". A "child" node can have one or more parent nodes in the DAG. Several other works have been proposed for predicting the biological functions of proteins according to the GO [31, 32, 33].

*Current Models*

*Towards Protein Function Prediction with AIS for Hierarchical Classification*
The vast majority of classification algorithms assign just one class to an example (a protein, in the case of protein function prediction). Such classification algorithms solve a so-called single-label classification problem. However, in the context of protein function prediction, it is often necessary that the algorithm be flexible enough to be able to assign multiple classes (functions) to a protein, characterizing a multi-label classification problem [51]. In addition, protein functions are often defined in a hierarchical fashion, such as the functions included in the Gene Ontology (GO) - briefly discussed earlier. IFRAIS is a single-label, "flat" (non-hierarchical) classification algorithm. Work is ongoing in modifying IFRAIS to be a multi-label hierarchical classification algorithm [36]. One of the extensions being incorporated in the algorithm is to make it consistent with the semantics of the protein function hierarchy in GO. More precisely, when a protein is annotated with a GO term, this means that it contains not only the function specified by that term, but also the functions specified by all other terms which are ancestors of the former term in the GO's function hierarchy. IFRAIS [27] is being modified to guarantee that such hierarchical semantics is preserved in the candidate classification rules throughout the training of the algorithm. Another modification being implemented is to allow the algorithm to solve a multi-label classification problem, so that a single classification rule can predict one or more classes at once. Another research direction being pursued is the development of an AIS for the hierarchical prediction of GPCR (G protein coupled receptors) functions [52]. The AIS being developed in this project is a hierarchical

classification system, but not a multi-label one, since the GPCR classes being predicted are mutually exclusive at each level of the class hierarchy. A distinctive characteristic of this project is that it uses a novel methodology for designing an AIS where, instead of just using the natural immune system as a source of inspiration at a high level of abstraction (as usual in the field of AIS), the design of the AIS is influenced by the computational modelling of some aspect of the natural immune system. Hence, this project tries to achieve a much closer integration between computer science and biology than in previous AIS projects. More precisely, the key aspect of the natural immune system being modelled in the above project is the concept of antigen receptor degeneracy, which, according to [53], is essentially the capacity of a single antigen receptor to bind and recognize many different ligands. Cohen's theory is based on the idea that the degeneracy of different receptors is combined in order to achieve immune specificity. Mendao et al in [52] developed an agent-based computational model of immune degeneracy, and derived from it a high-level degeneracy-based clonal selection algorithm. This algorithm is currently being refined and extended in order to produce a degeneracy-based AIS for hierarchical classification [78].

### 11.3.3   Results

In the first set of experiments, we apply the approach to six proteins sequences, five extracted from reference [73] and one from [77]: 1ZDD, 1ROP, 1CRN, 1UTG, 1R69 and 1CTF.

Discussion is as follows. First we compare the performance of different versions of the PAES and I-PAES algorithms on the first protein set. Then we study the stability of the approach with respect to the native and predicted secondary structure constraints. Finally, we show specific results for each protein in terms of the obtained observed Pareto optimal sets at different time steps, $\mathcal{P}_{obs}^{*,t}$, and various dynamics of the algorithm during the evolution.

Four different versions of the PAES algorithm have been used [76] featuring dynamic (exponential decay)

The best conformation obtained with I-PAES has $DME = 0.77\text{Å}$ and $RMSD = 1.92\text{Å}$ (see figure 11.5).

**Fig. 11.5.** Native (left plot) and predicted (right plot) for 2MLT protein ($DME = 0.77\text{Å}$, $RMSD = 1.92\text{Å}$)

**Table 11.1.** Comparative results between I-PAES$_s$, I-PAES$_m$, (1+1)-PAES$_1$ and (1+1)-PAES$_2$. For each protein we report the Protein Data Bank (PDB) identifier, the length (number of residues), the approximate class ($\alpha$-helix, $\beta$-sheet), and the energy values of the native structures. The last three columns show the best results obtained for each protein on 10 independent runs. The $DME$ and $RMSD$ values are measured on $C_\alpha$ atoms from the native structure. Energy values are calculated using the ANALYZE routine from TINKER.

| Protein | Algorithm | $DME_{min}$ (Å) | $RMSD_{min}$ (Å) | Min energy ($kcal/mol$) |
|---|---|---|---|---|
| **1ROP**(56 aa) | I-PAES$_s$ | 2.01 | 4.11 | −661.48 |
| class: $\alpha$ | I-PAES$_m$ | **1.684** | **3.70** | **−902.36** |
| energy: *-667.05 kcal/mol* | (1+1)-PAES$_1$ | 4.91 | 6.31 | 2640.77 |
| | (1+1)-PAES$_2$ | 5.99 | 8.665 | −409.95 |
| **1UTG**(70 aa) | I-PAES$_s$ | 4.49 | 5.11 | **282.24** |
| class: $\alpha$ | I-PAES$_m$ | **3.79** | **4.60** | 573.89 |
| energy: *-142.46 kcal/mol* | (1+1)-PAES$_1$ | 4.71 | 6.04 | 7563.07 |
| | (1+1)-PAES$_2$ | 4.82 | 5.56 | 397.12 |
| **1CRN**(46 aa) | I-PAES$_s$ | 4.13 | 4.73 | **232.29** |
| class: $\alpha + \beta$ | I-PAES$_m$ | **3.72** | **4.31** | 509.09 |
| energy: *202.73 kcal/mol* | (1+1)-PAES$_1$ | 4.67 | 6.18 | 1653.93 |
| | (1+1)-PAES$_2$ | 6.05 | 7.89 | 509.52 |
| **1R69**(63 aa) | I-PAES$_s$ | 5.93 | 8.42 | **211.26** |
| class: $\alpha$ | I-PAES$_m$ | **4.91** | **5.05** | 264.56 |
| energy: *-676.53 kcal/mol* | (1+1)-PAES$_1$ | 5.16 | 7.59 | 9037.89 |
| | (1+1)-PAES$_2$ | 6.88 | 8.52 | 659.49 |
| **1CTF**(68 aa) | I-PAES$_s$ | 8.08 | 10.69 | **71.55** |
| class: $\alpha + \beta$ | I-PAES$_m$ | **6.82** | **10.12** | 218.99 |
| energy: *230.08 kcal/mol* | (1+1)-PAES$_1$ | 9.61 | 12.09 | 1424.33 |
| | (1+1)-PAES$_2$ | 8.84 | 10.21 | 617.69 |

## 11.4  Proteomic Multiple Sequence Alignments: Refinement Using an Immunological Local Search

### 11.4.1  Proteomics Multiple Sequence Alignments

The Multiple Sequence Alignment (MSA) of proteins plays a central role in molecular biology, as it can reveal the constraints imposed by structure and function on the evolution of whole protein families [78]. MSA has been used for building phylogenetic trees, for the identification of conserved motifs, to find diagnostic patterns families, and for predicting secondary and tertiary structures of RNA and protein sequences. In order to be able to align a set of biosequences, a reliable objective function for the measurement of an alignment in terms of its biological plausibility through an analytical or computational function is needed.

One of the most important and popular computational sequence analysis problem is to determine if two, or more, biological sequences have common subsequences. However, to check the similarities between two or more sequences, there are two primary issues that need to be faced: the choice of an objective function that assesses the biological alignment quality and the design of an effective algorithm to optimize the given objective function. The alignment quality is often the limiting factor in biological analyses of amino-acid sequences; defining a proper objective function is a crucial task.

The classical objective function used to measure the biological alignment quality is the *weighted sums-of-pairs* with affine gap penalties [79]: each sequence receives a weight proportional to the amount of independent information that it contains [80] and the cost of the multiple alignment is equal to the sum of the costs of all the weighted pairwise substitutions:

$$\max_{\hat{S}} \left( \sum_{i=1}^{n-1} \sum_{j=i+1}^{n} WSS(\hat{S}_i, \hat{S}_j) + \sum_{i=1}^{n} AGPS(\hat{S}_i) \right). \qquad (11.1)$$

Sequence weights are determined by constructing a guide tree from known sequences.

### 11.4.2  *IMSA*, an Immunological Algorithm

In this chapter we present an immunological algorithm, IMSA, to tackle the multiple sequence alignment problem. It incorporates two different strategies to create the initial population, as well as new hypermutation operators, specific operators for solving MSA, which insert or remove gaps in the sequences. Gap columns which have been matched are moved to the end of the sequence. The remaining elements (amino acids in this work) and existing gaps are shifted into the freed space.

*IMSA* considers antigens (Ags) and B cells. The Ag is a given MSA instance, and B cells a set of alignments, that have solved (or approximated) the initial problem. In tackling the MSA Ags and B cells are represented by a sequence matrix. In particular, let

$$\Sigma = \{A, R, N, D, C, E, Q, G, H, I, L, K, M, F, P, S, T, W, Y, V\} \qquad (11.2)$$

be the twenty amino acid alphabet, and let $S = \{S_1, S_2, \ldots, S_n\}$ be the set of $n \geq 2$ sequences with length $\{\ell_1, \ell_2, \ldots, \ell_n\}$, such that $S_i \in \Sigma^*,$. Then an Ag is represented by a matrix of $n$ rows and $max\{\ell_1, \ldots, \ell_n\}$ columns, whereas each B cell is represented by an $(n \times \ell)$ matrix, with $\ell = (\frac{3}{2} \cdot max\{\ell_1, \ldots, \ell_n\})$. By using such a representation *IMSA* was able to develop more *compact alignments*.

### 11.4.3  Results and Conclusions

To evaluate the biological alignment quality produced by *IMSA*, we tested it using the classical benchmark BALiBASE.

The obtained results showed in the next tables were obtained using a robust experimental protocol : $d = 10, dup = 1, \tau_B = 33, T_{max} = 2 \times 10^5$ and 50 independent runs. Moreover, we used the following substitution matrices:

- BLOSUM45 for Ref1v1 and Ref 3, with $GOP = 14, GEP = 2$;

**Table 11.2.** Pseudo-code of the proposed hybrid immune algorithm for the MSA

```
IMSA (d, dup, τ_B, T_max)
t ← 0;
FFE ← 0;
N_c ← d × dup;
P^(t) ← Initialize_Population(d);
Strip_Gaps(P^(t));
Evaluate(P^(t));
FFE ← FFE + d;
while (FFE < T_max)do
    P^(clo) ← Cloning (P^(t), dup);
    P^(gap) ← Gap_operators (P^(clo));
    Strip_Gaps(P^(gap));
    Evaluate(P^(gap));
    FFE ← FFE + N_c;
    P^(block) ← BlockShuffling_operators (P^(clo));
    Compute_Weights();
    Normalize_Weights();
    Strip_Gaps(P^(block));
    Evaluate(P^(block));
    FFE ← FFE + N_c;
    (P_a^(t), P_a^(gap), P_a^(block)) = Elitist-Aging(P^(t), P^(gap), P^(block), τ_B);
    P^(t+1) ← (μ + λ)-Selection(P_a^(t), P_a^(gap), P_a^(block));
    t ← t + 1;
end_while
```

**Table 11.3.** SP values given by several methods on the BAliBase v.1.0 benchmark

| Aligner | Ref 1 (82) | Ref 2 (23) | Ref 3 (12) | Ref 4 (12) | Ref 5 (12) | Overall (141) |
|---|---|---|---|---|---|---|
| *IMSA* | 80.7 | **88.6** | **77.4** | 70.2 | 82.0 | **79.7** |
| DIALIGN [89] | 77.7 | 38.4 | 28.8 | **85.2** | **83.6** | 62.7 |
| CLUSTALX [83] | 85.3 | 58.3 | 40.8 | 36.0 | 70.6 | 58.2 |
| PILEUP8 [82] | 82.2 | 42.8 | 33.3 | 59.1 | 63.8 | 56.2 |
| ML_PIMA [86] | 80.1 | 37.1 | 34.0 | 70.4 | 57.2 | 55.7 |
| PRRP [91] | **86.6** | 54.0 | 48.7 | 13.4 | 70.0 | 54.5 |
| SAGA [94] | 70.3 | 58.6 | 46.2 | 28.8 | 64.1 | 53.6 |
| SB_PIMA [86] | 81.1 | 37.9 | 24.4 | 72.6 | 50.7 | 53.3 |
| MULTALIGN [81] | 82.3 | 51.6 | 27.6 | 29.2 | 62.7 | 50.6 |

- BLOSUM62 for Ref1v2, Ref 2, Ref 4 and Ref 5, with $GOP = 11, GEP = 1$;
- BLOSUM80 for Ref1v3, with $GOP = 10, GEP = 1$.

Table 11.3 shows the average SP score obtained by the described alignment tools on every instance set of BAliBASE v.1.0. As it can be seen in the table, *IMSA* performs well on the Reference 2 and Reference 3 sets. The values obtained aid to raise the overall score, which is higher compared to the results published by the Bioinformatic platform of Strasbourg.

## 11.5   Conclusions and Open Questions

In this chapter we have analysed some applications of Artificial Immune System based algorithms in bioinformatics. Of course this is only a partial outlook on the world of AIS based approaches: interested readers can check references in order to obtain more detailed information about specific aspects of the proposed topics. Furthermore, given their infancy, AIS are currently undergoing very fast changes resulting in a very dynamical field of reasearch where tens of novel and promising projects are proposed in the time of some months. These aspects forced the authors to select a set of significant experiences to be used as examples of how the algorithms described herein can be successfully used in the field of bioinfomatics. This led to exclude interesting projects like BIAS-PROFS coordinated by Freitas and colleagues; even in this case interested readers can find useful information in the references. After these necessary statements some conclusions. In this chapter we have learned how novel bio-inspired computational intelligence paradigms can be used in very diverse field of research in bioinformatics. As previously stated AIS are considered a novel paradigm but they have been already able to reach significant results in highly complex contexst like Knowledge Discovery in Data bases (section 11.2) and Protein Structure Prediction (section 11.1). Even if immune-inspired algorithms have been successfully employed in several diverse problems, there are still some strategic fields of research in which solutions seem to be far from being reached, just to name few:

* Large molecules folding prediction;
* Gene networks inference;
* Disease profiling and evolution modelling.

These are only some of the most active areas of AIS based research in bioinformatics. From a theoretical point of view it should be noted that some areas like *danger theory* and *hybrid systems* have been exploited with a limited systematic approach in bioinformatics: these areas deserve a comprehensive analytic approach. Readers interested in these promising aspects of the AIS research in bioinformatics can find useful information in [43, 49].

## References

1. de Castro, L.N., Timmis, J.: Artificial Immune Systems: A New Computational Approach. Springer, Heidelberg (2002)
2. Scott, R.: Keynote Speach. TNTYN, San Francisco (2000)
3. Economist, Life 2.0. The new science of synthetic biology is poised between hype and hope. But its time will soon come. August 31, 2006 (2006)
4. Grossman, R., Kamath, C., Kumar, V.: Data Mining for Scientific and Engineering Applications. Springer, Heidelberg (2001)
5. Immon, W.H.: Building the Data Warehouse. John Wiley and Sons, New York (1996)
6. Frawley, W., Piatetsky-Shapiro, G., Matheus, C.: Knowledge Discovery in Databases: An Overview. AI Magazine, 213–228 (Fall 1992)

7. Barrett, T., Troup, D.B., Wilhite, S.E., Ledoux, P., Rudnev, D., Evangelista, C., Kim, I.F., Soboleva, A., Tomashevsky, M., Edgar, R.: NCBI GEO: Mining tens of millions of expression profiles–database and tools update. Nucleic Acids Res. (November 11, 2006)
8. Demeter, J., Beauheim, C., Gollub, J., Hernandez-Boussard, T., Jin, H., Maier, D., Matese, J.C., Nitzberg, M., Wymore, F., Zachariah, Z.K., Brown, P.O., Sherlock, G., Ball, C.A.: The Stanford Microarray Database: Implementation of new analysis tools and open source release of software. Nucleic Acids Res. 35(Database Issue), D766–D770 (2007)
9. Brazma, A., Parkinson, H., Sarkans, U., Shojatalab, M., Vilo, J., Abeygunawardena, N., Holloway, E., Kapushesky, M., Kemmeren, P., Lara, G.G., Oezcimen, A., Rocca-Serra, P., Sansone, S.A.: ArrayExpress–a public repository for microarray gene expression data at the EBI. Nucleic Acids Res. 31(1), 68–71 (2003)
10. Rigby, D.K., Ledingham, D.: CRM Done Right. Harvard Business Review (November 1, 2004)
11. Brownlee, J.: Artificial Immune Recognition System (AIRS) - A Review and Analysis [Technical Report]. Centre for Intelligent Systems and Complex Processes (CISCP), Faculty of Information and Communication Technologies (ICT), Swinburne University of Technology, Victoria, Australia, Technical Report ID: 1-01 (2005)
12. Brownlee, J.: Clonal Selection Theory and CLONALG - The Clonal Selection Classification Algorithm (CSCA) [Technical Report]. Centre for Intelligent Systems and Complex Processes (CISCP), Faculty of Information and Communication Technologies (ICT), Swinburne University of Technology, Victoria, Australia, Technical Report ID: 2-01 (2005)
13. Brownlee, J.: Immunos-81 – The Misunderstood Artificial Immune System [Technical Report]. Centre for Intelligent Systems and Complex Processes (CISCP), Faculty of Information and Communication Technologies (ICT), Swinburne University of Technology, Victoria, Australia, Technical Report ID: 3-01 (2005)
14. Brownlee, J.: Weka Classification Algorithms, http://sourceforge.net/projects/wekaclassalgos
15. Siadaty, M.S., Knaus, W.A.: Locating previously unknown patterns in data-mining results: a dual data- and knowledge-mining method. BMC Medical Informatics and Decision Making 6(13) (2006) doi:10.1186/1472-6947-6-13
16. Pool, R., Esnayra, J.: Bioinformatics: Converting Data to Knowledge. Natl. Acad. Press, Washington (2003)
17. Fayyad, U.M., Piatetsky-Shapiro, G., Smyth, P., Uthurusamy, R.: Advances in Knowledge Discovery and Data Mining. AAAI/MIT, Cambridge (1996)
18. Witten, I.H., Frank, E.: Data Mining: Practical Machine Learning Tools and Techniques, 2nd edn. Morgan Kaufmann, San Mateo (2005)
19. Mierswa, I., Wurst, M., Klinkenberg, R., Scholz, M., Euler, T.: YALE: Rapid Prototyping for Complex Data Mining Tasks. In: Proc. 12th ACM SIGKDD International Conference on Knowledge Discovery and Data Mining (KDD 2006) (2006)
20. Michalski, R.S., Bratko, I., Kubat, M.: Machine Learning and Data Mining: Methods and Applications. Wiley, Chichester (1998)
21. Gewehr, J.E., Szugat, M., Zimmer, R.: BioWeka-extending the Weka framework for bioinformatics. Bioinformatics 23(5) (March 2007) ISSN:1367-4803

22. Talia, D., Trunfio, P., Verta, O.: Weka4WS: a WSRF-enabled Weka Toolkit for Distributed Data Mining on Grids. In: Jorge, A.M., Torgo, L., Brazdil, P.B., Camacho, R., Gama, J. (eds.) PKDD 2005. LNCS (LNAI), vol. 3721, pp. 309–320. Springer, Heidelberg (2005)
23. Freitas, A.A.: Data Mining and Knowledge Discovery with Evolutionary Algorithms. Springer, Berlin (2002)
24. Vapnik, V.N.: The Nature of Statistical Learning Theory. Springer, Berlin (1995)
25. Haykin, S.: Neural Networks – A Comprehensive Foundation, 2nd edn. Prentice Hall, Upper Saddle River (1999)
26. Fogel, G.B., Corne, D.W.: Evolutionary Computation in Bioinformatics. Morgan Kaufmann Publishers, San Franciso (2003)
27. Alves, R.T., Delgado, M.R., Lopes, H.S., Freitas, A.A.: An artificial immune system for fuzzy-rule induction in data mining. In: Yao, X., Burke, E.K., Lozano, J.A., Smith, J., Merelo-Guervós, J.J., Bullinaria, J.A., Rowe, J.E., Tiňo, P., Kabán, A., Schwefel, H.-P. (eds.) PPSN 2004. LNCS, vol. 3242, pp. 1011–1020. Springer, Heidelberg (2004)
28. Pedrycz, W., Gomide, F.: An Introduction to Fuzzy Sets: Analysis and Design. MIT Press, Cambridge (1998)
29. Alberts, B., Johnson, A., Lewis, J., Raff, M., Roberts, K., Water, P.: Molecular Biology of the Cell, 4th edn. Garland Science, New York (2002)
30. The Gene Ontology Consortium, The Gene Ontology (GO) Database and Informatics Resource. Nucleic Acids Research 32(1), 258–261 (2004)
31. Vinayagam, A., Konig, R., Moormann, J., Schubert, F., Eils, R., Suhai, S.: Applying Support Vector Machines for Gene Ontology based gene function prediction. BMC Bioinformatics 5, 116–129 (2004)
32. Eisner, R., Poulin, B., Szafron, D., Lu, P., Greiner, R.: Improving Protein Function Prediction using the Hierarchical Structure of the Gene Ontology. In: Proc. IEEE Symposium on Computational Intelligence in Bioinformatics and Computational Biology (2005)
33. Tu, K., Yu, H., Guo, Z., Li, X.: Learnability-Based Further Prediction of Gene Func-tions in Gene Ontology. Genomics 86, 922–928 (2004)
34. Menolascina, F., Alves, R.T., Tommasi, S., Chiarappa, P., Delgado, M., Bevilacqua, V., Mastronardi, G., Freitas, A.A., Paradiso, A.: Fuzzy Rule Induction and Artificial Immune Systems in Female Breast Cancer Familiarity Profiling. In: Apolloni, B., Howlett, R.J., Jain, L. (eds.) KES 2007, Part III. LNCS (LNAI), vol. 4694, pp. 830–837. Springer, Heidelberg (2007)
35. Menolascina, F., Tommasi, S., Paradiso, A., Cortellino, M., Bevilacqua, V., Mastronardi, G.: Novel Data Mining Techniques in aCGH based Breast Cancer Subtypes Proling: the biological perspective. In: Proc. 2007 IEEE Symposium on Computational Intelligence in Bioinformatics and Computational Biology, Honolulu, HI, USA, April 1-5, pp. 9–16 (2007)
36. Alves, R.T.: An Artificial Immune System to Hierarchical Multi-label Classification for Predicting Protein Function. Ph.D. Qualifying Exam 42, Federal University of Technology of Paraná -UTFPR, Curitiba, Brazil (2007)
37. Timmis, J., Knight, T., de Castro, L.N., Hart, E.: An Overview of Artificial Computation in Cells and Tissues: Perspectives and Immune Systems. In: Tools for Thought, Anonymous, pp. 51–86. Springer, Heidelberg (2004)
38. Hart, E.: Immunology as a Metaphor for Computational Information Processing: Fact of Fiction. University of Edinburgh (2002)
39. Twycross, J.: An Immune System Approach to Document Classification. University of Sussex (2002)

40. Zeeberg, B.R., et al.: GoMiner: A Resource for Biological Interpretation of Genomic and Proteomic Data. Genome Biology 4(4), R28 (2003)
41. Reich, M., Liefeld, T., Gould, J., Lerner, J., Tamayo, P., Mesirov, J.P.: GenePattern 2.0. Nature Genetics 38(5), 500–501 (2006)
42. Watkins, A.B.: A resource limited artificial immune classifier. Mississippi State University (2001)
43. Sahan, S., Polat, K., Kodaz, H., Gunes, S.: A new hybrid method based on fuzzy-artificial immune system and k-nn algorithm for breast cancer diagnosis. Computers in Biology and Medicine 37(3), 415–423 (2007)
44. Tsankova, D., Rangelova, V.: Cancer Outcome Prediction by Cluster-based Artificial Immune Networks. In: Proc. Biomedical Engineering (2007)
45. de la Nava, J.G., Santaella, D.F., Alba, J.C., Carazo, J.M., Trelles, O., Pascual-Montano, A.: Engene: The processing and exploratory analysis of gene expression data. Bioinformatics 19(5), 657–658 (2002)
46. Polat, K., Gunes, S.: Principles component analysis, fuzzy weighting preprocessing and artificial immune recognition system based diagnostic system for diagnosis of lung cancer. Expert Systems with Applications: An International Journal 34(1) (2008)
47. Polat, K., Gunes, S.: Computer aided medical diagnosis system based on principal component analysis and artificial immune recognition system classifier algorithm. Expert Systems with Applications: An International Journal 34(1) (2008)
48. Polat, K., Gunes, S., Tosun, S.: Diagnosis of heart disease using artificial immune recognition system and fuzzy weighted pre-processing. Pattern Recognition 39(11) (2006)
49. Polat, K., Sahan, S., Gunes, S.: A novel hybrid method based on artificial immune recognition system (AIRS) with fuzzy weighted pre-processing for thyroid disease diagnosis. Expert Systems with Applications: An International Journal 32(4) (2007)
50. Polat, K., Gunes, S.: Medical decision support system based on artificial immune recognition immune system (AIRS), fuzzy weighted pre-processing and feature selection. Expert Systems with Applications: An International Journal 33(2) (2007)
51. Tsoumakas, G., Katakis, I.: Multi-label classification: An overview. International Journal of Data Warehousing and Mining 3(3), 1–13 (2007)
52. Mendao, M., Timmis, J., Andrews, P.S., Davies, M.: The Immune System in Pieces: Computational Lessons from Degeneracy in the Immune System. In: Proc. 2007 IEEE Symposium on Foundations of Computational Intelligence (FOCI 2007), Honolulu, HI, USA (2007)
53. Cohen, I.R., Hershberg, U., Solomon, S.: Antigen-receptor degeneracy and immunological paradigms. Molecular Immunology 40, 993–996 (2004)
54. Carter, J.H.: The immune system as a model for classification and pattern recognition. Journal of the American Informatics Association 7 (2000)
55. de Castro, L.N., von Zuben, F.J.: The Clonal Selection Algorithm with Engineering Applications. In: GECCO 2000, Workshop on Artificial Immune Systems and Their Applications, Las Vegas, USA, pp. 36–37 (2000)
56. Larranaga, P., Gallego, M.J., Sierra, B., Urkola, L., Michelena, M.J.: Bayesian Networks, Rule Induction and Logistic Regression in the prediction of the survival of women suffering from breast cancer. In: Costa, E. (ed.) EPIA 1997. LNCS, vol. 1323, pp. 303–308. Springer, Heidelberg (1997)

57. Bevilacqua, V., Chiarappa, P., Mastronardi, G., Menolascina, F., Paradiso, A., Tommasi, S.: Identification of Tumour Evolution Patterns by Means of Inductive Logic Programming. Journal - Genomics Proteomics and Bioinformatics (in press, 2007)

58. Menolascina, F., Alves, R.T., Tommasi, S., Chiarappa, P., Delgado, M., Bevilacqua, V., Mastronardi, G., Freitas, A.A., Paradiso, A.: Improving Female Breast Cancer Prognosis by means of Fuzzy Rule Induction with Artificial Immune Systems. Journal of Dynamics of Discrete Continuous and Impulsive Systems (to appear, 2007) ISSN:1492-8760

59. Menolascina, F., Alves, R.T., et al.: Induction of Fuzzy Rules with Artificial Immune Systems in aCGH based ER Status Breast Cancer Characterization. In: Proc. GECCO 2007, ACM 978-1-59593-697-4/07/0007 (2007)

60. Menolascina, F., Tommasi, S., Chiarappa, P., Bevilacqua, V., Mastronardi, G., Paradiso, A.: Data mining techniques in aCGH-based breast cancer subtype profiling: an immune perspective with comparative study. BMC Systems Biology 1(suppl. 1), P70 (2007)

61. de Castro, L.N., von Zuben, F.J.: Learning and Optimization IEEE Transactions on Evolutionary Using the Clonal Selection Principle Computation. Special Issue on Artificial Immune Systems 6, 239–251 (2002)

62. de Sousa, J.S., de Gomes, C.T., Bezerra, G.B., de Castro, L.N., von Zuben, F.J.: An Immune-Evolutionary Algorithm for Multiple Rearrangements of Gene Expression Data. Genetic Programming and Evolvable Machines 5(2), 157–179 (2004)

63. Bezerra, G.B., Cançado, G.M.A., Menossi, M., de Castro, L.N., von Zuben, F.J.: Recent advances in gene expression data clustering: a case study with comparative results. Genet. Mol. Res. 4(3), 514–524 (2005)

64. Hruschka, E.R., Campello, R.J.G.B., de Castro, L.N.: Evolving clusters in gene-expression data. Inf. Sci. 176(13), 1898–1927 (2006)

65. Ando, S., Iba, H.: Artificial Immune System for Classification of Cancer. In: Raidl, G.R., Cagnoni, S., Cardalda, J.J.R., Corne, D.W., Gottlieb, J., Guillot, A., Hart, E., Johnson, C.G., Marchiori, E., Meyer, J.-A., Middendorf, M. (eds.) EvoWorkshops 2003. LNCS, vol. 2611, p. 219. Springer, Heidelberg (2003)

66. Castro, P.A.D., Coelho, G.P., Caetano, M.F., von Zuben, F.J.: Designing ensembles of fuzzy classification systems: An immune-inspired approach. In: Jacob, C., Pilat, M.L., Bentley, P.J., Timmis, J.I. (eds.) ICARIS 2005. LNCS, vol. 3627, pp. 469–482. Springer, Heidelberg (2005)

67. Alatas, B., Akin, E.: Mining fuzzy classification rules using an artificial immune system with boosting. In: Eder, J., Haav, H.-M., Kalja, A., Penjam, J. (eds.) ADBIS 2005. LNCS, vol. 3631, pp. 283–293. Springer, Heidelberg (2005)

68. Anfinsen, C.: Principles that govern the folding of protein chains. Science 181, 223–230 (1973)

69. Simons, K.T., Kooperberg, C., Huang, E., Baker, D.: Assembly of of protein tertiary structures from fragments with similar local sequences using simulated annealing and Bayesian scoring function. J. Mol. Biol. 306, 1191–1199 (1997)

70. Hansmann, U.H., Okamoto, Y.: Numerical comparisons of three recently proposed algorithms in the protein folding problem. J. Comput. Chem. 18, 920–933 (1998)

71. Bowie, J.U., Eisemberg, D.: An evolutionary approach to folding small alpha-helical proteins that uses sequence information and an empirical guiding fitness function. Proc. Natl. Acad. Sci. USA 91, 4436–4440 (1994)

72. Pendersen, J.T., Moult, J.: Protein folding simulations with genetic algorithms and a detailed molecular description. J. Mol. Biol. 169, 240–259 (1997)

73. Cui, Y., Chen, R.S., Wong, W.H.: Protein Folding Simulation using Genetic Algorithm and Supersecondary Structure Constraints. Proteins: Structure, Function and Genetics 31(3), 247–257 (1998)
74. Plotkin, S.S., Onuchic, J.N.: Understanding protein folding with energy landscape theory. Quarterly Reviews of Biophysics 35(2), 111–167 (2002)
75. Foloppe, N., MacKerell Jr., A.D.: All-Atom Empirical Force Field for Nucleic Acids: I. Parameter Optimization Based on Small Molecule and Condensed Phase Macromolecular Target Data. J. Comput. Chem. 21, 86–104 (2000)
76. Cutello, V., Narzisi, G., Nicosia, G.: A Class of Pareto Archived Evolution Strategy Algorithms Using Immune Inspired Operators for Ab-Initio Protein Structure Prediction. In: Rothlauf, F., Branke, J., Cagnoni, S., Corne, D.W., Drechsler, R., Jin, Y., Machado, P., Marchiori, E., Romero, J., Smith, G.D., Squillero, G. (eds.) EvoWorkshops 2005. LNCS, vol. 3449, pp. 54–63. Springer, Heidelberg (2005)
77. Dal Palu, A., Dovier, A., Fogolari, F.: Constraint Logic Programming approach to protein structure prediction. BMC Bioinformatics 5(11), 186 (2004)
78. Eidhammer, I., Jonassen, I., Taylor, W.R.: Protein Bioinformatics. Wiley, Chichester (2004)
79. Altschul, S.F., Lipman, D.J.: Trees stars and multiple biological sequence alignment. SIAM J. on App. Maths. 49, 197–209 (1989)
80. Altschul, S.F., Carroll, R.J., Lipman, D.J.: Weights for data related by a tree. J. on Mol. Biol. 207, 647–653 (1989)
81. Corpet, F.: Multiple sequence alignment with hierarchical clustering. Nuc. Acids Research 16, 10881–10890 (1998)
82. Genetics Computer Group, Wisconsin Package v.8 (1993), http://www.gcg.com
83. Thompson, J.D., Gibson, T.J., Plewniak, F., Jeanmougin, F., Higgins, D.G.: The ClustalX windows interface: flexible strategies for multiple sequence alignment aided by quality analysis tools. Nuc. Acids Research 24, 4876–4882 (1997)
84. Zhou, H., Zhou, Y.: SPEM: Improving multiple sequence alignment with sequence profiles and predicted secondary structures. Bioinformatics 21, 3615–3621 (2005)
85. Do, C.B., Mahabhashyam, M.S.P., Brudno, M., Batzoglou, S.: ProbCons: Probabilistic consistency-based multiple sequence alignment. Genome Research 15, 330–340 (2005)
86. Smith, R.F., Smith, T.F.: Pattern-induced multi-sequence alignment (PIMA) algorithm employing secondary structure-dependent gap penalties for use in comparative protein modelling. Prot. Engineering 5, 35–41 (1992)
87. Carrillo, H., Lipman, D.J.: The Multiple Sequence Alignment Problem in Biology. SIAM J. on App. Maths. 48, 1073–1082 (1988)
88. Stoye, J., Moulton, V., Dress, A.W.: DCA: An efficient implementation of the divide-and conquer approach to simultaneous multiple sequence alignment. Bioinformatics 13(6), 625–626 (1997)
89. Morgenstern, B., Frech, K., Dress, A., Werner, T.: DIALIGN: Finding local similarities by multiple sequence alignment. Bioinformatics 14, 290–294 (1998)
90. Morgenstern, B., Frech, K., Dress, A., Werner, T.: DIALIGN 2: improvement of the segment-to-segment approach to multiple sequence alignment. Bioinformatics 15, 211–218 (1999)
91. Gotoh, O.: Further improvement in methods of group-to-group sequence alignment with generalized profile operations. Bioinformatics 10(4), 379–387 (1994)
92. Eddy, S.R.: Multiple alignment using hidden Markov models. In: Proc. 3rd Int. Conference on Intelligent Systems for Molecular Biology (ISMB 1995), Cambridge, UK, pp. 114–120 (1995)

93. Edgar, R.C.: MUSCLE: Multiple sequence alignment with high accuracy and high throughput. Nuc. Acids Research 32, 1792–1797 (2004)
94. Notredame, C., Higgins, D.G.: SAGA: Sequence alignment by genetic algorithm. Nuc. Acids Research 24, 1515–1539 (1996)
95. Notredame, C.: COFFEE: An objective function for multiple sequence alignments. Bioinformatics 14, 407–422 (1998)
96. Simossis, V.A., Heringa, J.: PRALINE: A multiple sequence alignment toolbox that integrates homology-extended and secondary structure information. Nuc. Acids Research 33, 289–294 (2005)
97. Shyu, C., Sheneman, L., Foster, J.A.: Multiple Sequence Alignment with Evolutionary Computation. Gen. Prog. and Evol. Machs. 5, 121–144 (2004)
98. Nguyen, H.D., Yoshihara, I., Yamamori, K., Yasunaga, M.: Aligning Multiple Protein Sequences by Parallel Hybrid Genetic Algorithm. Genome Inf. 13, 123–132 (2002)
99. Cutello, V., Narzisi, G., Nicosia, G., Pavone, M.: Clonal selection algorithms: A comparative case study using effective mutation potentials. In: Jacob, C., Pilat, M.L., Bentley, P.J., Timmis, J.I. (eds.) ICARIS 2005. LNCS, vol. 3627, pp. 13–28. Springer, Heidelberg (2005)
100. Cutello, V., Nicosia, G., Pavone, M.: Exploring the capability of immune algorithms: A characterization of hypermutation operators. In: Nicosia, G., Cutello, V., Bentley, P.J., Timmis, J. (eds.) ICARIS 2004. LNCS, vol. 3239, pp. 263–276. Springer, Heidelberg (2004)

# 12

# Evolutionary Algorithms for the Protein Folding Problem: A Review and Current Trends

Heitor Silvério Lopes

Bioinformatics Laboratory
Federal University of Technology – Paraná
Av. 7 de setembro, 3165, 80230-901 Curitiba – Brazil
hslopes@utfpr.edu.br

## 12.1 Introduction

Proteins are complex macromolecules that perform vital functions in all living beings. They are composed of a chain of amino acids. The biological function of a protein is determined by the way it is folded into a specific tri-dimensional structure, known as native conformation. Understanding how proteins fold is of great importance to Biology, Biochemistry and Medicine. Considering the full analytic atomic model of a protein, it is still not possible to determine the exact tri-dimensional structure of real-world proteins, even with the most powerful computational resources. To reduce the computational complexity of the analytic model, many simplified models have been proposed. Even the simplest one, the bi-dimensional Hydrophobic-Polar (2D-HP) model (see Sect. 12.2.2), was proved to be intractable due to its NP-completeness. The current approach for studying the structure of proteins is the use of heuristic methods that, however, do not guarantee the optimal solution. Evolutionary computation techniques have been proved to be efficient for many engineering and computer science problems. This is also the case of unveiling the structure of proteins using simple lattice models.

In this work the nature of the models used for the protein folding problem is reviewed, with special emphasis on discrete models. Also, we analyze how evolutionary computation techniques have been applied to solve it. Amongst these techniques, there are many different variants of genetic algorithms, besides ant colony optimization, differential evolution and artificial immune systems.

This chapter is structured as follows: the remaining of this section introduces some basic aspects of amino acids and proteins, and presents the protein folding problem. Sect. 12.2 presents the several models for protein folding with special emphasis on a specific discrete model: the hydrophobic-polar. Sect. 12.3 is dedicated to the several computational approaches for the protein folding problem, from molecular dynamics and approximation algorithms to several evolutionary computation algorithms. Next, Sect. 12.4 presents challenging issues that limit current research. Finally, in Sect. 12.5 current trends for future research and the conclusion are presented.

T.G. Smolinski et al. (Eds.): Comp. Intel. in Biomed. & Bioinform., SCI 151, pp. 297–315, 2008.
springerlink.com

### 12.1.1    Amino Acids and Proteins

The basic structure of an amino acid consists of a carbon atom ($C_\alpha$) connected with an amino group ($NH_2$), a carboxyl group (COOH) and a side-chain. The only difference between amino acids is due to the composition of their side-chain. There are 20 standard amino acids. According to the physical properties of the side-chain, amino acids can be classified according its polarity and acidity/basicity. Such classification leads to a hydrophilic (polar) or hydrophobic (nonpolar) character of the amino acid. The distribution of hydrophilic and hydrophobic amino acids along the protein ultimately determines structure of the protein.

The sequence of amino group, $C_\alpha$ and carboxyl group of an amino acid bounded with the following is known as backbone of a protein. There are three main levels of organization of the structure of a protein: primary, secondary and tertiary structures. The primary structure of a protein or polypeptide chain is its linear sequence of amino acids, represented by a string of letters. Some specific regions of the primary structure can fold into known tri-dimensional structures, such as $\alpha$-helices or $\beta$-sheets. These structures are known as secondary structures. The spatial representation of the protein is called tertiary structure. The shape into which a protein naturally folds is known as its native state, or native conformation. For some particular proteins, tertiary structures can be combined to form a super-structure known as quaternary structure.

The tertiary structure of a protein, or the quaternary structure of its complexes, is of particular interest, since it defines the biological function of the protein. The most effective method for unveiling the structure of real proteins is using nuclear magnetic resonance spectroscopy or X-ray crystallography. It is estimated that the human body has around 100,000 different proteins, but a only a small portion of them have its structure known. The Protein Data Bank (PDB) [7] ($http : //www.pdb.org$) is the repository for structural data of proteins. Currently, it holds structural information of almost 50,000 proteins. However, the amount of known proteins which structure is unknown is much larger, thus justifying the use of computational methods for this purpose. Therefore, this is an important research area in Bioinformatics and Computational Biology.

### 12.1.2    Protein Folding

The Protein Structure Prediction (PSP) problem can be defined as determining the final tri-dimensional structure of a protein by using only the information about its primary structure. On the other hand, the Protein Folding Problem (PFP) is understood as being the discovery of the pathways by which a protein is folded into its natural conformation, during its synthesis [34]. However, in the current literature those two terms are frequently used with no distinction, usually meaning only the first issue. A computational approach to predict the structure of a protein demands a model that represents it abstractly, in a given level of details. Basing on well-established thermodynamical laws, the prediction of the structure of a protein is modelled as the minimization of the corresponding free-energy with respect to the possible conformations that a protein is able to attain.

The minimization of this free-energy is the most important factor that drives the construction of the structure of a protein. Formally, the native conformation of a protein is defined as that in which the free-energy is minimal. According to [61], a computational model that obeys this principle must have the following features:

- A model of the protein, defined by a set of entities representing atoms and connections among them;
- A set of rules defining the possible conformations of the protein;
- A computationally feasible function for evaluating the free-energy of each possible conformation.

The amount of details of the structure modelled depends on the choices done about the model (see Sect. 12.2). For instance, a protein could a have a spatial representation of all its atoms, all all its atoms but hydrogen, only the backbone without the side-chain, or as simple hydrophobic-polar elements embedded in a lattice.

## 12.2  Free-Energy Models

### 12.2.1  Analytical Models

An analytical model has a detailed description of the tri-dimensional structure of a protein, including information about all its individual atoms [61]. A protein can be viewed as a collection of atoms connected each other. Therefore, to specify the tertiary structure of a protein, it is possible to establish values for angles, lengths and torsions of the connections among atoms in the structure. To reduce the inherent complexity of this model, some atoms could be disregarded or even grouped into larger elements, and treated as equivalent single atoms by the model. Obviously, such reduction decreases the visual equivalence between the model and a real protein, for a given conformation.

The free-energy function in an analytical model is frequently specified by parameters representing the individual contributions of atoms. For atoms connected each other, such parameters depend on the length, angles and torsions of the connection. For those atoms that are not directly connected each other, the parameters depend on physical forces (i.e., Coulomb and van der Walls forces) or statistical information inferred from known structures. However, using a detailed description of the structure of a protein and many parameters in the free-energy function, it is computationally hard to find an optimized solution for the prediction of the structure of a protein. For instance, analytical models were presented by [10, 25, 56, 57].

### 12.2.2  Discrete Models

The difficulty in using analytical models motivated researchers to develop simpler discrete models that allow a large number of computational simulations necessary to find optimal or quasi-optimal solutions for the PFP [10, 23]. The easiest way to limit the complexity of an analytical model is to limit the range of lengths, angles and torsions allowed in the model, and use predefined sets of values. Usually, these allowed values are obtained from known real-world structures [60, 61]. The simplest class of models for the PFP is known as lattice models. In these models, a protein is modelled as a sequence of simple elements, representing the amino acids, embedded in a lattice. The connection angles between amino acids are restricted by the lattice structure in the plane (2D) or in the space (3D). In a valid conformation, a given position in the lattice can be occupied by, at most, one amino acid, and adjacent amino acids in the sequence must occupy adjacent positions in the lattice. The free-energy of a conformation is defined as a function of the number of adjacent amino acids in the structure which are non-adjacent in the sequence. This is known as non-local bonds [22, 61] or H-H contacts. Although square and cubic lattices are the most popular, there are implementations that use other type of lattices, such as triangular [46, 63] and hexagonal [37].

Despite the simplicity of lattice models, both 2D and 3D HP models have some behavioral equivalency with real-world proteins [22, 23, 24, 61]. Also, the computational treatment of such models are much more convenient, when compared with analytical models and, for some instances, the exhaustive enumeration of the possible conformations can be done. These properties have made lattice models very popular. However, the main drawback of lattice models (and, in special, of HP models - see Sect. 12.2.2) is the difficulty in representing clear secondary structures in the folding [33].

**The Hydrophobic-Polar Model**

This model was introduced by [42] and it is the most known and studied discrete model for the PFP. The Hydrophobic-Polar (HP) model is the simplest possible model and, in most cases, uses a square (2D) or cubic (3D) lattice. Notwithstanding, even being simple, the PSP was proved to be NP-hard using this model, that is, there is no polynomial-time algorithm to solve it, either the 2D version [16, 55, 57, 73] or the 3D one [3, 6]. This fact has motivated the development of many heuristic approaches, such as in [8, 10, 22, 50, 52, 61, 70].

The HP model is based on the assumption that the major contribution to the free-energy of the native conformation of a protein is due to interactions between hydrophobic amino acids, which tend to be grouped in the inner part of the spatial structure, while the hydrophilic (polar) amino acids tend to stand more outside, thus protecting the hydrophobic amino acids from contact with the environmental solvent. For simplicity, the 20 standard amino acids are divided in either hydrophobic (H) or polar (P), based on experimental results [45]. Therefore, the primary structure of a protein is a string defined over the binary alphabet $\{H, P\}$. Although several different hydrophobicity scales can be found in the current literature, there is still no consensus about a standard translation table between the 20-letters amino acids into a simple $\{H, P\}$ alphabet. To circumvent this problem, some studies suggest the use of extended alphabets,

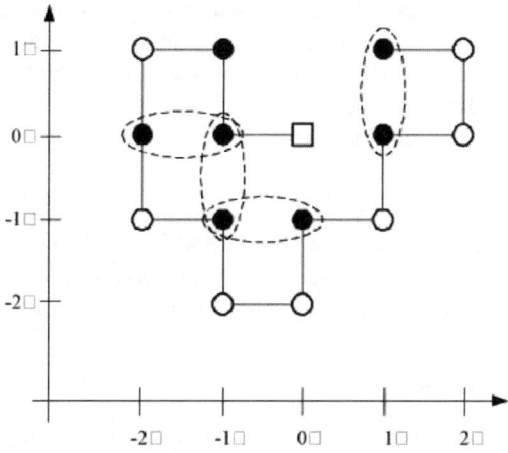

**Fig. 12.1.** A hypothetical conformation of 15 amino acids using the 2D-HP model

in which amino acids are converted to symbols more properly related to their physical and chemical properties [5, 48].

The $\{H, P\}$ string representing a protein is then embedded in a 2D or 3D lattice. Adjacent amino acids of the sequence are also adjacent in the lattice and, for a valid conformation, no point of the lattice can be occupied by more than one amino acid. The free-energy of a conformation is inversely proportional to the number of non-local bonds, as defined before. It is worth to mention that a non-local bond only takes place when a pair non-adjacent amino acids of the sequence lie in adjacent positions in the lattice. Consequently, minimizing the free-energy is equivalent to maximize the number of hydrophobic non-local bonds.

Figure 12.1 presents a conformation of polypeptide with 15 amino acids using the 2D-HP model. Black and white dots represent, respectively, the hydrophobic and hydrophilic amino acids. The square dot is the first amino acid of the sequence. The chain is connected by solid lines, and the bonds are represented by dotted lines. For this conformation there are 4 H-H contacts.

A simple free-energy function of a conformation, suggested by [44], is represented in Eq. (12.1):

$$E = \sum_{i<j} e_{r_i r_j} \Delta(r_i - r_j) \tag{12.1}$$

where: $\Delta(r_i - r_k) = 1$ , if amino acids $r_i$ and $r_j$ have a non-local bond, or $\Delta(r_i - r_k) = 0$, otherwise. Depending on the type of contact between amino acids, the energy will be $e_{HH}$, $e_{HP}$ or $e_{PP}$, corresponding to H-H, H-P or P-P contacts, respectively. According to [44], these energy parameters satisfy the following physical constraints:

1. Compact conformations have energy levels smaller than any other non-compact conformations;

2. Hydrophobic amino acids tend to be buried as inside as possible in the conformation. This is expressed by the relation $e_{PP} > e_{HP} > e_{HH}$ that decreases the energy of conformations in which the hydrophobic amino acids are hidden from the water solvent;

3. Amino acids of different types tend to get apart. This is expressed by the relation $2e_{HP} > e_{PP} + e_{HH}$.

In the standard HP model, values for those parameters are: $e_{HH} = -1.0$, $e_{HP} = 0$ and $e_{PP} = 0$ [42]. However, [44] suggested $e_{HH} = -2.3$, $e_{HP} = -1$ and $e_{PP} = 0$, since they satisfy the above conditions. According to them, results are not too sensitive to values of $e_{HH}$, provided those conditions are satisfied.

## Other Discrete Models

Besides the popular HP model, there are other discrete models for the PFP in which particular biological properties were explored:

- Lattice Polymer Embedding (LPE): this model was proposed by [73] and is based in a cubic lattice, similarly to the HP model. Each pair of amino acids have an affinity coefficient and the energy function to be minimized is the sum, over all possible pairs of amino acids, of the product of the affinity coefficient by the distance between amino acids.

- Charged Graph Embedding (CGE): This model was proposed by [27] and later used by [10]. It uses a 3D lattice and incorporates charges to the amino acids. Conformations allowed are not very realist because it considers bonds between every pair of amino acids in the chain and bonds are allowed to cross each other. On the other hand, the influence of a given amino acid on another one disappears when the Euclidean distance between them exceeds a critical value.

- Perturbed Homopolymer (PH): this model was suggested by [64] and reviewed by [67], and later used by [22]. This model does not take into account only the interactions between hydrophobic amino acids, but favors connections between amino acids of the same type (that is, H-H and P-P).

- Helical-HP model: it was presented by [72] and reviewed by [22]. This model considers only a 2D lattice and includes two types of interactions: non local interactions between hydrophobic amino acids, and local interactions represented by a propensity to form helices (helical propensity). [11] has extended this model taking into consideration the effect of hydrogen bonds in regular secondary structures (in both $\alpha$-helices and $\beta$-sheets).

- Tangent Spheres Side Chain model (HP-TSSC): introduced by [31], it uses the basic HP model, but does not embed amino acids in a lattice. In this model, a protein is represented by a graph that is transformed in a set of tangent spheres with equal radius, for both the backbone and the side chains. This model is an important contribution to off-lattice models for protein folding.

- HPNX model: this is a variation of the HP model in which the alphabet is extended to more letters [5, 9]. For instance, polar amino acids can be divided into three categories: positive charge (P), negative charge (N) and neutral (X). Therefore, the standard 20 amino acids is translated into the $\{H, P, N, X\}$ alphabet according to their physical and chemical properties and the energy of a conformation is computed using a matrix of energy potentials between every pair of contacts. Usually, $e_{HH} = -4.0$, $e_{PP} = e_{NN} = 1.0$, $e_{NP} = -1.0$ and for the remaining type of contacts the energy is null.

## 12.3 Computational Approaches for the PFP

### 12.3.1 Molecular Dynamics

This method, also known as *ab initio* [29, 58], seems to be the most realistic approach to simulate the folding of real proteins. The term *ab initio* means to start "from the beginning", without using previous knowledge about the structure of the protein. To do so, the basic idea is to simulate the movements of each atom of a protein, as well as of the water that surrounds it, as a function of time. The initial thermal energy of the system is established and atoms are enabled to move according to the rules of classical mechanics. The energy of a conformation is adapted to take into account all forces, accelerations and velocities to which each atom is submitted along time. Aiming at making the movement of the atoms the more realistic as possible, a very small time step is defined, such as $10^{-15}$ sec. For each time step the energy function is recomputed [75]. Even using supercomputers, the number of mathematical operations necessary to simulate the folding is so high that makes this methodology unfeasible even for very of small proteins. This type of simulation can be useful only for studying the behavior of the folding during very short periods of time, several orders of magnitude smaller than the time necessary to fold a real-world protein. However, it is believed that this method is potentially powerful to produce results according to the dynamic properties observed during the folding of real-world proteins [4], although it cannot be assured that it will converge to the native conformation. A full review of this methodology can be found in [43].

### 12.3.2 Approximation Algorithms

An approximation algorithm is a computational procedure capable of finding quasi-optimal solutions for specific problems (or specific instances), with a given predefined warranty of performance (regarding the optimal solution). According to [57], such class of algorithms can be useful for the PFP since they can find valid conformations somewhat near to the native conformation of a protein (provided the guaranteed maximum error is small). In a further step, the free-energy value of this approximated solution can be used as an upper-bound for another algorithm focused on local search. The main drawback of approximation algorithms is the need for a formal proof of its lower-bound performance.

Possibly, the first approximation algorithms devised for the PFP were proposed by [30, 31] using 2D- and 3D-HP models. Later, many other algorithms were proposed for different energy models and geometry of lattices [31, 32, 53, 54].

### 12.3.3 Genetic Algorithms

The minimization of the free-energy in discrete models frequently leads to a hard optimization problem. Despite the simplicity the usual free-energy function, based on the maximization of non-local H-H bonds, it leads to a multimodal search space. This search space is characterized by a large number of invalid solutions (corresponding to conformations in which more than an amino acid occupy a position in the lattice) and many local minima (corresponding to different conformations with the same number of non-local H-H bonds). These characteristics of the search space make it a hard problem for conventional optimization methods.

Amongst the many computational approaches for the PFP, certainly the most used is the genetic algorithm (GA), possibly due to its simplicity and efficiency in finding good solutions in large and complex search spaces. This of a GA in combining local features into a global solution makes it particularly appealing for the PFP.

A protein can have well-defined secondary structures, such as $\alpha$-helices or $\beta$-sheets. In most cases, important secondary structures can be identified in the primary structure as motifs. Some motifs have structures independent of its interaction with the remaining molecule, and they can be viewed as building blocks. In a similar way, the crossover operator of a GA works by recombining hopefully useful blocks to form solutions of increasing quality, thus providing a way to recycle partial solutions.

There are two basic issues for applying a GA for a given optimization problem: how the variables of the problem are encoded, and how the quality of a solution is measured. The first issue is the representation problem, and the latter, the evaluation problem. All other issues raised in an implementation, although important, are secondary.

### Encoding

When using a GA for the PFP, the way conformations of the protein are represented has a great importance on the dynamics and efficiency of the algorithm. Basically, one can devise three ways of representing a folding [40, 62]:

- Distance matrix: this encoding system describes a structure by means of a square matrix in which cells represent the distance between amino acids. This encoding system is rarely used in the literature [62].
- Cartesian coordinates: in this approach, a folding is described by a vector of elements representing the position of the amino acids of a sequence in the plane $\{x_i, y_i\}$ or in the space $\{x_i, y_i, z_i\}$. In general, this approach is not the most adequate for population-based algorithms (such as the evolutionary computation ones), since identical (or similar) structures can have completely representing vectors.
- Internal coordinates: a given conformation is represented as a set of movements of an amino acid relative to its predecessor in the chain. This is the

most usual representation approach found in evolutionary algorithms for the PFP, and two types of internal coordinates can be used:

- Absolute internal coordinates: they are based on the orientation of the axes the lattice in which the folding is embedded (either 2D or 3D). This encoding system is defined by the following set: $\{N, S, E, W, F, B\}$, corresponding to movements north, south, east and west (in the plane), and forward and backward (in the space).
- Relative internal coordinates: this encoding system defines the position of the next amino acid of the chain relative to the position of the preceding one in the lattice. The possible set of movements are: $\{F, L, R, U, D\}$, corresponding to forward, left, right, up and down, always having the previous position as reference. This encoding system has an important drawback: the initial population of a GA is randomly generated and, as a consequence, individuals will have an increased number of collisions in the structure (invalid conformations). This is specially true for proteins with an increased number of amino acids.

For instance, the conformation shown in figure 12.1 corresponds to the sequence $\{PHHPHPHPPHPHPPH\}$ and can be represented using:

- Relative internal coordinates: $\{LRLLFLRLLRLRLL\}$,
- Absolute internal coordinates: $\{WNWSSESENENENW\}$,
- Cartesian coordinates: $\{(0,0);(-1,0);(-1,1);(-2,1);(-2,0);(-2,-1); (-1,-1); (-1,-2);(0,-2);(0,-1);(1,-1);(1,0);(2,0);(2,1);(1,1)\}$.

A study of the two internal relative coordinates was done by [40], using different types of lattices. They concluded that, for square and cubic lattices, relative internal coordinates may lead a genetic algorithms to results much better than those that could be obtained using absolute internal coordinates. However, there are some authors who obtained satisfactory results using absolute coordinates for small chains [17]. For a triangular lattice, both types of coordinates have the same performance.

There are two restrictions to be satisfied for a valid conformation: there should be no collisions (a given point in the lattice should be occupied by at most one amino acid), and all adjacent amino acids of the sequence must be adjacent in the lattice. This last restriction is implicit in the encoding when using internal coordinates, but not when using Cartesian coordinates. To deal with the first restriction using internal coordinates there are two basic approaches:

- Delete invalid conformations that appear during the evolutionary cycle. This is the simplest way to deal with this issue, but, possibly, not the best one. When a protein is folded in a valid (but not optimal) conformation, the pathway to another valid conformation of smaller energy may be not achievable unless some invalid conformations are permitted in intermediary steps.
- Allow invalid conformations in the population and apply penalties. This approach is usual when using evolutionary algorithms for constrained problems. The genetic material present in some unfeasible solutions can be recombined further in the evolutionary cycle so as to form feasible and, hopefully, better

solutions. For the PFP there are two ways for applying penalties to invalid conformations: considering the number of pairs of amino acids that stand on the same point in the lattice, or considering the number of lattice points that have more than one amino acid in it. To date, it is not clear which of the two methods will give better results. Another somewhat different approach is due to [59] who suggest that hydrophobic amino acids that are in lattice points already occupied by other amino acids should not contribute to the free-energy function. that have more than two amino acids. This is an indirect way to apply a penalty to invalid conformations.

For off-lattice models, the encoding is somewhat straightforward. For instance, [20] represented a protein by means of internal angular coordinates of the atoms of the main chain. The torsion angles of the $C_\alpha$ (namely, $\phi$ and $\psi$) were restricted to small set of possible values, and were sufficient to represent the topology of the main chain for a large number of proteins with known structure. Therefore, using this kind of model, a chromosome can be encoded with integer [14] or binary [20, 21, 60] values representing those angles. On the other hand, [66] used a chromosome of real-valued genes for representing the same angles.

**Fitness Function**

There are many variations on the fitness function, and they are based on the model used (see Sect. 12.2.2).

For instance, [11] has proposed the use of an extra term to the Eq. (12.1), named secondary-structure-favored energy term, that considers the energy between hydrogen bonds formed by secondary structures. Also, [50] proposed a fitness function having three terms: the first is the regular free-energy function of the HP model and the other two are based on the concept of radius of gyration. The radius of gyration is computed separately for hydrophobic and for polar amino acids. Maximizing the radius of gyration of hydrophobic amino acids means that they are pushed towards the inner part of the conformation, while maximizing the radius of gyration of polar contacts means pushing them towards outside. This concept was used to force more compact and globular-like conformations.

Other variations can be found: [17] used a weighted sum of the number of H-H contacts, the number of H-P contacts and the number of hydrophobic-solvent contacts. They argue that this fitness function is more natural from the biological point of view, since it may be preferable for a hydrophobic amino acid to have a contact with a polar amino acid than to be in direct contact with the solvent.

Most approaches in the literature use some fitness function based on the number of H-H contacts, inherent to the HP model. However, the main criticism of this simple approach is that hydrophobic iteration alone is not sufficient to induce regular structures during folding, as pointed by [11].

**Genetic Operators**

For most implementations, the regular crossover and mutation operators have been used as part of a larger set of specialized operators.

Regarding the regular crossover, there are implementations using 1-point, 2-points and uniform variants. Although there is no consensus about which crossover type gives the best results, the traditional 1-point crossover is less disruptive and tends to keep larger schemata. Therefore, the more folded a conformation, the more the 1-point crossover seems to be appropriate. A different approach, known as systematic crossover, was proposed by [38]. In this case the best individual is always one of the parents selected for crossover and all possible crossover points are tried, generating a number of individuals. The two best offsprings are maintained in the population.

Some special types of mutation were also proposed. For instance, [15, 68, 74] proposed in-plane rotation, snake, out-of-plane rotation, crank shaft, kink and cornerchange, and [35] implemented diagonal move and tilt move. All these mutation operators aimed at producing different conformations by means of specific re-arrangement of the folding in the lattice.

Other researchers presented biologically-inspired operators such as the U-turn and Make-loops by [51]. These operators were meant to simulate the construction of stable secondary structures found in real folded proteins, such as $\alpha$-helices and $\beta$-sheets.

Two special genetic operators were proposed by [15]: duplicate predator and brood selection. The first is aimed at maintaining diversity in the populations throughout generations by means of deleting duplicate individuals and is similar to the pioneer search strategy introduced by [38]. The latter generates a brood of offsprings from two parents, and the best descendent is kept. This procedure is a kind of limited local search in the surrounding search space of the parents.

In some cases the use of the regular and specialized genetic operators is not sufficient to guarantee a proper fine-tuning of the conformation. This reflects the general knowledge that genetic algorithms are efficient for global search but do not display the same performance for local search. As a consequence, a number of different methods for local search have been proposed for the PFP. Many of such implementations are considered by authors as hybrid algorithms [51] or memetic algorithms [63, 39]. Possibly, the most popular procedures are Monte Carlo-based local search that has been used to improve solutions [15, 47, 74]. More sparsely, tabu search [36] and local hill-climbing [14, 71] are employed as genetic operators.

Another different approach is due to [51], who have proposed a local search procedure as a generalized version of the 2-opt method used for combinatorial optimization problems. This procedure starts by randomly selecting two non-consecutive amino acids in the chain and make their positions fixed in the lattice. Then, all possible conformations are evaluated, keeping the connectivity of the chain in the fixed points and changing the intermediate amino acids in between. The best conformation found in the procedure is kept. Although this procedure is computationally intensive (the number of possibilities increases exponentially as the distance between fixed points increase), it is useful to find best local conformation.

### 12.3.4   Ant Colony Optimization

Ant Colony Optimization (ACO) is an evolutionary technique inspired on the behavior of real ants searching for food. Possibly, [65] was the first to propose the use of ACO for the PFP. Their algoritm is based on three phases: construction, local search and pheromone updating. In the first phase, ants construct a folding over the lattice starting at a random point. Next, a greedy local search procedure is done, based on a long-range mutation method created by the authors. Then, the pheromone matrix is updated by ants, using two basic mechanisms: uniform evaporation ratio, and reinforcement of local folding motifs. They also used a mechanism of normalization of the pheromone matrix to prevent stagnation of the search. They have applied the ACO to several benchmark instances of using 2D and 3D-HP models, and results were compared with heuristic methods.

[26] also developed an ACO for the PFP using the 3D-HP model. The main difference between this implementation an that of [65] is the location of the polar amino acids, the form of the heuristic function that guides ant's decisions, and how the pheromone matrix is updated. Also, this implementation does not use any local search strategy. According the author, the implementation has achieved much better results than [65] and other heuristic methods.

Another implementation of ACO for the 3D-HP PFP is [69]. The differences of this approach to others is the use of a rapid coordinate transfer system to reduce computing time, as well a greedy local search procedure based on elementary moves, similar to the mutation operators proposed in [15, 68]. They also have devised a new method, inspired by Ethernet communication, for avoiding invalid foldings when an ant constructs a path in the lattice.

[13] has implemented single and multiple colony approaches of the ACO algorithm, with centralized and distributed processing. The main emphasis of the work was on distributed processing of multiple colonies, and they devised several methods for sharing information between evolving colonies. The several versions were tested with benchmarks of the 2D and 3D HP models. They have shown that the distributed multiple colony approach is scalable and has better performance over single colony approaches.

### 12.3.5   Differential Evolution

To date, the only work using Differential Evolution (DE) for the PFP is [8], using the 2D-HP model. Possibly, this is due to the fact that DE is a relatively recent evolutionary algorithm, and has been invented for continuous optimization problems. DE represents a possible solution for a problem using a vector of real numbers. The central idea of the DE algorithm is the use of difference vectors for generating perturbations in a population of vectors. This algorithm is conceptually simple, has few parameters do be tuned and, most times, converges fast to a good solution. In DE, the variables of the problem are encoded in a vector and, usually, the meaning of its elements to the real-world is straightforward.

Consequently, the concept of genotype, as in genetic algorithms, is not applicable to the original DE. However, for the PFP, authors devised an adaptation to represent possible solutions to the PFP by establishing a genotype-phenotype mapping. Individuals in DE are real-valued vectors which, in turn, are decoded into a specific fold of an amino acid chain in a square lattice. They also used special strategies for mixing vectors in DE and for initializing the population. Authors applied the proposed DE algorithm to benchmark instances up to 85 amino acids and reported consistent results better than genetic algorithms and other heuristic methods. Overall, the DE approach seems to be a promising option for finding good and fast solutions for the PFP.

### 12.3.6 Other Evolutionary Computation Methods

Only recently that the PFP has driven the attention of researchers of the Artificial Immune Systems (AIS) area. An AIS for 2D and 3D versions of the PFP using the lattice HP model was proposed by [18, 19]. In this work, they used two entities: antigens and B cells. The search space of the problem was efficiently partitioned by memory B cells with longer life span. Another work is due to [2] who proposed an AIS hybridized with tabu search and a fuzzy inference system. A fuzzy aging operator was introduced to decide which antibodies will be deleted from the population after the selection procedure. Also, they defined a mechanism of intensive affinity maturation that uses tabu search. The proposed AIS was tested with instances of the 3D-HP model.

A hybrid approach using operators from AIS and Pareto Archived Evolutionary Strategy was used by [18] for the PFP with an all-atom model. They have compared this approach with other evolutionary computation methods when applied to a set of small proteins up to 68 amino acids.

[28] has used Evolution strategies (ES) for a sub-problem of PFP: the side-chain packing problem. They used an all-atoms representation of the backbone plus the carbon atom of the side-chain that is bonded with the central $C_\alpha$. They used as energy function a measure of the deviation from a known structure. The encoding used was an array of integers representing the torsion angles for each amino acid of the chain. The evolutionary model used was a $(\mu + \lambda_t)$-ES, where $\mu$ parents generate $\lambda_t$ offsprings that compete with parents for survival.

### 12.3.7 Other Methods

There are several implementations of different neural network architectures for the PFP. For instance, [76] uses a self-organizing map (SOM) and the 2D-HP model. However, they obtained good results only for very small sequences, up to 36 amino acids. More traditional methods, such as the well-known branch-and-bound, were applied by [12] to a benchmark of sequences of up to 100 amino acids, using the 2D-HP model. Reported results were promising, but still lacks scalability.

## 12.4   Open Questions

### 12.4.1   Models and Implementations

As a matter of fact, the most studied models for protein folding are quite distant from reality, in special the HP model. However, as mentioned before, there are still no algorithm to solve this problem in polynomially-bounded time using simple lattice models. The more complex the models, certainly, the more difficult it will be to find an efficient computational algorithm for solving the PFP. This fact suggests that there is many room for development of models that, at the same time, have more realistic features and are computationally efficient. Possibly, both hybrid and evolutionary computation methods will be of great importance in this scenery.

Two basic issues come up when observing the implementations of evolutionary computation methods for PFP, as follows:

First, the way amino acids are encoded as a possible solution may be a serious drawback. If the encoding allows invalid conformations, the search space in which the evolutionary algorithm will look for solutions will have a large amount of invalid sites. Procedures for dealing with invalid conformations may be useful. However, a more efficient search could be done if the encoding itself did not allow invalid conformations. If so, the search space could be strongly reduced and then evolutionary (or other non-exact algorithms) could be more effective in searching for the optimal conformation. Also, with the current encoding methods it is possible that a very small change in a gene (that represents, for instance, a given move in the lattice) will cause a strong change in the conformation, thus indirectly affecting the role of other genes in the encoding. This effect is known as epistasis. Those drawbacks suggest that more studies are still necessary for finding less epistatic and intrinsically collision-free encodings.

Second, the fitness function dictates the fitness landscape, that is, the shape of the search space. Most models use the number of H-H contacts as the core of the fitness function. Consequently, the corresponding fitness landscape has many discontinuities and plateaus. The first is when the number of H-H contacts vary from one conformation to the next one, and the latter, when the number of H-H contacts is the same for many different neighbor conformations (possibly, the difference between these conformations is the position of amino acids that does not account for the number of H-H contacts). Due to the embedding in the lattice, a given conformation can be rotated and/or mirrored. The same holds for portions of the conformation that are not affected by the remaining amino acids. As a consequence, it is possible to have a lot of conformations, very different each other, that have the same number of H-H contacts.

The above-mentioned facts increase the difficulty of the PFP, thus leading to an increasing loss of performance of evolutionary computation methods, as they advance towards more realistic models and protein sizes.

## 12.4.2 Computational Power

The NP-hardness of the PFP with lattice models was one of the main motivations for using evolutionary computation, and other heuristic methods. To date, most works have approached only small sequences, usually chains with less than 100 amino acids. It is clearly observable that evolutionary computation methods display a decreasing performance as the number of amino acids increase. On the other hand, real-world proteins have an average of 300 amino acids, and some can have thousands. Since the number of possible solutions to the PFP tend to increase exponentially as the number of amino acids increase, the use of evolutionary computation methods for larger proteins seems to be unfeasible.

In the same way, simulations using all-atoms models (or some simplified version) have been done using very small chains, far away from real-world proteins.

Apart from the intrinsic loss of performance of evolutionary algorithms for the PFP, the main factor that has set bounds on their possible performances is the available computational power. Although the memory capacity and processing speed of modern desktop computers have increased extraordinarily in last years, they are still limited for large instances of the PFP. As a consequence, recent works have reported the use of distributed/grid computing [13, 21, 49, 71] or hardware-based techniques [1] for circumventing the computational power limitation. These seem to be the direction for future research to achieve the scalability necessary for studying the folding of real-world proteins.

## 12.4.3 Benchmarks

All the evolutionary computation methods proposed for the PFP use a kind of supervised learning procedure. In general, a set of amino acid sequences is used as training/test cases. The results of the algorithms are compared with some previous known results, regarding the free-energy of the conformation, the compactness of the structure, the processing time, etc.

Since the lattice HP models are the most widely studied, it can be found in the literature some sets of synthetically constructed amino acid chains (not real-world proteins) ranging from 20 to 100 elements for 2D-HP and up to 64 elements for 3D-HP [38, 40, 47, 59, 74]. For more realistic models, such as those that use all-atoms approach, biological data of short length has been used, provided the tri-dimensional structure is previously known.

However, there is a large gap between the available synthetic benchmarks and real-world proteins (this is especially true for the HP models). Even considering the limited representativeness of the model, synthetic instances do not capture important peculiarities of real-world proteins. Only recently, more realistic benchmarks were proposed, based on the translation of real-world proteins to the HP model [51, 65]. There are some issues to be solved regarding the translation procedure to construct such benchmarks, and they still do not have information about the native conformation, such as minimum free-energy and tri-dimensional structure. Notwithstanding, these benchmarks represent an important improvement for this research area, offering new challenges to the existing algorithms and methods.

## 12.5 Conclusion

Despite the progress done using evolutionary algorithms for protein folding prediction, this is still an open problem. To date, no technique has demonstrated acceptable scalability and accuracy for problem sizes comparable to those of real-world proteins. Notwithstanding, evolutionary computation methods have been intensively used for the PSP and are the most promising.

As mentioned in Sec. 12.4, currently, there are some important questions to be addressed in PFP. The most widely used models are far from reality, but, even so, computationally complex. Further research is necessary for inventing more adequate models, encodings and fitness functions for evolutionary computation methods.

Regarding the evolutionary computation methods themselves, it seems that genetic algorithms have achieved their limit of performance. More recent evolutionary computation methods, such as AIS, ACO and DE, seem to be more promising. However, the observation of the most successful evolutionary computation methods for PFP are those that use some kind of hybridism, mainly as a local search technique, and, certainly, this is a future trend.

Another important issue to be addressed is scalability, as research moves towards realistic models and the analysis of real-world proteins. The performance of computational systems for the PFP have to increase, at least, two orders of magnitude so as to deal efficiently with real-world problems. Therefore, future trends include distributed/grid processing and specialized hardware-based approaches.

## References

1. Armstrong Jr., N.B., Lopes, H.S., Lima, C.R.E.: Reconfigurable Computing for Accelerating Protein Folding Simulations. In: Diniz, P.C., et al. (eds.) ARCS 2007. LNCS, vol. 4419, pp. 314–325. Springer, Heidelberg (2007)
2. Almeida, C.P., Gonçalves, R.A., Delgado, M.R.B.S.: A Hybrid Immune-Based System for the Protein Folding Problem. In: Cotta, C., van Hemert, J. (eds.) EvoCOP 2007. LNCS, vol. 4446, pp. 13–24. Springer, Heidelberg (2007)
3. Atkins, J., Hart, W.E.: Algorithmica, 279–294 (1999)
4. Avbelj, F., Moult, J., Kitson, D.H., James, M.N.G., Hagler, A.T.: Biochemistry 29, 8658–8676 (1990)
5. Backofen, R., Will, S., Bauer, E.: Bioinformatics 15(3), 234–242 (1999)
6. Berger, B., Leighton, F.T.: J. Comput. Biol. 5, 27–40 (1998)
7. Berman, H.M., Westbrook, J., Feng, Z., Gilliland, G., Bhat, T.N., Weissig, H., Shindyalov, I.N., Bourne, P.E.: Nucl. Acids Res. 28, 235–242 (2000)
8. Bitello, R., Lopes, H.S.: A differential evolution approach for protein folding. In: Proc. IEEE Symp. on Computational Intelligence in Bioinformatics and Computational Biology, pp. 1–5 (2006)
9. Bornberg-Bauer, E.: In: Proc. 1st Ann. Int. Conf. on Computational Molecular Biology, pp. 47–55 (1997)
10. Chandru, V., Dattasharma, A., Kumar, V.S.A.: Discrete Appl. Math. 127, 145–161 (2003)

11. Chen, H., Zhou, X., Zhong-Can, O.-Y.: Phys. Rev. E 64, 041905–041910 (2001)
12. Chen, M., Huang, W.Q.: Genomics Proteomics Bioinformatics 3(4), 225–230 (2005)
13. Chu, D., Till, M., Zomaya, A.: Parallel ant colony optimizaiton for 3D protein structure prediction using the HP lattice model. In: Proc. 19th IEEE Int. Parallel and Distributed Processing Symp., pp. 193–199 (2005)
14. Cooper, L.R., Corne, D.W., Crabbe, M.J.C.: Comput. Biol. Chem. 27, 575–580 (2003)
15. Cox, G.A., Mortimer-Jones, T.V., Taylor, R.P., Johnston, R.L.: Theor. Chem. Acc. 112, 163–178 (2004)
16. Crescenzi, P., Goldman, D., Papadimitriou, C., Piccolboni, A., Yannakakis, M.: J. Comput. Biol. 5, 423–465 (1998)
17. Custódio, F.L., Barbosa, H.J.C., Dardenne, L.E.: Genet. Mol. Biol. 27(4), 611–615 (2004)
18. Cutello, V., Nicosia, G., Narzisi, G.: A Class of Pareto Archived Evolution Strategy Algorithms Using Immune Inspired Operators for Ab-Initio Protein Structure Prediction. In: Rothlauf, F., Branke, J., Cagnoni, S., Corne, D.W., Drechsler, R., Jin, Y., Machado, P., Marchiori, E., Romero, J., Smith, G.D., Squillero, G. (eds.) EvoWorkshops 2005. LNCS, vol. 3449, pp. 54–63. Springer, Heidelberg (2005)
19. Cutello, V., Nicosia, G., Pavone, M., Timmis, J.: IEEE T. Evol. Comput. 11(1), 101–117 (2007)
20. Dandekar, T., Argos, P.: J. Mol. Biol. 256, 645–660 (1996)
21. Day, R.O., Lamont, G.B., Pachter, R.: Protein structure prediction by applying an evolutionary algorithm. In: Proc. 2nd Int. Parallel and Distributed Processing Symp., pp. 155–162 (2003)
22. Dill, K.A., Bromberg, S., Yue, K., Fiebig, K.M., Yee, D.P., Thomas, P.D., Chan, H.S.: Protein Sci. 4, 561–602 (1995)
23. Dinner, A.R., Sali, A., Smith, L.J., Dobson, C.M., Karplus, M.: Trends Biochem. Sci. 25, 331–339 (2000)
24. Dobson, C.M., Karplus, M.: Curr. Opin. Struct. Biol. 9, 92–101 (1999)
25. Duan, Y., Kollman, P.A.: IBM Syst. J. 40, 297–309 (2001)
26. Fidanova, S.: 3D HP protein folding using ant algorithm. In: Proc. BioPS, pp. III.19–III.26 (2006)
27. Fraenkel, A.S.: Bull. Math. Biol. 55, 1199–1210 (1993)
28. Greenwood, G.W., Shin, J.M., Lee, B., Fogel, G.B.: A survey of recent work on evolutionary computation approaches to the protein folding problem. In: Proc. Congress on Evolutionary Computation, pp. 488–495 (1999)
29. Hardin, C., Pogorelov, T.V., Luthey-Schulten, Z.: Curr. Opin. Struct. Biol. 12, 176–181 (2002)
30. Hart, W.E., Istrail, S.: J. Comput. Biol. 3, 53–96 (1996)
31. Hart, W.E., Istrail, S.: J. Comput. Biol. 4(3), 241–259 (1997)
32. Heun, V.: Discrete Appl. Math. 127, 163–177 (2003)
33. Honig, B., Cohen, F.E.: Fold Des. 1, R17–R20 (1996)
34. Honig, B.: J. Mol. Biol. 293, 283–293 (1999)
35. Hoque, M.T., Chetty, M., Dooley, L.S.: A guided genetic algorithm for protein folding prediction using 3D hydrophobic-hydrophilic model. In: Proc. IEEE Congr. on Evolutionary Computation, pp. 2339–2346 (2006)
36. Jiang, T., Cui, Q., Shi, G., Ma, S.: J. Chem. Phys. 119, 4592–4596 (2003)
37. Jiang, M., Zhu, B.: J. Bioinform. Comput. Biol. 3(1), 19–34 (2005)
38. König, R., Dandekar, T.: Biosystems 50, 17–25 (1999)

39. Burke, E.K., Krasnogor, N., Blackburne, B.P., Hirst, J.D.: Multimeme Algorithms for Protein Structure Prediction. In: Guervós, J.J.M., Adamidis, P.A., Beyer, H.-G., Fernández-Villacañas, J.-L., Schwefel, H.-P. (eds.) PPSN 2002. LNCS, vol. 2439, pp. 769–778. Springer, Heidelberg (2002)
40. Krasnogor, N., Hart, W.E., Smith, J., Pelta, D.A.: Protein structure prediction with evolutionary algorithms. In: Proc. Int. Genetic and Evolutionary Computation Conf., pp. 1596–1601 (1999)
41. Krasnogor, N., Pelta, D., Lopez, P.E.M., Canal, E.: Genetic algorithm for the protein folding problem: a critical view. In: Proc. of Engineering of Intelligent Systems, pp. 353–360 (1998)
42. Lau, K., Dill, K.A.: Macromolecules 22, 3986–3997 (1989)
43. Lee, M.R., Duan, Y., Kollman, P.A.: J. Mol. Graph Model 19, 146–149 (2001)
44. Li, H., Helling, R., Tang, C., Wigreen, N.: Science 273, pp. 666–669 (1996)
45. Li, H., Tang, C., Wingreen, N.S.: Phys. Rev. Lett. 79, 765–768 (1997)
46. Li, Z., Zhang, X., Chen, L.: Appl. Bioinformatics 4(2), 105–116 (2005)
47. Liang, F., Wong, W.H.: J. Chem. Phys. 115(7), 3374–3380 (2001)
48. Liu, H.G., Tang, L.H.: Phys. Rev. E Stat Nonlin Soft Matter Phys. 74(5 Pt 1), 051918 (2006)
49. Liu, W., Schimidt, B.: Mapping of genetic algorithms for protein folding onto computational grids. In: Proc. IEEE Region 10 TENCON Ann. Conf., pp. 1–6 (2005)
50. Lopes, H.S., Scapin, M.P.: An Enhanced Genetic Algorithm for Protein Structure Prediction Using the 2D Hydrophobic-Polar Model. In: Talbi, E.-G., Liardet, P., Collet, P., Lutton, E., Schoenauer, M. (eds.) EA 2005. LNCS, vol. 3871, pp. 238–246. Springer, Heidelberg (2006)
51. Lopes, H.S., Scapin, M.P.: A hybrid genetic algorithm for the protein folding problem using the 2D-HP lattice model. In: Yang, A. (ed.) Success in Evolutionary Computation, Springer, Heidelberg (2007)
52. Lyngsø, R.B., Pedersen, C.N.S.: Protein folding in the 2D HP model. Technical Report RS-99-16, BRICS Bioinformatics Research Center, University of Aarhus (1999)
53. Mauri, G., Pavesi, G., Piccolboni, A.: Approximation algorithms for protein folding prediction. In: Proc. $10^{th}$ Ann. Symp. on Discrete Algorithms, pp. 945–946 (1999)
54. Newman, A.: A new algorithm for protein folding in the HP model. In: Proc. $13^{th}$ Ann. Symp. on Discrete Algorithms, pp. 876–884 (2002)
55. Nayak, A., Sinclair, A., Zwick, U.: Spatial codes and the hardness of string folding problems. In: Proc. $9^{th}$ Ann. Symp. on Discrete Algorithms, pp. 639–648 (1998)
56. Ngo, J.T., Marks, J.: Protein Eng. 5, 313–321 (1992)
57. Ngo, J.T., Marks, J., Karplus, M.: Computational complexity, protein structure prediction, and the Levinthal paradox. In: Merz Junior, K., LeGrand, S. (eds.) The Protein folding problem and terciary structure prediction. Birkhäuser, Boston (1994)
58. Osguthorpe, D.J.: Curr. Opin. Struct. Biol. 10, 146–152 (2000)
59. Patton, A.L., Punch III, W.F.: Goodman (eds) A standard GA approach to native protein conformation prediction. In: Proc. $6^{th}$ Int. Conf. on Genetic Algorithms, pp. 574–581 (1995)
60. Pedersen, C.N.S., Moult, J.: J. Mol. Biol. 269, 240–259 (1997)
61. Pedersen, C.N.S.: Algorithms in computational biology. PhD Thesis, Department of Computer Science. University of Aarhus, Denmark (2000)

62. Piccolboni, A., Mauri, G.: Application of evolutionary algorithms to protein folding prediction. In: Selected Papers from the $3^{rd}$ European Conference on Artificial Evolution, pp. 123–136 (1998)
63. Santos, E.E., Santos Jr., E.: Reducing the computational load of energy evaluations for protein folding. In: Proc. $4^{th}$ Symp. on Bioinformatics and Bioingineering, pp. 79–86 (2004)
64. Shakhnovich, E.I., Gutin, A.M.: Proc. Natl. Acad. Sci. USA 90, 7195–7199 (1993)
65. Shmygelska, A., Hoos, H.H.: BMC Bioinformatics 6, 30–52 (2005)
66. Shulze-Kremer, S., Tiedemann, U.: Parameterizing genetic algorithms for protein folding simulation. In: Proc. $27^{th}$ Ann. Hawaii Int. Conf. on System Sciences, pp. 345–354 (1994)
67. Socci, N.D., Onuchic, J.N.: J. Chen. Phys. 101, 1519–1528 (1994)
68. Song, J., Cheng, J., Zheng, T., Mao, J.: A novel genetic algorithm for HP model protein folding. In: Proc. $6^{th}$ IEEE Int. Conf. on Parallel and Distributed Computing, Applications and Technology, pp. 935–937 (2005)
69. Song, J., Cheng, J., Zheng, T.: Protein 3D HP model folding simulation based on ACO. In: Proc. $6^{th}$ Int. Conf. on Intelligent Systems Design and Applications, vol. 1, pp. 410–415 (2006)
70. Tang, C.: Physica. A 288, 31–48 (2000)
71. Tantar, A.-A., Melab, N., Talbi, E.-G., Parent, B., Horvath, D.: Future Gen. Comput. Syst. 23(3), 398-409 (2007)
72. Thomas, P.D., Dill, K.A.: Protein Sci. 2, 2050–2065 (1993)
73. Unger, R., Moult, J.: Bull Math. Biol. 55, 1183–1198 (1993b)
74. Unger, R., Moult, J.: J. Mol. Biol. 231, 75–81 (1993c)
75. Unger, R., Moult, J.: On the applicability of genetic algorithms to protein folding. In: $26^{th}$ Hawaii International Conference on System Sciences, vol. 1, pp. 715–725 (1993d)
76. Yanikoglu, B., Erman, B.: J. Comput. Biol. 9(4), 613–620 (2002)

# 13

# Flexible Protein Folding by Ant Colony Optimization

Xiao-Min Hu[1], Jun Zhang[1,3], and Yun Li[2]

[1] Department of Computer Science, SUN Yat-sen University, Guangzhou,
510275, China
[2] Department of Electronics and Electrical Engineering, University of Glasgow,
Glasgow G12 8LT, Scotland, UK
[3] Corresponding Author, Email: `junzhang@ieee.org`

**Summary.** Protein structure prediction is one of the most challenging topics in bioinformatics. As the protein structure is found to be closely related to its functions, predicting the folding structure of a protein to judge its functions is meaningful to the humanity. This chapter proposes a flexible ant colony (FAC) algorithm for solving protein folding problems (PFPs) based on the hydrophobic-polar (HP) square lattice model. Different from the previous ant algorithms for PFPs, the pheromones in the proposed algorithm are placed on the arcs connecting adjacent squares in the lattice. Such pheromone placement model is similar to the one used in the traveling salesmen problems (TSPs), where pheromones are released on the arcs connecting the cities. Moreover, the collaboration of effective heuristic and pheromone strategies greatly enhances the performance of the algorithm so that the algorithm can achieve good results without local search methods. By testing some benchmark two-dimensional hydrophobic-polar (2D-HP) protein sequences, the performance shows that the proposed algorithm is quite competitive compared with some other well-known methods for solving the same protein folding problems.

## 13.1 Introduction

With rapid development of bioinformatics, more and more about the molecular world becomes known. Following the completion of the human genome project in 2000, genetic sequencing is now made feasible by current technology [1]. However, there still exist challenges in analyzing relationships between protein structures and their related functions. As different structures reflect specifically different functions, predicting a protein structure to estimate its functions is one of the major goals of bioinformatics [2]. In nature, proteins fold spontaneously to their native structures very fast (on a time scale of milliseconds) when placed in an aqueous solution [3]. However, traditional methods for predicting the structures of proteins, such as the X-ray crystallography and the nuclear magnetic resonance (NMR) [4] [5] are expensive and time-consuming. More importantly, reflections that are gained by these methods may be blurry and incomplete. Since the remarkable discovery by Anfinsen et al [6] that many simple proteins have a unique native structure, which appears to depend on the sequence only,

T.G. Smolinski et al. (Eds.): Comp. Intel. in Biomed. & Bioinform., SCI 151, pp. 317–336, 2008.
springerlink.com

experimental results [7] [8] [9] have subsequently emerged to support this discovery. A commonly accepted hypothesis is that the protein sequence folds into the structure with the equilibrium minimum free energy (MFE) state (the thermodynamic hypothesis) [10] [11]. Given a two-dimensional square lattice board, the protein folding problem (PFP) is to place the protein sequence in the lattice to form a self-avoiding path. Thus, the aim of solving a PFP is to find the protein folding conformation that satisfies the MFE state.

Based on this hypothesis, the critical mission is to find a way to predict a protein structure fast and accurately from the protein sequence. The real structures of proteins are very complex for they are on the atomic level and in a relatively large three dimensional search space. Various protein folding models have been proposed to simplify the structure for better analysis. These models include the protein structure prediction (PSP) model [12], the lattice polymer embedding (LPE) model [10], the charged graph embedding (CGE) model [13], and the hydrophobic-hydrophilic (or hydrophobic-polar, HP) model [14]- [23], etc. In particular, the HP model can be further classified into three types - the square lattice model [14]- [18], the triangle lattice model [19]- [21], and the toy model [22] [23]. The algorithm proposed in this chapter is based on the HP square lattice model.

Since a number of simplified models have been proposed, various methods have been developed to solve the PFPs, such as the dynamic programming (DP), neural network (NN) [23], Monte Carlo (MC) [24]- [28], genetic algorithm (GA) [29]- [36], ant colony optimization (ACO) [37]- [41], particle swarm optimization (PSO) [22], and immune algorithm (IA) [42] [43] methods. The PFPs have been proven to be NP–complete [44], which cannot be solved by a deterministic polynomial algorithm. As a paradigm of swarm computation, ant colony algorithms [45] have shown great potential in solving NP-hard combinatorial problems.

This chapter develops a simple but effective ant algorithm to solve PFPs, termed the 'flexible ant colony (FAC) algorithm'. It has four special mechanisms, including the path construction, the path retrieval, the pheromone attraction, and the folding heuristics. These novel mechanisms make it behave differently from previous ant algorithms for solving PFPs with the HP square lattice model [37]- [41].

The ants of the FAC algorithm aim to find a 'conformation path' of protein in the lattice. The pheromones are deposited on the virtual connections between adjacent squares in the lattice. Such pheromone laying approach is similar to those on the arcs connecting cities in a traveling salesman problem (TSP). However, it is different from existing ant algorithms proposed for solving PFPs, whose pheromones are on three relative folding directions of the protein [37]- [41]. In fact, if the pheromones only indicate the relative directions of the protein folding (as the ones in [37]- [41] do), the ants may not grasp the folding situations entirely. On the contrary, if the pheromones guide the protein to fold on an absolute lattice, as the proposed algorithm does, each ant can sense the solutions which have been configured by the other ants.

A protein sequence in the HP square lattice model is a string of hydrophobic (H) and polar (P) amino acids. The amino acids are placed one by one by artificial ants. If the surrounding lattice squares of an amino acid are all occupied, the next amino acid cannot be placed. Such situation is termed stagnation. Then the path retrieving strategy should be applied. As all ants start to construct the folding path from the center of the lattice, diversity for ants to choose alternative squares to place the amino acid on the protein sequence is realized by decreasing the pheromone value on the arc that the ant has just passed. Such pheromone reduction method during solution construction is similar to the local pheromone update method used in the ant colony system (ACS) algorithm [45] [46].

In the proposed FAC algorithm, the heuristics and the pheromones cooperate to construct conformations. The heuristic information varies between the hydrophobic amino acids and the polar amino acids. An added local search method is optional and it is omitted in this chapter, as the performance of the FAC without the local search is good enough in most of the experimental tests. By comparing the performance with a genetic algorithm (GA), an immune algorithm (IA), and an ant algorithm in the literature, the proposed algorithm does present improvements.

The rest of this chapter is constructed as follows: Section 13.2 presents a brief review on the PFPs with the HP model and discusses the features of protein folding conformations. Then the characteristics of the ACO are briefly introduced. Section 13.3 details the ants' construction behaviors in the proposed FAC algorithm. Section 13.4 folds some benchmark protein sequences using the proposed algorithm and compares the performances with other well-known algorithms. For deeper analysis, this chapter also tests influences of the parameters of the FAC algorithm and highlights some prospects for enhancements. Finally, conclusions are drawn in Section 13.5.

## 13.2 Protein Folding and the Ant Colony Optimization

### 13.2.1 Characteristics of the Protein Folding

Some benchmark instances of protein sequences in the 2-dimensional square lattice HP (2D-HP) models are listed in Table 13.1, where $l$ is the number of amino acids and $E^*$ stands for the MFE level. The letter 'H' stands for the hydrophobic amino acid and 'P' stands for the polar amino acid, which is hydrophilic. There are 20 amino acids in nature. Using various classifications, they can be divided into acid, alkaline or neutral; positively or negatively charged or uncharged; and hydrophobic or hydrophilic, etc. To a globular protein in an aqueous solution, the hydrophilic amino acids tend to be on the surface of the globule as they are attracted to water molecules (note that the environment inside cells is primarily water). The hydrophobic amino acids are repelled by water so that most of them gather inside the globular protein to form a core except for some special hydrophobic regions on the surface of the protein.

Some conformations of the 2D-HP folding structures of the same protein sequence with 48 amino acids are presented in Fig. 13.1. For the square lattice HP model, the best conformation is judged by the number of hydrophobic-hydrophobic (H-H) bonds that hydrophobic amino acids are adjacent on the

**Table 13.1.** Standard HP benchmarks for 2-D square lattice

| No. | $l$ | $E^*$ | Protein Sequence |
|-----|-----|-------|------------------|
| 1 | 18 | -9 | PHPPHPHHHPHHPHHHHH |
| 2 | 18 | -8 | HPHPHHHPPPHHHHHPPHH |
| 3 | 20 | -10 | HHHPPHPHPHPPHPHPHPPH |
| 4 | 20 | -9 | HPHPPHHPHPPHPHHPPHPH |
| 5 | 24 | -9 | HHPPHPPHPPHPPHPPHPPHPPHH |
| 6 | 25 | -8 | PPHPPHHPPPPHHPPPPHHPPPPHH |
| 7 | 36 | -14 | PPPHHPPHHPPPPPHHHHHHHPPHH PPPPHHPPHPP |
| 8 | 48 | -23 | PPHPPHHPPHHPPPPPHHHHHHHHHH HPPPPPPHHPPHHPPHPPHHHHH |

lattice, but not consecutive in the sequence. The number of the H-H bonds in each conformation in Fig. 13.1 is 23 (e.g., the number of dashed lines in the first conformation), which forms the MFE state with $E^* = -23$. It can be seen that the hydrophobic amino acids do form a core inside the protein conformation, while the polar amino acids are surrounding the core and their placements are quite flexible.

Although the square lattice model is highly abstracted from the real protein folding model, some special conformations can reflect a possible secondary structure of a protein. Fig. 13.2 shows some special 2D-HP conformations and the corresponding three-dimensional protein structures of an $\alpha$-helix and $\beta$-sheets. However, such a model is unsatisfactory to many biologists. The 3-dimensional structure of a specific protein sequence is unique, but as we can see in Fig. 13.1,

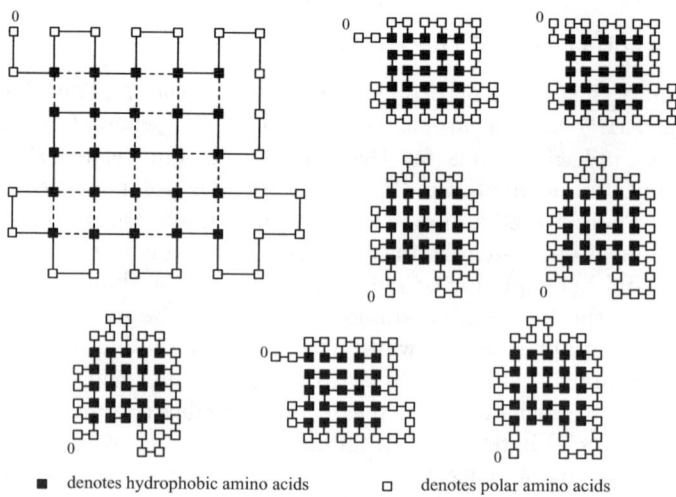

■ denotes hydrophobic amino acids    □ denotes polar amino acids

**Fig. 13.1.** Some conformations of sequence 8 (Length = 48)

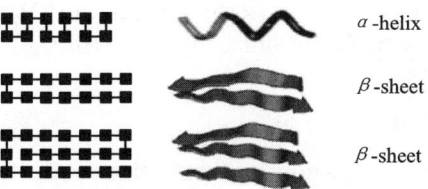

**Fig. 13.2.** Special HP conformations and the secondary structures of protein sequences

there may be several equivalent conformations by the same protein sequence. Therefore, the HP model is too simple to reflect the real protein structure completely. However, it is already a very challenging computational model.

### 13.2.2 Characteristics of the Ant Colony Optimization

To solve the traveling salesman problem, the first ant algorithm – the ant system (AS) – was proposed by Dorigo [45] [47] through stimulating the foraging behavior of real ants. Following this, several variants of ant algorithms have been developed, such as the elitist ant system (EAS), the max-min ant system (MMAS), and the ant colony system (ACS) [45], etc. They have been successfully applied to a wide range of application problems, such as the vehicle routing problem (VRP) [48], the job shop scheduling problem (JSP) [49], and the water distribution system (WDS) [50], etc. The AS and its successors at last form a kind of optimization paradigm termed the 'ant colony optimization (ACO) algorithms'. The basic framework for ACO includes:

Step 1: Construct ants' solutions (utilizing pheromone and heuristic information)
Step 2: Apply local search (optional)
Step 3: Update pheromones

A group of $m$ ants perform the above three steps to search for a better solution iteration by iteration. Firstly, based on the current density of pheromone in the environment and other heuristic information, each ant in the colony constructs a solution. Secondly, local search method can be applied to enhance the solutions found by the ants. Thirdly, the pheromone in the surrounding environment should be updated to guide more ants to the potentially best solution in the next iteration.

The FAC algorithm proposed in this chapter is based on the basic framework of the ant colony system (ACS) [45] [46], which includes mechanisms such as local pheromone update and global pheromone update. The implementation of these mechanisms is redefined in this chapter.

## 13.3 Ant Colony Search in Lattices

Given a two-dimensional square lattice board, the PFP is to place the protein sequence in the lattice to form a self-avoiding path. The mission of an ant colony is to discover a path, which maximizes the number of H-H bonds by two adjacent hydrophobic amino acids that are not consecutive in the protein sequence.

### 13.3.1  Path Construction

In order not to violate the region of the lattice, each ant starts building the path from the middle of the protein sequence in the center of the lattice. For a protein sequence with $n$ amino acids, which are denoted as $\{s_0, s_1, \cdots, s_{n-1}\}$ ($s_j \in \{P, H\}, j = 0, \cdots, n-1$), each ant starts from two horizontal squares in the middle of an $(n+2) \times (n+2)$ lattice board as depicted in Fig. 13.3. The squares in the lattice board are indexed from 0 to $(n+2)^2 - 1$, starting from the top left corner to the bottom right corner. The two squares with indexes $(\lfloor n/2 \rfloor + 1)(n+2) + \lfloor n/2 \rfloor$ and $(\lfloor n/2 \rfloor + 1)(n+2) + \lfloor n/2 \rfloor + 1$ are termed the 'left start square' and the 'right start square' respectively. The two squares are colored in the middle of the lattice shown in Fig. 13.3. The amino acid $s_{\lfloor n/2 \rfloor}$ is placed in the left start square while the amino acid $s_{\lfloor n/2 \rfloor + 1}$ is placed in the right start square. The sub-protein sequence $\{s_0, \cdots, s_{\lfloor n/2 \rfloor}\}$ that is built from the left start square is denoted as the 'left path', while the $\{s_{\lfloor n/2 \rfloor + 1}, \cdots, s_{n-1}\}$ is the 'right path'. Then an ant randomly chooses to go a step on the left part or on the right part of the protein sequence. After several construction steps, a protein conformation is built, similar to the dashed lines in Fig. 13.3. The squares that have been passed by the ant cannot be passed again by the same ant.

There are two advantages of indexing the squares in the lattice. One is that the coordinates of the squares are now one-dimensional. The other is that the four adjacent squares are convenient to obtain. For example, when an ant is now

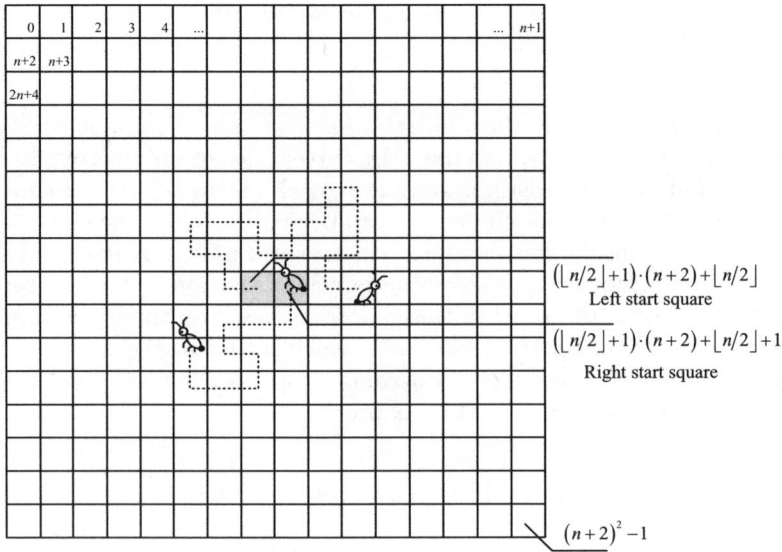

**Fig. 13.3.** Lattice board for a protein with $n$ amino acids

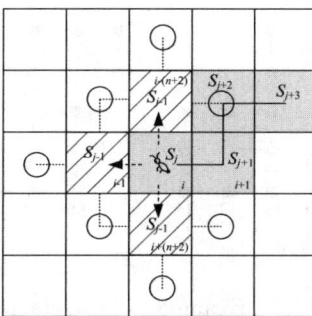

**Fig. 13.4.** An ant chooses a step to go

in square $i$ as shown in Fig. 13.4, which is not on the border of the lattice, its four adjacent squares are $i - (n+2)$ (going up $(U)$), $i + (n+2)$ (down $(D)$), $i - 1$ (left $(L)$), and $i + 1$ (right $(R)$). As the ant has passed the right square, it can only choose one of the other three directions to go.

### 13.3.2  Path Retrieval

If the ant has passed all the adjacent squares when placing a non-ending amino acid $s_j$ ($j \neq 0$ or $n - 1$), the protein cannot fold any more. Such situation is termed 'stagnation'. In this case, the folding needs to be retrieved. Consider an ant has constructed a sub-sequence $\{s_{left}, \cdots, s_{startL}, s_{startR}, \cdots, s_{right}\}$, where *left* is the index of the left most amino acid, *right* is the index of the right most amino acid, $startL = \lfloor n/2 \rfloor$, $startR = \lfloor n/2 \rfloor + 1$. To a 'right' retrieval, a random index $j$ is selected as

$$j = rand\%(right - startL - 1) + startL + 1 \tag{13.1}$$

where $rand$ is a random non-negative integer number. The amino acids from $s_{j+1}$ to $s_{right}$ are released as not been constructed by the ant and the corresponding squares in the lattice are set vacant. On the other hand, a 'left' retrieval point $j$ is selected as

$$j = rand\%(startL - left) + left + 1 \tag{13.2}$$

The amino acids from $s_{left}$ to $s_{j-1}$ are released and the corresponding squares are thus set vacant.

Although stagnation occurs on the right side of the protein, it doesn't mean that the right side of the protein is to be retrieved, because some stagnation situations cannot be cleared by simply retrieving the side where the stagnation happens. Fig. 13.5 illustrates two stagnation situations on the right path. The hollow beads stand for the left start amino acid and the right start amino acid, whereas the triangles are amino acids on the right path. In the example presented in Fig. 13.5(a), the stagnation can be released by the right retrieval when $j = 21$ and the ant is to go upward. However, in Fig. 13.5(b), the stagnation cannot

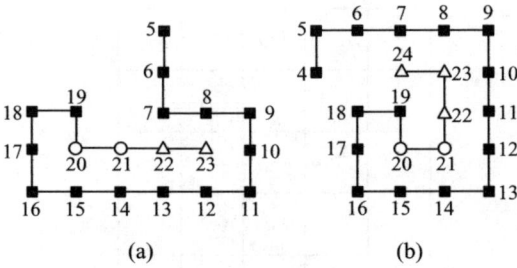

(a)                    (b)

**Fig. 13.5.** Examples of the stagnation

```
/* startL = ⌊n/2⌋, startR = ⌊n/2⌋+1
   left : the constructed left end, right: the constructed right end */
Procedure RightSideRetrieve({s_left, ⋯ , s_startL, s_startR, ⋯ , s_right})
   If startR < right && RightRetrieveBool == false
        j = rand % (right − startL − 1) + startL +1;
        RightRetrieveSequence({s_left, ⋯ , s_startL, s_startR, ⋯ , s_right}, j);
        LeftRetrieveBool = false;
        RightRetrieveBool = true;
   Else If startL != left
        j = rand % (startL − left) + left +1;
        LeftRetrieveSequence({s_left, ⋯ , s_startL, s_startR, ⋯ , s_right}, j);
        RightRetrieveBool = false;
   End


Procedure LeftSideRetrieve({s_left, ⋯ , s_startL, s_startR, ⋯ , s_right})
   If startL > left && LeftRetrieveBool == false
        j = rand % (startL − left) + left +1;
        LeftRetrieveSequence({s_left, ⋯ , s_startL, s_startR, ⋯ , s_right}, j);
        LeftRetrieveBool = true;
        RightRetrieveBool = false;
   Else If (startL+1) < right
        j = rand % (right − startL − 1) + startL +1;
        RightRetrieveSequence({s_left, ⋯ , s_startL, s_startR, ⋯ , s_right}, j);
        LeftRetrieveBool = false;
   End
```

**Fig. 13.6.** Outline of the retrieval process

be released by performing the right retrieval but only the left retrieval. So the Boolean values *RightRetrieveBool* and *LeftRetrieveBool* are used to judge such situations to make sure that a retrieval in the same direction cannot be performed twice consecutively for avoiding potential stagnation.

Whether to perform a right retrieval or a left retrieval is not only based on the location of the stagnation, but also the two Boolean values *RightRetrieveBool* and *LeftRetrieveBool*. If the stagnation happens on the right side of the protein, we term the retrieval procedure as 'RightSideRetrieve', while the procedure for the left stagnation is termed 'LeftSideRetrieve'. Fig. 13.6 illustrates the pseudo-code of the above process. The functions 'RightRetrieveSequence( )' and 'LeftRetrieveSequence( )' perform the respectively right/left retrieval. Take Fig. 13.5(b) as an example. The stagnation happens on the right path, so that the 'RightSideRetrieve' procedure is invoked. As $startR = 21$, $right = 24$, and $RightRetrieveBool = false$, a random integer $j$ is generated by (13.1). Suppose $j = 22$. Then the 'RightRetrieveSequence' function is invoked, so that the amino acids from 23 to 24 are released. The 'LeftRetrieveBool' and the 'RightRetrieve-Bool' are set as False and True respectively. It is known that this could not help to clear the stagnation. The construction of the path continues, until the stagnation happens again. Suppose the sequence is changed to be 2 to 24. At that time, the 'RightSideRetrieve' procedure is invoked again. As 'RightRetrieveBool' is true now, it can only perform 'LeftRetrieveSequence( )' to release some of the left path. The stagnation can be cleared if $j = 5$ to 20.

### 13.3.3    Pheromone Attraction

Pheromones are released on the directed arcs connecting the adjacent squares, which are denoted as $\tau_{id}$, where $i = 0, 1, 2, \cdots, (n+2)^2 - 1$ and $d = \{L, R, U, D\}$. Note that the protein sequence cannot exceed the lattice board, as the width of the board must be greater than the length of the protein.

### 1) Local Pheromone Update

As all ants start from the same left and right start squares in the lattice, an effective method for avoiding early convergence is to remove some pheromones between the two adjacent squares as (13.3)

$$\tau_{id} \leftarrow \delta \times \tau_{id} \tag{13.3}$$

where $d$ is the movement that the ant will go to place the next amino acid, $i$ is the index of the current square in which the ant locates. $\delta = (m-1)/m < 1$ is a 'local evaporation rate', and $m$ is the number of ants. If the pheromone on that arc is smaller than $\tau_{min}$, the pheromone is reset to $\tau_{min}$, which is the lower boundary of the pheromone value.

### 2) Global Pheromone Update

Once all ants have constructed a protein folding path, the pheromones on all arcs are 'evaporated' as defined by (13.4)

$$\tau_{id} \leftarrow \rho \times \tau_{id} \tag{13.4}$$

where $\rho$ is a 'global evaporation rate'. Then the best path found in the current iteration is reinforced by increasing the amount of pheromone as described by equation (13.5)

$$\tau_{i'd} \leftarrow \tau_{i'd} + \varepsilon/(-E^*_{min}) \tag{13.5}$$

where $i' \in$ {the squares that the iteration's best ant has just passed}, $d$ is the movement that the ant went to the adjacent square from square $i'$, $\varepsilon$ is the maximum number of H-H bonds in the current iteration, $E_{\min}^* < 0$ is the approximation of the MFE of the protein in the square lattice HP model.

### 13.3.4   Heuristics for Folding

While pheromones are the means for keeping the historical memories, heuristics are the strategies for current selection. Different from the heuristic information in [37]- [41] where only hydrophobic amino acids are considered, this chapter takes into account both the heuristic information for hydrophobic amino acids and polar amino acids.

#### 1) Heuristic for hydrophobic (H) amino acids

The goal for PFPs is to find the minimum energy conformation, which is reflected by the number of H-H bonds. Hence, if a conformation can yield more H-H bonds, it should have a higher probability to be constructed. Once the next amino acid $s_j$ for an ant $k$ to place is known as a hydrophobic (H) amino acid, the heuristic for it is determined by

$$\eta_{jd} = h_{jd} + 1 \tag{13.6}$$

where $h_{jd}$ is the number of the new obtained H-H bonds by placing the amino acid $s_j$ in the adjacent square, $d$ is the ant's movement.

Fig. 13.4 illustrates an ant that is currently locating in square $i$ with an amino acid $s_j$. The next step it chooses is to place an amino acid $s_{j-1}$. The slashed squares are the possible locations. For each of the slashed squares, the potential H-H bonds are the ones that connect the neighboring squares (shown as hollow spots in Fig. 13.4) where a hydrophobic amino acid has been placed.

#### 2) Heuristic for polar (P) amino acids

If the next amino acid $s_j$ to be placed is a polar amino acid, the heuristic value is the sum of the vacant squares (i.e., the squares that have not been passed by the ant) and polar amino acids (excluding consecutive polar amino acid) in the neighborhood of the possible location of the next amino acid plus one as given by

$$\eta_{jd} = v_{jd} + h'_{jd} + 1 \tag{13.7}$$

where $v_{jd}$ and $h'_{jd}$ are the numbers of vacant squares and polar amino acids in the neighborhood of the possible locations of the next amino acid repectively.

For a polar amino acid, more inclinations should be given to water molecules. As the protein folds in an aqueous solution, the vacant squares can be regarded as water molecules. The nearby polar amino acids imply that the edge of the protein is near.

### 13.3.5   Implementation of the Flexible Ant Colony Algorithm for PFPs

The roulette wheel selection method is used for each ant in the colony to choose the next step of path. If the ant currently locates in square $i$ and the next amino acid to be placed is $s_j$, then the probability of selecting the feasible movement $d$ is given by

$$
p_d = \frac{\tau_{jd} \eta_{jd}^{\beta}}{\sum_{q \in \{\text{feasible movements}\}} (\tau_{jq} \eta_{jq}^{\beta})}
\tag{13.8}
$$

where $\beta$ is the reinforcement to heuristic values.

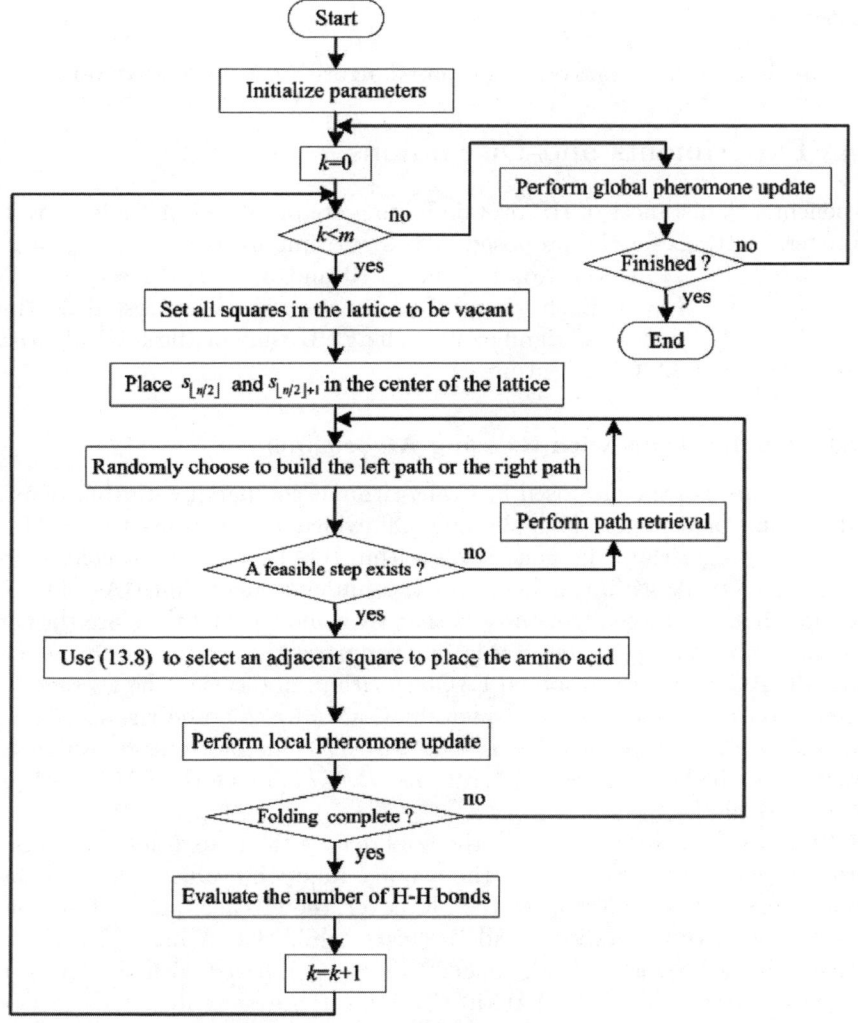

**Fig. 13.7.** Flowchart of the FAC algorithm

The implementation of the FAC algorithm can be realized as follows:

Step 1: Read in the protein sequence and initialize the parameters.
Step 2: Place all ants in the left start square and the right start square in the lattice.
Step 3: All ants construct feasible folding conformations to the input protein sequence. The local pheromone update is performed after every movement of ants.
Step 4: Evaluate the constructed folding paths and select the best ant in an iteration.
Step 5: Perform a global pheromone update.
Step 6: If the terminate condition is not met, go to Step 2; else terminate the algorithm.

A more detailed flowchart of the proposed algorithm is illustrated in Fig. 13.7.

## 13.4 Experiments and Discussions

The benchmark instances of HP protein folding are tabulated in Table 13.1. The parameters' settings for the proposed FAC algorithm are $\tau_0 = 1/3$, $\tau_{min} = 0.05$ and $\rho = 0.9$. For sequences No. 1–7, $m = 10$ and $\beta = 2$. For sequence No. 8, $m = 100$ and $\beta = 3$. Each group of parameters has been tested 30 times independently for statistical significance. The CPU time of the FAC algorithm was recorded on a 2.8 GHz Pentium IV PC.

### 13.4.1 Comparisons with Existing Algorithms

The performance of the proposed FAC algorithm is compared with that of existing algorithms presented in [36], [38] and [43], which are the conventional Monte Carlo (EMC) algorithm, the genetic algorithm (GA) [36], the ant colony optimization (ACO) algorithm in [38], and the immune algorithm (IA) [43]. The reason for choosing these algorithms is that their models and tests are the same as the ones used in this chapter. Table 13.2 compares the average performance of the IA, the ACO and the proposed FAC algorithm, in terms of the average time required ($AvgT$), the average energy evaluations ($A.E.E$), and the success rate (%ok). Table 13.3 compares the best time ($BestT$), the best energy evaluations ($B.E.E$), and the best number of iterations ($B.N.I$) among the FAC, the EMC, the GA, and the IA.

In Table 13.2, the mean values in the bold denote the best results of the three algorithms. Except for sequence 1, the average function evaluations of the FAC are much smaller than those of the IA. Moreover, the FAC has successfully found the best protein conformation in all the tests, while the IA has only managed to solve sequence 8 with a 56.67% success rate. Compared with the ACO, the average execution time of the FAC in obtaining the best protein for short protein sequences is not significantly longer, but it takes a shorter time for longer sequences such as Nos. 7 and 8.

**Table 13.2.** Comparison of the average performance in solving the 2D-HP problems

| No. | $l$ | $E^*$ | FAC | | | IA | | ACO | |
|-----|-----|-------|-----|-----|-----|-----|-----|-----|-----|
| | | | $AvgT$(sec.) | $A.E.E$ | %ok | $A.E.E$ | %ok | $AvgT$(sec.) | %ok |
| 1 | 18 | -9 | 3.17703 | 115384 | **100** | 69210 | **100** | – | – |
| 2 | 18 | -8 | 0.0967667 | **3149** | **100** | 41724 | **100** | – | – |
| 3 | 20 | -10 | 0.264167 | **8107** | **100** | 18086 | **100** | – | – |
| 4 | 20 | -9 | **0.103667** | **2981** | **100** | 23710 | **100** | <1 sec. | **100** |
| 5 | 24 | -9 | 1.28330 | **32159** | **100** | 69817 | **100** | <1 sec. | **100** |
| 6 | 25 | -8 | 3.90027 | **93883** | **100** | 269514 | **100** | <1 sec. | **100** |
| 7 | 36 | -14 | **1.25527** | 18683 | **100** | 2032504 | **100** | 4 sec. | **100** |
| 8 | 48 | -23 | **28.922** | 331103 | 100 | 6403985 | 56.67 | 1 min. | **100** |

–The corresponding values are unavailable in the reference [38].

**Table 13.3.** Comparisons of the best performances in solving the 2D-HP problems

| No. | $l$ | $E^*$ | FAC | | | EMC | GA | IA |
|-----|-----|-------|-----|-----|-----|-----|-----|-----|
| | | | $BestT$(sec.) | $B.E.E$ | $B.N.I$ | $B.E.E$ | $B.E.E$ | $B.E.E$ |
| 4 | 20 | -9 | 0.015 | **169** | 17 | 9374 | 30492 | 1925 |
| 5 | 24 | -9 | 0.078 | **1703** | 171 | 6929 | 30491 | 2479 |
| 6 | 25 | -8 | 0.234 | 5463 | 547 | 7202 | 20400 | **4212** |
| 7 | 36 | -14 | 0.031 | **234** | 24 | 12447 | 301339 | 43416 |
| 8 | 48 | -23 | 0.797 | **9102** | 92 | 165791 | 126547 | 37269 |

In Table 13.3, among the best values of all algorithms, the FAC is seen much faster than the EMC and the GA in solving the sequences listed. Only are the best energy evaluations to sequence 6 slightly larger than that of the IA. It can be seen that the FAC algorithm developed in this chapter can solve the given protein folding problems in a very shortest period of time.

### 13.4.2   Analysis on Different Parameter Values

The influence of parameters in the FAC algorithm is also tested in order to assess the best group of values of the parameters, including the number of ants $m$, the heuristic reinforcement value $\beta$, and the global pheromone evaporation rate $\rho$. Fig. 13.8 shows the trends of different parameter values for sequences No. 1 to 7.

### 1) The heuristic reinforcement value $\beta$

Fix the values of $m$ and $\rho$. When $\beta$ increases, the time needed to obtain solutions becomes shorter to sequences 1 to 5. Note that sequence 6 is distinctive in the sequences and it achieves the best result when $\beta = 1$.

### 2) The pheromone evaporation rate $\rho$

If the pheromone evaporation rate $\rho$ is about 0.9, the performance of the FAC is high in most test cases. Overall, the influence of $\rho$ is not so significant as $\beta$.

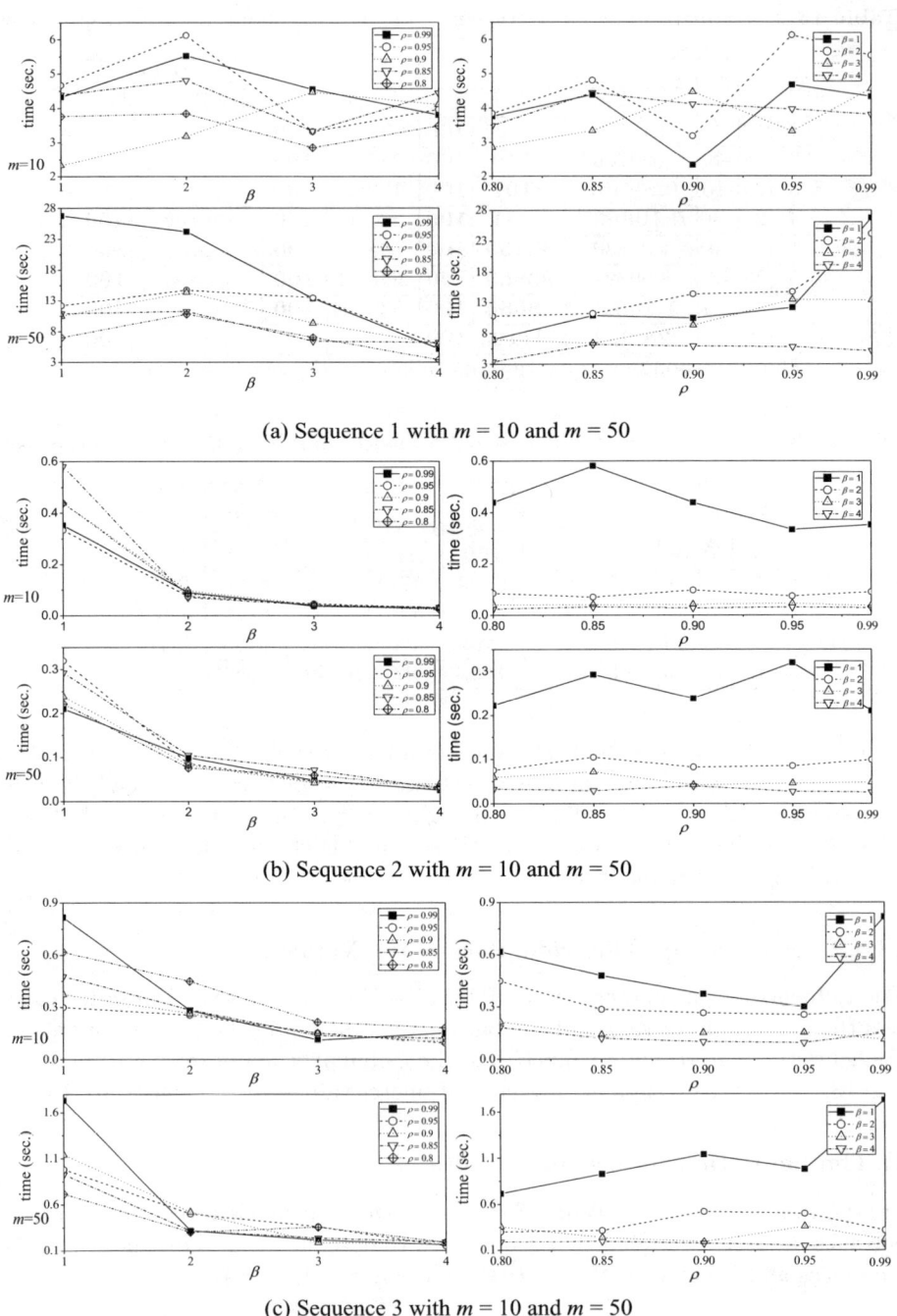

(a) Sequence 1 with $m = 10$ and $m = 50$

(b) Sequence 2 with $m = 10$ and $m = 50$

(c) Sequence 3 with $m = 10$ and $m = 50$

**Fig. 13.8.** Analysis of the FAC algorithm with various parameter values

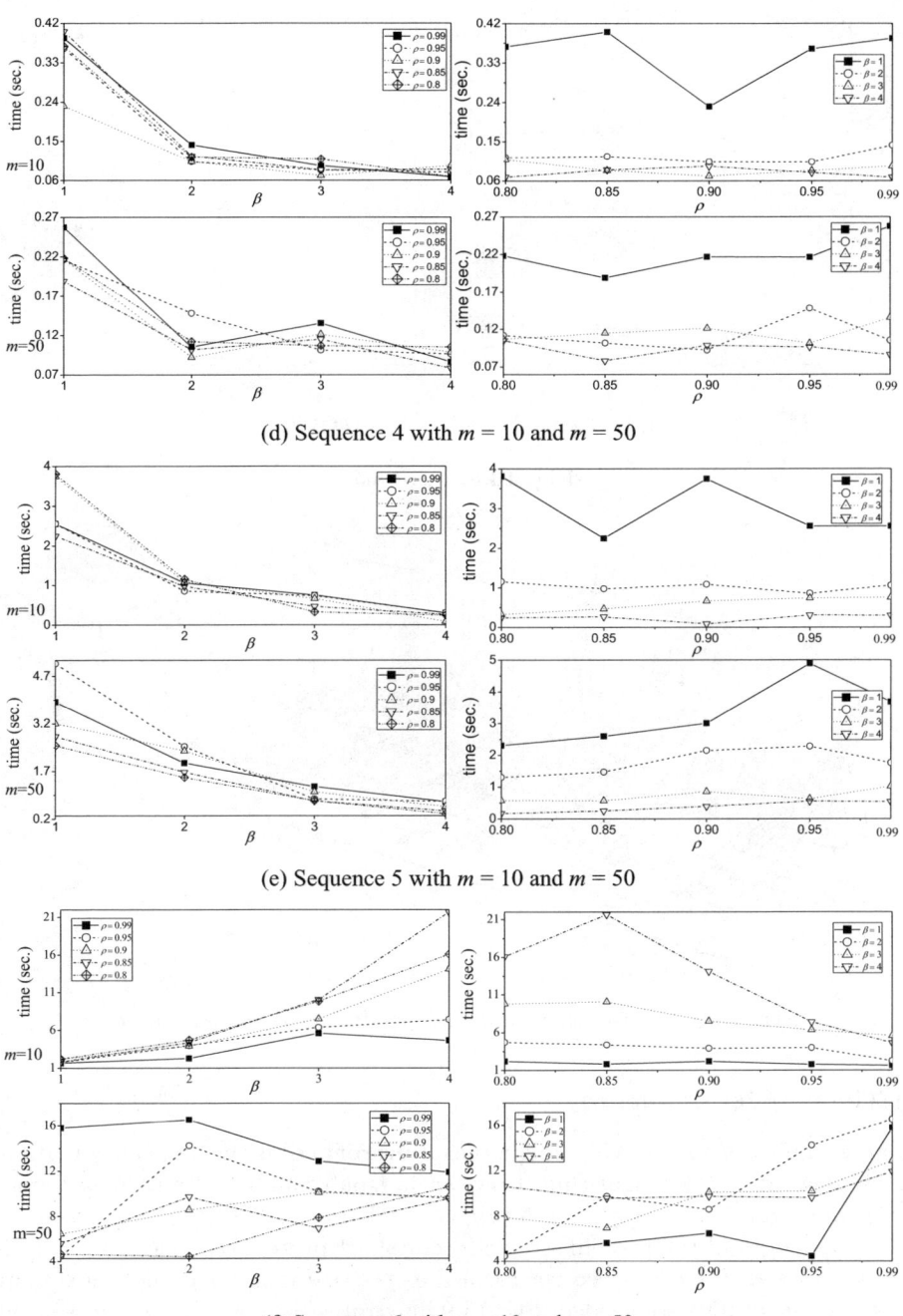

(d) Sequence 4 with $m = 10$ and $m = 50$

(e) Sequence 5 with $m = 10$ and $m = 50$

(f) Sequence 6 with $m = 10$ and $m = 50$

**Fig. 13.8.** (*continue*)

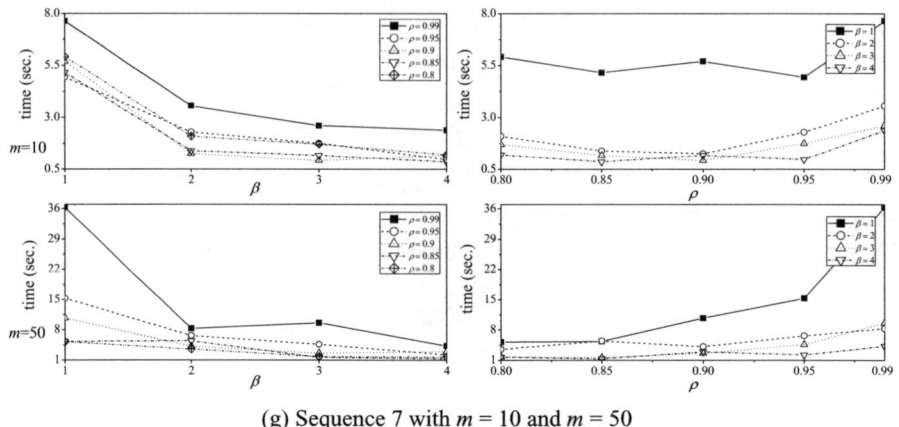

(g) Sequence 7 with $m = 10$ and $m = 50$

**Fig. 13.8.** (*continue*)

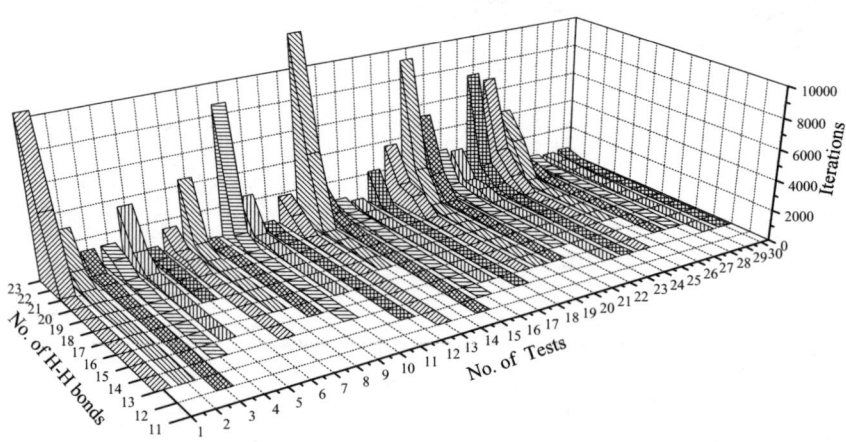

**Fig. 13.9.** Convergence in 30 independent tests to Sequence 8

## 3) The number of ants $m$

A large number of ants provide a higher insurance of finding the best conformation, but it slows down the algorithm. However, a small number of ants may induce early convergence to sub-optima. A proper number of ants is generally dependent upon the length of the protein sequence. For short protein sequences, $m = 10$ is enough. However, for long sequences such as the one with 48 amino acids, more ants (e.g., $m = 100$) are needed. Fig. 13.9 illustrates the convergent states in the 30 independent tests of sequence 8. Conformations with 12 or 13 H-H bonds are always found in the first iteration. As the optimization continues, it takes more time to improve. The fastest search for the optimum folding of sequence 8 in the 30 tests was 92 iterations, while the worst one needed more than 10,000 iterations.

**Table 13.4.** Comparisons on whether using heuristic information to polar amino acids

| No. | $l$ | $E^*$ | FAC (use) | | | FAC (not use) | | |
|-----|-----|-------|-----------|-----|-----|---------------|-----|-----|
|     |     |       | $AvgT$(sec.) | A.E.E | %ok | $AvgT$(sec.) | A.E.E | %ok |
| 1 | 18 | -9 | **3.17703** | **115384** | **100** | 7.69107 | 272580 | **100** |
| 2 | 18 | -8 | **0.0967667** | **3149** | **100** | 0.181733 | 5903 | **100** |
| 3 | 20 | -10 | **0.264167** | **8107** | **100** | 0.3943 | 11820 | **100** |
| 4 | 20 | -9 | **0.103667** | **2981** | **100** | 0.116833 | 3271 | **100** |
| 5 | 24 | -9 | **1.2833** | **32159** | **100** | 1.5146 | 36870 | **100** |
| 6 | 25 | -8 | **3.90027** | **93883** | **100** | 4.14727 | 95673 | **100** |
| 7 | 36 | -14 | **1.25527** | **18683** | **100** | 2.3422 | 33789 | **100** |
| 8 | 48 | -23 | **28.922** | **331103** | **100** | 334.755 | 3756947 | **100** |

### 13.4.3   Analysis of Heuristic Information to Polar Amino Acids

IIn the proposed FAC algorithm, there is heuristic information for folding polar amino acids, which is different from that used in the ACO algorithm [38]. The performance of the FAC algorithm is compared with or without heuristic information to polar amino acids. The results are tabulated in Table 13.4. With the same parameter settings, the algorithm without heuristic information to polar amino acids is slower than the one with the heuristic information in all test cases. The results demonstrate that the heuristic information proposed in this chapter is effective.

## 13.5   Conclusions

This chapter has presented a flexible ant colony algorithm for the protein folding problem. This FAC algorithm is based on the 2-dimensional square lattice hydrophobic-polar model, which is a highly abstract model for protein folding structures. Ants in the FAC algorithm start from the middle of the lattice and construct protein folding from the middle of the protein sequence. Pheromones are released to the directed arcs connecting adjacent squares in the lattice. Local pheromone update as well as global pheromone update mechanisms are also implemented. By using effective heuristic and pheromone method for selection, the proposed FAC algorithm can solve the PFP fast as shown by the test cases. Comparison with some well-known PFPalgorithms has highlighted superior performance of the proposed FAC algorithm.

## Acknowledgment

This work was supported by the Natioinal Science Foundation (NSF) of China (60573066) and the NSF of Guangdong (5003346), China and SRF for ROCS, SEM, China.

# References

1. Chandru, V., Dattasharma, A., Kumar, V.S.A.: The algorithms of folding proteins on lattices. Discrete Applied Mathematics 127, 145–161 (2003)
2. Mount, D.W.: Bioinformatics: sequence and genome analysis. Cold Spring Harbor Laboratory Press (2001)
3. Zaki, M.J., Jin, S., Bystroff, C.: Mining residue contacts in proteins using local structure predictions. IEEE Trans on Systems Man and Cybernetics - Part B 33(5), 789–810 (2003)
4. Chen, S.S.: A localized protein-folding problem. International Journal of Intelligent Systems 16, 449–457 (2001)
5. RCSB (Research Collaboratory for Structural Bioinformatics). Protein Data Bank, http://www.rcsb.org/pdb/
6. Anfinsen, C.B., Haber, E., Sela, M., White, F.H.: The kinetics of formation of native ribonuclease during oxidation of the reduced polypeptide chain. Proc Natl Acad Sci USA 47, 1309–1314 (1961)
7. Anfinsen, C.B.: Principles that govern the folding of proteins chains. Science 181, 223–230 (1973)
8. Woldawer, A., Miller, M., Jaskolski, M., Sathyanarayana, B.K., Baldwin, E., Weber, I.T., Selk, L.M., Clawson, L., Schneider, J., Kent, S.B.: Conserved folding in retroviral proteases: crystal structure of a synthetic HIV-1 protease. Science 245, 616–621 (1989)
9. Clore, G.M., Gronenborn, A.M.: Comparison of the solution nuclear magnetic resonance and X-ray crystal structures of human recombinant interleukin-1$\beta$. J. Mol. Biol. 221, 47–53 (1991)
10. Unger, R., Moult, J.: Finding the lowest free energy conformation of a protein is an NP-hard problem: proof and implications. Bulletin of Mathematical Biology 55, 1183–1198 (1993)
11. Govindarajan, S., Goldstein, R.A.: On the thermodynamic hypothesis of protein folding. Proc Natl Acad Sci USA 95, 5545–5549 (1998)
12. Ngo, J.T., Marks, J.: Computational complexity of a problem in molecular-structure prediction. Protein Engineering 5(4), 313–321 (1992)
13. Fraenkel, A.S.: Complexity of protein folding. Bulletin of Mathematical Biology 55(6), 1199–1210 (1993)
14. Guo, Y.Z., Feng, E.M.: Exploration of two-dimensional hydrophobic-polar lattice model by combining local search with elastic net algorithm. Journal of Chemical Physics 125, 154102 (2006)
15. Lesh, N., Mitzenmacher, M., Whitesides, S.: A complete and effective move set for simplified protein folding. In: RECOMB 2003 Berlin Germany, pp. 188–195 (2003)
16. Dill, K.A., Bromberg, S., Yue, K., Fiebig, K.M., Yee, D.P., Thomas, P.D., Chan, H.S.: Principles of protein folding - a perspective from simple exact models. Protein Science 4, 561–602 (1995)
17. Hart, W.E., Istrail, S.: Fast protein folding in the hydrophobic-hydrophilic model within three-eights of optimal. In: STOC 1995, USA, pp. 157–168 (1995)
18. Chan, H.S., Dill, K.A.: Transition states and folding dynamics of proteins and heteropolymers. J. Chem. Phys. 100(12), 9238–9257 (1994)
19. Hoque, M.T., Chetty, M., Dooley, L.S.: A hybrid genetic algorithm for 2D FCC hydrophobic-hydrophilic lattice model to predict protein foldings. In: Sattar, A., Kang, B.-h. (eds.) AI 2006. LNCS (LNAI), vol. 4304, pp. 867–876. Springer, Heidelberg (2006)

20. Agarwala, R., Batzoglon, S., Dančík, V., Decatur, S.E., Farach, M., Hannenhalli, S., Skiena, S.: Local rules for protein folding on a triangular lattice and generalized hydrophobicity in the HP model. In: Proceedings of the eighth annual ACM-SIAM symposium on discrete algorithms, USA, pp. 390–399 (1997)

21. Decatur, S.E.: Protein folding in the generalized hydrophobic-polar model on the triangular lattice. MIT LCS Technical memo: MIT-LCS-TM-559 (1996)

22. Liu, J., Wang, L.H., He, L.L., Shi, F.: Analysis of toy model for protein folding based on particle swarm optimization algorithm. In: Wang, L., Chen, K., S. Ong, Y. (eds.) ICNC 2005. LNCS, vol. 3612, pp. 636–645. Springer, Heidelberg (2005)

23. Stillinger, F.H., Head-Gordon, T., Hirshfeld, C.L.: Toy model for protein folding. Phys. Rev. E 48(2), 1469–1477 (1993)

24. Liang, F., Wong, W.H.: Evolutionary Monte Carlo for protein folding simulations. J. Chem. Phys. 115(7), 3374–3380 (2001)

25. Backofen, R., Will, S., Clote, P.: Algorithmic approach to quantifying the hydrophobic force contribution in protein folding. In: PSB 2000, pp. 92–103 (2000)

26. Bastolla, U., Frauenkron, H., Gerstner, E., Grassberger, P., Nadler, W.: Testing a new Monte Carlo algorithm for protein folding. Proteins: Structure Function and Genetics 32, 52–66 (1998)

27. Hsu, H.P., Mehra, V., Nadler, W., Grassberger, P.: Growth algorithm for lattice heteropolymers at low temperatures. J. Chem. Phys. 118, 444 (2003)

28. Ramakrishnan, R., Ramachandran, B., Pekny, J.F.: A dynamic Monte Carlo algorithm for exploration of dense conformational spaces in heteropolymers. J. Chem. Phys. 106(6), 2418–2425 (1997)

29. Tantar, A.A., Melab, N., Talbi, E.G., Parent, B., Horvath, D.: A parallel hybrid genetic algorithm for protein structure prediction on the computational grid. Future Generation Computer Systems 23, 398–409 (2007)

30. Arunachalam, J., Lanagasabai, V., Gautham, N.: Protein structure prediction using mutually orthogonal Latin squares and a genetic algorithm. Biochemical and Biophysical Research Communications 342, 424–433 (2006)

31. Hoque, M.T., Chetty, M., Dooley, L.S.: A new guided genetic algorithm for 2D hydrophobic-hydrophilic model to predict protein folding. CEC 2005 1, 259–266 (2005)

32. Bui, T.N., Sundarraj, G.: An efficient genetic algorithm for predicting protein tertiary structures in the 2D HP Model. In: Bui, T.N., Sundarraj, G. (eds.) GECCO 2005, USA, pp. 385–392 (2005)

33. Custódio, F.L., Barbosa, H.J.C., Dardenne, L.E.: Investigation of the three-dimensional lattice HP protein folding model using a genetic algorithm. Genetics and Molecular Biology 27(4), 611–615 (2004)

34. Krasnogor, N., Hart, W.E., Smith, J., Pelta, D.A.: Protein structure prediction with evolutionary algorithms. In: Proceedings of the Genetic and Evolutionary Computation Conference, pp. 1596–1601 (1999)

35. Pedersen, J.T., Moult, J.: Genetic algorithms for protein structure prediction. Current Opinion in Structural Biology 6(2), 227–231 (1996)

36. Unger, R., Moult, J.: Genetic algorithms for protein folding simulations. J. Mol. Biol. 231(1), 75–81 (1993)

37. Daeyaert, F., Jonge, M.D., Koymans, L., Vinkers, M.: An ant algorithm for the conformational analysis of flexible molecules. Journal of Computational Chemistry 28(5), 890–898 (2007)

38. Shmygelska, A., Hoos, H.H.: An ant colony optimisation algorithm for the 2D and 3D hydrophobic polar protein folding problem. BMC Bioinformatics 6, 30 (2005)

39. Chu, D., Till, M., Zomaya, A.: Parallel ant colony optimization for 3D protein structure prediction using the HP lattice model. In: IPDPS 2005, pp. 1–7 (2005)
40. Shmygelska, A., Hoos, H.H.: An improved ant colony optimisation algorithm for the 2D HP protein folding problem. In: Xiang, Y., Chaib-draa, B. (eds.) Canadian AI 2003. LNCS (LNAI), vol. 2671, pp. 400–417. Springer, Heidelberg (2003)
41. Shmygelska, A., Aguirre-Hernandez, R., Hoos, H.H.: An ant colony optimisation algorithm for the 2D HP protein folding problem. In: Dorigo, M., Di Caro, G.A., Sampels, M. (eds.) Ant Algorithms 2002. LNCS, vol. 2463, pp. 40–52. Springer, Heidelberg (2002)
42. Cutello, V., Nicosia, G., Pavone, M., Timmis, J.: An immune algorithm for protein structure prediction on lattice models. IEEE Trans. on Evolutionary Computation 11(1), 101–117 (2007)
43. Cutello, V., Nicosia, G., Pavone, M.: An immune algorithm with hyper-macromutations for the Dill's 2D hydrophobic-hydrophilic model. CEC 2004 1, 1074–1080 (2004)
44. Crescenzi, P., Goldman, D., Papadimitriou, C., Piccolboni, A., Yannakakis, M.: On the complexity of protein folding. J. Comp. Biol. 5(3), 423–466 (1998)
45. Dorigo, M., Stützle, T.: Ant Colony Optimization. MIT Press, Cambridge (2004)
46. Dorigo, M., Gambardella, L.M.: Ant Colony System: A Cooperative Learning Approach to the Traveling Salesman Problem. IEEE Trans. on Evol. Comput. 1(1), 53–66 (1997)
47. Dorigo, M., Maniezzo, V., Colorni, A.: Ant system: optimization by a colony of cooperating agents. IEEE Trans on Systems Man and Cybernetics–Part B 26(1), 29–41 (1996)
48. Gambardella, L.M., Taillard, É.D., Agazzi, G.: MACS-VRPTW: A Multiple Ant Colony System for Vehicle Routing Problems with Time Windows. In: Corne, D., Dorigo, M., Glover, F. (eds.) New Ideas in Optimization, pp. 63–76. McGraw-Hill, London (1999)
49. Zhang, J., Hu, X.M., Tan, X., Zhong, J.H., Huang, Q.: Implementation of an Ant Colony Optimization Technique for Job Shop Scheduling Problem. Transactions of the Institute of Measurement and Control 28(1), 1–16 (2006)
50. Zecchin, A.C., Simpson, A.R., Maier, H.R., Nixon, J.B.: Parametric Study for an Ant Algorithm Applied to Water Distribution System Optimization. IEEE Trans. Evol. Comput. 9, 175–191 (2005)

# Considering Stem-Loops as Sequence Signals for Finding Ribosomal RNA Genes

Kirt M. Noël[1] and Kay C. Wiese[2]

[1] McKesson Medical Imaging, Richmond, B.C., Canada, V6X 3G5
   kirtnoel@gmail.com
[2] School of Computing Science, Simon Fraser University,
   Surrey, B.C., Canada, V3T 0A3,
   Tel.: 778-782-7436; Fax: 778-782-8116
   wiese@cs.sfu.ca

**Summary.** Several factors make stem-loops an attractive sequence signal for a structural RNA gene-finder. Structural RNAs are virtually obligated to form stem-loops on their way to forming stable structures. Also, stem-loops can be identified along a sequence of length $n$ in $O(n)$ time. We postulate that stem-loops found in structural RNA genes may tend to be longer than those found in their genomic counterparts - coding sequences and noncoding DNA. We also postulate that stem-loops may occur in higher frequency in the structural RNA regions.

Methods: To examine these possibilities, rRNAs were selected as a test bed. An algorithm was developed to identify stem-loops along a genomic sequence which are similar to those found in rRNA secondary structures. This algorithm scanned the genomes in our training set to establish average metric values observed in rRNA genes. These values were subsequently used in an effort to identify rRNA genes in genomes outside of the training set.

Results: The values for the stem-loop metrics we tested are sensitive to G+C content. Two of the metrics reported here are able to identify rRNA genes when there is a marked difference in G+C content between rRNAs and their genomic counterparts. Another metric has demonstrated an ability to roughly target rRNA genes when there is a negligible difference in G+C content levels.

Conclusions: Our results are encouraging and demonstrate that stem-loops have the potential to act as sequence signals to discover rRNA genes. Our results also suggest that more study into stem-loops is warranted to further improve the performance of our algorithm and to examine the application to a wider population of structural RNA genes.

## 14.1 Background

Noncoding RNA (ncRNA) genes produce transcripts which function directly in RNA form without the need to generate proteins. The pool of noncoding RNAs, and microRNAs (miRNA) in particular, is growing at a brisk pace [1, 2]. A review is provided by Eddy [3]. This paper describes an examination into stem-loops as a sequence signal for identifying a subset of ncRNA genes - structural RNAs.

T.G. Smolinski et al. (Eds.): Comp. Intel. in Biomed. & Bioinform., SCI 151, pp. 337–357, 2008.
springerlink.com                                           © Springer-Verlag Berlin Heidelberg 2008

Inherent properties allow RNA sequences to form biological machinery with exclusive capabilities [2, 3, 4]. They form base pairs which allow them to fold into shapes capable of executing specialized catalytic processes [5, 6, 7]. RNAs are known to mimic the structure of nucleic acids to block translation [8, 9]. RNA gene products also interact with proteins to form complex structures [10]. The architecture and mechanism underlying RNA telomerase activity is a testimony to their complex ability [11, 12]. This diverse set of properties leads one to suspect that a wave of unknown structural RNA gene products may await discovery. The term 'structural RNA' explicitly refers to RNAs whose function is related to structure and not simply their primary nucleotide sequence.

To date, an effective and efficient means to computationally find novel structural RNA genes has been elusive [13, 14]. The difficulty primarily results from the lack of sequence conservation in RNA genes. Rather than scan genomes for consensus sequences, researchers have had to resort to other methods.

Genomic segments which code for structural RNAs tend to favor high G+C base compositions [15, 16, 17]. This is related to the structural integrity provided by the 3 hydrogen bonds which couple G-C base pairs. Exploiting differences in base composition between structural RNA genes and their genomic counterparts (i.e. coding sequences (CDS) and non-coding DNA (NC)) is a simple and effective means to locate structural RNA genes in many genomes [18, 19]. Unlike co-variance methods, this approach does not require precise information detailing the secondary structure of the structural RNA gene under pursuit [20, 21, 22, 23]. This is advantageous for pursuing *novel* structural RNA genes. However, the base composition approach has one key weakness. This approach is not effective when there is little disparity in the G+C content levels between the structural RNAs and their genomic counterparts.

Other attempts to develop a structural RNA gene-finder have relied on RNA folding algorithms to calculate the free energy ($\Delta G$) of sequence segments or windows along a given genomic sequence [24, 25, 26, 27]. Structural RNAs have an inherent ability to form stable secondary structures. Therefore, researchers suspected that windows which overlap with structural RNA genes may generate statistically significant $\Delta G$ values, which could act as a sequence signal. The effectiveness of applying $\Delta G$ to this problem has been brought into question by Rivas et al [14]. Part of the problem pertains to the size of the $\Delta G$ windows which have no biological justification or relevance. Instead, the window size is an input parameter necessitated by and optimized for the RNA folding algorithm [28, 29, 30]. The extraneous factors and complications brought about by selecting an optimized window size have not been addressed. Notably, RNA folding algorithms are burdened with a $O(n^3)$ computational complexity where $n$ is the length of the sequence to be folded. Hence, the $\Delta G$ sequence signal has inherent constraints which limit its applicability to the RNA gene-finding problem.

Herein we propose and examine an alternative sequence signal to pursue structural RNAs: Stem-loops. There are a number of factors which make stem-loops an intriguing sequence signal candidate worth studying. Pairing rules and sequence directionality (5'→3') virtually obligate stable RNA structures to form stem-loops. Therefore, it seems that stem-loops could be considered to RNA structures what $\alpha$-helices and $\beta$-sheets are to proteins. Another reason to focus on stem-loops pertains to their conduciveness to being searched. A set of search parameters can be tailored to pursue stem-loops which are characteristically found in structural RNAs. This element of control is lost when calculating the $\Delta G$ of arbitrarily sized windows. Lastly, scanning for stem-loops along a given sequence according to our search parameters can be accomplished in $O(n)$ time and space complexity where $n$ is the length of the sequence (see Sect. 14.2 for details).

The phrase stem-loop metric refers to a quantified feature describing a given stem-loop. In this work, stem-loop metrics are studied which measure the length of a given stem-loop and the distance between neighboring stem-loops. With the use of annotated genomic sequences, we can calculate and compare the average stem-loop metric values across genomic domains - coding sequences (CDS), non-coding DNA (NC), ribosomal RNA (rRNA), and transfer RNA (tRNA). The aim is to study the potential for stem-loop metrics to provide a statistical signal which sufficiently differentiates structural RNAs from their genomic counterparts - CDS and NC - across the G+C content spectrum.

Our training set includes 58 microbial complete genomic sequences in order to cover a wide range of G+C content levels. Details on the sequences and accession numbers are available from [31]. The known ncRNAs in most microbes are essentially limited to rRNA and tRNA. Several of the microbes in our test set have other unknown structural RNAs. We chose to focus on rRNAs to establish consistency in comparing the performance of stem-loop metrics between the various bacterial genomes with different *global* G+C content levels. This decision does create a bias. However, we suspect that the reported results reflect the potential applicability and the possible pitfalls of using this approach on other structural RNA gene families.

## 14.2 Methods

### 14.2.1 The Stem-Loop

A stem-loop is comprised of 2 elements - a hairpin loop and a stack or stem of base pairs which close or stabilize the hairpin loop (Figure 14.1A). Importantly, the base pairs in the stem are commonly interrupted by mismatched or unpaired nucleotides.

Below, stem-loops are defined with the help of logical expressions. Suppose an RNA sequence consists of $n$ nucleotides. An indexed sequence can be denoted from 5' to 3' as $(0, 1, 2, 3, e, i, k, p, t, n)$ where, $0 < e < i < k < p < t < n$.

Consider a stem-loop where the base pair on the end of the stem nearest the hairpin loop includes nucleotides $i$ and $p$, denoted $(i, p)$ (Figure 14.1B). The base pair, $(e, t)$, is the furthest from the hairpin loop; it forms the outer boundary of the stem-loop structure. The unpaired nucleotides in the hairpin loop are denoted $k$; hence, $\forall k \quad i < k < p$.

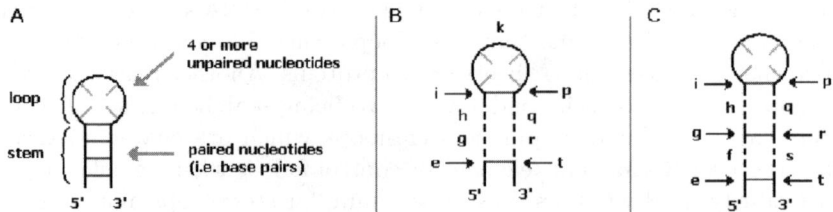

**Fig. 14.1.** (A) The stem-loop is comprised of 2 components: a hairpin loop and a stack or helix of base-pairs. (B) Annotated stem-loop. The (i,p) base pair lies nearest the hairpin loop. The (e,t) base pair marks the outer boundary of the stem-loop. (C) A pseudo-knot cannot be entirely comprised of nucleotides which lie between the (i,p) and (e,t) base pairs. See text for a detailed description.

For any stem-loop, the nucleotides which lie between $e$ and $i$ cannot base pair with one another nor can the nucleotides between $p$ and $t$ pair with one another (Figure 14.1B). This rule can be expressed as follows. Suppose a stem-loop is comprised of the following nucleotide sequence: $(e, g, h, i, p, q, r, t)$. If the hairpin is defined by $(i, p)$ and the stem-loop is bound by $(e, t)$ then $\forall g \forall h \; \neg(g, h) \wedge \forall q \forall r \; \neg(q, r)$.

Lastly, a pseudo-knot cannot be comprised entirely of nucleotides between base pairs $(i, p)$ and $(e, t)$. Suppose a stem-loop is comprised of the following nucleotide sequence: $(e, f, g, h, i, p, q, r, s, t)$. If $(e, t) \wedge (g, r) \wedge (i, p)$: then $\forall f \forall q \; \neg(f, q) \wedge \forall h \forall s \; \neg(h, s)$. See Figure 14.1C.

### 14.2.2   Search Parameters

This research is based on an algorithm designed to find stem-loops along a genomic sequence. The intent is to target stem-loops typically observed in RNA secondary structures. A cursory analysis of several rRNA secondary structures was performed to establish trends which could be used as search parameters. For instance, the number of nucleotides which typically comprise a hairpin loop falls within a certain range. Similarly, the minimum number of base pairs necessary to form a stem-loop can be deduced. In this manner, a set of search parameters for our algorithm was devised (Table 14.1). The secondary structures were obtained through online databases [32, 33].

### 14.2.3   Stem-Loop Finding Algorithm

The algorithm begins its search for stem-loops by finding the base pair nearest the hairpin loop (i.e. the $(i, p)$ base pair in Figure 14.1B). It then attempts to elongate the stem by finding adjacent base pairs as long as the search parameters are met. Suppose, the initial upstream and downstream nucleotides are denoted $i$ and $p$, respectively. The upstream nucleotide, $i$, acts as the anchor which moves the search through the sequence from 5' to 3'. The downstream nucleotide, $p$, is initialized to $p = i + (x_{min} + 1)$ where $x_{min}$ is the minimum

**Table 14.1.** This table presents the search parameters which guide the algorithm as it scans an input sequence for stem-loops. The hairpin loop closure refers to the base pairs which lie immediately below the hairpin loop. This parameter indicates that at least 3 adjacent base pairs must "close" the hairpin loop. Importantly, the search parameters also require a stem-loop consisting of at least 4 base pairs overall. Suppose only 3 adjacent base pairs close a given hairpin loop. The parameters dictate that more base pairs must exist after a bulge or internal loop to bring the total number of base pairs to 4 or more.

| Structural Element | Min./Max. Parameter |
|---|---|
| $x$ nucleotides in hairpin loop | $3 \leq x \leq 15$ |
| Max. bulge | 6 nucleotides |
| Max. internal loop | 6 nucleotides |
| Min. bulge or internal loop closure | 3 base pairs |
| Min. GC Base Pair Content | 30% |
| Max. GU Base Pair Content | 34% |
| Min. hairpin loop closure | 3 base pairs |
| Overall min. number of base pairs | 4 |

hairpin loop size. The nucleotides which separate $i$ and $p$ make up the hairpin loop (Figure 14.1A). When a base pair is found, the algorithm decrements $i$ and concurrently increments $p$. Then, the algorithm checks to see whether these nucleotides form a base pair. If they form a base pair, then the indices are decremented/incremented again. Importantly, the elongation process of the stem tolerates a certain number of unpaired or mismatched nucleotides in the stem (Table 14.1). If a stem-loop starting at index $i$ is found, it is stored and $i$ is incremented to $i + 1$. Suppose, a stem-loop is *not* found starting at index $i$ and $p = i + x_{min}$. The algorithm increases the size of the hairpin loop and then attempts to elongate a stem starting with the nucleotides at indices $i$ and $p = i + (x_{min} + 2)$. It will continue this until it reaches $x_{max}$ - the maximum number of nucleotides permitted in the hairpin loop. If $x_{max}$ is surpassed, then $i$ is incremented and $p = i + (x_{min} + 1)$; this moves the search along the sequence. At each iteration, the algorithm finds the stem-loop with the smallest hairpin loop that meets the search parameters. Some additional simplistic rules are applied to deal with the boundary regions close to the 5' and 3' ends.

### 14.2.4  Computational Complexity

As the algorithm scans along a genomic sequence of length $n$, it constructs the longest possible stem-loop within the search parameters. While it is possible to construct pathological artificial sequences where this could lead to a scan time of $O(n^2)$ with a maximum stem-loop size of $\frac{1}{2}n$, such sequences do not exist in nature. For real sequences the length of the largest possible stem-loop is typically negligible compared to the length of the sequence. Also, our search parameters further restrict the size of the stems and loops. For all practical purposes this length can be considered constant and depends more on the G+C content of the sequence than its length (for large enough sequences). In addition, at each position $i$, once a stem-loop is found $i$ is incremented by 1. This limits the number of iterations to $O(n)$. Considering the constant time to find a stem-loop at position $i$, we can assume that the scan of a sequence of length $n$ takes linear or $O(n)$ time. This was also confirmed in practical experiments (not presented here) where the doubling of the length of an input sequence led to roughly doubling the CPU time to scan the sequence assuming that the G+C content of both sequences was the same.

### 14.2.5   Resource for Genomic Sequences

A thorough analysis to examine the signal capacity of stem-loop metrics requires they be tested on sequences across the G+C content spectrum. Selecting a few A+T rich sequences could misrepresent the ability of stem-loop metrics to distinguish structural RNA domains from their genomic counterparts. Therefore, 58 bacterial sequences were arbitrarily chosen. Their G+C base compositions range from 25% to 68%. The genomic sequences were obtained from the NCBI website [34].

### 14.2.6   Stem-Loop Metrics

In this body of work, 4 stem-loop metrics are examined. They relate to the length of the stem-loops and to their frequency. The base pairs metric, denoted *bps*, gauges the length of stem-loops by simply totaling the number of base pairs present in a stem-loop. Two metrics measure the average distance from a given stem-loop to its nearest upstream and downstream neighbours. Note, these neighbouring stem-loops are not permitted to overlap (i.e. they cannot share

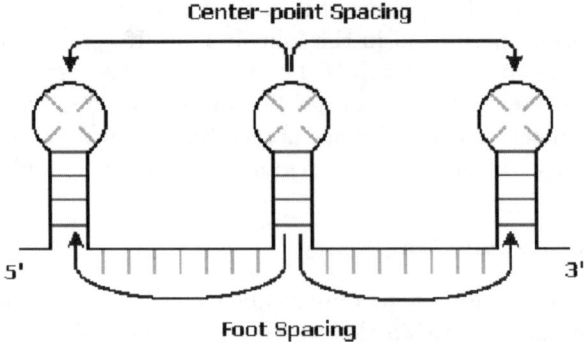

**Fig. 14.2.** The center-point spacing (*cSpacing*) metric is measured by averaging the distance between neighbouring hairpin loops. The Foot Spacing (*f Spacing*) metric is measured by averaging the distance between the paired nucleotides which are most distal to the hairpin loop.

nucleotides between them) (Figure 14.2). In the event that the immediate up-stream/downstream neighbour overlaps with the center stem-loop, the algorithm jumps to the next upstream/downstream stem-loop until a non-overlapping neighbour is found. Two methods were used to measure the spacing between a stem-loop and its 2 opposing neighbours (Figure 14.2). Center-point spacing, denoted *cSpacing*, measures the spacing between stem-loops as the distance (in nucleotides) between the center of their hairpin loops. The foot spacing metric, denoted *fSpacing*, is measured from the nucleotides which make-up the "foot" of the stem - the base pair most distal to the hairpin loop (Figure 14.2). Lastly, the *cSpacing* and *bps* metrics were combined by multiplying them to create the (*cSpacing* × *bps*) metric.

### 14.2.7   How Average Metric Values were Tabulated

Annotated sequences (from the NCBI website) are used to create a one dimensional map of the sequence. This map is equal in length to the input sequence and it labels each nucleotide in the sequence into 1 of 4 broad categories - CDS, NC, rRNA, tRNA. This allows us to map each stem-loop to its genomic category. As a result, the average stem-loop metric for each of the genomic domains can be calculated and compared.

One of our goals is to study how changes to G+C content affect the average stem-loop metric values in each genomic region - CDS, NC, rRNA, and tRNA. This, in turn, provides an indication of how useful stem-loop metrics may be in identifying structural RNAs along a given sequence.

### 14.2.8   Identifying rRNA Genes in Genomes Outside the Training Set

The next experiments apply these average metric values in search of rRNAs along bacterial genomes not included in our training set. Statistical inferences are made using hypothesis tests which rely on $Z$ Scores. This test determines whether a given region along the sequence has a metric value which falls within the 95% confidence interval (C.I.) of the mean rRNA value for the same metric. The mean rRNA metric value is denoted $\mu_{rRNA}$:

$$Z = \frac{\overline{x} - \mu_{rRNA}}{\sqrt{s^2/N}} \qquad (14.1)$$

Here, $\overline{x}$ denotes sample mean, $\mu_{rRNA}$ denotes the mean rRNA value derived from our training set, $N$ denotes sample size, and $s^2$ denotes the standard deviation.

The search for rRNAs considers all the stem-loops identified along the sequence. The algorithm moves through the list of stem-loops from 5' to 3'. In doing so, it calculates the average metric value between groups of 30 to 500 stem-loops. Changes to the sample size can affect the sample's standard deviation. This, in turn, affects the sensitivity of the statistical inference. Hence, the sample size was sometimes modified between various genomes to reduce the number of false positives. When the sample mean falls within the 95% C.I., we assume this suggests the stem-loop at the center of the sample may fall within a region which codes for rRNA.

The accuracy of the predictions is evaluated by comparing the location of the rRNA genes to the statistically inferred "hits". Ideally one would want to compare these results to the $\Delta G$ method. However, the $\Delta G$ approach cannot be practically applied to the sequences in the range of 500,000 to 5,000,000 nucleotides long [25, 26].

### 14.2.9   Hardware

The computer used to run this search algorithm is equipped with an Intel Pentium 4© processor, 1.5 gigabytes of RAM, and a Linux operating system. The algorithm is implemented in C++. Our program requires less than 10 seconds to scan a million nucleotide sequence for stem-loops.

## 14.3   Results

### 14.3.1   G+C Content

It was noted that the base composition method is not effective when the difference in base composition between structural RNAs and their genomic counterparts is negligible. This can be illustrated by plotting the G+C content of the genomic domains against the *global* G+C content. Figure 14.3 depicts the G+C content levels found in the various regions for each of the 58 sequences in our training set. Note, each genome has only one *global* G+C content value. Therefore, the local G+C values for the CDS, NC, rRNA, and tRNA domains within a given genome are all aligned vertically. The opposing plots intersect when their respective G+C content levels are equivalent. The proximity of the opposing plots when the *global* G+C content is roughly 50% illustrates why this method is at times ineffective at distinguishing structural RNAs from their genomic counterparts.

What follows are the results obtained with each of the stem-loop metrics and illustrated in a similar manner. It should be evident that a promising metric is one where the values of structural RNAs do not merge with the values of their genomic counterparts.

A Bezier curve has been drawn through the data points strictly to illustrate trends which occur with changing G+C content. This curve is not intended to act as a substitute for interpreting the actual data but rather as an interpretive guide.

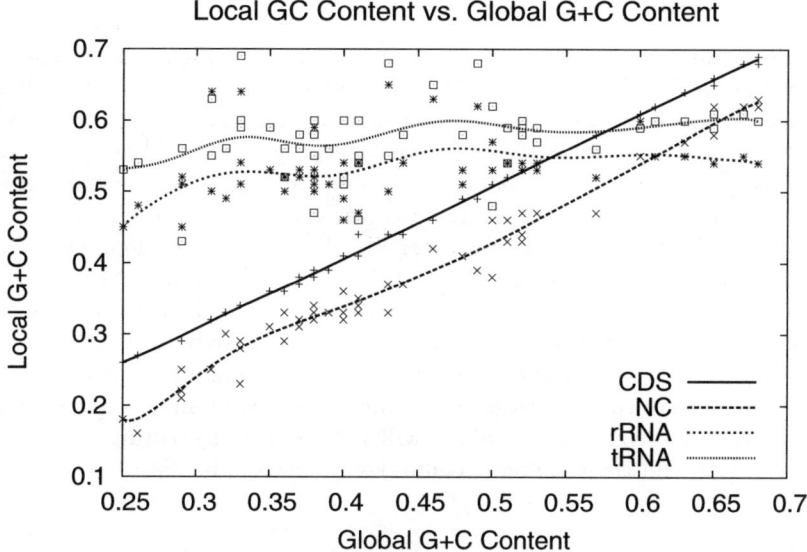

**Fig. 14.3.** Local G+C nucleotide content versus global G+C content (over the entire genome). The G+C content in each of the 58 test genomes is measured in 4 broadly defined genomic domains - CDS, NC, rRNA, and tRNA. The graph illustrates differences in base composition between these regions for each of the genomes in our test set.

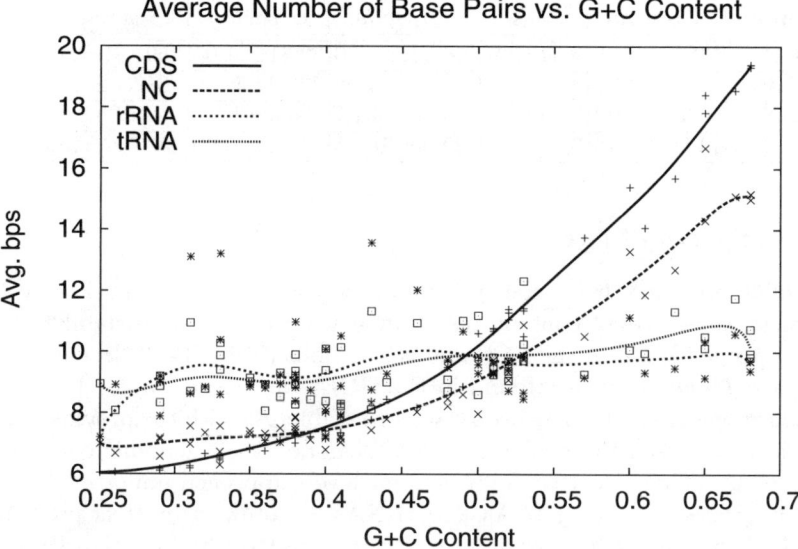

**Fig. 14.4.** Average *bps* values found in CDS, NC, rRNA, and tRNA regions over 58 bacterial genomes with varying *global* G+C content levels

## 14.3.2  *bps* Metric

The results attained with the *bps* metric are depicted in Figure 14.4. The average *bps* values in the CDS and NC regions appear strongly correlated to changes in *global* G+C content. This observation makes sense given that the CDS and NC regions comprise roughly 95% of the bacterial genome. At low *global* G+C content levels, there is a lesser likelihood of finding long stem-loops which meet the minimum GC base pair parameter. Conversely, at higher G+C content levels, the reverse is true - there is a greater likelihood of finding long stem-loops. Earlier, Figure 14.3 and Table 14.2 reveal that the G+C content is more stable in rRNAs and tRNAs than their genomic counterparts. This helps to explain the relatively consistent *bps* values observed in these structural RNAs across the training set. However, the average *bps* values of these structural RNAs and their genomic counterparts merge when their G+C content levels are equivalent. This suggests average *bps* values will not be able to distinguish between structural RNAs and their genomic counterparts across the entire G+C content spectrum.

**Table 14.2.** Summary of data collected on 58 bacterial genomes spanning a wide range of *global* G+C content levels. Note the values for the stem-loops metrics are more stable in rRNAs than their genomic counterparts. This is also true for tRNAs but to a lesser extent. This, presumably, is related to the relatively stable G+C content levels in rRNAs and tRNAs.

| Genomic Domain | G+C Content | Avg. *bps* | Avg. *fSpacing* | Avg. *cSpacing* | Avg. (*cSpacing* × *bps*) |
|---|---|---|---|---|---|
| CDS | 0.45±0.12 | 9.64±3.92 | 34.84±25.30 | 62.81±21.05 | 591.81±231.68 |
| NC | 0.39±0.12 | 8.92±2.63 | 46.33±31.98 | 72.72±28.86 | 612.14±176.68 |
| rRNA | 0.53±0.043 | 9.44±1.27 | 16.39±3.16 | 44.48±1.90 | 437.05±70.17 |
| tRNA | 0.58±0.048 | 9.51±1.01 | 25.95±16.45 | 53.43±16.00 | 537.42±158.74 |

## 14.3.3  *fSpacing* Metric

The results obtained from the *fSpacing* metric are shown in Figure 14.5. Changes in *global* G+C content result in a wide degree of variability in the *fSpacing* values found in the CDS and NC regions (Table 14.2). In comparison, the average *fSpacing* values found in the tRNAs are more stable. The average *fSpacing* values in the rRNAs are even more stable. As with the previous metric, the disparity between these rRNAs and tRNAs and their genomic counterparts disappears as their respective G+C content levels approach uniformity.

The difference between rRNAs and tRNAs is noteworthy. It is presumably related to how the search parameters of our stem-loop finding algorithm were devised. The Methods Section describes how the search parameters were developed by studying rRNA secondary structures.

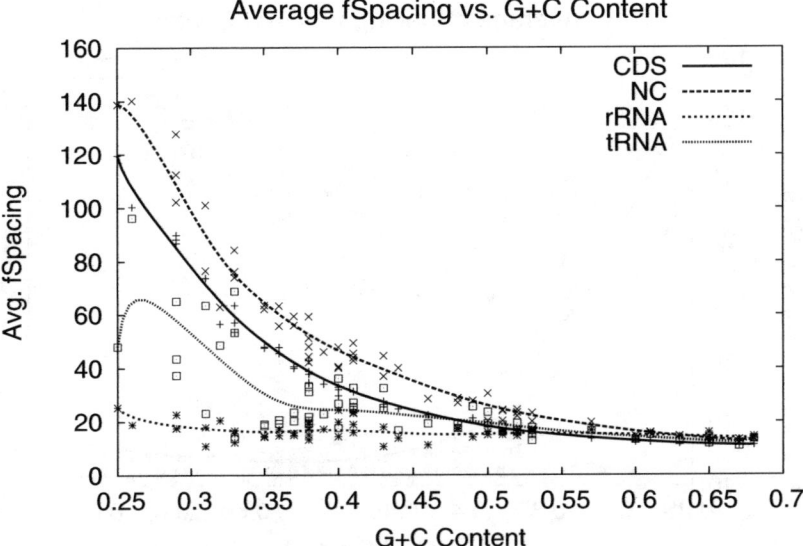

**Fig. 14.5.** Average *fSpacing* values found in CDS, NC, rRNA, and tRNA regions over 58 bacterial genomes with varying *global* G+C content levels

### 14.3.4  *cSpacing* Metric

The results obtained from the *cSpacing* metric are shown in Figure 14.6. There are several noteworthy observations. The disparity between the rRNAs and tR-NAs and their counterparts diminishes as the G+C content levels approach uniformity. Like the *fSpacing* metric, the disparity in *cSpacing* values between rRNAs and their counterparts is greater than the disparity observed between tRNAs and their counterparts.

The average *cSpacing* metric values found in rRNA are remarkably more stable than the values attained for tRNA, NC, and CDS (Table 14.2). Notably, the rRNA values do not merge with the CDS and NC values. With regards to our training set, this indicates that there is always a difference in *cSpacing* values between the rRNAs and their genomic counterparts. This claim cannot be made for the base composition approach on this training set in particular (compare Figures 14.3 and 14.6). Importantly, when the *global* G+C content is roughly 40-60%, the disparity in the average *cSpacing* values between the rRNAs and their genomic counterparts is insufficient for our statistical inference method to distinguish rRNAs from their genomic counterparts.

### 14.3.5  (*cSpacing* × *bps*) Metric

The previous results illustrate the difficulty in finding a sequence signal capable of identifying rRNAs when there is a negligible difference in G+C content

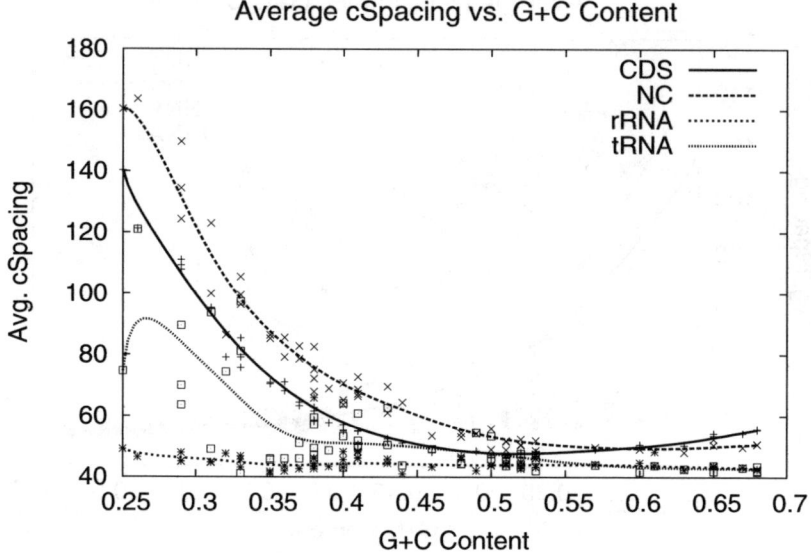

**Fig. 14.6.** Average *cSpacing* values found in CDS, NC, rRNA, and tRNA regions over 58 bacterial genomes with varying *global* G+C content levels

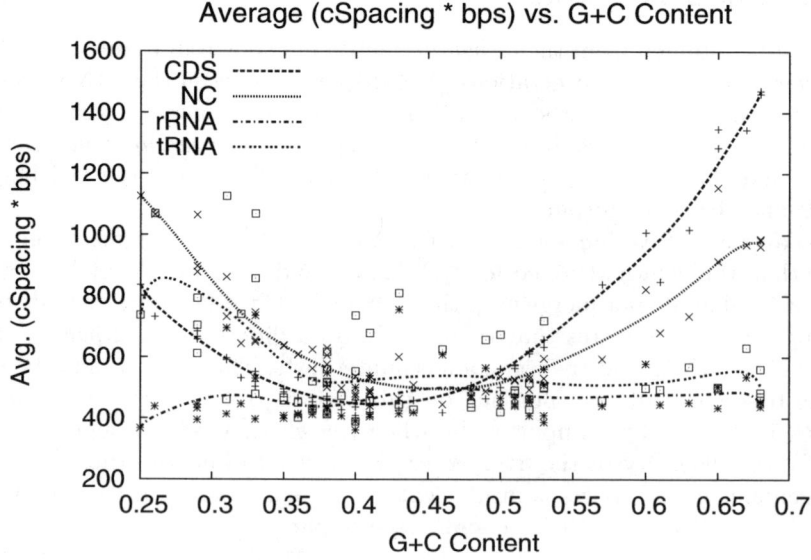

**Fig. 14.7.** Average (*cSpacing* × *bps*) values found in CDS, NC, rRNA, and tRNA regions over 58 bacterial genomes with varying *global* G+C content levels

between opposing genomic regions. We postulated that combining multiple stem-loop metrics could help to amplify their inherent differences. Figure 14.7 shows the results obtained when the (*cSpacing* × *bps*) metric was employed. Like with the previous metrics there are points along the G+C content spectrum where the (*cSpacing* × *bps*) values merge. Furthermore, the CDS and NC regions display a high degree of variability in the (*cSpacing* × *bps*) values (Table 14.2). In turn, this makes the metric ineffective across a wide portion of the G+C content spectrum. However, there is a noteworthy advantage to this metric. The average (*cSpacing* × *bps*) metric displays more disparity between rRNAs and their genomic counterparts when their G+C content levels are virtually equivalent (i.e. G+C = 52-54%). The significance of this observation is substantiated in the following experiments.

Our next task was designed to examine whether the information generated from the 58 genome training set is helpful in finding ever-present structural RNAs - namely rRNA genes - by using statistical inference. In several genomes, the rRNAs were successfully identified. Notably, these genomes had *not* been included in the training set.

*Rickettsia typhi* str. Wilmington, NC 006142, measures 1,111,496 base pairs (bps) and has a *global* G+C content of 29%. We used the average rRNA *cSpacing* value attained from our training set to identify the rRNA genes. Note the largest hit returned by the statistical inference corresponds well with the rRNA gene (Figure 14.8). Notably, the hit at index 609,030 corresponds with an RNA of unknown type (as indicated by the NCBI annotated sequence). The other "hits" do not correspond with other known structural RNAs and are presumed to be false positives. In the interest of space, the figures which follow depict a 500,000 nucleotide segment of the tested genome. The results over the reported segment are representative of the findings over the length of the entire sequence. As the *cSpacing* metric is used on sequences with higher and higher G+C content levels, an untenable number of false positives is eventually reached. This occurs when the *global* G+C content is roughly 40-60%.

In an attempt to overcome this hurdle, we devised the (*cSpacing* × *bps*) metric. The next 2 sequences convey what was commonly observed. *Synechococcus sp*, NC_005070, is 2,434,428 bps long and has a *global* G+C content of 59%. In this case the difference in G+C content between the opposing genomic domains (i.e. rRNAs and CDS and NC) is roughly 5-7%. Interestingly, the statistical inference is able to identify the rRNA genes (Figure 14.9). The number of false positives appears reasonable.

Next, we attempted a more difficult task - to identify the rRNAs in a genome when there is a negligible difference in the G+C content levels between rRNAs and their genomic counterparts. *Salmonella enterica*, NC_006905, is 4,755,700 bps long and has a *global* G+C content of 52%. Notably the G+C content in rRNA, tRNA, NC, and CDS is essentially equivalent. The results are depicted in Figure 14.10. The prevalence of false positives is immediately evident. Roughly 30% of the sequence is flagged - including the rRNAs.

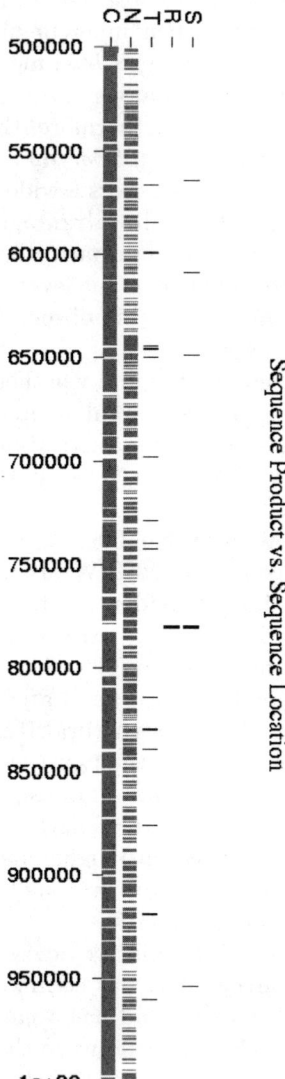

**Fig. 14.8.** *Rickettsia typhi* str. Wilmington, complete genome. Accession No. NC_006142. Sequence Length 1,111,496 bps. Global G+C is 29%. Metric *cSpacing*. Sample size 31. The segment depicted is denoted by the x-axis and it includes nucleotides 500,000 to 1,000,000. The y-axis outlines the annotated sequence and the statistically inferred structural RNAs. The abbreviations are as follows: C= CDS, N = NC, R = rRNA, T = tRNA, S = statistically inferred structural RNA. Ideally, the "hits" should correspond with the locations of the rRNA genes. In this case, the hits overlap with the rRNA gene. Also, there is a hit at index 609,030 which corresponds with an RNA of unknown type per the NCBI annotated sequence.

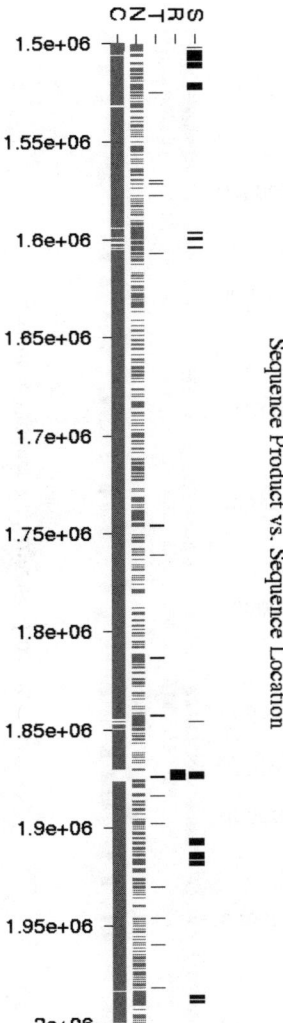

**Fig. 14.9.** *Synechococcus sp.* WH 8102, complete genome. Accession No. NC_005070. Sequence Length 2,434,428 bps. Global G+C is 59%. Metric (*cSpacing* × *bps*). Sample size 351. The segment depicted is denoted by the x-axis and it includes nucleotides 1,500,000 to 2,000,000. The y-axis outlines the annotated sequence and the statistically inferred RNAs. The abbreviations are as follows: C = CDS, N = NC, R = rRNA, T = tRNA, S = statistically inferred structural RNA. Note, the hits (S) correspond well with the rRNAs located in the region surrounding indices 1,870,000 to 1,874,000.

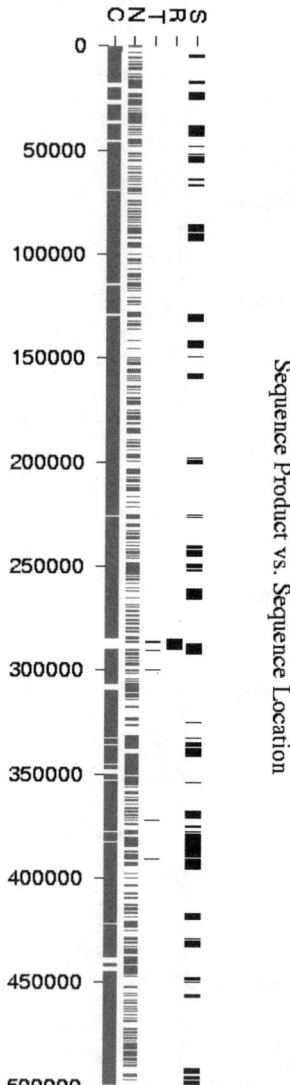

**Fig. 14.10.** *Salmonella enterica* subsp. enterica serovar Choleraesuis str. SC-B67, complete genome. Accession No. NC_006905. Sequence Length 4,755,700 bps. Global G+C is 52%. Metric (*cSpacing* × *bps*). Sample size 501. The segment depicted is denoted by the x-axis and it includes nucleotides 0 to 500,000. The y-axis outlines the annotated sequence and the statistically inferred RNAs. The abbreviations are as follows: C = CDS, N = NC, R = rRNA, T = tRNA, S = statistically inferred structural RNA. Note, the hits surrounding index 287,000 overlaps with the rRNAs in this segment. The other hits are presumed to be false positives.

## 14.4   Discussion

At the outset, our intuition led us to suspect that stem-loops identified in regions which code for structural RNAs would, on average, be longer than those
identified in their genomic counterparts. Our results indicate this is not always
true. Instead, the *bps* values are strongly related to G+C base composition (Figure 14.4). For instance, an increased G+C base composition is accompanied by a
higher *bps* value. As a result, the average *bps* value in structural RNA is greater
than those of their genomic counterparts only when the G+C content in the
structural RNAs is also greater. In comparing several genomes, it is clear that
the G+C base composition in the structural RNAs is more stable than the G+C
content in the CDS and NC domains (Figure 14.3). Consequently, the rRNA and
tRNA *bps* values are more stable from genome to genome than the average *bps*
values measured in the CDS and NC domains. In several genomes which make
up our test set, the rRNA and tRNA G+C content is essentially uniform across
the various genomic regions. In turn, the *bps* values calculated for structural
RNAs and their counterparts are indistinguishable.

Another initial suspicion led to the belief that stem-loops would occur more
frequently in regions which code for structural RNAs compared to their genomic
counterparts. To study this, two spacing metrics - *fSpacing* and *cSpacing* -
were devised. The following paragraphs explain to what extent our experiments
support our initial intuition.

Like the *bps* metric, the *fSpacing* and *cSpacing* metrics are affected by
changes to G+C content levels. Increases to G+C content are accompanied
by decreases to spacing values - i.e. a higher density of stem-loops. In A+T
rich genomes, the *fSpacing* and *cSpacing* values recorded for structural RNAs
differ significantly from those of their counterparts. This difference is largely
attributable to the notable disparity in their G+C content levels (Figure 14.3).
Recall that the search parameters require stem-loops to be comprised of at least
30% GC base pairs (Table 14.1). The probability of finding these base pairs
increases with increasing G+C content. As a result, the average spacing values
observed in A+T rich CDS and NC regions are significantly higher than those
recorded for structural RNAs.

There are some important differences in the *fSpacing* and *cSpacing* results.
When the G+C content levels between opposing genomic regions are equivalent,
the disparity in their *fSpacing* values is completely lost. This is conveyed by
the intersecting plots in Figure 14.5. In regards to *cSpacing*, only the tRNA
plot intersects with its counterparts when the G+C content levels are equivalent
(Figure 14.6). Notably, the rRNA *cSpacing* plot does *not* intersect with its
counterparts. However, the difference in the average *cSpacing* values between
the rRNAs and their counterparts is too small to distinguish between them
when the G+C content levels are roughly 45% to 55%.

One explanation why the rRNA *cSpacing* plot does not intersect with its
counterparts while the *fSpacing* metric does likely relates to inherent differences between these metrics. The *fSpacing* metric measures from the "foot" of

the stem. As a result, it does not incorporate attributes related to the length of the stem (i.e. the number of base pairs). In contrast, the *cSpacing* metric incorporates length attributes since it measures the distance between hairpin loop centers (Figure 14.2). Revisiting the results for the *bps* metrics we note that increases to the G+C content in the CDS and NC domains correspond to increases in the average stem length (Figure 14.4). This pattern, though more subtle, is also seen in the *cSpacing* metric. When the G+C content is roughly 50% or more the average *cSpacing* value in the CDS and NC regions begin an upward trend (Figure 14.6). This partially explains why the rRNA data points do not merge with the NC, and CDS data points.

In A+T rich genomes, the average rRNA spacing values are more discrepant from their counterparts - CDS and NC - than the average tRNA spacing values are (Figures 14.5 and 14.6). Similarly, the average rRNA spacing values are more stable relative to the tRNA spacing values. These differences are likely related to the fact that the search parameters were devised by studying rRNA secondary structures rather than tRNA structures. Notably, when our algorithm was equipped with only the mean rRNA *cSpacing* value calculated from our training set (Table 14.2), our approach attained a hit which overlapped with an RNA of unknown type in the *Rickettsia typhi* genome (Figure 14.8). This suggests that it would be advantageous to develop or model the stem-loop search parameters over a wider variety of structural RNA gene families.

The stability observed in the *cSpacing* and *fSpacing* stem-loop metrics provides for some interesting speculation. Structural RNAs undergo evolutionary pressures which strongly favour conserving base pairs rather than sequence motifs. Conserving base pairs allows RNA genes to conserve structure. The stability observed in the *cSpacing* and *fSpacing* metrics is likely related to this tendency to conserve structure. It might also be indicative of an equilibrium state. This equilibrium, it is postulated, may foster an environment which promotes the formation of the final structure. It seems plausible that RNA transcripts with an overabundance of stem-loops would increase the likelihood that an obstructive amount of disassembly may be necessary before reaching their final RNA structure. Conversely, RNA transcripts which resist folding may hinder distal segments from coming into close proximity. Presumably, an environment between these two extremes may be most conducive to efficient RNA structure folding. Cast in this light, stem-loops or base pair interactions which are not observed in the final structure may not be mere coincidence.

The results obtained with the (*cSpacing* × *bps*) metric are noteworthy. In the example of NC_006905, the difference between the opposing genomic regions is negligible. The presence of numerous apparent false positives in Figure 14.10 should not overshadow the fact that this relatively simple metric and statistical inference approach perform with a marked degree of success. It eliminates roughly two-thirds of the sequence which is *not* rRNA material without any knowledge other than the average rRNA (*cSpacing* × *bps*) value reported in Table 14.2. Although a direct comparison is not presented here, it is presumed that a G+C content approach would not function as effectively when such negligible differences in G+C content are present. Perhaps $(cSpacing \times bps)$ could be useful as a preliminary screen for rRNAs in large bacterial genomes. This could be followed-up by more computationally taxing comparative methods to more accurately pinpoint the location of the rRNA genes.

## 14.5  Conclusion

This research provides a consideration of stem-loop metrics as sequence signals for uncovering structural RNA genes. An algorithm was developed to identify stem-loops which share basic characteristics similar to those observed in rRNA secondary structures. Initially, it was hypothesized that stem-loops identified in regions which code for structural RNAs would, on average, consist of more base pairs than stem-loops found in their genomic counterparts. In addition, it was suspected that stem-loops would occur more frequently in structural RNA regions in comparison to their genomic counterparts. Our hypotheses were tested over 58 bacterial genomes encompassing a wide range of G+C content levels. The results support the above hypotheses to some extend.

The values attained for the stem-loop metrics are strongly tied to G+C content. For instance, a high G+C content level correlates with a high number of base pairs in a stem. Conversely, a low G+C content correlates with a lower number of base pairs in a stem. Similar observations were made with respect to the $fSpacing$ metric. A high G+C content correlates with a higher frequency of stem-loops. A low G+C content is linked to a lower frequency of stem-loops. When the G+C content levels in structural RNAs and their genomic counterparts are equivalent, their $bps$ and $fSpacing$ metric values are also equivalent. As a result, these metrics cannot distinguish between structural RNAs and their counterparts based on their average metric valuations.

The $cSpacing$ metric injects some promise into the possibilities embodied by stem-loops as a sequence signal. The rRNA $cSpacing$ values are less than those of their genomic counterparts for all the genomes in our test set. However, the $cSpacing$ metric is not able to distinguish rRNAs from their genomic counterparts when their G+C content levels are essentially equivalent.

When encountered with genomes where the G+C content is essentially uniform across all the genomic domains we employed the $(cSpacing \times bps)$ metric. Given the novelty of this approach, the results are very encouraging. Equipped with only the average rRNA $(cSpacing \times bps)$ value established with our training set, we were able to correctly eliminate roughly two-thirds of the sequences which do *not* code for rRNAs. This level of accuracy seems to suggest that the $(cSpacing \times bps)$ metric could act as a preliminary screen for rRNAs before more computationally taxing approaches are applied. In addition, in several instances real and putative rRNA sites were identified.

This article has successfully introduced a unique approach to identifying ribosomal RNA genes. It relies on a sequence signal which is based on the stem-loop. Stem-loops are an attractive target since RNA sequences are essentially obligated to form them enroute to generating stable structures. This work also reports some of the limitations of this approach. Suggestions for future endeavors include designing and testing alternative metrics, combining multiple metrics, and optimizing the search parameters.

# References

1. Storz, G.: An expanding universe of noncoding RNAs. Science 296, 1260–1262 (2002)
2. Erdmann, V.A., Szymański, M., Hochberg, A., Groot, Nd., Barciszewski, J.: Collection of mRNA-like non-coding RNAs. Nucleic Acids Research 27(1), 192–195 (1999)
3. Eddy, S.R.: Non-coding RNA genes and the modern RNA world. Nature Reviews Genetics 2, 919–929 (2001) (review)
4. Szymański, M., Erdmann, V.A., Barciszewski, J.: Noncoding regulatory RNAs database. Nucleic Acids Research 31(1), 429–431 (2003)
5. Kiss-László, Z., Henry, Y., Bachellerie, J., Caizergues-Ferrer, M., Kiss, T.: Site-specific ribose methylation of preribosomal RNA: A novel function for small nucleolar RNAs. Cell 85(7), 1077–1088 (1996)
6. Nicoloso, M., Qu, L., Michot, B., Bachellerie, J.: Intron-encoded, antisense small nucleolar RNAs: The characterization of nine novel species points to their direct role as guides for the 2'-O-ribose methylation of rRNAs. Journal of Molecular Biology 260(2), 178–195 (1996)
7. Tycowski, K.T., Smith, C.M., Shu, M., Steitz, J.A.: A small nucleolar RNA requirement for the site-specific ribose methylation of rRNA in *Xenopus*. Proceedings of the National Academy of Sciences 93(25), 14480–14485 (1996)
8. Lease, R.A., Cusick, M.E., Belfort, M.: Riboregulation in *Escherichia coli*: DsrA RNA acts by RNA: RNA interactions at multiple loci. Proceedings of the National Academy of Sciences 95(21), 12456–12461 (1998)
9. Lease, R.A., Belfort, M.: Riboregulation by DsrA RNA: *transactions* for global economy. Molecular Microbiology 38(4), 667–672 (2000)
10. Walter, P., Blobel, G.: Signal recognition particle contains a 7S RNA essential for protein translocation across the endoplasmic reticulum. Nature 299, 691–698 (1982)
11. Lustig, A.J.: Telomerase RNA: a flexible RNA scaffold for telomerase biosynthesis. Current Biology 14(14), 565–567 (2004)
12. Romero, D.P., Blackburn, E.H.: A conserved secondary stucture for telomerase RNA. Cell 67(2), 343–353 (1991)
13. Eddy, S.R.: Computational genomics of noncoding RNA genes. Cell 109(2), 137–140 (2002) (review)
14. Rivas, E., Eddy, S.R.: Secondary structure alone is generally not statistically significant for the detection of noncoding RNAs. Bioinformatics 16(7), 583–605 (2000)
15. Galtier, N., Lobry, J.: Relationships between genomic G+C content, RNA secondary structures, and optimal growth temperature in prokaryotes. Journal of Molecular Evolution 44, 632–636 (1997)
16. Hurst, L., Merchant, A.: High guanine-cytosine content is not an adaptation to high temperature in prokaryotes: a comparative analysis amongst prokaryotes. Proceedings of the Royal Society of London - Biological Sciences 268, 493–497 (2001)
17. Wang, H., Hickey, D.A.: Evidence for strong selective constraint acting on the nucleotide composition of 16s ribosomal RNA genes. Nucleic Acids Research 30(11), 2501–2507 (2002)

18. Klein, R.J., Misulovin, Z., Eddy, S.R.: Noncoding RNA genes identified in AT-rich hyperthermophiles. Proceedings of the National Academy of Sciences 99(11), 7542–7547 (2002)
19. Schattner, P.: Searching for RNA genes using base-composition statistics. Nucleic Acids Research 30(9), 2076–2082 (2002)
20. Eddy, S.R., Durbin, R.: RNA sequence analysis using covariance models. Nucleic Acids Research 22(11), 2079–2088 (1994)
21. Lowe, T.M., Eddy, S.R.: tRNAscan-SE: a program for improved detection of transfer RNA genes in genomic sequence. Nucleic Acids Research 25(5), 955–964 (1997)
22. Rivas, E., Klein, R.J., Jones, T.A., Eddy, S.R.: Computational identification of noncoding RNAs in *E. coli* by comparative genomics. Current Biology 11(17), 1369–1373 (2001)
23. Weinberg, Z., Ruzzo, W.L.: Exploiting conserved structure for faster annotation of non-coding RNAs without loss of accuracy. Bioinformatics 20(suppl. 1), i334–i341 (2004)
24. Carter, R., Dubchak, I., Holbrook, S.: A computational approach to identify genes for structural RNAs in genomic sequences. Nucleic Acids Research 29, 3928–3938 (2001)
25. Chen, J., Le, S., Shapiro, B., Currey, K., Maizel, J.: A computational procedure for assessing the significance of RNA secondary structure. Computer Applications in the Biosciences 6, 7–18 (1990)
26. Le, S., Chen, J., Currey, K., Maizel, J.: A program for predicting significant RNA secondary structures. Computer Applications in the Biosciences 4, 153–159 (1988)
27. Rivas, E., Eddy, S.R.: A dynamic programming algorithm for RNA structure prediction including pseudoknots. Journal of Molecular Biology 285, 2053–2068 (1999)
28. Zuker, M., Stiegler, P.: Optimal computer folding of large RNA sequences using thermodynamics and auxiliary information. Nucleic Acids Research 9, 133–148 (1981)
29. Zuker, M.: In: Griffin, A.M., Griffin, H.G. (eds.) Computer Analysis of Sequence Data, pp. 267–294. Humana Press Inc. (1994)
30. Zuker, M.: Mfold web server for nucleic acid folding and hybridization prediction. Nucleic Acids Research 31(13), 3406–3415 (2003)
31. Noel, K.: Examining Stem-Loops as a Sequence Signal for Identifying Structural RNA Genes. MA Thesis, Simon Fraser University, Burnaby, British Columbia, Canada (2005)
32. Cannone, J.J., Subramanian, S., Schnare, M.N., Collett, J.R., D'Souze, L.M., Du, Y., Feng, B., Lin, N., Madabusi, L.V., Muller, K.M., Pande, N., Shang, Z., Yu, N., Gutell, R.R.: The comparative RNA web (CRW) site: An online database of comparative sequence and structure information for ribosomal, intron, and other RNAs. BMC Bioinformatics 3(2) (2002), http://www.rna.icmb.utexas.edu
33. Van de Peer, Y., De Rijk, P., Wuyts, J., Winkelmans, T., De Wachter, R.: The European database on small subunit ribosomal RNA. Nucleic Acids Research 30(1), 183–185 (2002)
34. National Center for Biotechnology Information. Microbial Genomes. World Wide Web, http://www.ncbi.nlm.nih.gov/genomes/MICROBES/Complete.html

# 15

# Power-Law Signatures and Patchiness in Genechip Oligonucleotide Microarrays

Radhakrishnan Nagarajan

University of Arkansas for Medical Sciences, 4301 W. Markham,
Little Rock, AR 72205, USA
nagarajanradhakrish@uams.edu

**Summary.** Genechip oligonucleotide microarrays have been used widely for transcriptional profiling of a large number of genes in a given paradigm. Gene expression estimation precedes biological inference and is given as a complex combination of atomic entities on the array called probes. These probe intensities are further classified into perfect-match (PM) and mis-match (MM) probes. While former is a measure of specific binding, the latter is a measure of non-specific binding. The behavior of the MM probes has especially proven to be elusive. The present study investigates qualitative similarities in the distributional signatures and local correlation structures/patchiness between the PM and MM probe intensities. These qualitative similarities are established on publicly available microarrays generated across laboratories investigating the same paradigm. Persistence of these similarities across raw as well as background subtracted probe intensities is also investigated. The results presented raise fundamental concerns in interpreting Genechip oligonucleotide microarray data.

## 15.1 Introduction

Oligonucleotide Genechip microarrays [1, 35, 36] have been used widely for transcriptional profiling of large number of genes across distinct biological paradigms including (i) *stem cell differentiation* [27, 47], (ii) *molecular portraits and heterogeneity in tumors* [43, 50], (iii) *Aging and neurobiology* [13], (iv) *infectious disease research and environmental applications* [31]. Prevalence of such high throughput assays can especially be attributed to the rapid sequencing of genomes [11]. A recent multiple-laboratory and multi-platform study [26] established the superiority of oligonucleotide microarrays from accuracy and precision standpoints. Unlike classical biological approaches, microarrays can be used to model functional relationships between genes, hence provide *system-level* understanding [30] of the paradigm [14, 59]. There is also the possibility of oligonucleotide arrays being used as active screening tools in clinical settings in the near future [21].

Developing suitable computational techniques for meaningful interpretation of oligonucleotide gene expression data is one of the major challenges and precedes biological inference. Gene expression is estimated as a complex combination of atomic entities on the array called *probes* [45]. While several algorithms have been proposed for gene expression estimation and subsequent higher level analysis [2, 3, 24-26, 34, 46, 48], understanding the qualitative behavior at the probe level is still *incomplete*.

T.G. Smolinski et al. (Eds.): Comp. Intel. in Biomed. & Bioinform., SCI 151, pp. 359–377, 2008.
springerlink.com

Probes are broadly classified into *perfect match* (PM) and *mismatch* (MM). The former is a measure of *specific binding* whereas the latter is a measure of *non-specific binding* and used as an internal control (Sect. 15.1.1) [1, 35, 36]. While PM and MM probes are biologically distinct by very design they are spatially proximal on the array. Several statistical techniques have been proposed for gene expression estimation and subsequent higher-level analysis. While some techniques use perfect as well as mismatch probes [2, 3, 34], others have encouraged using the perfect match probes only [24, 25] in the estimation procedure. The choice of the latter was possibly inspired by [38], which pointed out that arithmetic subtraction of (PM, MM) probe intensities may not translate into biological subtraction. The qualitative behavior of the MM probes has especially proven to be elusive.

The objective of the present study is to investigate qualitative similarities in the distributional signatures and local correlation structure across the perfect-match and mismatch probe intensities. Qualitative similarities are demonstrated on the raw as well background subtracted (PM, MM) probe intensities in publicly available Genechip arrays generated across laboratories investigating the same biological paradigm [26]. These qualitative similarities to our knowledge have never been reported and raise fundamental concerns in interpreting oligonucleotide gene expression data and higher level analyses such as (*a*) gene expression estimation and normalization [2, 3, 6, 24, 25, 34, 46, 48, 58]. (*b*) inferring functional relationships and network structure [14, 59] (*c*) ontology [5] and (*d*) expression quantitative trait loci (eQTL) [28] The present study is especially encouraged by our (*i*) recent research on various aspects of microarray gene expression analysis [39, 40] and growing evidence of (*ii*) hybridization interactions/multiple targeting of the probes [42, 57, 60]; (*iii*) spatial artifacts [52] and (*iv*) redefinition of probe-transcript relationship [16, 33] in oligonucleotide Genechip arrays.

The chapter is organized as follows. In Sect. 15.1.1, a brief introduction to Genechip oligonucleotide microarrays along with the associated terminologies is provided. Qualitative similarities along with power-law and exponential approximations to the PM and MM probe intensity distributions is investigated in Sec. 15.2. Qualitative similarities in local correlations/patchiness across PM and MM probe intensity matrices is investigated in Sec. 15.3. The choice of multiscale decomposition for accomplishing the same is also explored. The impact of the findings in the present study on gene expression estimation and subsequent higher level analyses is discussed in Sect. 15.4.

### 15.1.1 Oligonucleotide Genechip Microarrays

Oligonucleotide Genechip microarray [1, 35, 36] comprise of a large number of atomic entities called *probes* [45] arranged as a rectangular matrix. Each probe is an *oligomer*, i.e. around ~25 nucleotides long, (e.g. 5'-GTGATCGTTTACTTCGGTGCCACCT-3'). A set of (~16 to 20) probes also called a *probeset*, represents a particular *transcript* on the array. The term transcript is generic and can represent either a *gene* or an *expressed sequence tag (EST)*. Probes can be broadly classified into *perfect-match* (PM) and *mismatch* (MM) probes. PM probes correspond to a short region of the transcript and are designed to be complementary to the *target sequence* [1, 35, 36], hence ideally a measure of *specific binding*. The nucleotide content of an MM probe is the same as that of the corresponding PM probe except for the middle most nucleotide, which is changed deliberately. Thus MM probes are used as an internal control to *assess non-specific binding*. Gene expression $g'$ of a transcript $t$ on the array is given as a complex combination of the corresponding (PM and MM) or PM only probe intensities [2, 3, 24, 25, 34, 48]. An example of PM, MM and their target probe is shown below for clarity.

**Example.** PM, MM and target probe:

| | |
|---|---|
| **PM** | (5' G T G A T C G T T T A C T $\boxed{\text{T}}$ C G G T G C C A C C T 3') |
| **MM** | (5' G T G A T C G T T T A C T $\boxed{\text{C}}$ C G G T G C C A C C T 3') |
| **Target** | (5' C A C T A G C A A A T G A  A G C C A C G G T G G A 3') |

Analysis of oligonucleotide microarrays begins by extracting the raw (PM, MM) probe intensities from the .CEL files [1] (Affymetrix Technical Manual, Santa Clara, CA). Subsequently, these are *background subtracted* [2, 3, 24-26, 34, 48, 58] to minimize contributions of non-biological factors to the probe intensity/gene expression. In the present study, we investigate the qualitative similarities of the raw as well as background subtracted [3, 24, 25] (PM, MM) probe intensities. Such an approach is useful in rejecting the claim that the observed qualitative similarities are an outcome of not subtracting the background. Background subtraction is accomplished with Bioconductor [17] implementtation of two popular algorithms namely: MAS 5.0 [3] and RMA [24, 25]. Consider the PM $\pi^{pmt} : \pi_1^{pmt} ... \pi_{20}^{pmt}$ and MM $\pi^{mmt} : \pi_1^{mmt} ... \pi_{20}^{mmt}$ probe intensities corresponding to a transcript $t$. The gene expression of that transcript is a mapping of $\pi^{pmt}$ and $\pi^{mmt}$ onto a single value $(g')$ by a chosen estimation procedure $f$, represented by $(\pi^{pmt}, \pi^{mmt}) \xrightarrow{f} g'$.

It is important to note that depending on the choice of the estimation procedure $f$, gene expression $(g')$ is either a *linear* or *nonlinear* combination of $(\pi^{pmt}, \pi^{mmt}, t = 1...20)$. An example of linear and a nonlinear estimation procedures assuming two $(\pi^{pmt}, \pi^{mmt})$ probes per transcript $(g')$ and their impact on the distributions is shown below for clarity.

**Example.** Mapping from probe intensity to gene expression
  *(a) Linear estimation procedure*

$$f: g' = (2\pi_1^{pmt} + 3\pi_2^{pmt}) - (0.5\pi_1^{mmt} + \pi_2^{mmt})$$

In (a), gene expression estimation is given as a difference of the corresponding PM and MM intensities. If $(\pi^{pmt}, \pi^{mmt})$ are normally distributed then $(g')$ is *normally* distributed.

  *(b) Nonlinear estimation procedure*

$$f: g^1 = (2\pi_1^{pm1} + 3\pi_2^{pm1})^2 - (0.5\pi_1^{mm1} + \pi_2^{mm1})^2$$

In (b), gene expression estimation is given as a difference of the square of corPM and MM intensities. Unlike (a), even if $(\pi^{pmt}, \pi^{mmt})$ are normally distributed $(g^t)$ is *not normally* distributed in (b).

**Remark 1.** *From the above section we note the following important points:*

(i)     *Gene expression is estimated as a complex combination of (PM and MM or PM only) probe intensities using an estimation procedure f. Thus conclusions drawn about the statistical properties such as distributional profiles at the gene expression level are dependent on the assumptions behind that particular estimation procedure f. However, conclusions drawn at the probe intensity level is independent of the estimation procedure f.*

(ii)    *PM and MM probe intensities although biologically distinct are located physically adjacent to each other on the array (i.e. spatially proximal).*

(iii)   *Spatial information preserved at the probe intensity level is lost at the gene expression level. Since one of the objectives of the present study is to understand the qualitative similarities in local non-random structure/patchiness across the array, retaining the spatial information is crucial.*

A significant number of studies [22, 32, 56] have argued in favor of power-law distributional approximation to microarray gene expression data and attributed the same to biological factors governing gene expression. The authors in [22] demonstrated power-law (pareto-like) distribution in gene expression across genomes. They attributed such a behavior to common probabilistic mechanism in the gene expression process conserved in eukaryotic evolution. The authors in [32] claimed that gene expression distributions across several microarray platforms show close similarities to power-law behavior. Their findings also claimed that the variance of the log spot intensities were proportional to the genome size. In [56], the authors demonstrated persistence of power-law signatures in microarray gene expression from bacteria (Escherichia Coli) to humans (Homo Sapiens) across distinct biological conditions. Such a behavior was attributed to universality in transcriptional organization across genomes.

In this section, we explore biological and non-biological factors that contribute to the distribution of probe intensities; hence gene expression estimates (Remark 1). A schematic diagram representing the microarray data acquisition process and subsequent higher level analysis is shown in Fig. 15.1 [41]. Specific details such as array layout, probe descriptions, hybridization protocols, laser scanning and image segmentation are intentionally excluded in Fig. 15.1 and can be found elsewhere [1, 35, 36]. Oligonucleotide microarrays can be regarded as *measurement devices* or transducers that map the true transcriptional activity (i.e. mRNA expression) onto a measurement value (i.e. raw probe intensities). The data acquisition (Fig. 15.1) is accompanied by considerable noisiness $(\eta_t, \in_t)$ and nonlinearities $(\varphi, \psi)$ at the transcriptional and the measurement levels, Fig. 15.1. Transcriptional noise is coupled to the dynamics of the system, hence *biological*. It can be attributed to uncertainty in gene expression

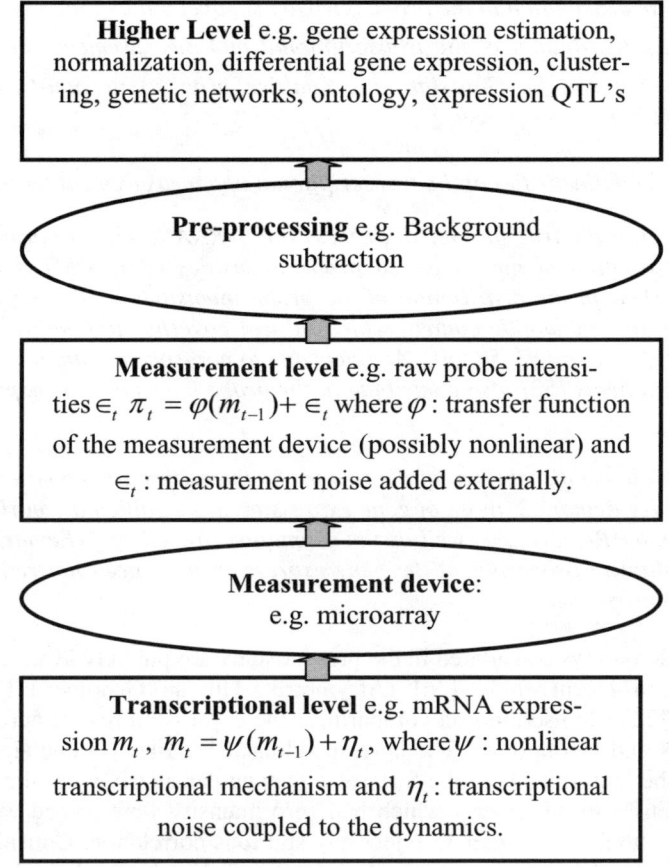

**Fig. 15.1.** Schematic diagram representing the contribution of various factors to probe intensity and gene expression estimates

[12, 29, 53]. However, measurement noise is uncoupled to the dynamics of the biological system, hence *non-biological*. Biological systems by their very nature are nonlinear feedback systems [15, 18, 51]. An example of nonlinearity ($\psi$) in the case of gene expression is that of transcriptional cooperativity [15, 18], where promoters work in tandem to facilitate transcription. The actual mRNA expression and those output by a measurement device such as an oligonucleotide microarray need not necessarily be linearly related. The measurement device is often accompanied by an associated transfer function ($\varphi$) possibly nonlinear, that maps the true biological activity (i.e. mRNA activity) onto the raw (PM, MM) probe intensities. It is important to appreciate the fact that ($\psi$) is *biological* whereas ($\varphi$) is *non-biological*.

***Remark 2.*** *From the above section we note the following important points:*

   *(i)*    *Biological as well as non-biological factors can contribute to the probe intensity/gene expression estimates, Fig. 15.1.*

(ii)    *The distribution at the probe intensity is governed by the*

*(a) distribution of the transcriptional and measurement noise* $(\eta_t, \in_t)$

*which can be Gaussian (i.e. additive process) or non-Gaussian (e.g. multiplicative process)*

*and*

*(b) nonlinearities at the transcriptional and measurement levels* $(\psi, \varphi)$.

*Therefore, even if the true biological process (i.e. mRNA levels) is normally distributed, the distribution of the measured probe intensities (PM, MM) is likely to be skewed. The skew in the distribution of the probe intensities is also accentuated by their non-uniform nucleotide content which in turn governs the binding efficiencies, hence their expression [57, 58, 60]. Artifacts due to non-specific binding [16, 33, 42] and spatial gradients [52] also contribute to the probe intensity/gene expression estimates.*

**Remark 3.** *While the distribution of the raw probe intensities are governed by the factors listed under Remark 2, those of gene expression has significant contribution from the factors under Remark 2 as well as the estimation procedure f (Remark 1). Therefore, the qualitative properties at the gene expression level need not reflect those at the probe intensity level.*

**Data.** The microarrays considered in the present study are publicly available and were generated in a recent study [26] (Affymetrix, Human Genome U133 set, i.e. HGU133A, 22283 transcripts) on comparing gene expression results across microarray platforms and laboratories. The corresponding .CEL files [1] containing the PM and MM probe intensities is in the form of a rectangular matrix with dimensions 356 x 712. All entries in this matrix which had zero intensity were forced with uncorrelated random numbers in order to reject any spurious correlation. Considering replicate arrays across laboratories rejects the claim that the observed results are not an outcome of experimental protocols adopted by a particular laboratory.

## 15.2  Power-Law Distributional Approximations to PM and MM Probe Intensities

Array-wide gene expression has been widely reported to exhibit a significant skew towards lower expression values and a decaying trend with increasing magnitude of expression. Several parametric distributions can be used to model such a decaying trend [9]. Static nonlinear transforms such as Box-Cox normality transforms $\xi(x) = (x^\lambda - 1)/\lambda$ [8] have been used widely in statistical literature to argue in favor of near-normality assumptions. The log-transform in conjunction with 2-fold cut-off used widely in microarray community for identifying differential gene expression is the limiting case of classical Box-Cox normality transforms, i.e. $\lim_{\lambda \to 0} \xi(x)$.

This in turn implicitly assumes log-normal distribution of the gene expression values. Two popular distributions used widely to model decaying trends include the exponential and power-law distributions. The parameters of both the distributions can be attuned so as to capture the decaying trend with increasing magnitude. However, these two classes of distributions have marked differences in their statistical properties. Unlike exponential distribution, the power-law distributions exhibit *scale-invariance*, where the basic shape of the distribution does not alter with scaling. Let $p(k) \sim k^{-\gamma}$ then we have $p(\theta.k) \sim \theta^{-\gamma}.k^{-\gamma} = \theta^{-\gamma}p(k)$ i.e. the distribution of $p(k)$ resembles that of $p(\theta.k)$ other than for a constant scaling factor. The constant scaling factor can also be viewed as the *global normalization* of the microarray, which is used as an important pre-processing step to remove systematic bias between arrays prior to inferring differential gene expression [46, 48]. Unlike exponential distribution, scale-invariance of power-law distributions ensures non-negligible probability of occurrence at large expression values (i.e. heavy tailed). In temporal data power-law distributions are associated with presence of memory whereas exponential distributions are deemed memoryless. These differences in the statistical properties between these two classes of distributions can have far-reaching consequences on biological interpretation.

Power-law distributions as noted earlier have been observed at the gene expression level [22, 32, 44, 56]. In the present study, we investigate the validity of exponential and power-law distributional approximations at the probe PM as well as MM probe intensity levels using three different criteria, namely: $R^2$, Akaike Information Criterion (AIC) and Schwarz information criterion (SIC) [4, 7, 20, 23]. The term *approximation* is deliberately used to accommodate outliers, saturated intensities and finite sample effects inherent in microarray data. A more rigorous analysis using maximum-likelihood approach [10] may provide further insight into the distributional signatures. It is important to note that model(s) with highest $R^2$ is preferred whereas model(s) with lowest AIC and SIC are preferred. Using a combination of model validation criteria minimizes spurious conclusion that is an outcome of inherent assumptions behind a single validation criterion.

Prior to model validation the distributions were log-transformed as follows:

(*i*) Transforming the exponential distribution $P(k) = \alpha_e e^{-\gamma_e k}$ yields

$$\log_2[P(k)] = \log_2(\alpha_e) - \gamma_e k, k > 0$$

(*ii*)Transforming the power-law distribution $P(k) = \alpha_p k^{-\gamma_p}$

$$\log_2[P(k)] = \log_2(\alpha_p) - \gamma_p \log_2(k), k > 0$$

Preliminary inspection of the log-log (power-law) and semi-log (exponential) plots at the probe intensity and gene expression levels revealed significant distortions for values greater than ($2^{13}$). Given the dynamic range (0, $2^{16}$-1) [1, 35, 36] of the probe intensities, it is likely that values greater than ($2^{13}$) may have significant contributions from saturated pixels. Therefore, gene expression and probe intensities above (> $2^{13}$) were filtered prior to model validation. The exponential and power-law approximations were validated using the three different criteria ($R^2$, AIC and BIC) on the filtered and background subtracted gene expression data generated across two different

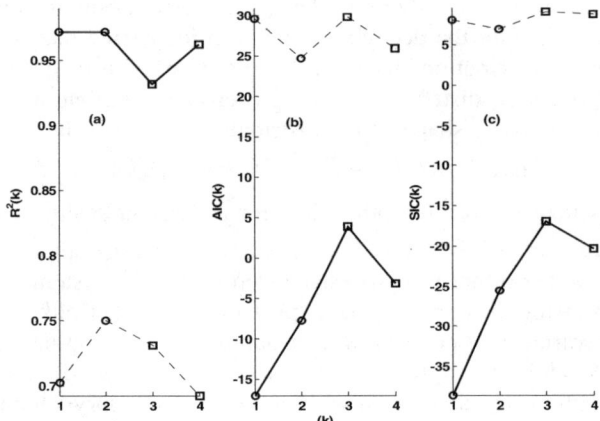

**Fig. 15.2.** Validation metrics ($R^2$, AIC and BIC) for the exponential (dotted lines) and power-law approximations (solid lines) across the four different distributions ($k = 1, 2, 3$ and $4$). The background subtracted gene expression data from two distinct laboratories ($L_1$, $L_2$) investigating the same paradigm [26] are represented by circles and squares respectively. While ($k = 1, 3$) correspond to MAS 5.0, ($k = 2, 4$) correspond to RMA. The results across the three validation metrics argue in favor of power-law approximations over exponential approximations across laboratories.

laboratories investigating the same paradigm generated in a recent study [26], Fig. 15.2. Background subtraction was accomplished by MAS 5.0 [3] and RMA [24] represented by ($k = 1, 3$) and ($k = 2, 4$) in Fig. 15.2. The two different laboratories are represented by (circles, $k = 1, 2$,) and (squares, $k = 3, 4$) in Fig. 15.2, respectively. The $R^2$ values corresponding to the power-law approximation was relatively higher than that of the exponential approximation, Fig. 15.2a. The AIC and the BIC estimates were relatively lower for the power-law as opposed to exponential. These findings were consistent across arrays between laboratories and across background subtraction techniques. Thus power-law approximations seem to better explain the gene expression distribution as opposed to exponential approximation. These results conform to earlier findings [22, 32, 56].

A similar analysis was carried out for the raw and background subtracted $\pi^{PM}$ and $\pi^{MM}$ probe intensities obtained from the same arrays across the same laboratories [26], Fig. 15.3. The raw PM and MM intensities across laboratories ($L_1$, $L_2$) are represented by ($k = 1$ and $4$), those obtained by background subtraction with MAS 5.0 and RMA are represented by ($k = 2$ and $5$) and ($k = 3$ and $6$) respectively, Fig. 15.3. The results obtained across the three validation criteria were consistent and argued in favor of power-law approximation over exponential approximations at the probe intensity levels.

***Remark 4.*** *Power-law and exponential approximations exhibit significant difference in their statistical properties.*

    *(i)       Analysis of the gene expression estimates across laboratories investigating the same paradigm using three validation criteria argued in favor of power-law over exponential approximations.*

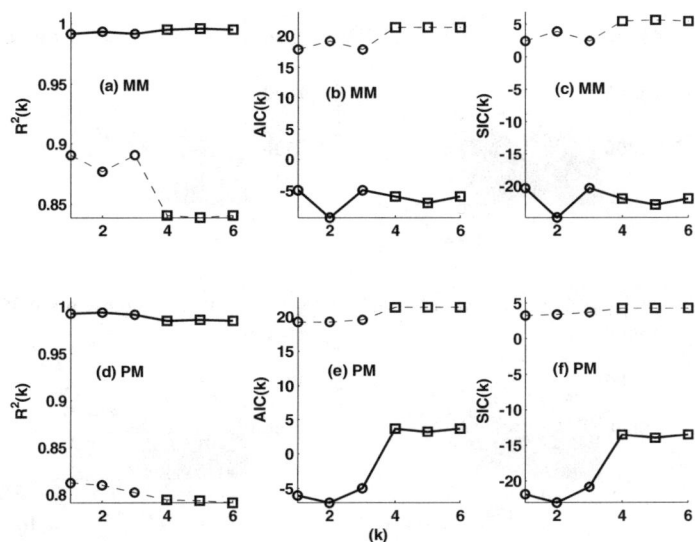

**Fig. 15.3.** Validation metrics ($R^2$, AIC and BIC) for the exponential (dotted lines) and power-law approximations (solid lines) across the raw and background subtracted $\pi^{MM}$ (a, b, c) and $\pi^{PM}$ (d, e, f) probe intensity distributions ($k = 1...6$) obtained across two laboratories $L_1$,(circles) and $L_2$(squares) investigating the same paradigm [26]. The x-labels ($k = 1$ and 4) correspond to the raw PM and MM intensities across ($L_1$, $L_2$); ($k = 2$ and 5) correspond to the background subtracted (MAS 5.0) PM and MM intensities across ($L_1$, $L_2$); ($k = 3$ and 6) correspond to background subtracted (RMA) PM and MM intensities across ($L_1$, $L_2$) respectively. The results across the three validation metrics argue in favor of power-law approximations over exponential approximations across PM as well as MM intensity distributions.

(ii)     *Analysis of the raw and background subtracted PM and MM probe intensities in arrays across laboratories investigating the same paradigm using three validation criteria argued in favor of power-law over exponential approximations. These qualitative similarities in the distributional properties across the PM as well as MM intensities is especially intriguing as the former is a measure of specific binding whereas the latter is a measure of non-specific binding. The persistence of power-law approximations across PM and MM intensities argue in favor of non-biological factors such as static nonlinear measurement function contributing the distributional signatures.*

(iii)    *Power-law distributions observed at the probe intensity levels may also imply inherent clustering/patchiness in the intensities [49].*

## 15.3  Patchiness in PM and MM Probe Intensity Matrices

Classical linear correlation coefficient is widely used for inferring statistically significant *linear* dependencies between a given pair of variables. Correlation coefficient between the raw and background subtracted (RMA) $\pi^{PM}$ and $\pi^{MM}$ intensities across

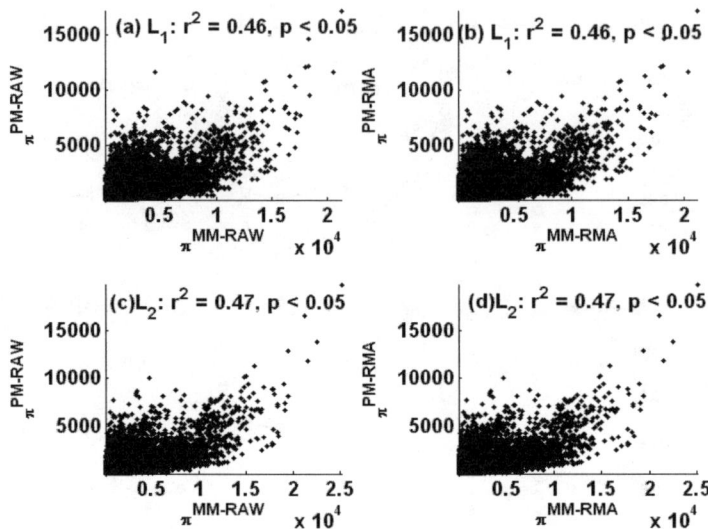

**Fig. 15.4.** Scatter plot of the raw and background subtracted (RMA) $\pi^{PM}$ and $\pi^{MM}$ probe intensities in arrays generated across laboratories $L_1$ (a, b) and $L_2$ (c, d) investigating the same paradigm [26]

laboratories $L_1$ (Figs. 15.4a and 15.4b) and $L_2$ (Figs. 15.4c and 15.4d) were ($r^2 \sim 0.46$, p-value $< 0.05$) and ($r^2 \sim 0.47$, p-value $< 0.05$).respectively However, visual inspection of the scatter plots, Fig. 15.4 revealed considerable noisiness with no apparent linear trend. Thus direct estimation of the correlation coefficient may not provide sufficient insight into their qualitative similarities and correlation structure.

Techniques such as *global singular-value decomposition* (SVD) [19] have been used widely in interpreting microarray gene expression data [6, 59]. Global SVD of a matrix $\Gamma$ is equivalent to eigen-decomposition of symmetric and $\Gamma^T\Gamma$ and $\Gamma\Gamma^T$, hence a measure of *linear correlation* between the probe intensities. While $\Gamma^T\Gamma$ is a measure of the row-wise correlation, $\Gamma\Gamma^T$ is a measure of the column-wise correlation. However, they both yield the same eigen-spectrum, hence equivalent.

***Remark 5.*** *Classical correlation coefficient and global SVD may be useful in establishing the non-random nature of the PM and MM probe intensity matrices. However, it is possible that only a subset of the probes on the array contribute to the observed correlation. Global assessment also does not provide insights into which probes on the array contribute significantly to the observed similarity in correlation signatures between the probe intensity matrices.*

In order to overcome some of the caveats listed under Remark 5, we chose local SVD as opposed to global SVD. The procedure to determine *statistically significant patchiness* using local SVD is described in the following section.

### 15.3.1 Local SVD of (PM, MM) Probe Intensity Matrices

**Algorithm I**

**Step 1.** *Partition the PM probe intensity matrix $PM^{R1 \times C1}$ into non-overlapping blocks each of size r x c. This maps $PM^{R1 \times C1}$ into $B^{R2 \times C2}$, such that, $R2 = \lfloor R1/r \rfloor$, $C2 = \lfloor C1/c \rfloor$ where $\lfloor y \rfloor$ stands for largest positive integer greater than or equal to y.*

**Step 2.** *Choose a block $B = B_{UV}$, $U = 1...R2$, $V = 1...C2$. Retrieve the eigen-spectra $\lambda_K$, $K = 1...min(R2,C2)$. Subsequently, normalize the eigen-values to obtain $\delta^i = \dfrac{\lambda_i^2}{\sum\limits_{i=1}^{K} \lambda_i^2} i = 1...K$.*

**Step 3.** *Complexity $(\eta^B)$ of block $B$ is given by*

$$\eta^B = -\frac{1}{\log K} \sum_{k=1}^{K} \delta^k \log(\delta^k)$$

*Complexity $\eta^B$ is inversely proportional to the linear correlation in $B$. Alternatively, increased redundancy/local correlation between neighboring probes in the block results in low complexity. Ideally, for a random structure the eigen-values will be uniformly distributed resulting in maximum complexity.*

**Step 4.** *Block $B$ is deemed as significantly correlated if the estimate of the covariance complexity on $B$ is significantly different from those obtained on its random shuffled counterparts $B_i^*, i = 1...n_s$ of $B$. Random shuffled counterparts/matrices were constructed by bootstrapping the elements of $B$ randomly without replacement [54, 55]. Such constrained realizations retain the distribution of the probe intensities in $B$ in the shuffled counterparts whereas the spatial information between neighboring probes is destroyed.*

**Step 5.** *In the presence of correlations, we expect the complexity of block $B$ ( $\eta^B$ ) to be lesser than that of its random shuffled counterparts $(\eta_i^{B*}, i = 1...n_s)$. Therefore, a one-side non-parametric test is sufficient to establish statistical significance. i.e. the null hypothesis that the given block is not significantly correlated can be rejected at a significance level $\alpha = 1/(1+n_s)$ if $\eta^B < \eta_i^{B*} \forall i = 1...n_s$ [54, 55]. In the present study, we fix ($n_s = 99$), which corresponds to $\alpha = 0.01$ [54, 55]. Parametric approaches [54, 55] are less stringent. However, their conclusions implicitly rely on implicit normality assumptions; hence can give rise to false positives when these assumptions are violated.*

**Step 6.** *For visualization a binary mask $\Phi$ is generated such that*
$$\Phi_{UV} = 1 \qquad \text{for a significantly correlated block } U = 1...R2,$$
$$V = 1...C2.$$
$$= 0 \text{ otherwise}$$
*Repeat steps 2 to 5 for each of the block $\boldsymbol{B} = B_{UV}$, $U = 1...R2$, $V = 1...C2$ of the PM matrix.*

**Step 7.** *Repeats Steps 1-6 independently for the (MM) probe intensity matrix.*

Global SVD is special case obtained by setting ($r = 1$, $c = 1$) in Step 1 of Algorithm I. As expected, complexity ($\eta$) obtained from global SVD of the PM and MM matrices with and without background subtraction were significantly lower than those of their random shuffled surrogates $\eta < \eta_i^s$, $i = 1... 99$, indicative of non-random structure in the PM and MM matrices. This was verified across replicate arrays generated across laboratories ($L_1$, $L_2$) investigating the same given paradigm. However, from Remark 5, we note that the correlation across the probes in the PM and MM matrices *need not necessarily be global*, i.e. the statistical properties can vary considerably across the probe intensity matrices. This is to be expected as the binding efficiencies of the probes can vary considerably by their very design, also

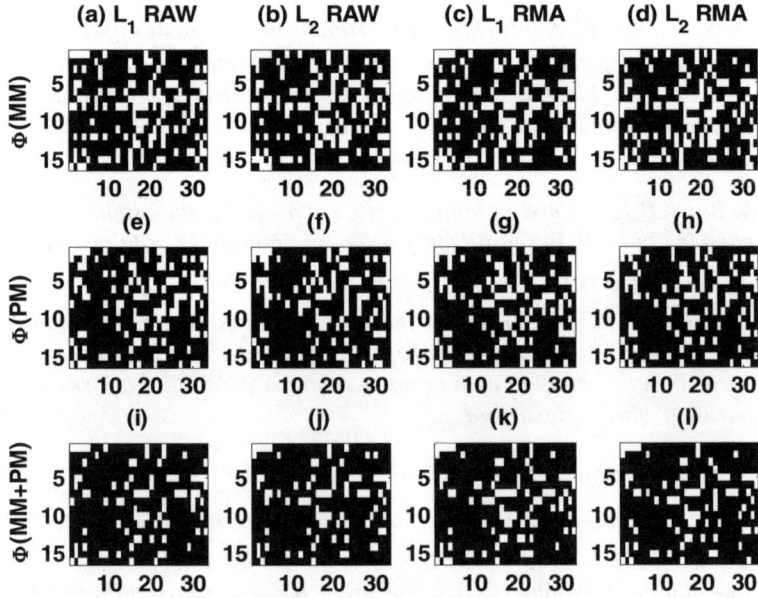

**Fig. 15.5.** Binary masks generated (Step 6, Algorithm I) with (r x c = 21 x 21, $n_s$ = 99) across the raw (RAW) and background subtracted (RMA) PM and MM probe intensity matrices across laboratories ($L_1$, $L_2$) investigating the same paradigm [26]. Correlated patches (white pixels) across MM, i.e. $\Phi(MM)$, and PM, i.e. $\Phi(PM)$, probe intensity matrices are shown in the top two rows (a-d and e-h), whereas those common to PM as well as MM, i.e. $\Phi(PM+MM)$, are shown in the bottom row (i-l). The size of the probe intensity matrices are (356 x 712), hence the dimension of the binary masks are (356/21 x 712/21), i.e. (16 x 33).

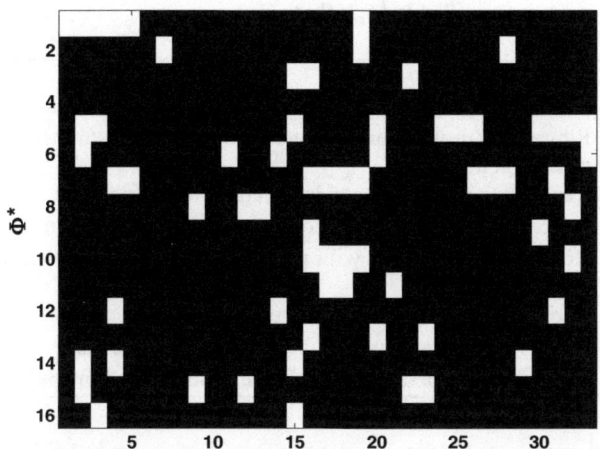

**Fig. 15.6.** Binary mask ($\Phi^*$) generated by intersection of the binary masks in the last row of Fig. 5, i.e.5i- 5l. The correlated patches (white pixels) in the above binary mask were common across PM and MM probe intensity matrices, across raw and background subtracted intensities and across replicate arrays generated across laboratories ($L_1$, $L_2$) [26]. The size of the probe intensity matrices are (356 x 712), hence the dimension of the masks are (356/21 x 712/21), i.e. (16 x 33).

reflected by the skewed distribution of the probe intensity matrices (Sec. 15.2). In order to capture the local variation in correlation structure, we analyzed the probe intensity matrices using local SVD with block size (r x c = 21 x 21) and the number of surrogates ($n_s$ = 99), Fig. 15.5. It is important to note that there are several significantly correlated patches that persists across PM as well as MM probe intensity matrices. This is especially interesting as the former is a measure of specific binding whereas the latter is a measure of non-specific binding.

Interestingly, there were correlated patches ($\Phi^*$) Fig. 15.6, that persisted (*i*) across PM and MM probe intensity matrices, (*ii*) across replicate arrays from two distinct laboratories and (*iii*) across the raw and background subtracted intensities. These patches were generated as intersection of the binary masks in Figs. 15.5i to 15.5l.

The probes on the Genechip microarrays are designated based on their sequence information (see Table 15.1 and [3]). A recent study [42], investigated the contributions of two specific probe designations (_s_at and _x_at) on hybridization interactions and spurious correlations. Probesets with suffix (_s_at) have the ability to target multiple transcripts (i.e. multiple targeting), on the other hand those with _x_at can contribute significantly to cross-hybridization and non-specific binding. Interestingly, ~70% of the probes comprising the patchy region, Fig. 15.6, were classified under _s_at whereas ~11% were classified as _x_at.

***Remark 6.*** *Local SVD can be useful in identifying significantly correlated patches. Preliminary results indicated patchiness that persists across PM as well as MM probe intensity matrices with and without background subtraction. Probes that were common across the PM and MM intensity matrices, across laboratories, across raw*

**Table 15.1.** Probe designations

| Suffix | Description |
| --- | --- |
| _f_at | Represents polymorphic probes which share considerable similarity |
| _s_at | Represents probes common across several genes/transcripts, i.e. multiple targeting |
| _g_at | Represents probes chosen in a region of overlap |
| _r_at | Represents probes picked comprising the selection rules. |
| _i_at | Represents transcripts with incomplete/fewer number of probes than required |
| _b_at | Represents ambiguous probe sets |
| _l_at | Represents transcripts with more than 20 probe pairs |
| _x_at | Represents probe-sets which share probes, i.e. non-specific binding |

*and background subtracted data consisted mainly of cross-hybridizing and multiple targeting probes.*

*It should be noted that ($\eta$) by definition is a measure of linear correlation, hence Algorithm I can give rise to false-negatives in the presence of nonlinear correlations among the probe intensities. However, it cannot give rise to false-positives (i.e. it cannot indicate presence of correlation in a seemingly random patch). For the same reason, results obtained with ($\eta$) represent the lower limit in identifying locally correlated regions. More sophisticated measures, possibly nonlinear may be used to gain further insight into the correlation structure. Algorithm I implicitly assumes a rectangular geometry, however the locally correlated regions can be irregular. This in turn may result in the inclusion/exclusions of probes which are not a member of the locally correlated region. Overlapping blocks is a suitable alternative and may be used in order to obtain finer representation of the correlation structure and minimize edge effects (i.e. accommodate all the probes on the array). The choice of block size can also affect the conclusions. A large block size provide better statistical description and especially encouraged when the probe intensity matrices are homogeneous, i.e. not much variation in the correlation properties. Small block sizes are preferred when the correlation properties show marked variations. However, smaller the block size, lesser the statistical information. There is no straightforward way to determine the optimal block size. An exhaustive approach would be to repeat Algorithm I for varying block sizes. A more elegant approach would be to use multiscale decomposition techniques such as wavelets that provide both spatial and frequency resolution.*

### 15.3.2 Multiscale Decomposition of (PM, MM) Probe Intensity Matrices

Multiscale approaches such as discrete wavelet transforms (DWT) are ideally suited for capturing varying statistical properties and correlation structure in 1D and 2D data. Unlike classical 2D Fourier transform (FT), DWT provides time/spatial as well as frequency resolution of the given data [37]. While high frequency components require better time resolution, low frequency components require better frequency resolution. The delicate balance between time and frequency resolutions in DWTs is dictated by the Heisenberg's uncertainty principle. DWT is a linear transform 2D FT and represents the given data as a linear combination of basis functions generated by dilating and shifting the scaling function and the mother wavelet. Dilating and shifting interrogates the correlation content in the 2D structure at various scales, hence termed as multiscale decomposition. This very aspect makes DWTs far more superior to techniques such as STFT and local SVD which captures the correlation structure at a single scale. DWT coefficients at lower-scales provide finer resolution (details) and high frequency (H) components in the given data. Those at higher-scales provide coarser resolution (approximations) or low-frequency (L) components in the given data. Since the objective of the proposed study is to understand local correlation structures and their variation across the (PM, MM) probe intensity matrices, the emphasis will be on the approximation coefficients in the DWTs.

**Example.** A $(k = 3)$ level hierarchical decomposition of $X$ into details and approximations using 1D DWT is shown below. The details and the approximations correspond to high-frequency (H) and low-frequency (L) components respectively. Thus at each stage one encounters two possibilities (H and L).

$$X \qquad \text{(given data)}$$
$$= L_1 \qquad + H_1 \quad (k = 1, \text{ first level decomposition})$$
$$L_1 = L_2 \qquad + H_2 \quad (k = 2, \text{ second level decomposition})$$
$$L_2 = L_3 + H_3 \quad (k = 3, \text{ third level decomposition})$$

At each level $(k)$, the relation $L_{k-1} = L_k + H_k$ holds. 2D DWT [37] is given as a tensor product of row-wise and column-wise 1D (separable) DWTs of the given matrix. Row-wise and column-wise decompositions give rise to approximations and details along either directions resulting in four possible outcomes namely: (LL, LH, HL and HH) respectively. Similar to 1D DWT, 2D DWT decomposition at the level $k$ satisfies the relation $LL_{k-1} = LL_k + LH_k + HL_k + HH_k$. The term $LL_k$ corresponds to the approximation (low frequency component) whereas ($LH_k$, $HL_k$ and $HH_k$) correspond to vertical, horizontal and diagonal details (high frequency components) respectively. The choice of a particular wavelet is dictated by important properties. These include (a) compact support (b) symmetry (c) orthogonality (d) regularity and (e) vanishing moments. A brief explanation of these terms are enclosed below. (a) *compact support*: wavelets with compact support correspond to FIR (finite impulse response) filters and useful in time localization. (b) *symmetry*: symmetric wavelets do not give rise to artifacts at the boundaries (c) *orthogonality*: orthogonality significantly reduces the computational burden, hence results in faster implementation (d) *regularity*: governs the degree of smoothness and usually proportional to the order of the filters. (e) *vanishing moments*: the maximum polynomial degree representation that can be generated by the scaling function. From the perspective of the proposed study, emphasis will be on (a) compact support, (b) symmetry (d) regularity and (e) vanishing moments. As noted earlier, DWT represents the correlation structure in the given matrix as a hierarchical decomposition that satisfies the recursive relation $LL_{k-1} = LL_k + LH_k + HL_k + HH_k$ at each level $k$.

A three level hierarchical decomposition (DWT) of a portion of the raw MM and the corresponding PM probe intensity matrices using Biorthogonal wavelet 2.6 (i.e. order of reconstruction = 2 and order of decomposition = 6) is shown in Figs. 15.7a and 15.7b

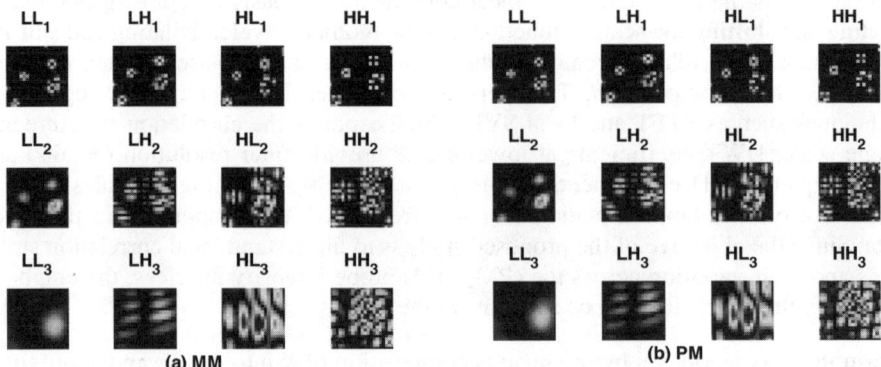

**Fig. 15.7.** Three-level hierarchical decomposition (DWT) of a small portion of the raw MM probe intensity matrix (a) and raw PM intensity matrix (b) using biorthogonal wavelet 2.6. The details and the approximation coefficients are color coded for visualization.

respectively. The choice of biorthogonal wavelet is encouraged by the fact that it is compact and symmetric. The approximations at the three levels are represented by $LL_i$, $i = 1$, 2 and 3, the horizontal, vertical and diagonal details are represented by $LH_i$, $HL_i$ and $HH_i$ respectively with $i = 1$, 2 and 3. The magnitudes of the coefficients are color coded to aid visualization of locally correlated regions. The corresponding color-coefficient mapping is also included. Brighter colors correspond to probes which exhibit significant local correlation/patchy regions. From Figs. 15.4a and 15.4b, there is a clear overlap in local correlation structures between the PM and MM probe intensity matrices. In the following section, we propose an approach to determine whether the correlation structures are statistically significant.

## 15.4  Discussion

Gene expression estimation in Genechip microarrays are governed by the qualitative behavior of atomic entities on the arrays called probes. These probes can be broadly classified into perfect and mismatch probes. While the former is a measure of specific binding, the latter is used an internal control to assess non-specific binding. Understanding the qualitative behavior at the probe level can have significant impact on gene expression estimation, higher level analyses and subsequent biological inference. Classical techniques estimate gene expression as a complex combination of PM or PM and MM intensities. The behavior of the mismatch probes has especially proven to be elusive. The present study elucidates qualitative similarities in the distributional signatures and local correlation structure/patchiness of the perfect match and mismatch probe intensity matrices. The results were established on publicly available microarray gene expression data generated across laboratories investigating the same biological paradigm. These results were also established on the raw and background subtracted PM and MM probe intensity data. Thus background subtraction using popular techniques seem to have negligible effect on the qualitative similarities between PM and MM probe intensities.

Power-law approximations attributed to inherent biological mechanisms were found to persist across the PM as well as MM probe intensities and across replicate arrays generated across laboratories investigating the same paradigm. These preliminary findings argue in favor of non-biological factors contributing to the observed power-law signatures including the transfer function of the measurement device (i.e. microarray) which maps the true biological phenomena onto the probe intensity value. Analysis of the PM and MM probe intensity matrices using local singular value decomposition revealed statistically significant locally correlated patches reflecting inherent heterogeneity and variation in statistical properties. Patchiness persisted across the PM and MM probe intensity matrices. The results were established across the raw as well as background subtracted probe intensity data and across replicate arrays between laboratories investigating the same paradigm. Majority of the probes comprising the patchy regions were found to be either multiple targeting or cross-hybridizing probes. The preliminary results reported in this study raise fundamental concerns in interpreting gene expression data and encourage possible exclusion of certain probes that are common to PM as well as MM probe intensity matrices from gene expression estimation and subsequent higher level analysis. A more detailed investigation using sophisticated approaches such as maximum likelihood and multiscale decomposition is necessary in order to completely understand the distributional signatures and local correlation structures at the probe intensities.

# References

1. Affymetrix Genechip Expression Analysis Technical Manual
2. Affymetrix Microarray Suite 3.0 (MAS 3.0), Affymetrix Santa Clara
3. Affymetrix Microarray Suite 5.0 (MAS 5.0), Affymetrix Santa Clara
4. Akaike, H.: Information theory and an extension of the Maximum Likelihood Principle. In: Proceedings of the 2nd International Symposium of Information Theory, Akadamiai Kiado, Budapest, pp. 267–281 (1973)
5. Alexa, A., Rahnenfuhrer, J., Lengauer, T.: Improved scoring functional groups from gene expression data by decorrelating GO graph structure. Bioinformatics 22(13), 1600–1607 (2006)
6. Alter, O., Brown, P.O., Botstein, D.: Singular Value Decomposition For Genome-Wide Expression Data Processing and Modeling. Proc. Natl. Acad. Sci. USA 97(18), 10101–10106 (2000)
7. Bogdan, M., Ghosh, J.K., Doerge, R.W.: Modifying the Schwarz Bayesian Information Criterion to Locate Multiple Interacting Quantitative Trait Loci. Genetics (167), 989–999 (2004)
8. Box, G.E.P., Cox, D.R.: An analysis of transformations. J. Roy. Stat. Soc. B 26, 211–252 (1964)
9. Castillo, E.: Extreme Value Theory in Engineering. Academic Press, Boston (1988)
10. Clauset, A., Shalizi, C.R., Newman, M.E.J.: Powerlaw distributions in empirical data. Rev. Mod. Physics (2007), http://arxiv.org/abs/0706.1062
11. Dhand, R.: The finished landscape. Nature S1, 7 (2006)
12. Elowitz, M.B., Levine, A.J., Siggia, E.D., Swain, P.S.: Stochastic Gene Expression in a Single Cell. Science 297(5584), 1183–1186 (2002)
13. Fraser, H.B., Khaitovich, P., Plotkin, J.B., Paabo, S., Eisen, M.B.: Aging and Gene Expression in the Primate Brain. PLoS Biology 3(9), e274 (2005)

14. Friedman, N.: Inferring Cellular Networks Using Probabilistic Graph Models. Science 303(5659), 799–805 (2004)
15. Gardner, T.S., Cantor, C.R., Collins, J.J.: Construction of a genetic toggle switch in Escherichia coli. Nature 403, 339–342 (2000)
16. Gautier, L., Moller, M., Friis-Hanse, L., Knudsen, S.: Alternative mapping of probes to genes for Affymetrix chips. BMC Bioinformatics 5, 111 (2004)
17. Gentleman, R.C., et al.: Bioconductor: open software development for computational biology and bioinformatics. Genome Biol. 5(10), R80 (2004)
18. Goldbeter, A., Dupont, G.: Allosteric regulation, cooperativity, and bio-chemical oscillations. Biophys. Chem. 37, 341–353 (1990)
19. Golub, G.H., van Loan, C.F.: Matrix Computations. Johns Hopkins University Press (1996)
20. Harrell Jr., F.E.: Regression Modeling Strategies. Springer, N.Y. (2001)
21. Hofmann, W.-K.: Gene Expression Profiling by Microarrays: Clinical Implications. Cambridge University Press, Cambridge (2006)
22. Hoyle, D.C., Rattray, M., Jupp, R., Brass, A.: Making sense of microarray data distributions. Bioinformatics 18(4), 576–584 (2002)
23. Hurvich, C.M., Tsai, C.L.: Regression and time series model selection in small samples. Biometrika 76, 297–307 (1989)
24. Irizarry, R.A., Bolstad, B.M., Collin, F., Cope, L.M., Hobbs, B., Speed, T.P.: Summaries of Affymetrix GeneChip probe level data. Nucleic Acids Res. 31(4), e15 (2003)
25. Irizarry, R.A., Hobbs, B., Collin, F., Beazer-Barclay, Y.D., Antonellis, K.J., Scherf, U., Speed, T.P.: Exploration, Normalization, and Summaries of High Density Oligonucleotide Array Probe Level Data. Biostatistics 4(2), 249–264 (2003)
26. Irizarry, R.A., et al.: Multiple-laboratory comparison of microarray platforms. Nature Methods 2, 345–350 (2005); This entire issue was dedicated to various aspects of microarray analysis
27. Ivanova, N.B., Dimos, J.T., Schaniel, C., Hackney, J.A., Moore, K.A., Lemischka, I.R.: A Stem Cell Molecular Signature. Science 298, 601–604 (2002)
28. Jansen, R.C., Nap, J.P.: Genetical genomics: the added value from segregation. Trends GenetICS (17), 388–391 (2001)
29. Kaern, M., Elston, T.C., Blake, W.J., Collins, J.J.: Stochasticity in Gene Expression: From Theories to Phenotypes. Nat. Rev. Genetics 6, 451–464 (2005)
30. Kitano, H.: Systems Biology: A Brief Overview. Science 295(5560), 1662–1664 (2002)
31. Kobayashi, M.D., et al.: Bacterial Pathogens modulate an apoptosis differentiation program in human neutrophils. Proc. Nat. Acad. Sci. (USA) 100(19), 10948–10953 (2003)
32. Kuznetsov, V.A., Knott, G.D., Bonner, R.F.: General statistics of stochastic process in Eukaryotic cells. Genetics 161(3), 1321–1332 (2002)
33. Leong, H.S., Yates, T., Wilson, C., Miller, C.J.: ADAPT: A database of affymetrix probesets and transcripts. Bioinformatics 21(10), 2552–2553 (2005)
34. Li, C., Wong, W.H.: Model-based analysis of oligonucleotide arrays: Expression index computation and outlier detection. Proc. Natl. Acad. Sci. (USA) 98, 31–36 (2001)
35. Lipshutz, R.J., Fodor, S., Gingeras, T., Lochart, D.: High density synthetic oligonucleotide array. Nature Genetics 21(suppl. 1), 20–24 (1999)
36. Lockhart, D.J., Dong, H., Byrne, M.C., Follettie, M.T., Gallo, M.V., Chee, M.S., Mittman, M., Wang, C., Kobayashi, M., Horton, H., Brown, E.L.: Expression monitoring by hybridization to high-density oligonucleotide arrays. Nature Biotechnology 14(13), 1675–1680 (1996)
37. Mallat, S.: A wavelet tour of signal processing. Academic Press, London (1998)

38. Naef, F., Lim, D.A., Patil, N., Magnasco, M.: DNA hybridization to mismatched templates: a chip study. Phys. Rev. E 65, 040902 (2002)
39. Nagarajan, R., Upreti, M.: Correlation Statistics for cDNA Microarray Image Analysis. IEEE/ACM Trans. Comp. Biology Bioinform. 3(3), 232–238 (2006)
40. Nagarajan, R., Upreti, M.: Qualitative assessment of gene expression in Affymetrix genechip arrays. Physica A 373(1), 486–496 (2007)
41. Nagarajan, R., Aubin, J.E., Peterson, C.A.: Modeling genetic networks from clonal analysis. J. Theor. Biology 230(3), 359–373 (2004)
42. Okoniewski, M.J., Miller, C.J.: Hybridization interactions between probesets in short oligo microarrays lead to spurious correlations. BMC Bioinformatics 7, 276 (2006)
43. Perou, C.M., et al.: Molecular portraits of human breast tumors. Nature 406, 747–752 (2000)
44. Pavelka, N., et al.: A power law global error model for the identification of differentially expressed genes in microarray data. BMC Bioinformatics 5, 203 (2004)
45. Phimister, B.: Going global Nature Genetics 21, 1 (1999)
46. Quackenbush, J.: Microarray data normalization and transformation. Nature Genetics 32, 496–501 (2002)
47. Ramalho-Santos, M., Yoon, S., Matsuzaki, Y., Mulligan, R.C., Melton, D.A.: Stemness: Transcriptional Profiling of Embryonic and Adult Stem Cells. Science 298, 597–600 (2002)
48. Speed, T.: Statistical Analysis of Gene Expression Microarray Data. CRC Press, Boca Raton (2003)
49. Stanley, H.E.: Phase Transitions: Power Laws and Universality. Nature (1995) 378, 554 (2002)
50. Staudt, L.M.: It's ALL in the diagnosis. Cancer Cell 1, 109–110 (2002)
51. Strogatz, S.H.: Nonlinear dynamics and chaos: With applications to physics, biology, chemistry, and engineering. Perseus Books, Reading (2001)
52. Suarez-Farinas, M., Haider, A., Wittowski, K.M.: Harshlighting small blemishes on microarrays. BMC Bioinformatics 6, 65 (2005)
53. Thattai, M., van Oudenaarden, A.: Intrinsic noise in gene regulatory networks. Proc. Natl. Acad. Sci. (USA) 98, 8614 (2001)
54. Theiler, J., Eubank, S., Longtin, A., Galdrikian, B., Farmer, J.D.: Testing for nonlinearity in time series: the method of surrogate data. Physica D 58, 77–94 (1992)
55. Theiler, J., Prichard, D.: Constrained-realization MonteCarlo method for hypothesis testing. Physica D 94(4), 221–235 (1996)
56. Ueda, H.R., Hayashi, S., Matsuyama, S., Yomo, T., Hashimoto, S., Kay, S.A., Hogenesch, J.B., Lino, M.: Universality and flexibility in gene expression from bacteria to human. Proc. Natl. Acad. Sci. USA 16(101), 3765–3769 (2004)
57. Wu, C., Carta, R., Zhang, L.: Sequence dependence on cross-hybridization on short oligo microarrays. Nucl. Acids Res. 33(9), e84 (2005)
58. Wu, Z., Irizarry, R.A.: Preprocessing of oligonucleotide array data. Nat. Biotech. 22, 656–658 (2004)
59. Yeung, M.K., Tegner, J., Collins, J.J.: Reverse engineering gene networks using singular value decomposition and robust regression. Proc. Natl. Acad. Sci. USA 30, 6163–6168 (2002)
60. Zhang, L., Miles, M.F., Aldape, K.A.: A model of molecular interactions on short oligonucleotide arrays. Nature Biotech. 21(7), 818–821 (2003)

# 16

# Case Study: Structure and Function Prediction of a Protein with No Functionally Characterized Homolog

Vijayaraj Nagarajan and Mohamed O. Elasri

Department of Biological Sciences, The University of Southern Mississippi, Hattiesburg, MS 39406, USA
{vijayaraj.nagarajan,mohamed.elasri}@usm.edu

**Summary.** The post-genomic era has seen a significant increase in the use of computational prediction methods to gain insights into structure and function of proteins. Prediction tools are used to guide the experimental design to test various hypotheses about structure and function of known proteins. However, these tools are particularly useful when studying putative protein sequences with no known function. The genomic era produced a large number of sequences that are described as either hypothetical proteins or as proteins with unknown function. Current molecular biology techniques are not adequate to efficiently study this vast reservoir of genetic information. However, computer algorithms can process large amounts of sequence data to predict structure and function. These knowledge-based computational tools use available experimental data and are regularly updated to improve their predictive power. The simplest form of function prediction is achieved by comparison of the query sequence to all available sequences using BLAST. If the query sequence is highly similar to previously characterized proteins, then it is likely that the query sequence has similar functions. However, if the query sequence does not have any homologous sequence with known function, then more sophisticated computational tools are necessary to gain insight into structure and function. Various methods have been developed to search for known domains, motifs, patterns, or profiles. The quality of predictions is dependent on the type of tools used and is limited to the closeness of the query sequence to known proteins.

In this chapter, we will describe and discuss methods and tools we used to predict structure and function of a putative protein sequence (Msa) with unknown function. We will address the advantages and limitations of all these approaches by using the Msa protein from the human pathogen *Staphylococcus aureus* as a case study. Msa is a novel protein that is involved in regulation of virulence. Since Msa has no known homolog, computational tools are being used to predict its structure and mechanism of action. These predictions are used to design experiments to study Msa and explore its use as a therapeutic target to combat antibiotic-resistant infections.

## 16.1 Background

The post-genomic era has seen a significant increase in the use of computational prediction methods designed to gain insights into structure and function of proteins. Prediction tools are useful guide for experimental design. They can be

T.G. Smolinski et al. (Eds.): Comp. Intel. in Biomed. & Bioinform., SCI 151, pp. 379–395, 2008.
springerlink.com                                            © Springer-Verlag Berlin Heidelberg 2008

used to test various hypotheses about structure and function of known proteins. They are also particularly useful when studying protein sequences with no known function.

### 16.1.1   Genomic Era

The genomic era produced a large number of sequences that are described as either hypothetical proteins or as proteins with unknown function. Indeed, the NCBI Gene database contains 571,064 bacterial genes with hypothetical function (34.27% of known bacterial genes) and 180,062 bacterial genes with unknown function (10.80% of known bacterial genes) (statistics as on June 28, 2007). Knowledge of the functional properties of these unknown proteins would greatly enhance our understanding of the molecular mechanisms of biology with great payoffs in several areas of research such as medicine, agriculture, environment and biotechnology.

Traditional techniques of molecular biology like mutation, cloning and over expression are not adequate to efficiently study this vast reservoir of genetic information. Therefore, reliance on computer algorithms which can process large amounts of sequence data to predict structure and function is necessary to gain knowledge. There are two types of prediction tools: knowledge-based and *ab initio*. knowledge-based computational tools use available experimental data and are regularly updated to improve their predictive power. On the other hand, *ab initio* tools use only the physico-chemical properties of the biological molecules to predict structure and function.

### 16.1.2   Function Prediction

Function prediction could be done based on a variety of aspects like physico-chemical properties, sequence similarity, structure similarity, gene expression data, biomolecular interaction information, gene ontology, phylogeny and text mining. Depending on how much prior information we know about the query sequence, one or more of these prediction methods could be used. The simplest form of function prediction is done by comparison of a query sequence to all available protein sequences and/or structures using a similarity search tool such as BLAST [1]. If the query sequence is similar to sequences of characterized proteins, then function is inferred from the known proteins. However, if the query sequence does not have any homology ('homology' is defined as the similarity that is inherited by virtue of common ancestry) with characterized proteins, then the problem becomes more challenging. In the latter case, specialized and more sophisticated computational tools are necessary to gain insights into the structure and function of novel proteins. A wide variety of methods have been developed to predict and search for known domains, motifs, patterns, or profiles. The predictive quality of these methods depend on the type of tools used and how close the query sequence is to known proteins.

### 16.1.3   An Embarrassment of Prediction Tools

There is no shortage of prediction tools; in fact the large and growing number of computational tools available makes it more difficult to keep up to date on the latest algorithms in the field of Bioinformatics. This is particularly problematic for "bench" scientists (like molecular biologists and microbiologists) who do not use these tools on a regular basis. There are literally hundreds of online/offline tools, packages, modules, scripts and automated pipelines for large scale predictions. There are even operating systems (linux based) that comes preconfigured with several open source Bioinformatics software. For a recent comprehensive survey of protein function prediction tools see Gaurav *et al.* [2].

The efficiency and performance of these prediction tools are constantly enhanced; however, function assignment is still a daunting task. Indeed searching NCBI protein database with the term "unknown function" returned about half a million (431,070) protein sequences. We have faced the problem of function assignment after the discovery of a novel protein Msa, in *Staphylococcus aureus* [3].

### 16.1.4   Msa, a Case Study

*Staphylococcus aureus* is an important human pathogen causing several diseases ranging from common skin infection to serious life-threatening diseases. The emergence of antibiotic resistant strains of *S. aureus* necessitate discovery of novel drugs to address this important public health problem. An important prerequisite to developing new treatment is the identification of novel therapeutic target. We have discovered a novel protein (Msa) which plays a critical role in regulation of virulence in *S. aureus* [3]. Since Msa has no homolog with known function, we used computational tools to predict its structure and mechanism of action [4]. These predictions allowed us to develop a hypothetical model for Msa which served as the basis for our wet lab experimental design to study this protein and explore its use as a therapeutic target.

In this chapter we will describe and discuss the methods and tools that we used to study the Msa protein *in silico*. This chapter is not a comprehensive review of all the available tools for function prediction, rather we focused on a set of tools that we found most useful to study a protein with no known homolog. This set of tools can be used, with some variations, to study other proteins.

## 16.2   The Starting Point: BLAST

Basic Local Alignment Search Tool (BLAST) [1] is the first tool that one could start with in protein function prediction. BLAST is a powerful similarity searching tool hosted by National Center for Biotechnology Information (NCBI). Given a query sequence (either nucleotide or protein), BLAST searches for similarity against several databases that contain sequence, structure, domain or expression information.

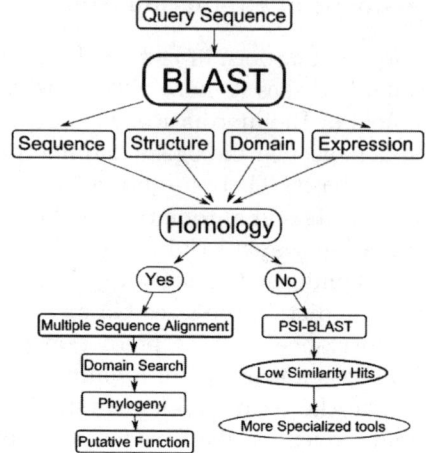

**Fig. 16.1.** The Overview of function prediction based on initial BLAST results

If the BLAST results contain sequences with statistically significant similarity (high score and low E-value) to the query sequence, then one can infer function from homology. From this point additional sequence analysis can be done such as multiple sequence alignment, conserved domain search, phylogeny etc. (Figure 16.1). One can then proceed to experimental design to test the predictions. In some cases however, no significant hits are produced using the default parameters. This was the case of Msa. In this event one can change the BLAST parameters such as by increasing the E-value threshold or using other version of BLAST like PSI-BLAST to find distant similarities.

### 16.2.1    BLAST Results for Msa

Our simple BLAST search against the non-redundant protein database, using the 133 amino acid length Msa protein sequence (Locus : Q7A5P4) did not return

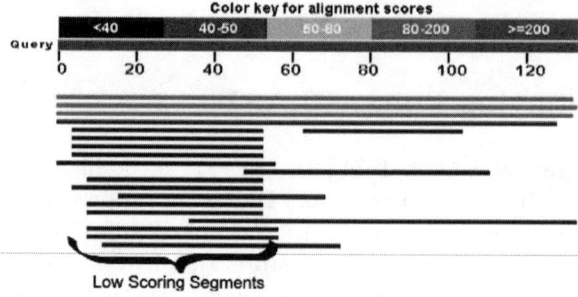

**Fig. 16.2.** The BLAST similarity search result showing very few similar sequences. Several low scoring segments are also found in the N-terminal region.

any significant homolog. Searching additional databases that are not included in the BLAST default setting also did not yield any significant homolog. We did not find any homolog in the conserved domain database (CDD) or by using PSI-BLAST. However, BLAST and PSI-BLAST gave us several hits that had low scores and high E-values. The similarity of these low scoring hits were confined to the N-terminal region (~40 residues) (Figure 16.2). Interestingly, despite the fact that these low scoring sequences came from a diverse group of organisms, they were all integral membrane proteins.

## 16.3   The Next Step

Similarity searches that do not yield clear homologs can still produce some clues that can help in deciding on the next step in analysis. Indeed, in the absence of homologs one can utilize prediction tools that rely primarily on the physico-chemical properties or structural properties (primary, secondary, or tertiary). Tools that predict other properties like localization, solubility, or other features could also be used. The decision of the most appropriate tool to use can be based on the findings from the BLAST analysis.

In the case of Msa, BLAST results suggested that Msa might be a membrane protein. Multiple sequence alignment and phylogeny at this point did not yield any more useful information since Msa sequences from various strains of *S. aureus* are highly conserved (~98% similar). So, we started with sub-cellular localization prediction tools to find out if Msa localizes to the membrane.

Several tools are available to predict protein sub-cellular localization [5]. These tools make localization predictions for either prokaryotic or eukaryotic sequences. They use a wide variety of methods from simple amino acid composition to complex machine learning approach. Each tool has advantages and disadvantages associated with it. The best way to approach these problems is by using as many tools as possible with the hope to build a consensus.

### 16.3.1   Localization of Msa

When studying the localization of Msa, the first tool that we used was PRED-CLASS [6], since it claimed to produce near 100% accuracy for classifying proteins as membrane, fibrous, globular or mixed. PRED-CLASS predicted Msa to be a membrane protein. We then used PSORT [7] for gram positive bacteria, which also predicted Msa as a membrane protein, with significant "certainty". In addition, PSORT predicted the presence of a cleavable N-terminal signal sequence and two transmembrane regions. ProtCompB, which specializes in gram positive bacteria also predicted Msa as an integral membrane protein with an N-terminal signal sequence. We also used PSLpred [8] which predicted localization for prokaryotic proteins, based on five different methods. Only two of the PSLpred methods (the amino acid composition based method and the properties based method) predicted Msa as a membrane protein. PSLpred however does not distinguish between gram positive and gram negative bacteria in its predictions. CELLO [9] ; one of the recent tools also predicted Msa as a membrane protein, with a significant score.

Finally, two other programs SVMProt [10] and ProtFun [11], both of which predict the functional category of a protein sequence, also predicted Msa as belonging to the membrane protein category. Results from most of the tools used showed a consensus prediction of Msa as a membrane protein. The GRAVY (grand average of hydropathicity) index [12] value of 1.021 also suggests that Msa is probably an insoluble protein. This allowed us to focus our analysis on aspects of Msa that pertain to integral membrane proteins.

## 16.4    Advanced Analyses of a Putative Membrane Protein

Given the fact that Msa was predicted to be a membrane protein by a variety of localization tools, there is a number of analysis that could be done to further study Msa *in silico*. For instance, we could search for a signal peptide, predict the transmembrane topology, determine the hydropathy index as well as make secondary structure predictions. Interestingly, two of the sub-cellular localization tools predicted a cleavable signal sequence in the N-terminal region of Msa. This prompted us to use specialized signal prediction tools.

### 16.4.1    Signal Peptide Prediction

Signal peptides are essential components for many membrane proteins. Several tools are available to predict the presence and identity of a signal peptide in a putative membrane protein. These tools can guide experimental design to determine the mechanism of targeting of membrane proteins. One such tool is SignalP [13] which predicts the presence of signal peptides as well as a putative cleavage site, using a hidden Markov model (HMM) and a neural network (NN).

Both HMM and NN of SignalP showed a high probability for the presence of an N-terminal signal peptide in the Msa protein. However, while HMM predicted a cleavage site between $29^{th}$ and $30^{th}$ residue, NN predicted the cleavage site between $30^{th}$ and $31^{st}$ residue. The HMM prediction had a higher probability than the NN prediction. Another signal peptide prediction tool iPSORT [7] predicted the presence of a signal peptide in the first 30 residues with a score several fold higher than the threshold value (Figure 16.3). It is noteworthy to state that PSORTb [14], a member of the PSORT family, failed to predict a signal peptide for Msa, but the more generic version of PSORT [7] predicted a signal peptide cleavage site at position 20. Indeed PSORTb is mainly recommended for signal peptide predictions in gram negative organisms (those bacteria that fail to retain the dye (crystal violet) during the Gram staining protocol).

Surprisingly SIG-Pred for gram positive organisms did not find a signal peptide in Msa. In contrast, the eukaryotic version of the SIG-Pred predicted a cleavable signal peptide in Msa at position 20. This could be due to the fact that SIG-Pred has a false negative rate that is several fold higher in comparison

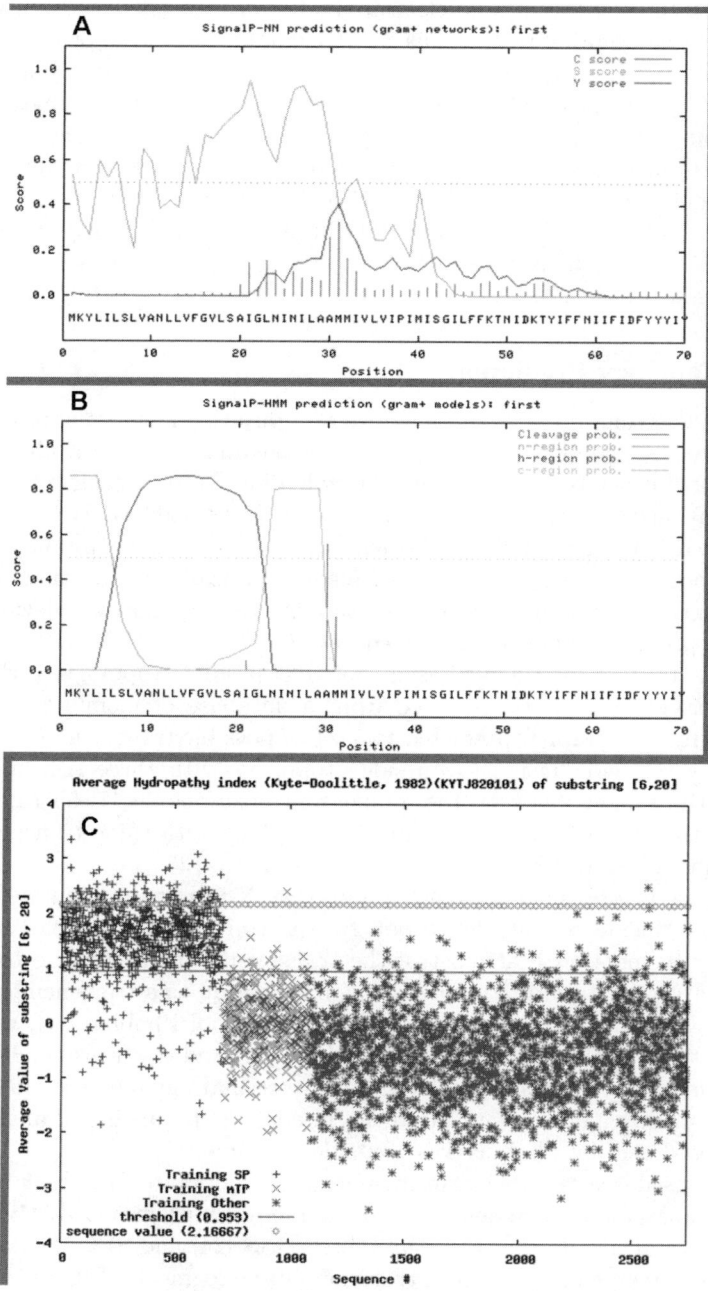

**Fig. 16.3.** Signal peptide prediction results from A: SignalP-NN (with largest C-score near the 30 $^{th}$ amino acid), B: SignalP-HMM (with highest cleavage probability between $29^{th}$ and $30^{th}$ amino acid) and C: iPSORT (with first 30 amino acids showing high probability for being the signal peptide sequence)

to the other signal peptide prediction tools. Finally, another program that did not find a signal peptide in Msa was SOSUIsignal [15]. This is probably due to fact that SUIsignal does not perform well for prokaryotic sequences in comparison to eukaryotic ones. Indeed, this is acknowledged by SOSUIsignal's authors who attribute this to the lack of enough data for prokaryotic signal anchors. sigcleave [16] predicted the cleavage position at 20. Similarly, Phobius [17] also predicted a cleavage site at position 20. Based on the preponderance of evidence (predicted cleavage site by sigcleave, PSORT, Phobius and SIG-Pred), we drew a consensus prediction that Msa has a putative signal peptide that is cleavable at amino acid position 20.

### 16.4.2    Topology Prediction

Having made strong predictions that Msa is a putative membrane protein with an N-terminal signal peptide, the next step in the analysis was the characterization of the transmembrane topology. This will allow us to determine the number and orientation of the transmembrane segments. Transmembrane topology which is characterized by the number and orientation of the transmembrane segments, is very important for determination of membrane protein's function. We used several transmembrane prediction tools to analyze Msa. For a review of transmembrane prediction tools refer to Ikeda *et al.*, [18].

TMpred [19] is a knowledge-based tool that predicts topology of the query sequence based on statistics derived from a database of membrane spanning protein segments. TMpred predicted two set of possible transmembrane helices. One of this set had "inside to outside" topology with three transmembrane helices in the regions between the amino acid positions 29-47, 55-75 and 107-123. The other set had "outside to inside" topology with three transmembrane segments in the regions between 3-23, 27-47 and 107-125.

Since Msa is predicted to have a putative N-terminal signal peptide sequence, the "inside to outside" topology was considered plausible. It is also noteworthy to mention that most of the transmembrane prediction tools have the inherent property of predicting the signal peptide as a transmembrane segment [17]. To address this issue, we used another tool Phobius [17], which has the ability to distinguish between the signal peptide and the transmembrane segments. As expected, Phobius predicted the N-terminal signal peptide followed by three transmembrane segments. However, in contrast to our initial analysis, the topology predicted by Phobius was "outside to inside".

We then used several other transmembrane prediction tools to look for a consensus topology for Msa. When we used the following programs, TMHMM [20], SPLIT [21], HMMTOP [22], MEMSAT [23], DAS [24] and SEG [25], a consensus topology emerged showing four transmembrane segments (Figure 16.4), with "inside to outside" orientation. Based on previous analysis, we concluded that the first transmembrane segment predicted by all these tools is probably the signal peptide sequence (see Sec. 16.4.1). This prompted us to conclude that the consensus topology actually consists of three transmembrane segments with "inside to outside" orientation. We also tested the possibility of other models for

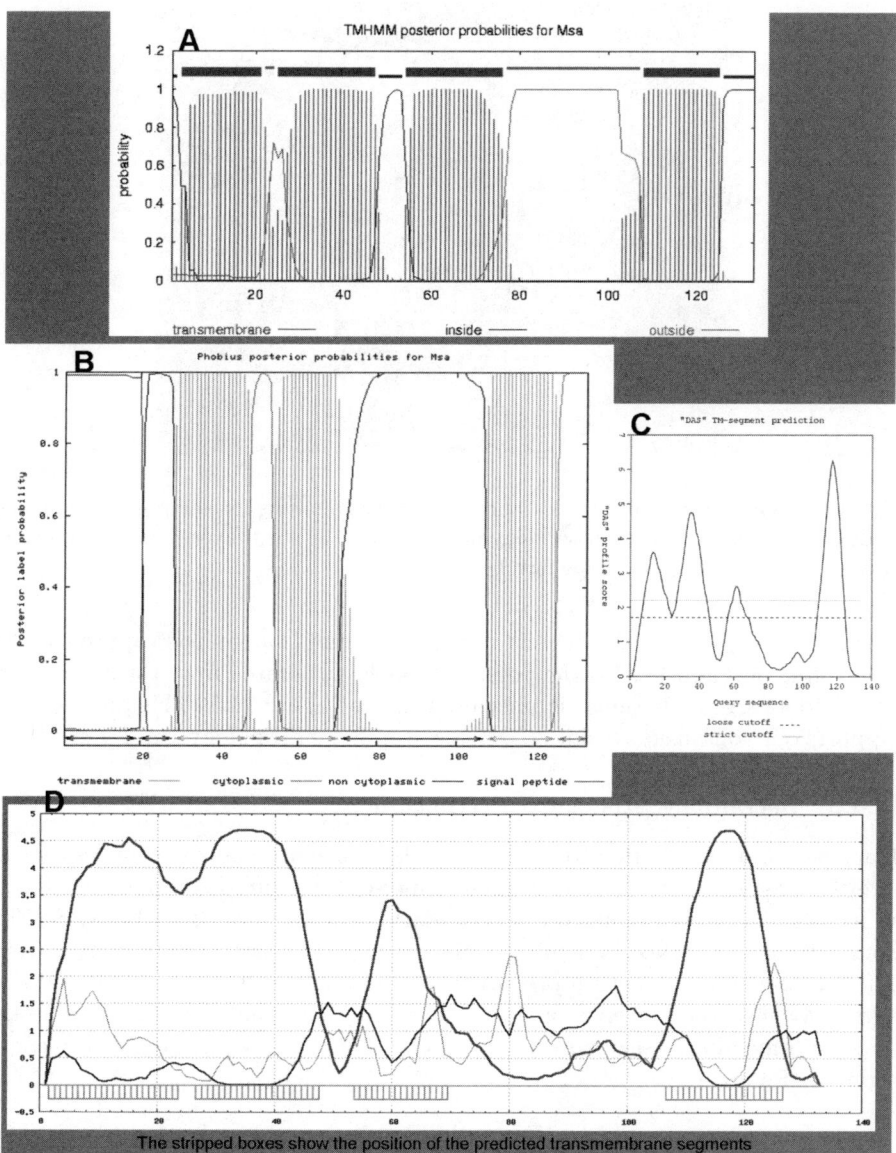

**Fig. 16.4.** The transmembrane predictions by A: TMHMM (showing four transmembrane segments), B: Phobius (clearly showing three distinct transmembrane segments), C: DAS (showing four transmembrane segments as four peaks above the cutoff line), D: SPLIT (shows four transmembrane segments). The first segment in the TMHMM, DAS and SPLIT predictions actually correspond to the signal peptide sequence.

388     V. Nagarajan and M.O. Elasri

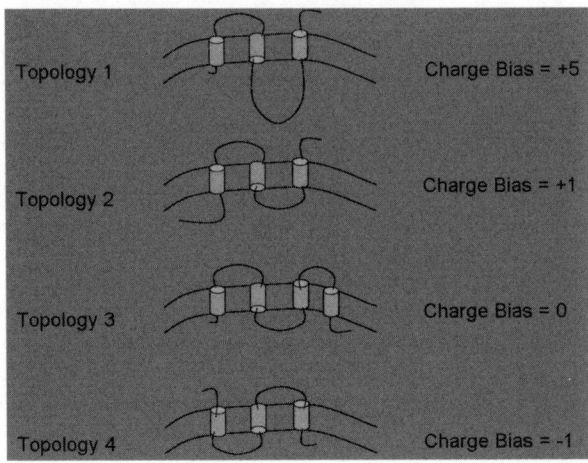

**Fig. 16.5.** The Positive-inside rule and Charge bias analysis for the possible topology models of Msa. Our consensus topology model is also the most favored model according to the positive-inside rule (Topology 1).

the Msa topology using the "Positive-inside rule and Charge bias" approach [26]. The charge bias analysis for the possible models (generated from the above mentioned topology predictions) is presented in the Figure 16.5. This analysis also supports our consensus "inside to outside" topology (Topology 1 in Figure 16.5).

### 16.4.3 Secondary Structure Prediction

Transmembrane prediction tools also provide information about secondary structure. For instance, the three transmembrane segments in Msa were predicted to be helices. To further analyze the secondary structure, we used the NPS (Network Protein Sequence Assembly) [27] server. NPS uses results from several different secondary structure prediction tools to derive at a consensus. NPS results confirmed the presence of three helices in Msa which corresponded to the three transmembrane segments and one helix that corresponded to a cytoplasmic segment (Figure 16.6).

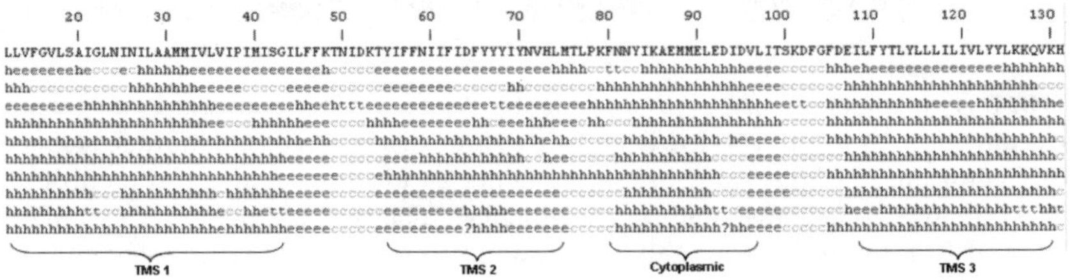

**Fig. 16.6.** The Consensus secondary structure prediction. TMS - Transmembrane Segment [4]

### 16.4.4 Domains, Patterns, Motifs

Domains and Motifs are distinct structural and functional units which confer specific function to a protein sequence. For example, presence of a DNA binding domain in a protein sequence is suggestive of the proteins function as a transcriptional factor or a polymerase enzyme . Domains, patterns and motifs are important in assigning putative function to proteins. Several tools are available to predict domains, patterns or motifs in protein sequences. A search against the protein families and domains database Pfam [28] did not yield any known functional domains for Msa. Search against another protein domain database ProDom [29] also did not yield any known functional domain for Msa. Using a meta-server InterProScan [30], an N-terminal signal peptide was the only prediction made on Msa.

We then used SMART (a Simple Modular Architecture Research Tool) [31], which predicted the presence of a PreATP-grasp domain in Msa. This PreATP-grasp domain (Structural Classification of Proteins (SCOP) entry: d1gsa_1) usually precedes the ATP-grasp domain and could contain a substrate-binding function. SMART located this domain between the $85^{th}$ and $116^{th}$ residue. Interestingly this location is predicted to be in a cytoplasmic loop region of Msa.

To search for patterns we used the PROSITE [32] database. We used different pattern searching tools such as PPSearch, PSITE [33] and ScanProsite [34]. Predictions from all these tools showed the presence of four putative phosphorylation sites in Msa. These phosphorylation sites were located at residues 48 (Tyrosine Kinase), 49 and 99 (Casein Kinase II). Residue 99 was also predicted to be a phosphorylation site for Protein Kinase C. Interestingly, two of the putative phosphorylation sites are located outside the membrane while one is located in the cytoplasmic region. Moreover, multiple sequence alignment of the Msa protein sequences from different strains of *S. aureus* showed that these putative

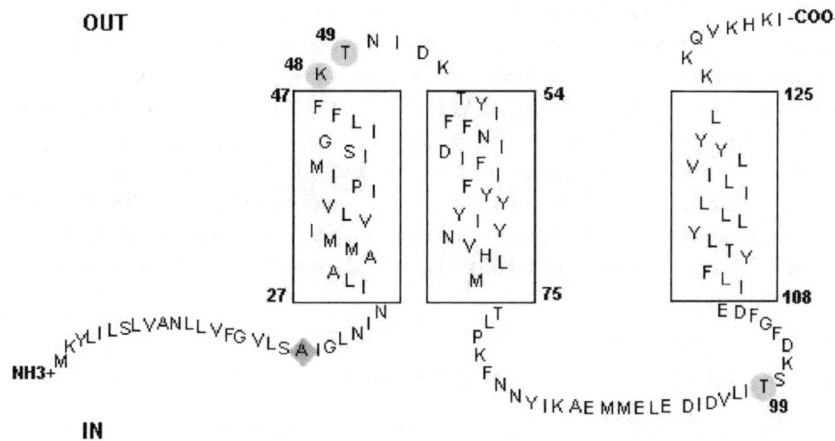

**Fig. 16.7.** The hypothesized model for Msa with three putative transmembrane regions, phosphorylation sites (circled) and signal peptide cleavage site (diamond)

phosphorylation sites are highly conserved. These predictions suggests that Msa is phosphorylated by kinases in the cytoplasm as well as on the outside of the membrane (e.g. from the host cells).

These findings led to the hypothesis that Msa functions as a signal transducer (Figure 16.7). However, Msa did not share homology with any known transmembrane signal transducers. These predictions will be tested by mutagenesis experiments. If the hypothesis is true, then Msa will be considered as a novel signal transducer.

## 16.5    3-D Structure Prediction and Analysis

Ultimately, tertiary structure of a protein is the most important determinant of its function. Determination of tertiary structure of proteins can be difficult, especially for membrane proteins. Several methods and tools are available for the prediction of 3-Dimensional (3-D) tertiary structure. Tertiary structure can be predicted using homology modeling, fold recognition or *ab initio* based methods. Once a tertiary structure is predicted, it could be used for structure based function predictions.

### 16.5.1    Structure Quality

Since Msa did not have any homologous structures we were not able to use homology modeling. We used the fold-recognition-based tool Phyre [35] to predict a 3-D structure for Msa. We visualized the predicted structure using Swiss-PDB Viewer (SPDBV) [36]. Preliminary analysis showed that the predicted 3-D structure, correlated with the predicted structural features of Msa. The 3-D structure showed the three transmembrane helices in similar positions as predicted previously. We analyzed the quality of the structure using both SPDBV as well as WHATIF [37] program. We refined the preliminary structure by energy minimization, side-chain fixing, removal of amino acid clashes, and fixing the problematic loops under the SPDBV environment. The refined structure was again checked for its quality using WHATIF and SPDBV. The refinement reduced the total energy of the structure slightly (Total energy of predicted model: -2359.390, Total energy of energy minimized model: -3368.924) . We also examined the protein backbone structure using Ramachandran plot [38] (the plot is used to visualize the dihedral angles of amino acid residues in proteins backbone conformation) which suggested that the predicted structure did not need any major improvement.

### 16.5.2    Structure Based Function Prediction

We used the 3-D model of Msa to predict properties like clefts and binding sites. ProFunc [39], Q-SiteFinder [40], PINUP [41] and SuMo [42] are the tools that we used to analyze the predicted tertiary structure of Msa. Three of them, ProfFunc, PINUP and Q-SiteFinder predicted a putative binding site in the

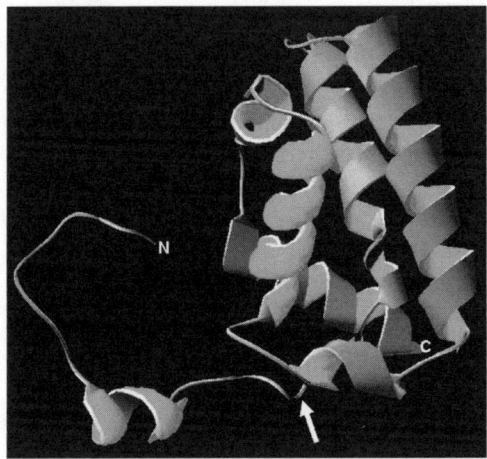

**Fig. 16.8.** The predicted tertiary structure of Msa protein, showing the three putative transmembrane helices and the N-terminal signal peptide

cytoplasmic loop between the second and the third transmembrane segments. This site also corresponds to the binding site previously predicted by SMART (see 16.4.4) based on the sequence similarity. In addition, based on tertiary structure, ProFunc predicted a "nest" (residues 47-50) in Msa. This nest showed features of an anion-binding site and is characteristic of functional motifs found in ATP or GTP binding proteins. The predicted tertiary structure strengthens our hypothesis that Msa is a novel signal transducer.

## 16.6  Summary

Prediction tools are extremely valuable as a first step in building hypotheses and designing experiments to study novel proteins. There is a wide variety of tools that utilize different algorithms to draw predictions on protein function. The best approach would be to become familiarized with as many tools as possible with a good understanding of their features and use several of them for the predictions. This might yield a consensus that strengthens the predictions. In the end, these predictions have to be tested experimentally before drawing any conclusions about the function of a protein.

## 16.7  URL's of Tools and Applications Used

### 16.7.1  Sequence Analysis

BLAST: http://www.ncbi.nlm.nih.gov/BLAST/
NPS: http://npsa-pbil.ibcp.fr/

## 16.7.2   Localization

PRED-CLASS http://athina.biol.uoa.gr/PRED-CLASS/
PSORT: http://psort.nibb.ac.jp/
ProtCompB: http://www.softberry.com
PSLpred: http://www.imtech.res.in/raghava/pslpred/
CELLO: http://cello.life.nctu.edu.tw/
SVMProt: http://jing.cz3.nus.edu.sg/cgi-bin/svmprot.cgi
ProtFun: http://www.cbs.dtu.dk/services/ProtFun-2.2/
ProtParam: http://ca.expasy.org/tools/protparam.html

## 16.7.3   Signal Peptide Prediction

SignalP: http://www.cbs.dtu.dk/services/SignalP/
iPSORT: http://hc.ims.u-tokyo.ac.jp/iPSORT/
psortB: http://www.psort.org/psortb/
SIG-Pred: http://www.bioinformatics.leeds.ac.uk/prot_analysis/Signal.html
SOSUIsignal: http://bp.nuap.nagoya-u.ac.jp/sosui/

## 16.7.4   Topology Prediction

TMpred: http://www.ch.embnet.org/software/TMPRED_form.html
Phobius: http://phobius.cgb.ki.se/
TMHMM: http://www.cbs.dtu.dk/services/TMHMM/
SPLIT: http://split.pmfst.hr
HMMTOP: http://www.enzim.hu/hmmtop
MEMSAT: http://saier-144-37.ucsd.edu/memsat.html
DAS: http://www.sbc.su.se/~miklos/DAS/
SEG: http://www.genome.jp/SIT/tsegdir/

## 16.7.5   Domains/Patterns/Motifs

Pfam: http://www.sanger.ac.uk/Software/Pfam/
ProDom: http://prodom.prabi.fr/
InterProScan: http://www.ebi.ac.uk/InterProScan/
SMART: http://smart.embl-heidelberg.de/
PPSearch: http://www.ebi.ac.uk/ppsearch/
PSITE: http://www.softberry.com/
ScanProsite: http://www.expasy.ch/tools/scanprosite/

## 16.7.6   3-D Structure Prediction and Analysis

Phyre: http://www.sbg.bio.ic.ac.uk/~phyre/
SWISS-MODEL: http://swissmodel.expasy.org
WHATIF: http://swift.cmbi.kun.nl/WIWWWI/
ProFunc: http://www.ebi.ac.uk/thornton-srv/databases/ProFunc/

Q-SiteFinder: http://www.bioinformatics.leeds.ac.uk/qsitefinder/
PINUP: http://sparks.informatics.iupui.edu/PINUP/
SuMo: http://sumo-pbil.ibcp.fr/cgi-bin/sumo-welcome

# References

1. Altschul, S.F., Madden, T.L., Schaffer, A.A., Zhang, J., Zhang, Z., Miller, W., Lipman, D.J.: Gapped blast and psi-blast: a new generation of protein database search programs. Nucleic Acids Res 25(17), 3389–3402 (1997)
2. Pandey, G., Kumar, V., Steinbach, M.: Computational approaches for protein function prediction: A survey. Tech. Rep. TR 06-028, Department of Computer Science and Engineering, University of Minnesota (2006)
3. Sambanthamoorthy, K., Smeltzer, M.S., Elasri, M.O.: Identification and characterization of msa (sa1233), a gene involved in expression of sara and several virulence factors in staphylococcus aureus. Microbiology 152(Pt 9), 2559–2572 (2006)
4. Nagarajan, V., Elasri, M.O.: Structure and function predictions of the msa protein in staphylococcus aureus. BMC Bioinformatics 8(suppl 7), S5 (2007)
5. Matsuda, S., Vert, J.P., Saigo, H., Ueda, N., Toh, H., Akutsu, T.: A novel representation of protein sequences for prediction of subcellular location using support vector machines. Protein Sci. 14(11), 2804–2813 (2005)
6. Pasquier, C., Promponas, V.J., Hamodrakas, S.J.: Pred-class: cascading neural networks for generalized protein classification and genome-wide applications. Proteins 44(3), 361–369 (2001)
7. Nakai, K., Horton, P.: Psort: a program for detecting sorting signals in proteins and predicting their subcellular localization. Trends Biochem Sci 24(1), 34–36 (1999)
8. Bhasin, M., Garg, A., Raghava, G.P.: Pslpred: prediction of subcellular localization of bacterial proteins. Bioinformatics 21(10), 2522–2524 (2005)
9. Yu, C.S., Chen, Y.C., Lu, C.H., Hwang, J.K.: Prediction of protein subcellular localization. Proteins 64(3), 643–651 (2006)
10. Cai, C.Z., Han, L.Y., Ji, Z.L., Chen, X., Chen, Y.Z.: Svm-prot: Web-based support vector machine software for functional classification of a protein from its primary sequence. Nucleic Acids Res. 31(13), 3692–3697 (2003)
11. Jensen, L.J., Gupta, R., Staerfeldt, H.H., Brunak, S.: Prediction of human protein function according to gene ontology categories. Bioinformatics 19(5), 635–642 (2003)
12. Gasteiger, E., Hoogland, C., Gattiker, A., Duvaud, S., Wilkins, M.R., Appel, R.D., Bairoch, A.: In: Walker JM (ed.) The Proteomics Protocols Handbook, pp. 571–607. Humana Press (2005)
13. Bendtsen, J.D., Nielsen, H., von Heijne, G., Brunak, S.: Improved prediction of signal peptides: Signalp 3.0. J. Mol. Biol. 340(4), 783–795 (2004)
14. Gardy, J.L., Laird, M.R., Chen, F., Rey, S., Walsh, C.J., Ester, M., Brinkman, F.S.: Psortb v.2.0: expanded prediction of bacterial protein subcellular localization and insights gained from comparative proteome analysis. Bioinformatics 21(5), 617–623 (2005)
15. Gomi, M., Sonoyama, M., Mitaku, S.: High performance system for signal peptide prediction: Sosuisignal. Chem-Bio. Informatics Journal 4(4), 142–147 (2004)
16. von Heijne, G.: A new method for predicting signal sequence cleavage sites. Nucleic Acids Res. 14(11), 4683–4690 (1986)

17. Kall, L., Krogh, A., Sonnhammer, E.L.: A combined transmembrane topology and signal peptide prediction method. J. Mol. Biol. 338(5), 1027–1036 (2004)
18. Ikeda, M., Arai, M., Lao, D.M., Shimizu, T.: Transmembrane topology prediction methods: a re-assessment and improvement by a consensus method using a dataset of experimentally-characterized transmembrane topologies. Silico Biol. 2(1), 19–33 (2002)
19. Hofmann, K., Stoffel, W.: Tmbase - a database of membrane spanning protein segments. Biol. Chem. Hoppe-Seyler 374, 166 (1993)
20. Krogh, A., Larsson, B., von Heijne, G., Sonnhammer, E.L.: Predicting transmembrane protein topology with a hidden markov model: application to complete genomes. J. Mol. Biol. 305(3), 567–580 (2001)
21. Juretic, D., Zoranic, L., Zucic, D.: Basic charge clusters and predictions of membrane protein topology. J. Chem. Inf. Comput. Sci. 42(3), 620–632 (2002)
22. Tusnady, G.E., Simon, I.: The hmmtop transmembrane topology prediction server. Bioinformatics 17(9), 849–850 (2001)
23. Jones, D.T., Taylor, W.R., Thornton, J.M.: A model recognition approach to the prediction of all-helical membrane protein structure and topology. Biochemistry 33(10), 3038–3049 (1994)
24. Cserzo, M., Wallin, E., Simon, I., von Heijne, G., Elofsson, A.: Prediction of transmembrane alpha-helices in prokaryotic membrane proteins: the dense alignment surface method. Protein Eng. 10(6), 673–676 (1997)
25. Kihara, D., Shimizu, T., Kanehisa, M.: Prediction of membrane proteins based on classification of transmembrane segments. Protein Eng. 11(11), 961–970 (1998)
26. Heijne, G.v.: Membrane protein structure prediction. hydrophobicity analysis and the positive-inside rule. J. Mol. Biol. 225(2), 487–494 (1992)
27. Deleage, G., Blanchet, C., Geourjon, C.: Protein structure prediction. Implications for the biologist. Biochimie 79(11), 681–686 (1997)
28. Finn, R.D., Mistry, J., Schuster-Bockler, B., Griffiths-Jones, S., Hollich, V., Lassmann, T., Moxon, S., Marshall, M., Khanna, A., Durbin, R., Eddy, S.R., Sonnhammer, E.L., Bateman, A.: Pfam: clans, web tools and services. Nucleic Acids Res. 34(Database issue), D247–D251 (2006)
29. Bru, C., Courcelle, E., Carrere, S., Beausse, Y., Dalmar, S., Kahn, D.: The prodom database of protein domain families: more emphasis on 3d. Nucleic Acids Res. 33(Database issue), D212–D215 (2005)
30. Quevillon, E., Silventoinen, V., Pillai, S., Harte, N., Mulder, N., Apweiler, R., Lopez, R.: Interproscan: protein domains identifier. Nucleic Acids Res. 33(web server issue), W116–W120 (2005)
31. Letunic, I., Copley, R.R., Schmidt, S., Ciccarelli, F.D., Doerks, T., Schultz, J., Ponting, C.P., Bork, P.: Smart 4.0: towards genomic data integration. Nucleic Acids Res. 32(Database issue), D142–D144 (2004)
32. Hulo, N., Bairoch, A., Bulliard, V., Cerutti, L., Castro, E.D., Langendijk-Genevaux, P.S., Pagni, M., Sigrist, C.J.: The prosite database. Nucleic Acids Res. 34(Database issue), D227–D230 (2006)
33. Solovyev, V.V., Kolchanov, N.A.: Search for functional sites using consensus. In: Kolchanov, N.A., Lim, H.A. (eds.), pp. 16–21. World Scientific, Singapore (1994)
34. Castro, E.de., Sigrist, C.J., Gattiker, A., Bulliard, V., Langendijk-Genevaux, P.S., Gasteiger, E., Bairoch, A., Hulo, N.: Scanprosite: detection of prosite signature matches and prorule-associated functional and structural residues in proteins. Nucleic Acids Res. 34(Web Server issue), W362–W365 (2006)

35. Kelley, L.A., MacCallum, R.M., Sternberg, M.J.: Enhanced genome annotation using structural profiles in the program 3d-pssm. J. Mol. Biol. 299(2), 499–520 (2000)
36. Schwede, T., Kopp, J., Guex, N., Peitsch, M.C.: Swiss-model: An automated protein homology-modeling server. Nucleic Acids Res. 31(13), 3381–3385 (2003)
37. Vriend, G.: What if: a molecular modeling and drug design program. J. Mol. Graph 8(1), 29, 52–56 (1990)
38. Ramachandran, G.N., Ramakrishnan, C., Sasisekharan, V.: Stereochemistry of polypeptide chain configurations. J. Mol. Biol. 7, 95–99 (1963)
39. Laskowski, R.A., Watson, J.D., Thornton, J.M.: Profunc: a server for predicting protein function from 3d structure. Nucleic Acids Res. 33(web server issue), W89–W93 (2005)
40. Laurie, A.T., Jackson, R.M.: Q-sitefinder: an energy-based method for the prediction of protein-ligand binding sites. Bioinformatics 21(9), 1908–1916 (2005)
41. Liang, S., Zhang, C., Liu, S., Zhou, Y.: Protein binding site prediction using an empirical scoring function. Nucleic Acids Res. 34(13), 3698–3707 (2006)
42. Jambon, M., Imberty, A., Deleage, G., Geourjon, C.: A new bioinformatic approach to detect common 3d sites in protein structures. Proteins 52(2), 137–145 (2003)

# 17

# From Biomedical Literature to Knowledge: Mining Protein-Protein Interactions

Deyu Zhou[1], Yulan He[1], and Chee Keong Kwoh[2]

[1] Informatics Research Centre, The University of Reading, Reading, RG6 6BX, UK
d.zhou@reading.ac.uk, y.he@reading.ac.uk
[2] School of Computer Engineering, Nanyang Technological University,
Singapore 639798
asckkwoh@ntu.edu.sg

**Summary.** To date, more than 16 million citations of published articles in biomedical domain are available in the MEDLINE database. These articles describe the new discoveries which accompany a tremendous development in biomedicine during the last decade. It is crucial for biomedical researchers to retrieve and mine some specific knowledge from the huge quantity of published articles with high efficiency. Researchers have been engaged in the development of text mining tools to find knowledge such as protein-protein interactions, which are most relevant and useful for specific analysis tasks. This chapter provides a road map to the various information extraction methods in biomedical domain, such as protein name recognition and discovery of protein-protein interactions. Disciplines involved in analyzing and processing unstructured-text are summarized. Current work in biomedical information extracting is categorized. Challenges in the field are also presented and possible solutions are discussed.

## 17.1 Introduction

In post genomic science, proteins are recognized as elements in complex protein interaction networks. Hence protein-protein interactions play a key role in various aspects of the structural and functional organization of the cell. Knowledge about them unveils the molecular mechanisms of biological processes. However, most of this knowledge hides in published articles, scientific journals, books and technical reports. To date, more than 16 million citations of such articles are available in the MEDLINE [1] database. In parallel with these plain text information sources, many databases, such as DIP [2], BIND [3], IntAct [4] and STRING [5], have been built to store various types of information about protein-protein interactions. Nevertheless, data in these databases were mainly hand-curated to ensure their correctness and thus limited the speed in transferring textual information into searchable structure data. Retrieving and mining such information from the literature is very complex due to the lack of formal structure in the natural-language narrative in these documents. Thus, automatically extracting information from biomedical text holds the promise of easily discovering large amounts of biological knowledge in computer-accessible forms.

T.G. Smolinski et al. (Eds.): Comp. Intel. in Biomed. & Bioinform., SCI 151, pp. 397–421, 2008.
springerlink.com                                    © Springer-Verlag Berlin Heidelberg 2008

**Table 17.1.** Online databases, systems, tools relating to the extraction of protein-protein interactions

| Description | URL |
|---|---|
| **Online databases storing protein-protein interactions** | |
| BIND | Biomolecular Interaction Network Database contains over 200,000 human-curated interactions. | www.bind.ca/ |
| DIP | Database of Interacting Proteins catalogs experimentally determined interactions between proteins. Until now, it contains 55,732 interactions, combining information from various sources to create a single, stable set of protein-protein interactions. | dip.doe-mbi.ucla.edu/ |
| HPRD | The Human Protein Reference Database [14] contains interaction networks for each protein in the human proteome. All the information in HPRD has been manually extracted from the literature by expert biologists who read, interpret and analyze the published data. | www.hprd.org/ |
| HPID | Human Protein Interaction Database integrates the protein interactions in BIND, DIP and HPRD. | www.hpid.org/ |
| IntAct | IntAct consists of a open source database system and analysis tools for protein interaction data. It now contains more than 100,000 curated binary molecular interactions. | www.ebi.ac.uk/intact/ |
| MINT | Molecular INTeraction database [15] is a database storing interactions between biological molecules. It focuses on experimentally verified protein interactions with special emphasis on proteomes from mammalian organisms. | mint.bio.uniroma2.it/mint/ |
| STRING | STRING, a database consisting of known and predicted protein-protein interactions, quantitatively integrates interaction data from several sources for a large number of organisms. It currently contains 736,429 proteins in 179 species. | string.embl.de/ |
| **Online protein-protein interaction information extraction systems** | |
| BioRAT | BioRAT is a search engine and information extraction tool for biological research. | bioinf.cs.ucl.ac.uk/biorat/ |
| GeneWays | GeneWays is a system for automatically extracting, analyzing, visualizing and integrating molecular pathway data from the literature. It focuses on interactions between molecular substances and actions, providing a graphical consensus view on these collected information. | geneways.genomecenter.columbia.edu/ |
| MedScan | MedScan is a commercial system based on natural language processing technology for automatic extraction of biological facts from scientific literature such as MEDLINE abstracts, and internal text documents. | www.ariadnegenomics.com/products/medscan.html |
| **Online tools for biomedical literature mining** | |
| iProLINK | iProLINK is a resource to facilitate text mining in the area of literature-based database curation, named entity recognition, and protein ontology development. It can be utilized by computational and biomedical researchers to explore literature information on proteins and their features or properties. | pir.georgetown.edu/iprolink/ |
| PreBIND | PreBIND is a tool helping researchers locate biomolecular interaction information in the scientific literature. It identifies papers describing interactions using a support vector machine (SVM). | prebind.bind.ca/ |
| PubGene | PubGene is constructed to identify the relationships between genes and proteins, diseases, cell processes, and so on based on their co-occurrences in the abstracts of scientific papers, their sequence homology, and statistical probability of their co-occurrences. | www.pubgene.org/ |
| Chilibot | Chilibot [16] is a search software for the MEDLINE literature database to rapidly identify relationships between genes, proteins, or any keywords that the user might be interested. | www.chilibot.net/ |
| iHOP | Information Hyperlinked over Proteins [17] constructs a gene network by converting the information in MEDLINE into one navigable resource using genes and proteins as hyperlinks between sentences and abstracts. | www.ihop-net.org/UniPub/iHOP/ |

Many systems [6, 7, 8, 9, 10], such as EDGAR [11], BioRAT [12], GeneWays [13] and so on, have been developed to accomplish this goal, but with limited success. Table 17.1 lists some popular online databases, systems, and tools relating to the extraction of protein-protein interactions.

In general, to automatically extract protein-protein interactions, a system needs to consist of three to four major modules [13, 18], which is illustrated in Figure 17.1.

- *Zoning module.* It splits documents into basic building blocks for later analysis. Typical building blocks are phrases, sentences, and paragraphs. In special cases, higher-level building blocks such as sections or chapters may be

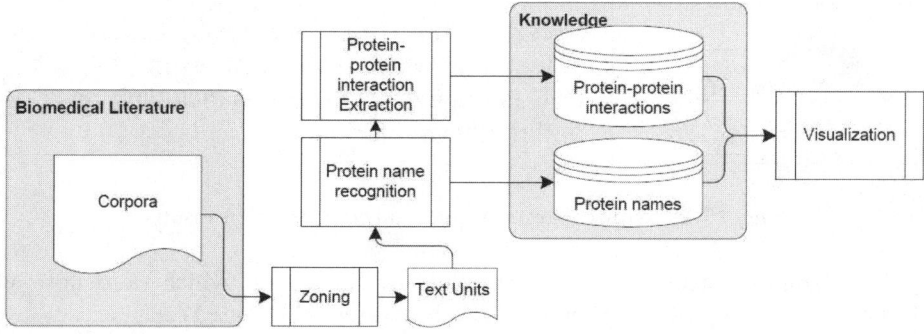

**Fig. 17.1.** A general architecture of an information extraction system for protein-protein interactions

chosen. Ding [19] compared the results of employing different text units such as phrases, sentences, and abstracts from MEDLINE to mine interactions between biochemical entities based on co-occurrences. Experimental results showed that abstracts, sentences, and phases all can produce comparative extraction results. However, with respect to effectiveness, sentences are significantly better than phrases and are about the same as abstracts.

- *Protein name recognition module.* Before the extraction of protein-protein interactions, it is crucial to facilitate the identification of protein names, which still remains a challenging problem [20]. Although experimental results of high recall and precision rates have been reported, several obstacles to further development are encountered while tagging protein names for the conjunctive natural of the names [21]. Chen [22] and Leser [23] provided a quantitative overview of the cause of gene-name ambiguity, and suggested what researchers can do to minimize this problem.

- *Protein-protein interaction extraction module.* As the retrieval of protein-protein interactions has attracted much attention in the field of biomedical information extraction, plenty of approaches have been proposed. The solutions range from simple statistical methods relying on co-occurrences of genes or proteins to methods employing a deep syntactical or semantical analysis.

- *Visualization module.* This module is not as crucial as the aforementioned three modules, but it provides a friendly interface for users to delve into the generated knowledge [24]. Moreover, it allows users to interact with the system for ease of updating the system's knowledge base and eventually improve its performance.

To evaluate the performance of an information extraction system, normally recall and precision values are measured. Suppose a test dataset has $T$ positive information (for example, protein-protein interactions), and an information extraction system can extract $I$ "positive" information. In $I$, only some information is really positive which we denote as $B$ and the remaining information is negative, however the system falsely extracts as positive which we denote as

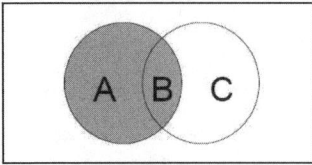

A+B (the blue area) is the positive information in test data which need to be extracted
B+C is the extracted results including positive and negative information

**Fig. 17.2.** Venn Diagram of information extraction results

$C$. In $T$, some information is not extracted by the system which we denote as $A$. The relationships of $A$, $B$, and $C$ are illustrated in Figure 17.2.

Based on the above definitions, recall and precision can be defined as:

$$\text{Precision} = \frac{\|B\|}{\|B\| + \|C\|} \tag{17.1}$$

$$\text{Recall} = \frac{\|B\|}{\|A\| + \|B\|} \tag{17.2}$$

For example, a test dataset has 10 protein-protein interactions $T$. An information extracting system extracts 11 protein-protein interactions $I$. In $I$, only 6 protein-protein interactions ($B$) can be found in $T$, which are considered as true positive (TP). The remaining 5 protein-protein interaction ($C$) can not be found in $T$, which are considered as false positive (FP). In $T$, 4 protein-protein interactions ($A$) are not extracted by the system, which are considered as false negative (FN). Thus, the recall of the system is $6/(6 + 4) = 60\%$ and the precision is $6/(6 + 5) = 54.5\%$.

Obviously, an ideal information extracting system should fulfil $\|A\| \longrightarrow 0, \|C\| \longrightarrow 0$. To reflect these two conditions, F-measure is defined by the harmonic (weighted) average of precision and recall [25] as :

$$\begin{aligned} F_\beta &= \frac{(1 + \beta^2) \cdot \text{Precision} \cdot \text{Recall}}{\beta^2 \cdot \text{Precision} + \text{Recall}} \\ &= \frac{(1 + \beta^2)\|B\|}{(1 + \beta^2)\|B\| + \beta^2\|A\| + \|C\|} \end{aligned} \tag{17.3}$$

where $\beta$ indicates a relative weight of precision. For further details of the state of the science in text mining evaluations, please refer to Hersh [26].

In this chapter, we focus on the protein name recognition and the protein-protein interaction extraction module. A brief survey and classification on the developed methodologies is provided. In general, the methods proposed so far rely on the techniques from one or more areas [27, 28, 29, 30] including Information Retrieval (IR) [25, 31], Machine Learning (ML) [32, 33], Natural Language Processing (NLP) [34, 35, 36], Information Extraction (IE) [37, 38, 39, 40], and Text Mining [41, 42, 43, 44, 45, 46, 47]. The surveyed work illustrates the progress of the field and shows the increasing complexity of the proposed methodologies.

The rest of the chapter is organized as follows. Firstly, systems and methods implemented for protein name recognition are discussed in Section 17.2.

Section 17.3 presents a survey of various methods applied in automatical extraction of protein-protein interactions from literature. In succession, challenges are identified and possible solutions are suggested.

## 17.2    Protein Name Recognition

As mentioned in section 17.1, recognizing protein names in biomedical literature is crucial for protein-protein interactions extraction. An example of a sentence with its protein names in italic is given as follows:

> *Interleukin-2 (IL-2)* rapidly activated *Stat5* in fresh PBL, and *Stat3* and *Stat5* in preactivated PBL. [PMID: 7719938]

There are various methods for recognizing protein name. Traditionally, these methods can be divided into four categories namely the dictionary based approaches, rule-based approaches, machine learning approaches, and hybrid approaches.

### 17.2.1    Dictionary Based Approaches

In dictionary based approaches, protein names are identified from text by using a provided list of protein names. These names can be identified using substring matching techniques such as exact matching and approximate string matching.

Egorov et al. [48] implemented a protein name identification system, ProtScan, using a carefully constructed dictionary. This dictionary was built based on the LocusLink database and enriched by the GenBank, GoldenPath and HUGO database entries. The system was evaluated on a gold standard, which consists of 1,000 randomly selected MEDLINE abstracts and achieved 88.6% recall and 98% precision. When evaluated on a more general set of biomedical documents other than MEDLINE abstracts, 98.5% recall and 84% precision were reported. Krauthammer et al. [49] proposed a dictionary-system based on BLAST [50], a tool for DNA and protein sequence comparison. An exhaustive list of gene and protein names was extracted from GenBank and translated into DNA sequences to form a dictionary. Names in the dictionary and input texts were converted into nucleotide sequences and then BLAST was implemented. The system achieved a recall of 78.8% and precision of 71.7%, of which 4.4% of names not included in the dictionary are fully recognized when evaluated on a gold standard review article marked by 2 experts.

Dictionary-based approaches in general can not identify protein names that are not listed in the pre-constructed dictionary. Their performance is highly dependent on the quality of their base dictionaries.

### 17.2.2    Rule-Based Approaches

Rule-based approaches identify protein names based on a set of manually defined rules. These rules usually employ surface clues and the syntactic and semantic properties of the gene and protein names.

Fukuda et al. [51] proposed a rule based system, PROPER, for identifying protein names using surface clues on character strings. These clues include capital letters, numerical figures and non-alphabetical letters. Evaluation was conducted based on 30 abstracts from MEDLINE in the SH3 protein domain and a recall of 98.84% and a precision of 94.70% was achieved. The Yapex system, based on hand-written rules was implemented by Franzen et al. [52]. Lexical analysis of single word tokens, syntactic analysis of noun phrases was performed to identify new protein names. 99 abstracts were randomly selected from MEDLINE to form the training corpus and 101 MEDLINE abstracts formed the test corpus. Yapex achieved a recall of 66.4% and a precision of 67.8%. In GPmarkup [53], abbreviations were first mapped to full names using a set of guidelines and protein symbols were mapped to the names by a set of pattern-matching rules. The mappings were performed on 11 million MEDLINE records and the abbreviation-name or symbol-name pairs were stored in a knowledge database. Non-protein abbreviation-name pairs in the database were then filtered out based on a set of heuristic rules. 50 abstracts from MEDLINE were randomly selected to form the test set and it achieved a recall of 73% and a precision of 93%.

Rule-based systems have the advantage that rules are able to be defined and extended when needed. However, the construction of rules has to be done manually and can be very time-consuming.

### 17.2.3   Machine Learning Approaches

Machine Learning approaches use various algorithms to automatically identify protein names. There are three commonly used approaches, Naive Bayes (NB), Support Vector Machine (SVM), and Hidden Markov Model (HMM).

### Naive Bayes

The NB Classifier is the most commonly used approach to identify protein names. It is a simple probabilistic model based on the Bayes' rule. It assumes that the effect of one feature on a given class is independent from that of another feature.

Nobata et al. [54] developed a system by calculating the similarity between a string and a class. NB was used to estimate the probability of a word occurring in a particular class. 100 abstracts were tagged by a human expert using Genia Ontology. Out of the 100 abstracts, 20 were used for testing and the remaining 80 were used for training. The system achieved an F-measure of 65.8%. Wilbur [55] considered several approaches based on NB. As the NB algorithm assumes that values are independent of each other and each term can be weighted separately based on its distribution in the training set. Documents in the test set are then scored by summing the weights of the terms they contain. The test set consisted of 100 documents and the training set consisted of 3,021 documents. The precision obtained was 71.4%. The staged NB algorithm was also be implemented. NB was first trained on the entire training set, then tested using both the training and test sets. A second training involved the positive examples and the negative ones that were unable to be separated in the first training. The precision achieved for this algorithm was 78.9%.

The main advantages of using the Naive Bayes approach is that it is fast to train and evaluate.

**Support Vector Machine**

A support vector machine (SVM) is a supervised learning technique for classification and regression. Mika and Rost [56] proposed a system, NLProt, that combines dictionary and rule-based filtering together with SVMs to tag protein names. The system used two dictionaries to perform pre-filtering. The first dictionary is a protein name dictionary with names generated from SWISS-PROT and TrEMBL [57] and the second is a common dictionary containing non-protein names. Input text is then tagged and run on four trained SVMs. When tested on the Yapex corpus, the system achieved an F-measure of 75% compared to the 67.1% on the Yapex system. GAPSCORE [58] identifies protein names based on their syntax, appearance, morphology, context and abbreviations. Features in it were developed on a Yapex independent corpus in order to obtain an accurate evaluation of the performance. A training set of 735 abstracts from MEDLINE was used for training the NB, Maximum Entropy (ME), and SVM classifiers. When evaluated on the Yapex training set, SVM outperformed the other two classifiers with a recall, precision and F-measure of 79.3%, 77%, and 78.1% respectively. When evaluated on the Yapex test set, SVM achieved a recall of 70.3%, a precision of 81.4%, and an F-measure of 75.4%. Hakenberg et al. [59] developed a system to solve the Name Entity Recognition (NER) problem posed by BioCreAtIve task 1A[1]. In the system, words are separated into tokens and an SVM is used to identify features that describe gene and non-gene names. Then, post-processing is performed by passing the tokens through a POS (Part-of-Speech)-tagger to find complete gene phrases. On a given test corpus of 5,000 previously unseen and untagged sentences, the system attained a recall of 72.8%, precision of 71.4% and an F-measure of 72.1% on the closed-division. The system is later enhanced by performing Recursive Feature Elimination (RFE) where features with the lowest weights are removed until 150 features remain. Post-processing is then done. Recall and precision of 82.8% and 83.4% respectively were achieved with this enhanced system.

SVM is an extremely powerful binary non-linear classifier. However, it is computationally demanding to train and run when the dataset is very large. It is also sensitive to noisy data and prone to overfitting, which in turn leads to generalization failures.

**Hidden Markov Model**

A Hidden Markov Model (HMM) is a variant of a finite state machine where the modelled system is assumed to be a Markov process with unknown parameters.

---

[1] BioCreAtIvE task 1A [60] focuses on extracting gene names and participants involved were given 10,000 MEDLINE sentences with tagged protein and gene names.

It consists of a finite set of states, in which each state is associated with a probability value. However, the states in an HMM are not directly visible to an external observer, only the observations are visible.

Collier et al. [61] proposed a model to find the most probable class which a word belongs to. A first-order HMM is used to implement the model. 1,000 MEDLINE abstracts related to molecular biology were selected and marked up by an expert. Out of these 1,000 abstracts, 800 were used for training and 200 were used for testing. The system reported an F-measure of 72.8%. PowerBioNE [62] is a system that is implemented using HMM and an HMM-based name entity recognizer. A pattern-based post processing was also done to extract rules from the training data so as to deal with the cascaded entity name phenomenon. The GENIA Corpus v3.0 which contains 2,000 MEDLINE abstracts of 360 thousand words was used. 200 abstracts were selected as the testing data and the remaining 1800 formed the training data. On twenty-three classes of the GENIA corpus, the system achieved an F-measure of 66.6%. On the "protein" class, the system achieved an F-measure of 75.8%.

In general, machine learning methods are useful when the annotated training set is available, as it requires experts to determine the protein names and this can be very time-consuming. In addition, the training set must also contain enough amounts of data in order to prevent the data sparseness problem.

### 17.2.4   Hybrid Approaches

Hybrid approaches use a combination of the above mentioned methods in the extraction of protein and gene names.

Tanabe and Wilbur [63] presented a method that uses a combination of rules and machine learning strategies. Rules are automatically generated by the Brill POS Tagger and then augmented with hand-crafted rules. The Brill tagger is trained on a corpus of 7,000 MEDLINE sentences to produce rules for tagging the texts. Next, rule-based post-processing rules are applied to identify potential gene names and then NB learning is used to rank documents based on their likelihood to contain a gene name. A test corpus of 56,469 MEDLINE abstracts was used to identify gene and protein names and results showed that higher performance can be achieved on documents with a higher Bayesian score. PROTEX [64] is a system employing a set of heuristic rules, a probabilistic model and a protein name dictionary to identify protein names. In the approach, heuristic rules reported in [51, 52] were first used to detect protein names, and then a probabilistic model was used to identify complete protein names. Finally, a dictionary compiled from the SWISS-PROT and TrEMBL protein databases were used to detect protein names that were not identified earlier. The Yapex gold standard was used for training and testing respectively. The system was then compared to the system by Franzen et al. [52]. Based on the exact matching evaluation criteria, PROTEX reported a recall, precision, and F-measure of 67.7%, 60.2%, and 63.7% respectively, outperforming Yapex.

A table summarizing the various algorithms and their performance is tabulated in Table 17.2.

**Table 17.2.** Performance of existing protein name recognition methods and the data corpora used

| Category | Result (%) | | Corpus | Ref |
|---|---|---|---|---|
| | Recall | Precision | | |
| Dictionary-based | 88.6 | 98 | 1000 randomly selected abstracts from MEDLINE | [48] |
| | 78.8 | 71.7 | Gold standard review articles marked by 2 experts | [49] |
| Heuristic rule-based | 98.8 | 94.7 | Test Set: 30 abstracts on the SH3 protein domain from MEDLINE | [51] |
| | 66.4 | 67.8 | Training Set: 99 random abstracts from MEDLINE; Test Set: 101 abstracts from MEDLINE. | [52] |
| | 73 | 93 | Test Set: 50 abstracts from MEDLINE | [65] |
| Naive Bayes | - | - | Training Set: 80 abstracts; Test Set: 20 abstracts | [54] |
| | 71.4 | - | Training Set: 3,021 documents; Test set: 100 documents | [55] |
| | 78.9 | - | Training Set: 3,021 documents; Test set: 3,121 documents | [55] |
| SVM | - | - | Training and Test Set: Yapex Corpus | [56] |
| | 58.5 | 56.7 | Training set: 735 abstracts from MEDLINE; Test Set: Yapex Test Set | [58] |
| | 83.4 | 82.8 | Training set: 10,000 sentences from MEDLINE. Test set: 5000 sentences | [59] |
| | 74.2 | 75.7 | Training Set: 1,600 abstracts from Genia Corpus; Test Set: 400 abstracts from Genia Corpus | [66] |
| HMM | - | - | Training Set: 800 abstracts from MEDLINE; Test Set: 200 abstracts from MEDLINE | [61] |
| | - | - | Training Set: 1,800 abstracts from GENIA Corpus; Test Set: 200 abstracts from GENIA Corpus | [62] |
| Hybrid Systems | 67.7 | 60.2 | Training and Test Set: Yapex Corpus | [64] |

## 17.2.5 Challenges

Despite the availability of many well-known nomenclatures for biomedical entities, there is no community-wide agreement on how a particular gene should be named. One name can stand for a particular gene, may include homologue of this gene in other organisms, may also encompass the protein the gene encodes. As a consequence, recognition of protein names automatically in the biomedical literature is not straightforward. In this section, we list the several open issues.

- *Ambiguous Names.* An ambiguous name denotes different entities. Some protein names are not distinguished from common English words, such as "white", "shaggy" and son on. Some names may denote biomedical entities of different classes. Other names may refer to certain entities before, but refer to another entities now.
- *Multi-word names.* Multi-word names are names consisting of more than one word (or token). For gene and protein names, multi-word names are rather than an exception. Multi-word names are not only harder to find, but in many cases there is no agreement on the exact borders of such names.
- *Synonyms and acronyms.* In synonymy relation, a protein name can be denoted by multiple names. Acronyms are abbreviation of names and are very popular in scientific writing because they allow for shorter texts. However, acronyms are difficult to resolve to their true names because they are often homonyms.
- *Names of newly discovered genes and proteins.* The overwhelming growth rate and the constant discovery of novel genes and proteins make protein name recognition more complex. Methods based on dictionary can not figure out these new names because registering the new names of genes and proteins is time-consuming and occurs much later.

## 17.3    Protein-Protein Interaction Extraction

This section presents a brief discussion on the existing techniques and methods for extracting protein-protein interactions. In general, current approaches can be divided into three categories:

- *Computational linguistics-based methods.* To discover knowledge from unstructured text, it is natural to employ computational linguistics and philosophy, such as syntactic parsing or semantic parsing to analyze sentence structures. Methods of this category define grammars to describe sentence structures and use parsers to extract syntactic information and internal dependencies within individual sentences. Approaches in this category can be applied to different knowledge domains after being carefully tuned to the specific problems. But, there is still no guarantee that the performance in the field of biomedicine can achieve comparable performance after tuning. Until recently, methods based on computational linguistics still could not generate satisfactory results.
- *Rule-based methods.* Rule-based approaches define a set of rules for possible textual relationships, called patterns, which encode similar structures in expressing relationships. When combined with statistical methods, scoring schemes depending on the occurrences of patterns to describe the confidence of the relationship are normally used. Similar to computational linguistics methods, rule-based approaches can make use of syntactic information to achieve better performance, although it can also work without prior parsing and tagging of the text.
- *Machine learning and statistical methods.* Machine learning refers to the ability of a machine to learn from experience to extract knowledge from data corpora. As opposed to the aforementioned two categories that need laborious effort to define a set of rules or grammars, machine learning techniques are able to extract protein-protein interaction patterns without human intervention.
  Statistical approaches are based on word occurrences in a large text corpus. Significant features or patterns are detected and used to classify the abstracts or sentences containing protein-protein interactions, and characterize the corresponding relations among genes or proteins.

It has to be mentioned that many existing systems in fact adopt a hybrid approach for better performance by combining methods from two or more of the aforementioned categories.

Figure 17.3 illustrates the process of information extraction on an example sentence by employing the typical methods in the above three categories.

### 17.3.1    Computational Linguistics-Based Methods

In general, computational linguistics-based methods employ linguistic technology to grasp syntactic structures or semantic meanings from sentences.

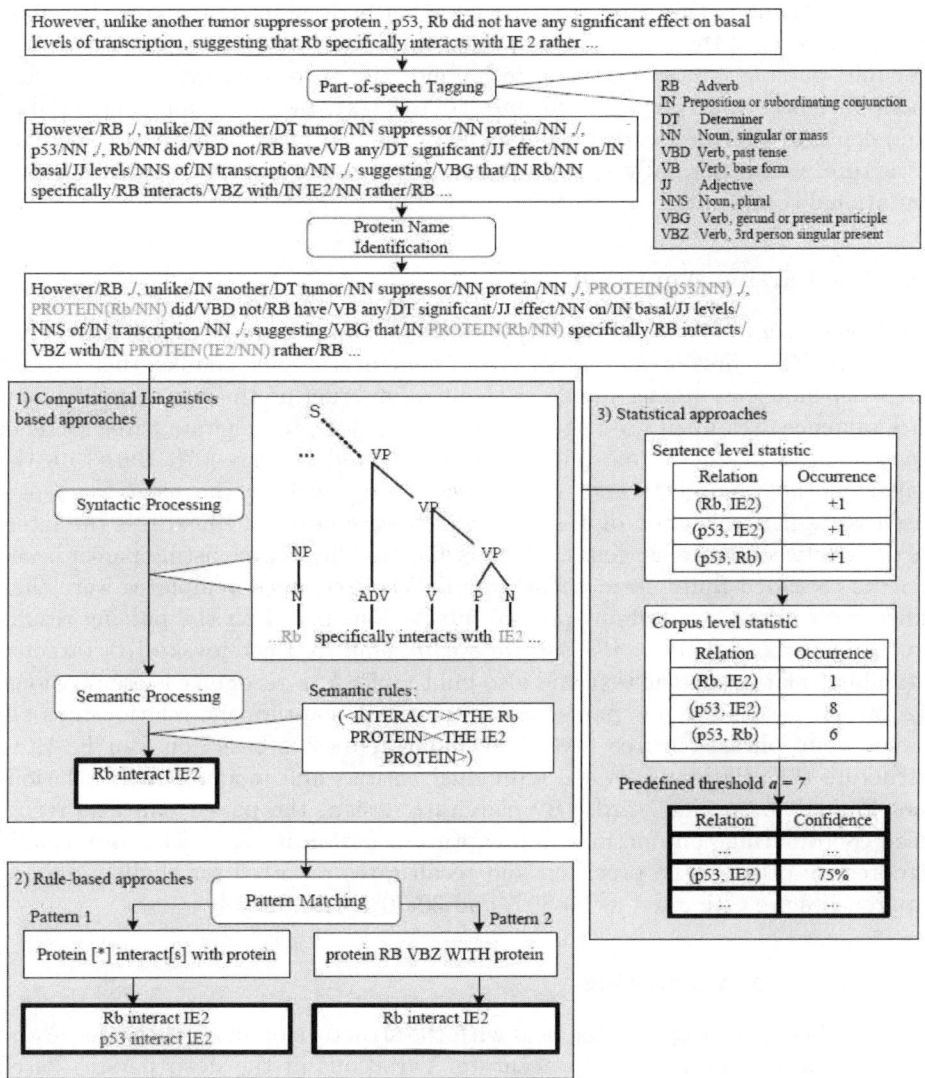

**Fig. 17.3.** General dataflow of information extraction system employing different methodologies

Techniques for analyzing a sentence and determining its structure in computational linguistics are called parsing techniques. Parsing the corpus firstly to obtain the morphological and syntactic information for each sentence is extremely important, and probably only after that, it would be possible to fulfill sophisticated tasks such as identifying the relationship between proteins and gene products in a fully automatic way. However, it is well known that parsing unrestricted texts, such as those in the biomedical domain, is extremely difficult.

The methods in this category can be further divided into two types, based on the complexity of the linguistics methods, as shallow (or partial) parsing or deep (or full) parsing. Shallow parsing techniques aim to recover syntactic information efficiently and reliably from unrestricted text, by sacrificing completeness and depth of analysis, while deep parsing techniques analyze the entire sentence structure, which normally achieve better performance but with increased computational complexity.

**Shallow Parsing Approaches**

Shallow parsers [67, 68, 69, 70, 71] perform partial decomposition of a sentence structure. They first break sentences into none-overlapping chunks, then extract local dependencies among chunks without reconstructing the structure of an entire sentence. Sekimizu used shallow parser, EngCG, to generate three kinds of tags, such as syntactic, morphological, and boundary tags [67]. Based on the tagging results, subjects and objects were recognized for the most frequently used verbs in a collection of abstracts which were believed to express the interactions between proteins, genes. Thomas [69] modified a preexisting parser based on the cascaded finite state automata (FSA). Predefined templates were then filled with information about protein interactions based on the parsing results for three verbs: *interact with, associate with, bind to.* Pustejovsky [70] targeted "inhibit" relations in the text and also built an FSA to recognize these relations. Leroy [71] used a shallow parser to automatically capture the relationships between noun phrases in free text. The shallow parser is based on four FSAs to structure the relations between individual entities and model generic relations not limited to specific words. By elaborate design, the parser can also recognize coordinating conjunctions and capture negation in text, a feature usually ignored by others. The precision and recall rates reported for shallow parsing approaches are estimated at 50-80% and 30-70%, respectively.

**Deep Parsing Approaches**

Systems based on deep parsing deal with the structure of an entire sentence and therefore are potentially more accurate. Variations of the deep parsing-based approach have been proposed [10, 72, 73, 74, 75, 76, 77, 78, 79, 80, 81]. Based on the way of constructing grammars, deep parsing-based approaches can be divided into two types: rationalist methods and empiricist methods. Rational methods define grammars by manual efforts, while empiricist methods automatically generate the grammar by some observations.

*Rationalist Methods*

Yakushiji [75] used a general full parser with grammars for biomedical domain to extract interaction events by filling sentences into slots of semantic frames. Information extraction itself is done using pattern matching on the canonical structure. Park [74] proposed bidirectional incremental parsing with combinatory categorial grammar (CCG). This method first localized target verbs, and then scanned the left and right neighborhood of the verb respectively. The lexical and grammatical rules of CCG are more complicated than those of a general context-free grammar (CFG). The recall and precision rate of the system were reported to be 48% and 80%. Temkin [78] introduced a lexical analyzer and a CFG to extract protein, gene and small molecule interactions with a recall rate of 63.9% and precision rate of 70.2%. Ding [79] investigated link grammar parsing for extracting biochemical interactions. It can handle many syntactic structures and is computationally relatively efficient. A better overall performance was achieved compared to those biomedical term co-occurrence based methods. Ahmed [10] split complex sentences into simple clausal structures made up of syntactic roles based on a link grammar. Complete interactions were then extracted by analyzing the matching contents of syntactic roles and their linguistically significant combinations. In GENIES [76], a parser and a semantic grammar consisting of a large set of nested semantic patterns (incorporating some syntactic knowledge) are used. Unlike other systems, GENIES is capable of extracting a wide variety of different relations between biological molecules as well as nested chains of relations. However, the downside of the semantic grammar-based systems such as GENIES is that they may require complete redesign of the grammar in order to be tuned for used in different domain.

*Empiricist Methods*

Many empiricist methods [77, 80] have been proposed to automatically generate the language model to mimic the features of unstructured sentences. For example, Seymore [72] used Hidden Markov Model (HMM) for extracting important fields from the headers of computer science research papers. Following the trend, Souyma [73] applied HMM to the biomedical domain to describe the structure of sentences. More recently, Skounakis [82] proposed an approach that is based on hierarchical HMMs to represent the grammatical structure of the sentences being processed. Firstly, shallow parser to construct a multi-level representation of each sentence being processed was used. Then hierarchical HMMs to capture the regularities of the parses for both positive and negative sentences were trained. In [83], a broad-coverage probabilistic dependency parser was used to identify sentence level syntactic relations between the heads of the chunks. The parser used a hand-written grammar combined with a statistical language model that calculates lexicalized attachment probabilities.

## 17.3.2 Rule-Based Approaches

In rule-based approaches [6, 7, 9, 12, 84, 85, 86, 87, 88, 89, 90, 91, 92], a set of rules need to be defined which may be expressed in forms of regular expressions over words or POS tags. Based on the rules, relations between entities that are relevant to tasks such as proteins, can be recognized.

Ng [84] defined five rules based on the word form, such as <A> ... <fn> ... <B> in which the symbols A, B refer to protein names while the symbol fn refers to the verb which describes the interaction relationship. Obviously, such rules are too simple to produce satisfactory results. Ono [87] manually defined a set of rules based on syntactic features to preprocess complex sentences, with negation structures considered as well. It achieves good performance with a recall rate of 85% and precision rate of 84% for *Saccharomyces cerevisiae (yeast)* and *Escherichia coli*. Blaschke [7] induced a probability score to each predefined rule depending on its reliability and used it as a clue to score the interaction events. Sentence negations and the distance between two protein names were also considered. In [89], gene-gene interactions were extracted by scenarios of patterns which were constructed manually. For example, "gene product acts as a modifier of gene" is a scenario of the predicate act, which can cover a sentence such as: "Egl protein acts as a repressor of BicD". Egl and BicD can be extracted as an argument of an event for the predicate acts. Shatkay and Leroy [88] employed preposition-based parsing to generate templates. It achieved a template precision of 70% when processing literature abstracts.

Using predefined rules can generate nice results. It is however not feasible in practical applications as it requires heavy manual processing to define patterns when shifting to another domain.

Huang [90] tried to automatically construct the protein-protein interaction patterns. At first, part-of-speech tagging was employed. Then dynamic programming to automatically extract similar patterns from sentences based on POS tags was used. Based on the automatically constructed patterns, protein-protein interactions can be identified. Their results gave precision of 80.5% and recall of 80.0%. Phuong [93] used some sample sentences, which were parsed by a link grammar parser, to learn extraction rules automatically. By incorporating heuristic rules based on morphological clues and domain specific knowledge, the method can remove the interactions that are not between proteins.

Rule-based approaches have been found to be overall limiting in the set of interactions that can be extracted by the extent of the recognition rules that were implemented, and also by the complexity of sentences being processed. Specifically, complicated cases such as interaction descriptions that span several sentences of text are often missed by these approaches. The shortcoming of such approaches is their inability to correctly process anything other than short, straightforward statements, which are quite rare in information-saturated biomedical literature. They also ignore many important aspects of sentence construction such as mood, modality, and sometimes negation, which can significantly alter or even reverse the meaning of the sentence.

### 17.3.3   Machine-Learning and Statistical Approaches

Many machine-learning (ML) methods have been proposed ranging from simple methods such as deducing relationship between two terms based on their co-occurrences to complicated methods which employ NLP technologies. Approaches combing machine learning and NLP have been discussed in section 17.3.1. Here we focus on the methods without employing NLP techniques.

A variety of machine-learning and statistical techniques based on the discovery of co-occurrence of protein names have been applied for protein-protein information extraction [8, 94, 95, 96, 97, 98, 99, 100, 101, 102, 103, 104, 105, 106]. They can be further divided into different types based on the mining units, such as abstracts, sentences and so on.

Approaches proposed in Miguel and Marcottle [94, 100] aim to extract protein-protein interactions from a set of abstracts. Miguel [94] used a group of relevant documents against a set of random documents to extract domain specific information such as gene functions and interactions. Marcottle [100] was only interested in retrieving a large number of documents that probably contained information about protein-protein interactions.

The first machine-learning sentence-based information extraction system in molecular biology was described in Craven and Kumlien [96]. They developed a Bayesian classifier which, given a sentence containing mentions of two items of interest, returns a probability that the sentence asserts some specific relations between them. Later systems have applied other technologies, including hidden Markov models and support vector machines, to identify sentences describing protein-protein interactions.

Other approaches [8, 97, 98, 99] focus on a pair of proteins and detect the relations between them using probability scores. Stapley [97] used fixed lists of gene names and detected relations between these genes by means of co-occurrences in MEDLINE abstracts. A matrix that contains distance dissimilarity measurement of every pair of genes based on their joint and individual occurrence statistics was constructed based on a user-defined threshold. Stephens [98] furthered the method to discover relationships using more complicated computation on co-occurrences. Jenssen [99] used a similar approach to find relations between human gene clusters obtained from DNA array experiments. Donaldson [8] constructed PreBIND and Textomy - an information extraction system that uses support vector machines to evaluate the importance of protein-protein interactions.

Simple statistical methods such as those based on protein co-occurrence information can not precisely describe the relations between proteins and therefore tend to generate high false negative error rate. On the contrary, complex statistical models need a large amount of training data in order to reliably estimate model parameters, which is usually difficult to obtain in practical applications. To strike the balance, we applied the hidden vector state model (HVS) which was previously used in spoken language understanding to extract protein interactions [107]. Unlike other statistical parsers which need fully-annotated treebank data for training, the HVS model explores the embedded sentence structures using only lightly annotated corpus. The details of how this is done can be found in [108].

### 17.3.4  Discussion

The performance of the existing protein-protein interaction extraction methods along with the data corpora they used are listed in Table 17.3.

**Table 17.3.** Performance of existing protein-protein interaction extraction methods and the data corpora used

| Category | Result (%) Recall | Precision | Corpus | Ref |
|---|---|---|---|---|
| Shallow Parsing | - | 73 | 34343 sentences from abstracts retrieved from MEDLINE using keywords "leucine zipper", "zinc finger", "helix loop helix motif" | [67] |
| | 29 | 69 | 2565 unseen abstracts extracted from MEDLINE with the keywords molecular, interaction and protein for year 1,998 (560k words) | [69] |
| | 57 | 90 | Training set consists of 500 abstracts from MEDLINE. Evaluation set consists of 56 abstracts collected using search strings "protein" and "inhibit" | [70] |
| | 62 | 89 | 26 abstracts | [71] |
| Deep Parsing | 48 | 80 | 492 sentences out of 250,000 abstracts on cytosine in MEDLINE | [74] |
| | 63.9 | 70.2 | The test corpus consists of 100 randomly selected scientific abstracts from MEDLINE. | [78] |
| | - | 96 | Articles from cell containing 7,790 words revealing 51 binary relations | [76] |
| | 21 | 91 | 3.4 million sentences from approximately 3.5 million MEDLINE abstracts dated after 1,988 containing at least one notation of a human protein | [80] |
| | 26.94 | 65.66 | 229 abstracts from MEDLINE correspond to 389 interactions from the DIP database. | [10] |
| Rule Based | 47 | 70 | 474 sentences from 50 abstracts retrieved using "E2F1" | [88] |
| | 86.8 Yeast 82.5 Escherichia | 94.3 Yeast 93.5 Escherichia | 834 and 752 sentences containing at least two protein names and one relation keyword for yeast and E.coli obtained by a MEDLINE search using the following keywords, "protein binding" as a MESH term and "yeast", "E coli", "protein", and "interaction" | [87] |
| | 39.7 | 44.9 | Five different sets of abstracts were used: 1. 1,435 MEDLINE abstracts directly referenced from each of the Drosophila Swiss-prot entries. 2. 4,109 MEDLINE abstracts referenced directly from Fly Base. 3. 111,747 abstracts retrieved by extending the set (2) with the Neighbors utility. 4. 518 MEDLINE abstracts containing any of the protein names (related with cell cycle control) and Drosophila in the MESH list of terms. 5. 6,278 MEDLINE abstracts by expanding set (4) using Neighbors to identify all related abstracts. | [7, 85] |
| | 60 | 87 | 3,343 abstracts were obtained by querying MEDLINE with the following keywords: "Saccharomyces cerevisiae", "protein", and "interaction". The abstracts were filtered and 550 sentences were retained containing at least one of four keywords "interact", "bind", "associate", "complex" or one of their inflections. | [93] |
| | 80.0 | 80.5 | The top 50 biomedical papers were retrieved from the Internet by querying using the keyword "protein-protein interaction". Full texts were segmented into 65,536 sentences and the sentences with fewer than two protein names were discarded. The final corpus consists of about 1,200 sentences. | [90] |

As in the area of extracting information about protein-protein interactions, competitive evaluations have played important roles in pushing the field of IE and NLP. Several evaluations have been held in recent years. Procreative challenge (Critical Assessment of Information Extraction in Biology) [109] began in 2004 and provided two common evaluation tasks to assess the state of the art methods for text mining applied to biological problems. The first task dealt with extraction of gene or protein names from text, and their mappings into standardized gene identifiers for three model organism databases (fly, mouse, yeast). The second task [110] addressed issues of functional annotation, requiring systems to identify specific text passages that supported Gene Ontology annotations for specific proteins, given full text articles. Genic Interaction Extraction Challenge [111] was associated with Learning Language in Logic Workshop (LLL05). The challenge focuses on information extraction of gene interactions in *Bacillus subtilin*, a model bacterium. It was reported that the best F-measure achieved with the balanced recall and precision is around 50%.

**Table 17.4.** Online annotated corpora for the extraction of protein-protein interactions.

| Corpus Name | Description | URL |
|---|---|---|
| GENA | GENA corpus version 3.0 consists of 2,000 MED-LINE abstracts with more than 400,000 words and almost 100,000 annotations for biological terms. | www-tsujii.is.s.u-tokyo.ac.jp/GENIA/ |
| Apex | It consists of two collections, training collection consisting of 99 abstracts with 1,745 protein names, test collection consisting of 101 abstracts with 1,966 protein names. The protein names in all the abstracts were annotated manually. | www.sics.se/humle /projects/prothalt/ |
| Penninite | The corpus consists of 2,258 MEDLINE abstracts in two domains: 1) the molecular genetics of oncology (1,158 abstracts); 2) the inhibition of enzymes of the CYP450 class (1,100 abstracts). | bioie.ldc.upenn.edu/ |
| LLL05 challenge Corpus | There are 80 sentences in the training set, including 106 examples of genic interactions without coreferences and 165 examples of interactions with coreferences. | genome.jouy.inra.fr /texte/LLLchallenge/ |

As annotated corpora are important to the development as well as the evaluation of protein-protein extraction systems, some online available annotated corpora are listed in Table 17.3.4.

## 17.4  Challenges and Possible Solutions

The continuing growth and diversification of the scientific literature, a prime resource for accessing worldwide scientific knowledge, will require tremendous systematic and automated efforts to utilize the underlying information. In the near future, tools for knowledge discovery will play a pivotal role in systems biology. The increasing fervor on the field of biomedical information extraction gives the evidence. IE in biomedicine has been studied for approximately ten years. Over these years, IE systems in biomedicine have grown from simple rule-based pattern matcher to sophisticated, hybrid parser employing computational linguistics technology. But, until now, there are still several severe obstacles to overcome.

Firstly, biomedical IE methods generate poorer results compared with other domains such as newswire. In general, biomedical IE methods are scored with F-measure, with the best methods scoring about 0.85 without considering the limitation of test corpus, which is still far from users' satisfaction. The main reason is that information from ontologies[2] or terminologies is not well used. Until

---

[2] Ontologies, structured lists of terms, are often used by NLP technologies to establish the semantic function of a word in a document. The simplest form of ontology is a lexicon or a list of terms that belong to a particular class. A lexicon usually consists of specialized terms and (optionally) their definitions. Another form of ontology is a thesaurus, a collection of terms and their synonyms which are of immense utility for NLP. A popular ontology in biomedicine is Gene Ontology (GO) [112, 113].

recently, most biomedical IE systems do not make use of information from ontologies or terminologies. Hence, ontologies together with terminological lexicons are prerequisites for advanced biomedical IE. Since different ontologies are employed in different systems currently, unification seems necessary and impendent. Also, biomedical text needs to be semantically annotated and actively linked to ontologies.

Secondly, relations between biological entities, such as proteins or genes are conditional and may change when the same entities are considered in a different functional context. As a consequence, every relation between entities should be linked with the functional context in which the relation was observed. Moreover, without considering the observed context, it is meaningless and impossible to make general statements whether a relation detected by literature mining is a "yes" or a "no" relation. Obviously, to overcome this obstacle, in-depth analysis based on more elaborately constructing grammars or rules in sentence or phrase level is requisite. Hopefully, it will result in the increase of performance.

Thirdly, it seems to be crucial to the success of biomedical IE to bridge the gap between biologists and computational scientists. Currently, this field is dominated by researchers with computational background; however, the biomedical knowledge is only possessed by biologists. That is crucial for defining standards for evaluation; for identification of specific requirements, potential applications and integrated information system for querying, visualization and analysis of data on a large scale; for experimental verification to facilitate the understanding of biological interactions. Hence, to attract more biologists into the field, it is important to design simple and friendly user interfaces that make the tools accessible to non-specialists.

Fourthly, the knowledge extracted from the literature may contradict itself under different environment, conditions, or because of author's errors, experimental errors or other issues. Although the contradictory knowledge may occupy minor part of the whole interaction network, it is worth more attention. To handle this challenge, one way is to categorize the corpora and define the confidence value for each category. For contradictory knowledge, the decision can be made based on these confidence values. The solution can also be applied to handling different parts of an article, such as the abstract, introduction, references and so on, which obviously are of different confidences.

Fifthly, some problems exist not only in the field of biomedical IE, but also in the field of NLP. Two of them are: (1) Dealing with negative sentences, which constitutes a well-known problem in language understanding [114]. (2) Resolving coreferences, the recognition of implicit information in a number of sentences may contain key information, e.g. protein names, that later are used implicitly in other sentences. Results in LLL challenge 05 show that F-measure can only achieve 25% when considering coreferences.

Finally, the development of gold standard for evaluation systems is still under way, far from maturity, which requires more concerted efforts. The experience in the newswire domain shows that the construction of evaluation benchmarks in the face of common challenges contribute greatly to the rapid development of IE. Thus it is crucial to attach importance to evaluate systems development in biomedicine. Also, efforts will be required to focus on linking the knowledge in the databases with text sources available. It is believed that in the future, biomedical IE might provide new approaches for relation discovery that exploit efficiently indirect relationships derived from bibliographic analysis of entities contained in biological databases.

# References

1. Pubmed-overview,
   http://www.ncbi.nlm.nih.gov/entrez/query/static/overview.html
2. Xenarios, I., Salwinski, L., Duan, X., Higney, P., Kim, S., Eisenberg, D.: DIP, the Database of Interacting Proteins: a research tool for studying cellular networks of protein interactions. Nucleic Acids Research 30(1), 303–305 (2002)
3. Bader, G., Betel, D., Hogue, C.: BIND: the Biomolecular Interaction Network Database. Nucleic Acids Research 31(1), 248–250 (2003)
4. Hermjakob, H., Montecchi-Palazzi, L., Lewington, C.: IntAct: an open source molecular interaction database. Nucleic Acids Research 1(32(Database issue)), 452–455 (2004)
5. von Mering, C., Jensen, L., Snel, B., Hooper, S., Krupp, M.: STRING: known and predicted protein-protein associations, integrated and transferred across organisms. Nucleic Acids Research 33(Database issue), 433–437 (2005)
6. Wong, L.: PIES, a protein interaction extraction system. In: Proc. Pacific Symposium on Biocomputing, Hawaii, U.S.A., pp. 520–531(2001)
7. Blaschke, C., Valencia, A.: The Frame-Based Module of the SUISEKI Information Extraction system. IEEE Intelligent Systems 17(2), 14–20 (2002)
8. Donaldson, I., Martin, J., de Bruijn, B., Wolting, C.: PreBIND and Textomymining the biomedical literature for protein-protein interactions using a support vector machine. BMC Bioinformatics 4(11) (2003)
9. Chiang, J.H., Yu, H.C., Hsu, H.J.: GIS: a biomedical text-mining system for gene information discovery. Bioinformatics 20(1), 120–121 (2004)
10. Ahmed, S.T., Chidambaram, D., Davulcu, H., Baral, C.: IntEx: A Syntactic Role Driven Protein-Protein Interaction Extractor for Bio-Medical Text. In: Proc. ACL-ISMB Workshop on Linking Biological Literature, Ontologies and Database 2005, pp. 54–61 (2005)
11. Rindflesch, T., Tanabe, L., Weinstein, J., Hunter, L.: EDGAR: extraction of drugs, genes and relations from the biomedical literature. In: Proc. Pacific Symposium Biocomputing, pp. 517–528 (2000)
12. Corney, D.P.A., Buxton, B.F., Langdon, W.B., Jones, D.T.: BioRAT: extracting biological information from full-length papers. Bioinformatics 20(17), 3206–3213 (2004)
13. Rzhetsky, A., Iossifov, I., Koike, T., Krauthammer, M., Kra, P., Morris, M., Yu, H., Duboué, P., Weng, W., Wilbur, W., Hatzivassiloglou, V., Friedman, C.: GeneWays: a system for extracting, analyzing, visualizing, and integrating molecular pathway data. Journal of Biomedical Informatic 37(1), 43–53 (2004)
14. Peri, S., Navarro, J.D., Amanchy, R.: Development of Human Protein Reference Database as an Initial Platform for Approaching Systems Biology in Humans. Genome Research 13, 2363–2371 (2003)

15. Zanzoni, A., Montecchi-Palazzi, L., Quondam, M., Ausiello, G., Helmer-Citterich, M., Cesareni, G.: MINT: a Molecular INTeraction database. FEBS letters 513(1), 135–140 (2002)
16. Chen, H., Sharp, B.M.: Conten-trich biological network constructed by mining PubMed abstracts. BMC Bioinformatics 8(5), 147 (2004)
17. Hoffmann, R., Valencia, A.: A gene network for navigating the literature. Nature Genetics 36, 664 (2004)
18. Mathiak, B., Eckstein, S.: Five Steps to Text Mining in Biomedical Literature. In: Proc. Data Mining and Text Mining for Bioinformatics European Workshop (2004)
19. Ding, J., Berleant, D., Nettleton, D., Wurtele, E.: Mining MEDLINE: Abstracts, sentences or phrases. In: Proc. Pacific Symposium on Biocomputing, Hawaii, U.S.A., pp. 326–337 (2002)
20. Krauthammer, M., Nenadic, G.: Term identification in the biomedical literature. Journal of biomedical informatics 37(6), 512–526 (2004)
21. Pearson, H.: Biology's name game. Nature 411(6838), 631–632 (2001)
22. Chen, L., Liu, H., Friedman, C.: Gene name ambiguity of eukaryotic nomenclatures. Bioinformatics 21(2), 248–256 (2005)
23. Leser, U., Hakenberg, J.: What makes a gene name? Named entity recognition in the biomedical literature. Briefings in Bioinformatics 6(4), 257–269 (2005)
24. Drabkin, H.J., Hollenbeck, C., Hill, D.P., Blake, J.A.: Ontological visualization of protein-protein interactions. BMC Bioinformatics 6(29) (2005)
25. van Rijsbergen, C.: Information Retrieval (1999), http://www.dcs.gla.ac.uk/Keith/Preface.html
26. Hersh, W.: Evaluation of biomedical text-mining systems: Lessons learned from information retrieval. Briefings in Bioinformatics 6(4), 344–356 (2005)
27. Andrade, M., Bork, P.: Automated extraction of information in molecular biology. FEBS Lett. 476(1-2), 12–17 (2000)
28. Hirschman, L., Park, J.C., Tsujii, J., Wong, L., Wu, C.H.: Accomplishments and challenges in literature data mining for biology. Bioinformatics 18(12), 1553–1561 (2002)
29. Shatkay, H., Feldman, R.: Mining the biomedical literature in the genomic era: an overview. Journal of Computational Biology 10(6), 821–855 (2003)
30. Jensen, L.J., Saric, J., Bork, P.: Literature mining for the biologist: from information retrieval to biological discovery. Nature Reviews Genetics 7, 119–129 (2006)
31. Hersh, W.R.: Information Retrieval: A Health and Biomedical Perspective (2003)
32. Chang, J.T.: Using Machine Learning to Extract Drug and Gene Relationships from Text. Ph.D. Thesis, Stanford University (2003)
33. Marquez, L.: Machine learning and natural language processing. In: Tech. Rep. LSI-00-45-R, Departament de Llenguatges i Sistemes Informatics (LSI), Universitat Politecnica de Catalunya (UPC), Barcelona, Spain (2000)
34. Cohen, K.B., Hunter, L.: Natural language processing and systems biology. In: Dubitzky, W. (ed.) Computational Biology. Azuaje, Francisco, vol. 5 (2004)
35. Yandell, M., Majoros, W.: Genomics and natural language processing. Nature Reviews Genetics 3(8), 601–610 (2002)
36. Hunter, L., Cohen, K.B.: Biomedical Language Processing: What's Beyond PubMed? Molecular Cell 21(5), 589–594 (2006)
37. Cardie, C.: Empirical Methods in Information Extraction. AI. Magazine 18(4), 65–80 (1997)

38. Blaschke, C., Hoffmann, R., Oliveros, J., Valencia, A.: Extracting information automatically from biological literature. Comparative and Functional Genomics 2(5), 2, 310–313 (2001)
39. Cunningham, H.: Information Extraction, Automatic, Encyclopedia of Language and Linguistics, 2nd edn. (2005)
40. Skusa, A., Rüegg, A., Köhler, J.: Extraction of biological interaction networks from scientific literature. Briefings in Bioinformatics 6(3), 263–276 (2005)
41. Text mining in the life sciences. Tech. rep. (2004)
42. Bruijn, B.D., Martin, J.: Literature Mining in Molecular Biology. In: Proc. EFMI Workshop on Natural Language Processing in Biomedical Application, pp. 1–5 (2002)
43. Krallinger, M., Erhardt, R., Valencia, A.: Text-mining approaches in molecular biology and biomedicine. Drug Discovery Today 10(6), 439–445 (2005)
44. Spasic, I., Ananiadou, S., McNaught, J., Kumar, A.: Text mining and ontologies in biomedicine: Making sense of raw text. Briefings in Bioinformatics 6(3), 239–251 (2005)
45. Cohen, A.M., Hersh, W.R.: A survey of current work in biomedical text mining. Briefings in Bioinformatics 6(1), 57–71 (2005)
46. Ananiadou, S., Mcnaught, J.: Text mining for biology and biomedicine (2006)
47. Shatkay, H., Craven, M.: Biomedical text mining. MIT Press, Cambridge (2007)
48. Egorov, S., Yuryev, A., Daraselia, N.: A simple and practical dictionary based approach for identification of proteins in medline abstracts. Journal of the American Medical Informatics Association 11, 174–178 (2004)
49. Krauthammer, M., Rzhetsky, A., Morozov, P., Friedman, C.: Using blast for identifying gene and protein names in journal articles. Gene 259(1), 245–252 (2000)
50. Altschul, S., Gish, W., Miller, W., Myers, E., Lipman, D.: Basic local alignment search tool. Journal of molecular biology 215, 403–410 (1990)
51. Fukuda, K., Tsunoda, T., Tamura, A., Takagi, T.: Toward information extraction: Identifying protein names from biological papers. In: Proc. Pacific Symposium on Biocomputing, Hawaii, USA, pp. 707–718 (1998)
52. Franzen, K., Eriksson, G., Olsson, F., Asker, L., Liden, P., Coster, J.: Protein names and how to find them. International Journal of Medical Informatics 67(1), 49–61 (2002)
53. Yu, H., Hatzivassiloglou, V., Rzhetsky, A., Wilburc, W.J.: Automatically identifying geneprotein terms in medline abstracts. Journal of Biomedical Informatics 35, 322–330 (2002)
54. Nobata, C., Collier, N., Tsujii, J.: Automatic term identification and classification in biology texts. In: Proc. 5th Natural Language Processing Pacific Rim Symposium, Beijing, China (1999)
55. Wilbur, W.: Boosting naive Bayesian learning on a large subset of MEDLINE. In: Proc. AMIA Symposium, Beijing, China, pp. 918–922 (2000)
56. Mika, S., Rost, B.: Nlprot: extracting protein names and sequences from papers. Nucleic Acids Research 32, 634–637 (2004)
57. Boeckmann, B., Bairoch, A., Apweiler, R., Blatter, M., Estreicher, A., Gasteiger, E., Martin, M., Michoud, K., ODonovan, C., Phan, I.: The Swiss-Prot protein knowledgebase and its supplement TrEMBL in 2003. Nucleic Acids Research 31(1), 365–370 (2003)
58. Chang, J., Schutze, H., Altman, R.: Gapscore: finding gene and protein names one word at a time. Bioinformatics 20(2), 216–225 (2004)

59. Hakenberg, J., Bickel, S., Plake, C., Brefeld, U., Zahn, H., Faulstich, L., Leser, U., Scheffer, T.: Systematic feature evaluation for gene name recognition. BMC Bioinformatics 6, S9 (2005)
60. Yeh, A., Morgan, A., Colosimo, M., Hirschman, L.: Biocreative task 1a: gene mention finding evaluation. BMC Bioinformatics 6, S2 (2005)
61. Collier, N., Nobata, C., Tsujii, J.: Extracting the names of genes and gene products with a hidden Markov model. In: Proc. 18th Conference on Computational linguistics, Saarbrucken, Germany, pp. 201–207 (2000)
62. Zhou, G., Zhang, J., Su, J., Shen, D., Tan, C.: Recognizing names in biomedical texts: a machine learning approach. Bioinformatics 20(7), 1178–1190 (2004)
63. Tanabe, L., Wilbur, W.: Tagging gene and protein names in biomedical text. Bioinformatics 18(8), 1124–1132 (2002)
64. Seki, K., Mostafa, J.: A hybrid approach to protein name identification in biomedical texts. Information Processing and Management 41, 723–743 (2005)
65. Yu, H., Hatzivassiloglou, V., Rzhetsky, A., Wilbur, W.J.: Automatically identifying gene/protein terms in MEDLINE abstracts. Biomedical Informatics 35(5/6), 322–330 (2002)
66. Yamamoto, K., Kudo, T., Konagaya, A., Matsumoto, Y.: Protein name tagging for biomedical annotation in text. In: Proc. ACL 2003 Workshop on Natural Language Processing in Biomedicine, Sapporo, Japan, pp. 65–72 (2003)
67. Sekimizu, T., Park, H., Tsujii, J.: Identifying the interaction between genes and gene products based on frequently seen verbs in MEDLINE abstracts. In: Proc. Workshop on Genome Informatics, vol. 9, pp. 62–71 (1998)
68. Rindflesch, T., Hunter, L., Aronson, A.: Mining molecular binding terminology from biomedical text. In: Proc. AMIA Symposium, pp. 127–131 (1999)
69. Thomas, J., Milward, D., Ouzounis, C., Pulman, S.: Automatic extraction of protein interactions from scientific abstracts. In: Proc. Pacific Symposium on Biocomputing, Hawaii, U.S.A., pp. 541–552 (2000)
70. Pustejovsky, J., Castano, J., Zhang, J., Kotecki, M., Cochran, B.: Robust Relational Parsing Over Biomedical Literature: Extracting Inhibit Relations. In: Proc. Pacific Symposium on Biocomputing, Hawaii, U.S.A., pp. 362–373 (2002)
71. Leroy, G., Chen, H., Martinez, J.D.: A Shallow Parser Based on Closed-Class Words to Capture Relations in Biomedical Text. Journal of Biomedical Informatics 36(3), 145–158 (2003)
72. Seymore, K., McCallum, A., Rosenfeld, R.: Learning Hidden Markov Model Structure for Information Extraction. In: Proc. AAAI 1999 Workshop on Machine Learning for Information Extraction (1999)
73. Ray, S., Craven, M.: Representing Sentence Structure in Hidden Markov Models for Information Extraction. In: Proc. 17th International Joint Conference on Artificial Intelligence (IJCAI 2001), pp. 1273–1279 (2001)
74. Park, J., Kim, H., Kim, J.: Bidirectional incremental parsing for automatic pathway identification with combinatory categorical grammar. In: Proc. Pacific Symposium on Biocomputing, Hawaii, U.S.A., vol. 6, pp. 396–407 (2001)
75. Yakushiji, A., Tateisi, Y., Miyao, Y., Tsujii, J.: Event extraction from biomedical papers using a full parser. In: Proc. Pacific Symposium on Biocomputing, Hawaii, U.S.A., vol. 6, pp. 408–419 (2001)
76. Friedman, C., Pauline Kra, H.Y., Krauthammer, M., Rzhetsky, A.: GENIES: a natural-language processing system for the extraction of molecular pathways from journal articles. Bioinformatics 17, S74–S82 (2001)

77. Novichkova, S., Egorov, S., Daraselia, N.: MedScan, a natural language processing engine for MEDLINE abstracts. Bioinformatics 19(13), 1699–1706 (2003)
78. Temkin, J.M., Gilder, M.R.: Extraction of protein interaction information from unstructured text using a context-free grammar. Bioinformatics 19(16), 2046–2053 (2003)
79. Ding, J., Berleant, D., Xu, J., Fulmer, A.W.: Extracting Biochemical Interactions from MEDLINE Using a Link Grammar Parser. In: Proc. 15th IEEE International Conference on Tools with Artificial Intelligence (ICTAI 2003) (2003)
80. Daraselia, N., Yuryev, A., Egorov, S., Novichkova, S., Nikitin, A., Mazo, l.: Extracting human protein interactions from MEDLINE using a full-sentence parser. Bioinformatics 20(5), 604–611 (2004)
81. Tan, S., Kwoh, C.K.: Cytokine Information System and Pathway Visualization. In: Proc. International Joint Conference of InCoB, AASBi and KSBI (BIOINFO 2005) (2005)
82. Skounakis, M., Craven, M., Ray, S.: Hierarchical Hidden Markov Models for Information Extraction. In: Proc. 18th International Joint Conference on Artificial Intelligence (2003)
83. Rinaldi, F., Schneider, G., Kaljurand, K., Dowdall, J.: Mining relations in the GENIA corpus. In: Proc. 2nd European Workshop on Data Mining and Text Mining for Bioinformatics (2004)
84. Ng, S.K., Wong, M.: Toward Routine Automatic Pathway Discovery from Online Scientific Text Abstracts. In: Proc. 12th National Conference on Artificial Intelligence (1999)
85. Blaschke, C., Andrade, M.A., Ouzounis, C., Valencia, A.: Automatic Extraction of Biological Information from Scientific Text: Protein-Protein Interactions. In: Proc. 7th International Conference on Intelligent Systems for Molecular Biology, pp. 60–67. AAAI Press, Menlo Park (1999)
86. Blaschke, C.V.A.: The potential use of SUISEKI as a protein interaction discovery tool. In: Proc. Workshop on Genome Informatics, vol. 12, pp. 123–134 (2001)
87. Ono, T., Hishigaki, H., Tanigam, A., Takagi, T.: Automated extraction of information on protein-protein interactions from the biological literature. Bioinformatics 17(2), 155–161 (2001)
88. Leroy, G., Chen, H.: Filling preposition-based templates to capture information from medical abstracts. In: Proc. Pacific Symposium Biocomputing, pp. 350–361 (2002)
89. Proux, D., Rechenmann, F., Julliard, L.: A Pragmatic Information Extraction Strategy for Gathering Data on Genetic Interactions. In: Proc. 8th International Conference on Intelligent Systems for Molecular Biology, pp. 279–285. AAAI Press, Menlo Park (2000)
90. Huang, M., Zhu, X., Hao, Y.: Discovering patterns to extract protein-protein interactions from full text. Bioinformatics 20(18), 3604–3612 (2004)
91. Chun, H.W., Hwang, Y.S., Rim, H.C.: Unsupervised Event Extraction from Biomedical Literature using Co-occurrence Information and Basic Patterns. Lecture Notes in Artificial Intelligence, pp. 777–786 (2005)
92. Hao, Y., Zhu, X., Huang, M., Li, M.: Discovering patterns to extract protein-protein interactions from the literature: Part II. Bioinformatics 21(15), 3294–3300 (2005)
93. Phuong, T.M., Lee, D., Lee, K.H.: Learning rules to extract protein interactions from biomedical text. In: Whang, K.-Y., Jeon, J., Shim, K., Srivastava, J. (eds.) PAKDD 2003. LNCS (LNAI), vol. 2637. Springer, Heidelberg (2003)

94. Andrade, M.A., Valencia, A.: Automatic extraction of keywords from scientific text: application to the knowledge domain of protein families. Bioinformatic 14(7), 600–607 (1998)

95. Craven, M.: Learning to extract relations from medline. In: Proc. AAAI 1999 Workshop on Machine Learning for Information Extraction, pp. 25–30 (1999)

96. Mark, C., Johan, K.: Constructing biological knowledge bases by extracting information from text sources. In: Proc. 7th International Conference on Intelligent ·Systems for Molecular Biology, Heidelberg, Germany, pp. 77–86 (1999)

97. Stapley, B., Benoit, G.: Biobibliometrics: information retrieval and visualization from co-occurrences of gene names in Medline abstracts. In: Proc. Pacific Symposium on Biocomputing, Hawaii, U.S.A., pp. 529–540 (2000)

98. Stephens, M., Palakal, M., Mukhopadhyay, S., Raje, R., Mostafa, J.: Detecting gene relations from MEDLINE abstracts. In: Proc. Pacific Symposium on Biocomputing, Hawaii, USA, vol. 6, pp. 483–495 (2001)

99. Jenssen, T., Laegreid, A., Komorowski, J., Hovig, E.: A literature network of human genes for high-throughput analysis of gene expression. Nature Genetics 28(1), 21–28 (2001)

100. Marcotte, E.M., Xenarios, I., Eisenberg, D.: Mining literature for protein-protein interactions. Bioinformatics 17(4), 359–363 (2001)

101. Hahn, U., Romarker, M.: Rich knowledge capture from medical documents in the MEDSYNDIKATE system. In: Proc. Pacific Symposium on Biocomputing, Hawaii, U.S.A., pp. 338–349 (2002)

102. Eom, J.H., Zhang, B.T.: PubMiner: Machine Learning-Based Text Mining System for Biomedical Information Mining. In: Bussler, C.J., Fensel, D. (eds.) AIMSA 2004. LNCS (LNAI), vol. 3192. Springer, Heidelberg (2004)

103. Rosario, B., Hearst, M.: Multi-way Relation Classification: Application to Protein-Protein Interaction. In: Proc. HLT-NAACL 2005, Vancouver (2005)

104. Mooney, R.J., Bunescu, R.: Mining knowledge from text using information extraction. SIGKDD Explor. Newsl. 7(1), 3–10 (2005)

105. Bunescu, R., Ge, R., Kate, R.J., Marcotte, E.M., Mooney, R.J., Ramani, A.K., Wong, Y.W.: Comparative experiments on learning information extractors for proteins and their interactions. Journal of Artificial Intelligence in Medicine, 139–155 (2005)

106. Chun, H.w., Tsuruoka, Y., Kim, J.D., Shiba, R., Nagata, N., Hishiki, T., Tsujii, J.: Extraction of Gene-Disease Relations from MedLine using Domain Dictionaries and Machine Learning. In: Proc. Pacific Symposium on Biocomputing (PSB), pp. 4–15 (2006)

107. Zhou, D., He, Y., Kwoh, C.K.: Extracting Protein-Protein Interactions from the Literature using the Hidden Vector State Model. In: Proc. International Workshop on Bioinformatics Research and Applications, Reading, UK (2006)

108. He, Y., Young, S.: Semantic processing using the hidden vector state model. Computer Speech and Language 19(1), 85–106 (2005)

109. Hirschman, L., Yeh, A., Blaschke, C., Valencia, A.: Overview of BioCreAtIvE: critical assessment of information extraction for biology. BMC Bioinformatics 6(supp. 1) (2004)

110. Christian, B., Alexander, Y., Evelyn, C., Marc, C., Rolf, A., Lynette, H., Alfonso, V.: Do you do text? Bioinformatics 21(23), 4199–4200 (2005)

111. Nédellec, C.: Learning Language in Logic - Genic Interaction Extraction Challenge. In: Proc. Learning Language in Logic workshop (LLL 2005), pp. 31–37 (2005)

112. Ashburner, M., Ball, C., Blake, J., Botstein, D.: Gene ontology: tool for the unification of biology. The Gene Ontology Consortium. Nature Genetics 25(1), 25–29 (2000)
113. Lomax, J.: Get ready to GO! A biologist's guide to the Gene Ontology. Briefings in Bioinformatics 6(3), 298–304 (2005)
114. Salton, G.: Automatic Text Processing. Addison-Wesley series in Computer Science (1989)

# Index

# Author Index